BACKGROUND TO PALAEOHYDROLOGY

INTERNATIONAL GEOLOGICAL CORRELATION PROGRAMME

PROJECT 158

BACKGROUND TO PALAEOHYDROLOGY

A Perspective

Edited by
K. J. GREGORY
Department of Geography, University of Southampton

A Wiley-Interscience Publication

JOHN WILEY & SONS
Chichester · New York · Brisbane · Toronto · Singapore

Library of Congress Cataloging in Publication Data:

Main entry under title:

Background to palaeohydrology.

'A Wiley–Interscience publication.'
Includes index.
1. Palaeohydrology. I. Gregory, K. J. (Kenneth John)
QE39.5.P27B33 1983 551.48 83-5929
ISBN 0 471 90179 2

British Library Cataloguing in Publication Data:

Gregory, K. J.
Background to Palaeohydrology
1. Palaeohydrology
I. Title
551.4 GB659.6
ISBN 0 471 90179 2

Typeset by Activity, Salisbury, Wiltshire.
Printed in Great Britain by The Pitman Press, Bath, Avon.

List of Contributors

V. R. Baker	Department of Geosciences, University of Arizona, Tucson, Arizona 85721, USA
B. E. Berglund	Department of Quaternary Geology, Tornavägen 13, S-22363 Lund, Sweden
A. G. Brown	Department of Geography, The University, Southampton SO9 5NH, UK
I. Bryant	Department of Geology, The University, Nottingham NG7 2RD, UK
G. H. Dury	46 Woodland Close, Risby, Bury St Edmunds, Suffolk IP28 6QN, UK
I. Fenwick	Department of Geography, No 2 Earley Gate, Whiteknights Road, Reading RG6 2AU, UK
V. Gardiner	Department of Geography, The University, Leicester LE1 7RH, UK
W. L. Graf	Department of Geography, Arizona State University, Tempe, Arizona 85721, USA
K. J. Gregory	Department of Geography, The University, Southampton SO9 5NH, UK
M. Hayward	Department of Geography, No 2 Earley Gate, Whiteknights Road, Reading RG6 2AU, UK
S. Kozarski	Adam Mickiewicz University, ul. Fredry 10, 61–701 Poznan, Poland
J. Lewin	Department of Geography, University College of Wales, Llandinam Building, Penglais, Aberystwyth, Dyfed SY23 3DB, UK
S. Limbrey	Department of Ancient History and Archaeology, University of Birmingham, PO Box 363, Birmingham B15 2TT, UK
J. G. Lockwood	School of Geography, The University, Leeds LS2 9JT, UK
J. K. Maizels	Department of Geography, University of Aberdeen, Old Aberdeen AB9 2UF, UK
P. D. Moore	Department of Plant Sciences, King's College, London SE24 9JF, UK

F. Oldfield	Department of Geography, The University, PO Box 147, Liverpool L69 3BX, UK
K. Rotnicki	Adam Mickiewicz University, Quaternary Research Institute, ul. Fredry 10, 61–701 Poznan, Poland
L. Starkel	Inst Geografii PAN, ul. Sw. Jana 22, 31–018 Krakow, Poland
J. B. Thornes	Department of Geography, Bedford College, Regents Park, London NW1 4NS, UK
D. E. Walling	Department of Geography, The University, Exeter, Devon EX4 4RJ, UK
B. W. Webb	Department of Geography, The University, Exeter, Devon EX4 4RJ, UK
P. E. J. Wiltshire	Department of Plant Sciences, King's College, London SE24 9JF, UK

Contents

Preface

Unlike many edited volumes this one emerged as an idea well before rather than after a conference. An international conference in the UK in 1983 was planned as one of a succession of meetings during the ten year research programme (1977–1986) for IGCP Project 158, and it seemed desirable to collect together a number of general papers reviewing the state of the art in the several fields which contribute to palaeohydrology. The flavour of a number of chapters reflects the inception of the sequence of essays against the background canvas of recent developments in British research but, as is usual in a rapidly-advancing field, it has also been possible to include contributions from other internationally recognized specialists. A number of research papers and volumes stimulated by IGCP Project 158 have already been produced, and I hope that this present volume will be a worthy addition and testimony to the vision of the initiators of the international project, Professor Leszek Starkel and Björn Berglund, who have been kind enough to contribute a foreword to the volume. The arrangement of the chapters is explained in the Introduction (pp. 12) but it is necessary to emphasize that in a newly-developing field such as palaeohydrology several alternative sequences of chapter organization can be visualized. It was decided to separate those chapters providing background from others affording a further perspective because a more conventional organization, according to climate, hydrology, vegetation history, river channels, etc., would emphasize that some subject areas are not reviewed as completely as others. This is indeed deliberate because palaeohydrology being at the cross-roads, or spaghetti junction, of a range of disciplines, cannot call upon reviews that extend too far into the subject areas of hydrology, climatology, and Quaternary vegetation history, all of which have commanded volumes of their own. To continue the transport analogy, a location at the confluence of a number of routes means that, not only does flow of information and techniques occur to the intersection, but there can also be a transmission of ideas from palaeohydrology back to the contributing disciplines. Research in the early 1980's has already embraced some very innovative developments and this volume will be worthwhile if it can convey a little of the excitement that palaeohydrological investigations can instill and if it stimulates further innovations.

It is a great pleasure to acknowledge, in addition to the advice generously provided by Professor Leszek Starkel and Professor Björn Berglund, the discussions with Professor J. B. Thornes; the work of all the contributors; the secretarial assistance provided by Mrs R. Flint and Miss Liane Bailey; the cartographic advice given generously by Mr A. S. Burn and last but not least the patience of my family.

KEN J. GREGORY

Foreword

LESZEK STARKEL AND BJÖRN E BERGLUND

The reconstruction of climatic variations is the goal of all environmental reconstructions for the period since the deglaciation of the last inland ice. Important evidence on palaeoclimates comes through the reconstruction of different palaeohydrological parameters such as sea and lake level changes, variations in river discharge and sediment load, vegetational changes traced locally in mires and regionally in ecotone displacements, and finally the rate of glacial retreat.

It is symptomatic, too, that in all IGCP (International Geological Correlation Programme) projects covering the last 15000 years this palaeohydrological component is in the centre of attention. This applies to the longer established projects devoted to Quaternary glaciations (No. 24) and to sea level changes (No. 61) as well as to the younger ones, still in progress, which are devoted to lakes and floods in lower latitudes (No. 146) and to the palaeohydrology of the temperate zone (No. 158). It is a pleasure for us that this volume saw the light of day in the atmosphere of the last-mentioned project. Project definitions and research methods were described in an extensive guidebook for Project 158 subproject B dealing with lake and mire environments (Berglund, 1979–1982) and in a shorter guidebook for Project 158 subproject A dealing with fluvial environments (Starkel and Thornes, 1981), both containing many international contributions. It should be mentioned that most European countries, and the North American ones too, are taking an active part in the project. Since 1978 annual symposia have been organized in Finland (1978), France (1979), West Germany and Austria (1980), Poland (1981), and the Soviet Union (1982), and these have been a great help in the development of new methods and ideas. Of special value was the latest symposium, organized jointly with other IGCP project leaders (mentioned above) and, as is usual every year, together with the Eurosiberian Subcommission for studies of the Holocene of INQUA. This latest symposium was held in Moscow in August 1982, during the XI INQUA Congress. Some of the papers presented at that session are included in this volume, but the greater part of the basic papers in this book are the contributions of our British colleagues who, under the guidance of Ken J. Gregory, decided to put together present knowledge on palaeohydrology and present it to all participants at the symposium in England and Wales, September 1983. We hope that this volume will contribute to the further development of a new branch of science, namely palaeohydrology, and be an important step in the realization of our own and of other IGCP projects.

As the leaders of the IGCP project 158 we express our cordial thanks to all contributors and especially to Ken for his editorial work.

Cracow and Lund, December 1982

LESZEK STARKEL BJÖRN E BERGLUND

BACKGROUND

Background to Palaeohydrology
Edited by K. J. Gregory

1
Introduction

K. J. Gregory

Department of Geography, University of Southampton

DEFINING PALAEOHYDROLOGY

Palaeohydrology as a term has been conspicuously absent from many earth science textbooks published since 1950. The 1982 Supplement to the Oxford English dictionary has received inclusions of palaeoclimatology, palaeoecology, and palaeogeomorphology, but not palaeohydrology. This deficiency is echoed in the content of many recently-published books and monographs which have avoided using the term palaeohydrology, so that it is not difficult to summarize the brief history of the usage of the word. One of the earliest uses of the word in the English literature is in a study of the post-glacial chronology for alluvial valleys in Wyoming (Leopold and Miller, 1954). In deducing the alluvial chronology, a series of questions arose concerned with the interaction of climate, vegetation, stream regimen, and runoff which obtained under several climates, each different from that of the present and so led to the use of the word palaeohydrology (Table 1.1). Evidence was presented to compare present and past gradients of channels of the Powder river and one of its tributaries, Clear Creek, and it was concluded that a trend towards a colder and wetter climate would have produced a differential increase in runoff from the plains, compared with runoff from the mountains, and that such conditions would have produced aggradation as was shown to have occurred in Kaycee time. Further deductions about the palaeohydrology of the Powder river basin were based upon consideration of possible rates at which the alluvial fills underlying the river terraces were deposited. Computation of the volume of fill was used to give estimates of sediment movement which could be compared with sediment transport in contemporary streams, and it was concluded that during alluviation under the alternating climate of the recent period, rates of sediment production were of the same order of magnitude as in present streams (Leopold and Miller, 1954).

Table 1.1 Some Definitions of Palaeohydrology

Author	Definition	Subject
L. B. Leopold and J. P. Miller, 1954	Consideration of the alluvial chronology brings to mind a host of questions concerning a comparison of the present with conditions which prevailed in the past. These questions are concerned not so much with past climate itself, but with the interaction of climate, vegetation, stream regimen, and runoff, which obtained under climates different from that of the present. That is, with hydrologic rather than merely climatic factors. To describe this general subject the word 'palaeohydrology' is introduced.	Postglacial chronology for alluvial valleys in Wyoming.
S. A. Schumm, 1965	The science of hydrology encompasses the behaviour of water as it occurs in the atmosphere, on the surface of the ground, and underground. Palaeohydrology treats these phenomena but has reference to the past. ... The term palaeohydrology will be restricted to that portion of the hydrologic cycle that involves the movement of water over the surface of the earth, because runoff and its sediment load are of major importance in determining the non-glacial erosional and depositional features of the Quaternary.	Quaternary Palaeohydrology.

K. J. Gregory and D. E. Walling, 1973	... the increased interest in palaeohydrology ... the study of water and sediment dynamics in the past, introduces a promising field for future investigation.	Drainage basin form and process.
G. H. Cheetham, 1976	Palaeohydrology, to follow the definition provided by Schumm (1965), is the study of fluvial processes which have operated in the past and their hydrologic implications.	Geoarchaeology.
S. A. Schumm, 1977	If hydrology is the study of the waters of the earth, their distribution, composition and movements, then palaeohydrology can be defined as the science of the waters of the earth, their composition, distribution and movement on ancient landscapes from the occurrence of the first rainfall to the beginning of historic hydrologic records. It is difficult to consider water movement on a landscape without becoming involved with the various problems of sediment transport, and therefore the inclusion of the word composition in the definition stipulates that the quantity and types of sediments moved through palaeochannels must be considered a part of palaeohydrology.	The fluvial system.
G. H. Dury, 1977	As has been clear for some time, the investigation of underfit streams forms part of the general enquiry into palaeoclimatology and palaeohydrology.	Underfit streams: retrospect, perspect and prospect.

Whereas this use of palaeohydrology was in relation to a very specific area, a general approach to Quaternary palaeohydrology was devised by Schumm in 1965. The foundation of the approach was developed from earlier work by Langbein and others (1949) and by Langbein and Schumm (1958), and was based on three major relationships. These were expressed as curves relating firstly, mean annual precipitation and mean annual runoff for five specific values of mean annual temperature; secondly, mean annual sediment yield and mean annual (effective) precipitation for four values of mean annual temperature; and thirdly, mean annual suspended sediment concentration and mean annual precipitation for four values of mean annual temperature.

This important paper included a useful definition of palaeohydrology (Table 1.1); it has proved to be a cornerstone for the development of palaeohydrology, and it included a series of examples of the effects of changes of climate upon interfluves and drainage density, upon rivers, especially channel patterns, and upon climatic river terraces. The way in which effects of climatic changes were visualized was by reference to possible changes the three groups of relationships presented. This allowed the estimation of effects on hydrologic variables; the deduction of possible changes in fluvial processes, and examples of these estimates as devised by Schumm (1965), are illustrated in Tables 1.2 and 1.3. Subsequently the way in which the approach required assistance from the hydrological profession was reviewed by Schumm (1967) and the approach was then extended (Schumm, 1968) in a more geologic sense to provide speculations about the palaeohydrologic controls of terrestrial sedimentation. These speculations centred upon the changing nature of the land phase of the hydrological cycle before and during colonization of the landscape by vegetation and it was concluded that, with the appearance of grasses in the Cenozoic, the relations between climate, vegetation, erosion, and runoff became much as today except for the subsequent influence of man.

An approach to Quaternary palaeohydrology depended upon extending knowledge of spatial variations of present processes to temporal variations, and this ergodic hypothesis has found support in a number of time-based hydrological studies. The relationships originally proposed in 1958 were necessarily based upon the limited amount of data then available and it has been noted (Dury, 1967) that the two graphs of the climatic variation and sediment yield were initially based upon 6 and 9 coordinate points. The basic relationships have subsequently been modified slightly as more data has become available in the 1970's (Gregory and Walling, 1973) and they are capable of further revision and development (see Walling and Webb, Chapter 4, pp. 69). More recently the approach has been elucidated in the more general context of the fluvial system (Schumm, 1977), it has been reflected in a number of studies of river channel change (Gregory, 1977), and it has been particularly useful in the study of underfit streams (Dury, 1977). Work in a number of papers by Dury (e.g. 1964, 1965) showed how the interpretation of

Table 1.2 Examples of Estimated Effects of Climatic Change on Hydrologic Variables (Based on Schumm, 1965)

Weighted mean annual temperature degrees F (degrees C)		Mean annual precipitation in (mm)		Ratio of changed to present mean annual runoff	Ratio of changed to present mean annual sediment yield	Ratio of changed to present mean annual sediment concentration
Present	Changed	Present	Changed			
A. Change to climate during glaciation						
50(10)	40(4.4)	10(254)	20(508)	>20	0.6	<0.03
50(10)	40(4.4)	40(1016)	50(1270)	2	0.9	—
60(15.6)	50(10)	10(254)	20(508)	>20	>2.0	—
60(15.6)	50(10)	40(1016)	50(1270)	2	0.9	>0.5
B. Change to climate during interglaciation or postglacial hypsithermal interval						
50(10)	55(12.8)	20(508)	15(381)	0.2	1.2	>6
50(10)	55(12.8)	40(1016)	35(889)	0.7	1.1	2
60(15.6)	65(18.3)	20(508)	15(381)	0.1	0.7	—
60(15.6)	65(18.3)	40(1016)	35(889)	0.5	1.4	3

TABLE 1.3 EXAMPLE OF POSSIBLE CHANGES OF PLEISTOCENE RIVER ACTIVITY (After Schumm, 1965) (D, deposition; E, erosion; S, stable)

	Climate change			River activity during designated phase					
Location or type of drainage basin	Interglacial	→	Glacial	Late interglacial	Early glacial	Glacial	Late glacial	Early interglacial	Interglacial
Basin partly occupied by ice sheet	Warm, moist		Cooler, drier	S	D	D	E	E	S
Periglacial, no glaciers in basin	Warm, moist		Cooler, drier	S	D	D	E	E	S
Closed interior basin	Hot, dry		Cooler, wetter	D	D	S	E	E	S
Unglaciated continental interior (humid, perennial streams)	Warm, moist		Cooler, wetter	S	E	E	D	D	S
Unglaciated continental interior (semiarid, ephemeral, and intermittent streams)	Warm, dry		Cooler, wetter	S	E	E	D	D	S

underfit streams and meandering valleys could be placed in a context of climatic change, and this involved palaeohydrological estimates of former peak discharges and of the characteristic runoff regimes, although the approach was not characterized as palaeohydrology until 1977 (Dury, 1977).

Since the late 1970's palaeohydrology has begun to feature in two types of contribution. First it has become more apparent in research papers as these have included palaeohydrology as a keyword in their objectives and titles. Thus an outline review was provided of the palaeohydrology of Wales (Lewin, 1977), the use of slack-water flood sediments for reconstruction of the frequency of palaeohydrologic events has been demonstrated for southwestern Texas (Kochel, Baker, and Patton, 1982; Kochel and Baker, 1981), the morphological variation of the Colorado River in central Texas has been interpreted in terms of climatic change and palaeohydrology (Baker and Penteado-Orellana, 1977), and palaeohydrology of the Radunia basin, North Poland has been envisaged in relation to landscape development of central and southern Poland (Koutaniemi and Rachocki, 1981). These studies all illustrate how palaeohydrology has featured increasingly in specific papers but in addition there have been a number of books or collections of papers produced. In 1970 a special issue of Acta Universitatis Ouluensis collected the proceedings of meetings held in Finland in 1978, associated with International Geological Correlation Programme 158, namely *The Palaeohydrology of the Temperate Zone in the Last 15 000 Years* (Vasari, Saarnisto, and Seppala, 1979). Also stimulated by IGCP Project 158, which is referred to in the foreword to this volume, were several other volumes of conference proceedings derived from Amiens in 1979 and Frankfurt in 1980, there have also been publications associated with techniques to be employed in palaeohydrological research for IGCP Project 158 (Starkel and Thornes, 1981; Berglund, 1979) and volumes devoted to specific areas (Starkel, 1981).

In a number of research papers and written contributions, palaeohydrology is not specifically defined although the general sense of usage is evident, as illustrated by the study of the palaeohydrology of southern Israel where Issar (1979) suggested that 'A synthesis of hydrogeological, hydrogeochemical and palaeogeographical considerations enables a reconstruction of the palaeohydrologic regime' (Issar, 1979, pp.302). The definitions summarized in Table 1.1 now indicate a broad measure of agreement and many derive from the proposal of Schumm (1977) which follows his earlier definitions (Schumm, 1967, 1968). To clarify the definitions of palaeohydrology as the science of the waters of the earth, their composition, distribution and movement on ancient landscapes, it is necessary to stipulate, as Schumm did (1968, 1977), that composition embraces sediment movement in relation to palaeochannel morphology and furthermore that palaeohydrology could be visualized as being prior to continuous hydrologic records because some hydrologic records date back more than 2 000 years, whereas continuous hydrologic records seldom pre-date the last century. The objectives of palaeohydrology may therefore be envisaged as including reconstruction of the components of the hydrological cycle, of the water balance, and of sediment budgets for the time before continuous hydrological records, and necessarily embracing an understanding of the way in which changes in the hydrological cycle occurred and establishing how they differed from the contemporary hydrological picture.

PALAEOHYDROLOGY IN AN INTERDISCIPLINARY POSITION

Approaches to the study of the past were partially focused by the development of a new journal *Palaeogeography, Palaeoclimatology and Palaeoecology* which first appeared in 1965 (Elsevier, 1965 *et seq*.). In the issues in 1982 the subject matter is described as palaeoenvironmental geology and this should encompass palaeohydrology. However, in the journal, and in the field of environmental geology and environmental sciences in general, it is evident that palaeohydrology has emerged from, and been stimulated by, developments in many related fields associated with climatology, hydraulics, geomorphology, and sedimentology. This situation has arisen because, just as hydrology requires a knowledge of climatology, of physical characteristics of the land-surface and of the soils, vegetation, land use, and rock types, so palaeohydrology is founded upon a knowledge of palaeoclimatology, of palaeogeomorphology or palaeomorphology, of palaeopedology, and of palaeoecology. Each of these lines of enquiry (Table 1.4) was initially concerned to make its own internal progress. Thus for reconstructing the past record of climate as a necessary ingredient of palaeoclimatology, Lamb (1982) listed historical evidence including diaries and records, sedimentary evidence including varves, radiocarbon dating, pollen analysis, and vegetation history, and evidence from faunal remains such as coleoptera, from archaeology, from tree rings, and from ocean deposits. Similarly, an equally wide-ranging variety of techniques (Berglund, 1979) can be employed for palaeoecological reconstruction. Once the climatology, vegetation, soils and surface form have been reconstructed, then it is desirable to see how the components of the palaeoenvironment integrated and operated, and it is at this stage that palaeohydrological reconstruction may begin. The necessary dating has meant that a number of contributions have come from geoarchaeology (e.g. Davidson and Shackley, 1976) and additional evidence to facilitate palaeohydrological reconstruction can be adduced from palaeohydraulics and from reconstructions based upon parameters of sedimentary deposits. The major strategies available are reviewed by Allen (1977), and the success of fluvial sedimentology and of the development of facies models to assist understanding of geologic and recent sedimentary environments is demonstrated by a collection of papers published after a symposium in 1977 (Miall, 1978a). In an historical review of

Table 1.4 Definitions of Fields Related to Palaeohydrology

Palaeoclimatology	The study or investigation of palaeoclimates (which are) the climates at periods in the geological past.	Oxford English Dictionary (1982)
Palaeogeomorphology	The geomorphology of ancient landscapes especially as represented today by features that are buried or newly exhumed.	Supplement to the Oxford English Dictionary (1982)
Palaeohydraulics	Palaeohydraulics is the study of the quantitative relationships between the hydraulic parameters of a river (depth, width, slope, discharge, sediment type, etc.) and its preserved deposits.	Miall, 1978b
Palaeopedology	Palaeopedology is the study of palaeosols. A palaeosol (from Greek, *palaio* ancient and Latin *solum* soil) is a soil that formed on a landscape of the past.	Ruhe, 1965

the development of fluvial sedimentology Miall (1978b) argued that it was astonishing how much development had been possible in recent years but he also cautioned that, despite an increasing dialogue between sedimentologists, geomorphologists, and hydraulic engineers, conceptual differences remained. This is exemplified by the use of different timescales as a frame of reference, because, whereas geologists are interested in buried fluvial deposits as well as in surface deposition and processes of sediment preservation, geomorphologists and engineers have a primary concern with surface processes and with events that a sedimentologist may regard as of ephemeral significance.

Palaeohydrology presents a situation which is broadly analogous to that described for sedimentology by Miall (1978b), in that it is an interdisciplinary field which can progress only when receiving contributions from the component disciplines, and it must endeavour to reduce the differences in perspective that exist between the component disciplines. Reconstruction of the palaeohydrological cycle is achieved by an ever-increasing range of techniques and approaches, and it is notable that some specialists tend to concentrate upon particular components of the palaeohydrological cycle such as the input from palaeoclimatology or from palaeoecology, and some researchers focus dominantly upon particular time periods. For investigations of the palaeohydrology of time periods as recent as the last 15 000 years, the achievements of the International Geological Correlation Programme Project 158 initiated by Starkel and Berglund and described in the foreword (pp. iv) will be to make considerable interdisciplinary and methodological progress. However, because of increasing interest in the field of palaeohydrology it seemed that there was a need for a volume which summarized the background to the evolving discipline of palaeohydrology. This volume is therefore intended to illustrate some aspects of progress already achieved including work in research areas that are potentially of significance to palaeohydrology, and to provide a background perspective for contemporary developments. Background is provided by reviewing subject or disciplinary areas, techniques, and comparative achievements, and therefore studies of climatic change (Chapter 2), of discharge (Chapter 3), of sediment yield (Chapter 4) and of contributions from soil analysis (Chapter 7) and from archaeology (Chapter 8) make up much of the first section. It is not possible to include extensive coverage of hydrological techniques or of sedimentological and palaeohydraulic approaches which may be useful in palaeohydrologic reconstruction, because these could be the subject of further volumes. However, it has been possible to include a general review of the reconstructions based upon gravel deposits (Chapter 5), an outline of magnetic studies which have already demonstrated considerable potential for palaeohydrology (Chapter 6), an outline of channel geometry studies (Chapter 1), and a consideration of changes in the fluvial environment (Chapter 9) and in vegetation (Chapter 10) during the last 15 000 years based upon data already accummulated from a number of areas throughout the temperate zone. Some perspectives are provided to supplement the background of the first ten chapters and these are chosen to range over the diverse ways in which palaeohydrology may be assembled. Although the balance of review and example differs amongst the perspective chapters, they all present data for one or more specific areas which have an extra tropical emphasis. These contributions provide a perspective by showing how morphological evidence from channel dimensions and planform (Chapters 13,14,15,17) and from networks (Chapters 11,12) can illuminate palaeohydrology. It is important to recall that gully development is an important feature of palaeoenvironments in many areas and so should be represented (Chapter 12). Reconstruction based upon sedimentary sequences can also contribute (Chapters 16,18) and much has been learnt from use of techniques such as those employing remains of coleoptera (e.g. Coope, 1977). Studies of vegetation history have contributed substantially to the reconstruction of palaeohydrology and this is illustrated (Chapter 19) for central Wales and so complements the contribution based upon climatic changes in central Britain.

Just as palaeohydrology has developed rapidly in the last decade and has catalysed our ability to make prognostications about components of former hydrological cycles and about controls upon these components, so the amalgamation of approaches and techniques should further enhance progress during the next decade. Indeed, because the hydrology of today is the palaeohydrology of tomorrow, we must be aware of the need to retrodict past from present and also to predict future hydrologies based upon present and palaeohydrology as indicated in Chapter 20.

THE EXAMPLE OF CHANNEL GEOMETRY

Palaeohydrology has to reconstruct and to explain the hydrological cycle and the pathways of energy transfer in the hydrological cycle under previous conditions. Such reconstruction can be achieved utilizing morphological information such as that based upon fragments of former river channels and illustrated for river planform and floodplains (Chapter 9,13,14,17), and for drainage networks (Chapter 11); sedimentological information whereby quantitative information of characteristics of sediments are employed to assist interpretation of former processes (e.g. Chapters 5,15,16); and in addition information can be obtained from measurements of contemporary processes (e.g. Chapter 4) and of recent changes to show how knowledge of the magnitude of spatial variations can illuminate the magnitude of past variation. In addition the dating of events, together with the range of techniques available for environmental reconstruction, can facilitate the identification of the pathways of former hydrological cycles. The contemporary hydrological cycle is of course a palimpsest which embraces parts of the pathways of earlier cycles under different hydrological controls. The river channel affords a major hydrological pathway and the contemporary channel may incorporate evidence of earlier processes and it may also be adjacent to remnants of earlier phases. This brief review is approached by proceeding from an outline of the recent development of studies of channel geometry, to the present state of knowledge, and finally to the recent developments which may influence our way of developing channel geometry further.

Studies of channel geometry cannot easily be separated from studies of channel planform (Chapters 13,15). In an early review Day Kimball (1948) proposed that historical geology could be divided into two parts: stratigraphy which deals with what is there and denudation chronology which is concerned with what isn't! In arguing that denudation chronology was concerned with the vestiges of past fluviatile processes, Kimball (1948) discussed the fundamental variables, the controlling factors and the orders of magnitude of change and he employed flow equations including the Manning equation to indicate how changing discharges could lead to alterations in channel form. An approach with a similar objective was provided by E.W. Lane (1955) when he proposed a basis of analysis in quantitative stream morphology founded upon the approximation $Q_s d \approx Q_w S$ where Q_s is the quantity of sediment, d the particle diameter or size of sediment, Q_w the water discharge and S the slope of the stream. This approximation was used to assist the interpretation of channel adjustments by identifying six classes of change such as decrease of discharge (Class 1) which could involve an increase in slope if Q_s and d remained constant. The approach was then used to interpret the changes following the effects of engineering works upon river channels.

As is evident from the citations in a number of the subsequent chapters, a very significant contribution has been made by the research of S.A. Schumm. This developed the foundation provided by E.W. Lane (1955) and involved the production of a classification of alluvial channels, the development of empirical relationships which could afford a refined method of analysing channel adjustments, and the study of a number of specific areas such as the Murrumbidgee on the Riverine Plain of New South Wales where changes of channel geometry parameters were demonstrated (Schumm, 1977). Empirical relationships established for stable alluvial channels in semiarid and subhumid areas included (Schumm, 1969) the approximation

$$Q_w \simeq \frac{w.d.M_L}{S}$$

where Q_w is water discharge, w is channel width, d channel depth, M_L meander wavelength, and S stream slope. In the interpretation of river metamorphosis advocated by Schumm (1969), it was possible to envisage a series of possible changes such as the diversion of water out of the system or reduction of discharge which would then be expressed as

$$Q_w^- \simeq \frac{w^- d^- M_L^-}{S^+}$$

and several other approximations of this type were proposed and are developed in Chapter 9. Developing in parallel with this research was work by Dury (1964, 1965, 1977) which utilized relationships between present meander wavelength and bankfull discharge as a basis for estimating former bankfull discharges from the wavelengths of meandering valleys and a specific instance of the Osage type is elaborated by Dury in this volume (Chapter 17). The relationships were further elaborated (Dury, 1976) by including calibre of sediment and a factor combining a slope factor with the area–velocity relationship of a stereotype stream for humid areas.

Approaches of this kind have been developed and utilized by sedimentologists who have begun to develop equations to employ palaeochannel cross-section data to estimate morphologic and hydrologic characteristics of former channels (Miall, 1978a). However, such estimations must be treated carefully (Ethridge and Schumm, 1978), not only because a limited amount of data was used as the basis for the original empirical relationships but also because measurements of palaeochannel morphology and sediments may be incomplete or difficult to obtain. The relationships available in 1978 allowed the estimation of palaeochannel and palaeohydraulic values (Ethridge and Schumm, 1978) using two methods, one based on width, depth, and per cent silt–clay, and the second requiring only width–depth ratio measurements of inferred palaeochannels. The latter was recommended because it was simpler to use although further studies were thought to be required on the changing morphology and sediment character of aggrading channels and on the effect of sediment load or channel morphology.

Recent research on the controls of river channel morphology has helped to illuminate the ways in which the reaction of river channels to recent process events may provide clues to adjustments in the longer term and the controls that have been investigated include discharge, sediment, slope and vegetation. In a number of early studies water discharge was regarded as of paramount importance in determining channel geometry, especially channel size, and any major modification of the discharge regime could be reflected in channel dimensions. However three groups of contributions have been made which can refine understanding of the discharge–river channel relationship. First has been the realization that there is a range of discharges which affects the cross sectional size and shape of the river channel. Although studies previously used a single value such as bankfull discharge, it has been shown that a range of flows affects the bedforms and that a further slightly less frequently occurring range of flows is responsible for the size of the channel cross-section (Pickup and Warner, 1976). Knowledge of the effects of discharge can be enhanced by a closer examination of the frequency of occurrence of flows and morphology and in the case of the Yampa river basin of Colorado and Wyoming effective discharges were defined as the increment of discharge that transports the largest fraction of the annual sediment load over a period of years (Andrews, 1980). In such analysis it is necessary to consider which method of flood frequency analysis should be used and Gregory and Madew (1982) indicated that whereas the annual series was employed to provide recurrence intervals in at least 17 studies out of 25 that were cited, the annual exceedance series or a partial duration series can be more appropriate. A second contribution to the understanding of the influence of discharge has been the study of the significance of large discharge events which allows recovery times of channels to be studied as well as the initial effects that the large flows may have upon channel size. Thus eight techniques have been used to convert the particle size of the coarse fraction of alluvium of Boulder Creek, Colorado, to competent velocities which were then compared with data calculated for major floods including the Big Thompson flood of 1976 (Bradley and Mears, 1980). Ability to identify former events in the discharge record pertinent to channel morphology may now be possible from the investigation of the dendrochronology of trees lining the channels (Yanosky, 1982) to supplement the ways in which slack water deposits (Kochel, Baker, and Patton, 1982) and dating methods associated with Holocene alluvial sequences (Costa, 1978) may be employed to erect long-term flood frequencies. A third development of potential significance is the use of catastrophe theory in explanation of the relationship between channel morphology and discharge and this has been introduced specifically (Thornes, 1980) in relation to processes producing spatial and temporal variations in the sediment transport and width of ephemeral channels in Spain, and more generally for situations where control variables can be specified (Graf, 1980).

Control by sediment upon channel morphology is notable through the classification of channels as of suspended load, bedload, or mixed load type (e.g. Schumm, 1972) and also for the development of empirical relations incorporating a per cent silt–clay parameter as noted above. In addition to recent work on the relationship of flow regime to bedforms and sediment transport which has indicated that empirical relationships should only be used within their limits of definition (Bridge, 1978), it has been shown that the relation between channel width–depth ratio (F) and per cent silt–clay (M) does not hold for clay-bedded streams in the Namoi–Gwydir distributary system of eastern Australia (Riley, 1975). In this area of southeastern Australia, it was therefore concluded that other relations amongst sediment, discharge and morphometry must be used in conjunction with the F–M relationship in order to interpret palaeochannels with some degree of certainty. In a later study of the upper Darling river system it was shown (Riley and Taylor, 1978) that palaeochannel sedimentology and morphology indicate fluvial regimes different from those of today but that these may not have been the result of climatic or tectonic change but could have been associated with the progressive degradational–aggradational development of the fluvial system under constant external conditions.

The slope and vegetation controls upon channel morphology have also been investigated in studies of channel process and channel adjustment, and both are pertinent to the use of flow equations for flow estimation; slope is pertinent directly and vegetation through its effect upon the roughness factor. In the case of contemporary channels it is difficult to decide over which distance water-surface slope should be measured for the purpose of flow estimation and so for former events or for reconstruction of the flows of palaeochannels, slope is even less clearly determined. Therefore some studies have concluded that slope calculated from the contour spacing on large-scale topographic maps can be an adequate approximation but one suspects that this is because of our inadequate appreciation of the real significance of the slope factor. Many river channels today are modified in many direct and indirect ways as a result of channelization procedures and in the case of England and Wales it has been demonstrated (Brookes, Gregory, and Dawson, 1983) that there is a density of 0.06 km km^{-2} of channelized river in England and Wales and along these rivers the vegetation influence is very much less than prior to channelization. Therefore when considering palaeochannels, the former significance of vegetation in affecting boundary conditions should not be forgotten. The significance can be appreciated from a number of studies which have demonstrated the significance of organic debris as debris jams upon the stream channel (e.g. Keller and Swanson, 1979; Gurnell and Gregory, 1983) and in New Zealand it has been shown (Mosley, 1981) that debris jams can be a major constraint on the application of physical laws and theories to explain sediment movement. The pattern of sediment accumulation can also be affected and along the Beatton river and other rivers in British Columbia, Canada, it has been shown that scroll bars concentrate around dead trees (Nanson, 1981) and in Arizona and Colorado gravel bars formed transverse to the flow incorporated downed timber derived from streamside forests (Heede, 1981).

Consideration of the controls upon stream channels should not be considered separately but rather in an integrated manner and similarly it is desirable to relate changes in channel geometry to changes in planform and to character of the entire network so that a three-dimensional view of the fluvial system may be obtained. In this way Gregory and Ovenden (1979) demonstrated how drainage network volumes could be related to indices of precipitation in Britain and this gives an idea of the range of channel size values that exist at the present time. A number of other recent developments have been possible which have stressed interrelationships in the fluvial system and the way in which downstream variations in channel parameters related to slope may be an important consideration (Nanson and Young, 1981) in relation to the formation of floodplains and to overbank deposition. Related to such

TABLE 1.5 SOME DIFFICULTIES CONFRONTING INTERPRETATION OF PALAEOCHANNELS FROM CHARACTERISTICS OF CONTEMPORARY RIVER CHANNELS

(Further general discussion in relation to sediments appears in Ethridge and Schumm, 1978, and in Chapters 5, 14, and 15 of this volume)

Controls of present river channels	Implications for channel morphology		Discussion included in
	Contemporary channels	Inferences concerning palaeochannels	
Human activity —direct effects	Channelized river channels	Present channels cannot easily be used as analogues	Brookes, Gregory, and Dawson, 1983
— indirect effects	Changes of catchment areas especially of land use leads to adjustments in channel process and channel morphology	Contemporary channels not typical of palaeochannels	Gregory and Madew, 1982
Discharge	No simple relationship between discharge and channel morphology	Interpretation of former bankfull problematical	Pickup and Warner, 1976
	Method of flood frequency analysis used can be significant	Flow frequency not easily established	Gregory and Madew, 1982
	Significance of large rare events. Channel morphology may have not recovered from major events	Significance of rare events not known	Bradley and Mears, 1980 Baker, Chapter 20 this volume
Sediment	Sediment transported may differ from sediment in bed and banks	Relationships between morphology and percent silt clay for present channels may not apply to palaeochannels	Riley, 1975 Riley and Taylor, 1978

	Channel size and shape may be affected by armouring and paving of bed	Present channels may not be typical, armouring deposits may not be detected	Maizels, Chapter 5 this volume
Slope	Distance over which slope should be measured not clear	Approximation to water surface slope for particular stage difficult to reconstruct	
	Morphology downstream related to slope	Relative position of palaeochannel in fluvial system not easily reconstructed	Nanson and Young, 1981
Vegetation — riparian	Riparian vegetation may be cut or reduced	Contemporary channels not typical of palaeochannels	Mosley, 1981
— over drainage area	Debris dams could have been more significant in past	Pattern of sediment accumulation and movement different from present	Keller and Swanson, 1979
		Accumulation around dead trees	Nanson, 1981
Integrated controls	Morphology must be seen in relation to drainage network		Gregory and Ovenden, 1979
	Morphology, slope and sediment accumulation related to position in fluvial system		Nanson and Young, 1981
	Discharge and sediments must be combined		Andrews, 1980

integrated considerations of several aspects of the channel system have been a number of theoretical developments including identification of thresholds (Schumm, 1980) and landscape sensitivity (Brunsden and Thornes, 1979). Recent work on controls on channel geometry is summarized in Table 1.5 and particular aspects further discussed in Chapters 5,9,14 and 17.

Although there are still uncertainties (Burkham, 1981) in estimating the behaviour of contemporary river channels, nevertheless a number of stimulating developments can be identified in the present research effort and some are reviewed by Gregory (1982). These developments, together with the contributions from the several fields contributing to palaeohydrology, should be the basis for further progress and should assist in the clearer elucidation of palaeohydrological systems.

REFERENCES

Allen, J. R. L. 1977. Changeable rivers: Some aspects of their mechanics and sedimentation. In Gregory, K. J. (Ed.) *River Channel Changes*, Wiley, Chichester, 15–45.

Andrews, E. D. 1980. Effective and bankfull discharges of streams in the Yampa river basin, Colorado and Wyoming. *Journal of Hydrology*, **46**, 311–330.

Baker, V. R., and Penteado-Orellana, M. M. 1977. Adjustment to Quaternary climatic change by the Colorado River in central Texas. *Journal of Geoglogy*, **85**, 395–422.

Berglund, B. E. (Ed.). 1979. *Palaeohydrological Changes in the Temperate Zone in the Last 15000 Years*, Subproject B Lake and Mire Environments. Project Guide Volume I General Project Description 123 pp; Volume II Specific Methods 340 pp. Department of Quaternary Geology, Lund.

Bradley, W. C., and Mears, A. I. 1980. Calculations of flows needed to transport course fraction of Boulder Creek Alluvium at Boulder, Colorado. *Geological Society of America Bulletin*, **91**, 1057–1090.

Bridge, J. S. 1978. Palaeohydraulic interpretation using mathematical models of contemporary flow and sedimentation in meandering channels. In Miall, A. D. (Ed.) *Fluvial Sedimentology*, Canadian Society of Petroleum Geologists, Memoir 5, 723–742.

Brookes, A., Gregory, K. J., and Dawson, F. H. 1983. An assessment of river channelization in England and Wales. *Science of the Total Environment*, in press.

Brunsden, D., and Thornes, J. B. 1979. Landscape sensitivity and change. *Transactions Institute of British Geographers*, **NS4**, 463–484.

Burkham, D. E. 1981. Uncertainties resulting from changes in river form. *Proceedings American Society of Civil Engineers, Journal Hydraulics Division*, **107**, 593–610.

Cheetham, G. H. 1976. Palaeohydrological investigations of river terrace gravels. In Davidson, D. A. and Shackley, M. L. (Eds.) *Geoarchaeology*, Duckworth, London, 335–344.

Coope, G. R. 1977. Fossil coleopteran assemblages as sensitive indicators of climatic changes during the Devensian (last) cold stage. *Philosophical Transactions Royal Society London Series B*, **280**, 313–340.

Costa, J. E. 1978. Holocene stratigraphy in flood frequency analysis. *Water Resources Research*, **14**, 626–632.

Davidson, D. L., and Shackley, M. L. (Eds.) 1976. *Geoarchaeology*. Duckworth, London, 408 pp.

Dury, G. H. 1964. Principles of underfit streams. *U.S. Geological Survey Professional Paper 452A*; Subsurface explorations and chronology of underfit streams. *U.S. Geological Survey Professional Paper 452B*.

Dury, G. H. 1965. Theoretical implications of underfit streams. *U.S. Geological Survey Professional Paper 452C*.

Dury, G. H. 1967. Climatic change as a geographical backdrop. *Australian Geographer*, **10**, 231–242.

Dury, G. H. 1976. Discharge prediction, present and former, from channel dimensions. *Journal of Hydrology*, **30**, 219–245.

Dury, G. H. 1977. Underfit streams: retrospect, perspect and prospect. In Gregory, K. J. (Ed.), *River Channel Changes*, Wiley, Chichester, 281–293.

Ethridge, F. G., and Schumm, S. A. 1978. Reconstructing palaeochannel morphologic and flow characteristics: methodology, limitations and assessment. In Miall, A. D. (Ed.), *Fluvial Sedimentology,* Canadian Society of Petroleum Geologists, Memoir 5, 703–721.

Graf, W. L. 1980. Catastrophe theory as a model for change in fluvial systems. In *Adjustments of the Fluvial System*, Kendal Hunt, Dubuque, Iowa, 13–32.

Gregory, K. J., and Walling, D. E. 1973. *Drainage Basin Form and Process*. Edward Arnold, London, 450 pp.

Gregory, K. J. (Ed.) 1977. *River Channel Changes*, Wiley, Chichester, 450 pp.

Gregory, K. J. 1982. Fluvial geomorphology: less uncertainty and more practical application? *Progress in Physical Geography*, **6**, 427–438.

Gregory, K. J., and Madew, J. R. 1982. Land use change, flood frequency and channel adjustments. In Hey, R. D., Bathhurst, J. C., and Thorne, C. R. (Eds.) *Gravel-bed Rivers*, Wiley, Chichester, 757–781.

Gregory, K. J., and Ovenden, J. C. 1979. Drainage network volumes and precipitation in Britain. *Transactions Institute of British Geographers*, **NS4**, 1–11.

Gurnell, A. M., and Gregory, K. J. 1983. The influence of vegetation on stream channel processes. In Walling, D. E. and Burt, T. P. (Eds.) *Catchment Experiments in Fluvial Geomorphology*, Geo Books in press.

Heede, B. M. 1981. Dynamics of selected mountain streams in the western United States of America. *Zeitschrift für Geomorphologie*, **25**, 19–32.

Isaar, A. 1979. The palaeohydrology of southern Israel and its influence on the flushing of the Kurnub and Arad groups (Lower Cretaceous and Jurassic). *Journal of Hydrology*, **44**, 289–303.

Keller, E. A., and Swanson, F. J. 1979. Effects of large organic material on channel form and fluvial processes. *Earth Surface Processes*, **4**, 361–380.

Kimball, D. 1948. Denudation chronology The dynamics of river action. *Occasional Paper Number 8*, University of London, *Institute of Archaeology*, 21 pp.

Kochel, R. C., and Baker, V. R. 1981. Palaeoflood hydrology. *Science*, **215**, 353–361.

Kochel, R. C., Baker, V. R., and Patton, P. C. 1982. Palaeohydrology of southwestern Texas. *Water Resources Research*, **18**, 1165–1183.

Koutaniemi, L., and Rachocki, A. 1981. Palaeohydrology and landscape development in the middle course of the Radunia basin, north Poland. *Fennia*, **159**, 335–342.

Lamb, H. H. 1982. *Climate, History and the Modern World*, Methuen, London, 387 pp.

Langbein, W. B. *et al.* 1949. Annual runoff in the United States. *U.S. Geological Survey Circular*, **52**, 14 pp.

Langbein, W. B., and Schumm, S. A. 1958. Yield of sediment in relation to mean annual precipitation. *Transactions American Geophysical Union*, **39**, 1076–1084.

Lane, E. W. 1955. The importance of fluvial morphology in hydraulic engineering. *Proceedings American Society of Civil Engineers*, **81**, paper 745, 1–17.

Leopold, L. B., and Miller, J. P. 1954. Postglacial chronology for alluvial valleys in Wyoming. *United States Geological Survey Water Supply Paper*, **1261**, 61–85.

Lewin, J. 1977. Palaeohydrology. *Cambria*, **4**, 112–123.

Miall, A. D. (Ed.) 1978a. *Fluvial Sedimentology*, Canadian Society of Petroluem Geologists, Memoir 5, Calgary, 859 pp.

Miall, A. D. 1978b. Fluvial sedimentology: an historical review. In *Fluvial Sedimentology*, Canadian Society of Petroleum Geologists, Memoir 5, Calgary, 1–47.

Mosley, M. P. 1981. The influence of organic debris on channel morphology and bedload transport in a New Zealand forest stream. *Earth Surface Processes*, **6**, 571–580.

Nanson, G. C. 1981. New evidence of scroll bar formation in the Beatton river. *Sedimentology*, **28**, 889–891.

Nanson, G. C., and Young, R. W. 1981. Overbank deposition and floodplain formation on small coastal streams of New South Wales. *Zeitschrift für Geomorphologie*, **25**, 332–347.

Pickup, G., and Warner, R. F. 1976. Effects of hydrologic regime on magnitude and frequency of dominant discharge. *Journal of Hydrology*, **29**, 51–75.

Riley, S. J. 1975. The channel shape–grain size relation in eastern Australia and some palaeohydrologic implications. *Sedimentary Geology*, **14**, 253–258.

Riley, S. J., and Taylor, G. 1978. The geomorphology of the upper Darling river system with special reference to the present fluvial system. *Proceedings Royal Society of Victoria*, **90**, 81–102.

Ruhe, R. V. 1965. Quaternary palaeopedology. In Wright, H. E., and Frey, D. G. (Eds.) *The Quaternary of the United States*, Princeton University Press, Princeton, 755–764.

Schumm, S. A. 1965. Quaternary palaeohydrology. In Wright, H. E., and Frey, D. G. (Eds.) *The Quaternary of the United States*, Princeton University Press, Princeton, 783–794.

Schumm, S. A. 1967. Palaeohydrology: application of modern hydrologic data to problems of the ancient past. *International Hydrology Symposium (Fort Collins) Proceedings*, **1**, 185–193.

Schumm, S. A. 1968. Speculations concerning palaeohydrologic controls of terrestrial sedimentation. *Bulletin Geological Society of America*, **79**, 1573–1588.

Schumm, S. A. 1969. River metamorphosis. *Proceedings American Society of Civil Engineers, Journal of Hydraulics Division*, **6352**, HY1, 255–273.

Schumm, S. A. 1972. Fluvial palaeochannels. In Rigby, J. K., and Hamblin, Wm. K. (Eds.), *Recognition of Ancient Sedimentary Environments, Society of Economic Palaeontologists and Mineralogists, Special Publication*, **16**, 98–107.

Schumm, S. A. 1977. *The Fluvial System*, Wiley, Chichester, 356 pp.

Schumm, S. A. 1980. Some applications of the concept of geomorphic thresholds. In Coates, D. R., and Vitek, D. (Eds.), *Thresholds in Geomorphology*, George Allen and Unwin, London, 473–485.

Starkel, L. (Ed.) 1981. *The Evolution of the Wisłoka Valley near Debiça during the Late Glacial and Holocene*, Folia Quaternaria, **53**, 91 pp.

Starkel, L., and Thornes, J. B. 1981. Palaeohydrology of river basins. *British Geomorphological Research Group*, Technical Bulletin No. 28, 107 pp.

Thornes, J. B. 1980. Structural instability and ephemeral channel behaviour. *Zeitschrift für Geomorphologie*, Supplementband, 36, 233–244.

Vasari, Y., Saarnisto, M., and Seppala, M. (Eds.) 1979. Palaeohydrology of the temperate zone: Proceedings of the working session of Commission on Holocene–Inqua (Eurosiberian Subcommission) Hailuoto–Oulanka–Kevo 1978.

Acta Universitatis Ouluensis Series A, Scientiae Rerum Naturalium No 82. Geologica No. 3, University of Oulu, Oulu, 176 pp.

Yanosky, T. M. 1982. Hydrologic inferences from tree ring widths of flood-damaged trees, Potomac river, Maryland. *Environmental Geology*, **4**, 43–52.

Background to Palaeohydrology
Edited by K. J. Gregory

2

Modelling climatic change

JOHN G. LOCKWOOD

School of Geography, University of Leeds

THE CLIMATIC SYSTEM AND CLIMATIC MODELS

The climatic system consists of the atmosphere (comprising the earth's gaseous envelope and its aerosols), the hydrosphere (comprising the liquid water distribution on or beneath the earth's surface), the cryosphere (comprising the snow and ice on and beneath the surface), the surface lithosphere (comprising the rock, soil and sediment of the earth's surface), and the biomass (comprising the earth's plant and animal life). Each of these components has quite different physical characteristics, and are linked to each other and to conditions external to the system by a variety of physical processes. The atmosphere is the central component of the climatic system, and displays a spectrum of conditions ranging from microclimates to the climate of the entire planet. Because of the ease with which the atmosphere can be heated and set in motion, it may generally be expected to respond to an imposed change more rapidly than do the other components of the climatic system. A close second to the atmosphere in terms of its overall importance in the climatic system is the hydrosphere. The extent and bulk of the world's oceans and the prevalence of surface water on the land ensures a potentially plentiful supply of water substance for the global hydrological cycle of evaporation, condensation, precipitation, and runoff. The cryosphere, like the hydrosphere, consists of a portion closely associated with the sea (sea-ice) and portions associated with the land (snow, glaciers and ice sheets). The importance of the cryosphere to the climatic system lies in the high albedo and low thermal conductivity of snow and ice. The surface lithosphere, in contrast to the atmosphere, hydrosphere and cryosphere, is a relatively passive component of the climatic system. An exception to this is the amount of soil moisture which is closely related to the local surface and ground hydrology. Soil moisture exerts a marked influence on the local surface balance of moisture and heat through its influence on the surface evaporation rate and

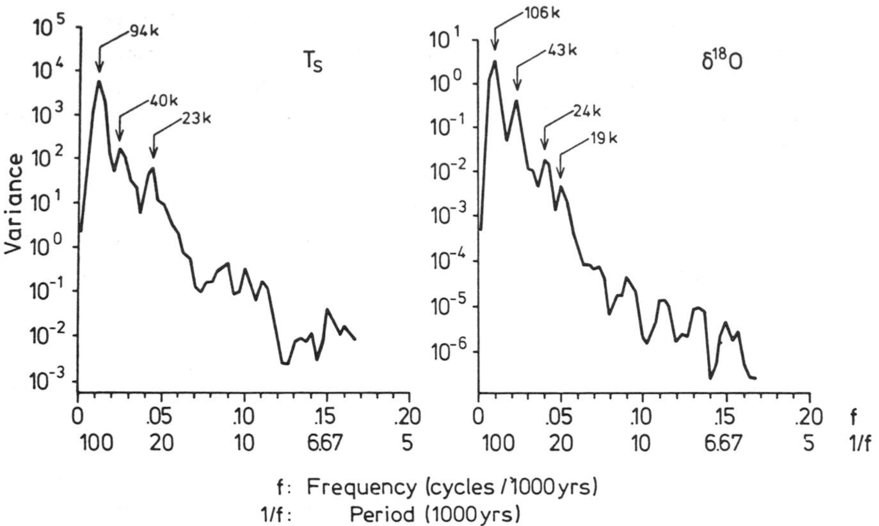

FIGURE 2.1 VARIANCE SPECTRA OF PALEOCLIMATIC TIME SERIES (inferred sea surface temperature T_S and $\delta^{18}O$). From Hays *et al.* (1976). Copyright 1976 by the American Association for the Advancement of Science

on the soil's albedo and thermal conductivity. The remaining component of the climatic system, the surface biomass, interacts with the other components on timescales that are characteristic of the life cycles of the earth's vegetative cover. The trees, plants, and ground cover modify the surface radiation balance and surface heat flux, and play a major role in the seasonal variations of local surface hydrology.

It is evident from all kinds of observations that there have been fluctuations in the climate on all time scales with large amplitudes on the very long time scales and small amplitudes on short time scales. Some general guidelines on the type of climate model required to simulate climate variability may be derived from the statistical properties of observed climate fluctuations. Figure 2.1 shows examples of the variance spectra of climatic fluctuations observed over wide timescale ranges. According to Hasselmann (1981) the distributions may be regarded as representative of climatic variance spectra on all time scales. The characteristic feature of all spectra is a continuous red distribution, increasing towards low frequencies according to some power law. An exception are the palaeoclimatic spectra of Figure 2.1, which are terminated on the low-frequency side by a peak at a period near 100 K years. This peak has been the subject of various speculations but has still not been conclusively explained. Also apparent in these spectra are weak peaks, barely statistically significant above the continuum, near periods of 20 K years and 40 K years, which have been attributed to variations of the earth's orbit in accordance with

Milankovitch's theory. However, Hasselman (1981) considers that, apart from these features, the dominant characteristic of essentially all climate–time series, covering timescales from tens of thousands of years down to a few weeks, is the continuous red distribution of the variance spectrum without the occurrence of prominent peaks.

The problem of climatic change on different timescales requires differing modelling approaches. On very short timescales only the rapidly varying components of the climate system need to be taken into account, such as the atmosphere and the top layer of the oceans. On these timescales atmospheric dynamics is in general treated explicitly, which leads to very complicated general circulation models (GCMs). On long timescales dynamical processes are parameterized partly (statistical dynamical models) or completely (energy balance models). These may include slowly varying climatic components such as ice sheets, the deep oceans, or vegetation changes.

The more important climatic models used in climatic change studies are briefly described below. Their use is illustrated in more detail later in the chapter.

Energy balance models

In energy balance models (EBMs) atmospheric dynamics are neglected completely or are very highly parameterized. In one-dimensional EBMs surface temperature is a function of latitude. At each latitude a balance is computed between incoming and outgoing radiation and horizontal transport of heat. This transport is parameterized as a diffusion term and an advection term. The diffusion constant can be derived by tuning the solution for the present climate. Various forms of EBMs have been used to study climatic variations on a geological timescale.

General circulation models

General circulation models (GCMs) are based on the fundamental physical laws that govern atmospheric behaviour. The equations are: the momentum balance equations and the hydrostatic relationship; the first law of thermodynamics; the continuity equation; the equation of state; and the water balance equation. The dynamics of pressure systems with a scale of the order of 1000 km or longer are treated explicitly, while small-scale exchanges are parameterized in terms of large-scale variables. All adiabatic heating components are taken into account with more or less detailed heating schemes. The equations are integrated in time over the whole globe or part of the globe.

Statistical dynamical models

A GCM generates via the process of baroclinic instability its own baroclinic eddies, that transport heat and momentum. Although individual synoptic eddies cannot be predicted at the right time or the right place, their statistical behaviour is almost correct. A statistical dynamical model (SDM) is based on the idea that the statistical behaviour of the transient eddies can be expressed in terms of the mean flow via parameterization relations. Thus it is possible to deal with the mean flow directly.

CLIMATIC MODELS AND SURFACE BOUNDARY CONDITIONS

Short timescale variations in weather are basically a problem in atmospheric hydrodynamics, and are nearly independent of surface conditions, therefore the movement of weather systems can be forecast over a period of a few days given only the known initial condition of the atmosphere. Unfortunately the memory of the atmosphere is short (about two weeks), so atmospheric conditions at any given instance in time give little information about longer term variations in climate. The memories of nearly all other components of the climatic system are considerably longer than that of the atmosphere, ranging from a few weeks for soil moisture systems to thousands of years for major ice sheets. On the climatic timescale the longer memories of the non-atmospheric components of the climatic system tend to dominate that of the atmosphere. Thus the atmospheric boundary conditions are very important in determining the long-term variations in climate.

In the study of long-term climatic variations it is very important to take into account surface boundary conditions. It will be demonstrated later that variations in surface conditions can greatly influence both local and regional climate. Indeed, many numerical models of climate are largely dependent on the surface boundary conditions and the predicted climate varies with changes in these conditions. Important surface boundary conditions include albedo, sea surface temperature, soil moisture, vegetation type, snow and ice distribution, relief, etc. Some of the more important surface conditions are discussed in detail below.

Vegetation

In the extratropics, with its large seasonal changes, the soil plays a role analogous to that of the ocean (Shukla and Mintz, 1982). The ocean stores some of the radiational energy it receives in summer and uses it to heat the atmosphere over the ocean in winter. The soil stores some of the precipitation it receives in winter and uses it to humidify the atmosphere in summer. According to Shukla and Mintz, vegetation and clouds play complementary roles. The clouds convert atmospheric water vapour into liquid water, which is transferred to the soil; the vegetation converts soil water into water vapour which is transferred to the atmosphere.

When bare soil is thoroughly wetted, the soil surface behaves like water insofar as the relative humidity of air in contact with the surface is 100 per cent. According to Monteith (1981) the rate of evaporation is usually very close to the rate for adjacent short vegetation, despite differences in radiative and aerodynamic properties. Loss of water from the soil surface establishes a gradient of water potential that drives water towards the surface from deeper, wetter layers. This process cannot continue indefinitely because the conductivity of soil for water decreases very rapidly as it dries and it is usually only a few days before the rate of evaporation becomes limited by the upward diffusion of liquid water towards the surface.

There are three basic requirements for significant land-surface evapotranspiration: moisture in the soil; vegetation, to transfer the moisture from the soil to the interface with the atmosphere; and energy, to convert that moisture to water vapour. During the temperate summer and in warm climates, much of the energy for evaporation comes from radiational heating of the surface and therefore depends on surface albedo. In cool seasons and climates much of the energy for evaporation comes from sensible heat advection from the atmosphere. During rain, radiation totals may be small and again much of the energy for evaporation may come from atmospheric heat advection. Therefore evaporation under low radiation conditions is going to depend on the rate of atmospheric heat advection which in turn will depend on the turbulence and stability structure of the lower atmosphere. The amount of turbulence depends partly on the height and roughness length of the vegetation. Indeed, recent work (e.g. Thom and Oliver, 1977) has indicated that interception loss from wet tall crops can be a large term in the total water balance of a small catchment.

Figure 2.2 summarizes the water balance of a small catchment covered by a specified vegetation type over a long time period. The water balance of the catchment may be expressed by the two equations:

$$P - I = pP + D + \Delta C \quad (1)$$

$$pP + D - J - E_s = R_o + \Delta M \quad (2)$$

where P is the rainfall rate
- I is the interception loss rate
- p is the throughfall proportion
- D is the drainage rate of the whole canopy
- C is the canopy water storage
- J is the transpiration loss
- E_s is the soil evaporation
- R_o is the runoff
- M is the soil moisture storage

Following the Penman–Monteith equation for evaporation, transpiration

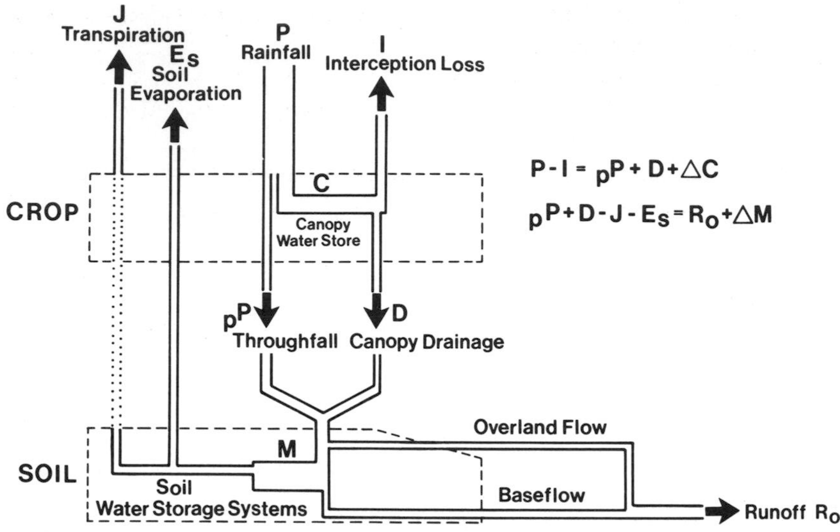

FIGURE 2.2 GENERALIZED WATER BALANCE OF A SMALL VEGETATED CATCHMENT

from vegetation may be expressed by:

$$LJ = \frac{\Delta R_N + \rho C_p \delta e / r_a}{\Delta + \gamma(1 + r_c/r_a)} \quad (3)$$

where Δ is the slope of the saturated vapour pressure versus temperature curve
R_N is the net radiation
δe is the vapour pressure deficit
ρ and C_p are the density and specific heat of air respectively
γ is the psychrometric constant
r_c is the bulk canopy resistance
r_a is the aerodynamic resistance
L is the latent heat of vaporization

For the evaporation rate I of intercepted water when the canopy is wet, r_c becomes zero. Assuming that the albedos of wet and dry canopies are not materially different, then

$$\frac{J}{I} = \frac{\Delta + \gamma}{\Delta + \gamma(1 + r_c/r_a)} \quad (4)$$

when r_c and r_a are of similar size, as they may be in herbaceous communities, then the rate of evaporation of intercepted water will not much exceed the potential transpiration rate in the same conditions. Indeed, for grass it is often assumed that they are the same. In contrast, when r_c is an order of magnitude greater than r_a, as it appears to be in coniferous forest, then intercepted water will be evaporated at 2 to 5 times the current transpiration rate. For well watered arable crops $r_c \simeq r_a \simeq 50\ \mathrm{s\,m}^{-1}$.

The term r_a expresses the 'aerodynamic' resistance to the diffusion of water vapour from the surface itself, where vapour pressure has an unknown value e_o, to some reference level above, when the vapour pressure is e. In ideal conditions of neutral atmospheric stability and large fetch, the resistance to the transfer of momentum to the effective surface of vegetation from a distance z above its zero plane displacement level d (see e.g. Thom, 1975) is:

$$r_a = \{\ln(z/z_o)\}^2/(k^2u) \tag{5}$$

in which z_o is the aerodynamic roughness parameter of the vegetation of the order $h/10$ (where h is vegetation height), u is wind speed, and k is von Karman's constant. However, according to Thom and Oliver (1977), with non-neutral conditions and for water vapour exchange, the theoretically rigorous expression

$$r_a = \{\ln(z/z_o) - \psi\}\{\ln(z/z_o') - \psi'\}/(k^2u) \tag{6}$$

applies, in which each ψ (see Paulson, 1970) is a specified function of the stability of the surface boundary layer, most conveniently expressed in terms of the gradient Richardson number R_i, and where a prime distinguishes a parameter of water vapour exchange from one of momentum exchange. Because of the enhancement of momentum transfer to a rough surface by the action of bluff-body forces (pressure drag) on its individual elements, z_o generally exceeds z_o' (Thom and Oliver, 1977). In particular, for vegetation z_o' is usually about one fifth as large as z_o (Thom, 1972, 1975). Nevertheless, in view of the many sources of error in estimating regional evaporation, Thom and Oliver consider that the simplified expression

$$r_a = \{\ln(z/z_o) - \psi\}\{\ln(z/z_o) - \psi'\}/(k^2u) \tag{7}$$

and equation (5) itself, adequately specify resistances to water vapour exchange. Clearly the theory could be developed further, but the important point that emerges is that both the height and the surface wetness of the vegetation will influence evaporation rates. Indeed, Thom and Oliver consider that where tall vegetation predominates, attempts to unravel the water balance of a vegetated region will be unrewarding unless there is adequate recognition of the effects of surface roughness and surface wetness.

Sellers and Lockwood (1981a, 1981b) have used a multilayer crop model to simulate in a realistic manner the interception and evaporation of rainfall as well as transpiration loss by grass, wheat, pine and oak canopies. In Sellers and Lockwood (1981b) and Lockwood and Sellers (1983) simulations using one year of real data from lowland Britain were presented and the data from these papers can be used in the present analysis. The same weather data are applied to each crop model in turn. Total transpiration losses for pine and grass are very similar, that from grass is some 9 per cent greater than that from pine (see Figure 2.3). The total interception loss (including soil evaporation) from pine is over four times as great as that from grass (252 mm as compared to 63 mm) however, making the annual total evapotranspiration loss some 40 per cent greater than that from grass (560 mm and 398 mm respectively). The effect of this difference on the runoff regimes of the two crops is substantial. The total runoff from the hypothetical pine catchment was predicted to be 152 mm, a little over half the 303 mm predicted for the grass catchment. Furthermore, the peak streamflows generated under grass are often more than 30 per cent higher than those produced by the same storms over pine. Figure 2.3 shows similar variations in interception loss and runoff over oak and wheat.

Interception loss is high throughout the year for pine but shows marked seasonal variations for oak and wheat. The annual courses of both transpiration and interception loss from an oak-covered area are closely related to the state of the crop's foliage. The leafless crop gives rise to a much reduced rate of interception loss due to minimal water storage in the canopy, high throughfall coefficient, and small maximum exposed wetted area. The interception loss from wheat foliage was just over 5 per cent of the annual precipitation. This low value (compared to the equivalent of 29 per cent for pine) is the result of the relatively short period of full plant cover in the summer and the complete absence of foliage during other wetter months. The low interception loss from wheat is also reflected in the very high annual runoff totals.

Though the transpiration rate from grass is relatively large, the interception loss is small compared with forest. This means that considerably more of the rainfall penetrates through the vegetation canopy and into the soil moisture store as compared with forest areas. Therefore the predicted soil moisture levels are generally higher under grass than under forest, and particularly than under coniferous forest. The oak model predicts the most extreme runoff regime of the four crops as there is no transpiration and little interception loss during the winter months and, by contrast, the highest predicted transpiration loss of the four crops during the summer. As a result, the oak model yields a soil moisture content roughly equivalent to that of wheat and slightly higher than that for grass over the first 90 days of the year. Thereafter, the soil moisture content drops rapidly as the canopy develops and falls below that of pine at day 189. The oak forest is then predicted to maintain the lowest level of moisture until day 302 when the effects of leaf fall reduce interception loss and halt transpiration. As a result of the unbalanced annual distribution of evapotranspiration loss, the oak forest's runoff regime is characterized by high peak flows and high levels of soil moisture in winter and an extended period of low flows during the summer. The numerical model predicts the highest runoff loss of all four crops from wheat. This is because of the long periods of the year during which the soil is bare or nearly bare. Thus the soil moisture content is maintained at a high level during the whole year and only drops below that of grass for a brief period during the late summer. From day 250 to day 365 the bare soil model used to represent the post-harvest state of the crop predicts the wettest soil moisture conditions.

Shukla and Mintz (1982) suggest that the soil/vegetation complex acts to transfer water from winter to summer and to humidify the summer atmosphere. This is illustrated in Figure 2.4, where predicted atmospheric water balances from the Sellers and Lockwood model for the spring and summer months of May, June, July, and part of August (days 120–225) are presented. Each vegetation type is exposed to exactly the same weather conditions, so differences in water balance are due to differing vegetation properties. At the start of the period soil moisture contents were high, and then fell with evaporation over the period. Heavy rain fell at the end of August, recharging the soil to field capacity and causing the negative atmospheric balances observed for July and August in Figure 2.3. All vegetation surfaces show positive atmospheric moisture balances, indicating that moisture is indeed being transferred from the winter to the summer atmosphere via the soil/vegetation complex. The properties of the grass surface are very different from those of the trees. The interception loss (including soil evaporation) from grass is only about 12 per cent as compared with about 34 per cent from oak and 42 per cent from pine. This implies a large degree of cycling of water between the tree canopies and the storms, in contrast to the case of grass where the coupling is relatively small. This is again reflected in the atmospheric water balances, that of grass being only about half of those for trees. The implication is that trees are considerably more effective in humidifying the atmosphere during the summer than grasslands. This is also reflected in the soil moisture storage at the end of the period on day 225, when the grass soil moisture store contains about 40 mm more moisture than the stores under oak or pine (Table 2.1). Correspondingly the predicted runoff from grass during the period of heavy rain at the end of August is up to 52 mm greater than that from oak or pine. This suggests that grass covered landscapes may generate higher peak flows after rainfall in summer than forest covered landscapes.

In summer the differences between the interception loss from various vegetation types may be less than in winter. Figure 2.5 shows the evaporation from wet canopies under meteorological conditions typical of Britain in winter and probably also under cloudy rainy conditions in summer. It is seen that the differences in interception loss between tall and short vegetation types increases rapidly with increasing geotrophic wind speed. Figures 2.6 and 2.7 are probably more typical of conditions under sunnier conditions in summer and show that under low wind speed conditions the differences in interception loss between short and tall vegetation types are much reduced. In contrast,

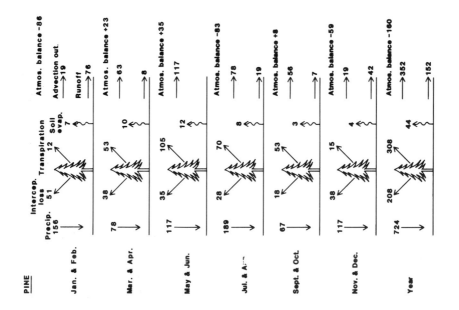

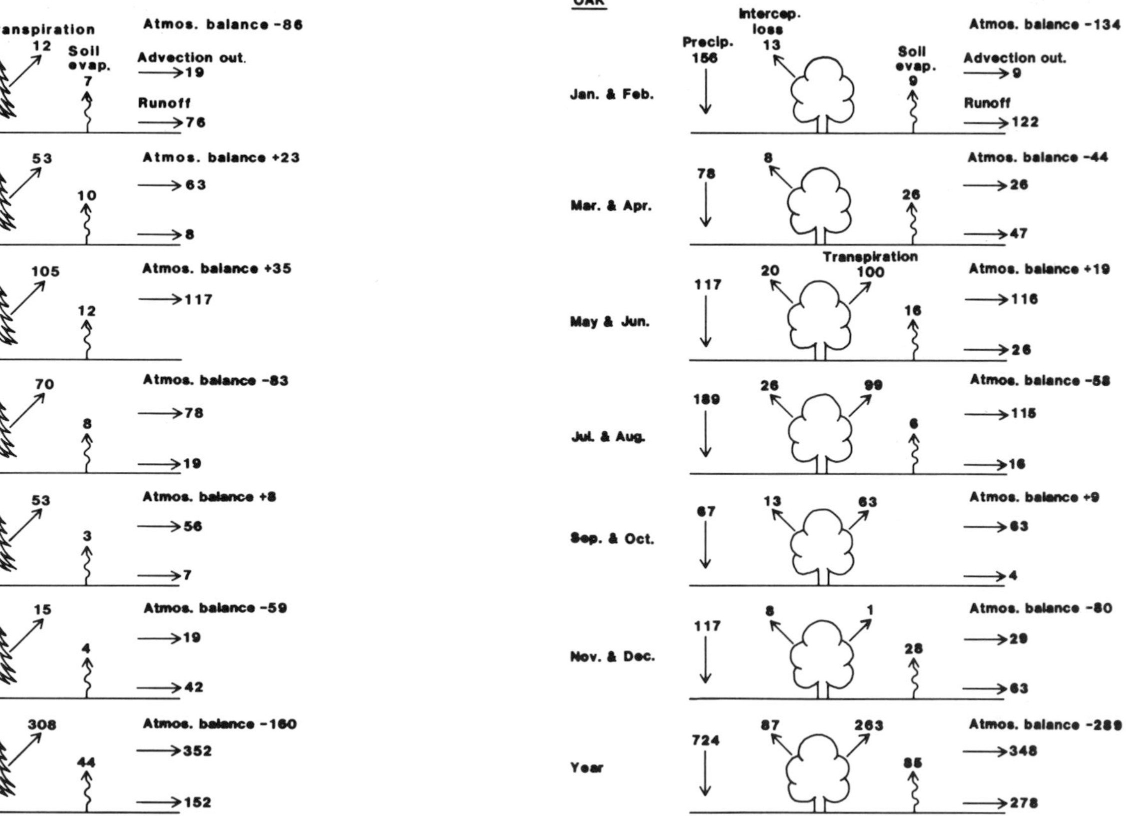
PINE
Jan. & Feb.
Precip. 156
Intercep. loss 51
Transpiration 12
Soil evap. 7
Atmos. balance -86
Advection out. 19
Runoff 76
Mar. & Apr.
78
38
53
10
Atmos. balance +23
63
8
May & Jun.
117
35
105
12
Atmos. balance +35
117
Jul. & A
189
28
70
8
Atmos. balance -83
78
19
Sept. & Oct.
67
18
53
3
Atmos. balance +8
56
7
Nov. & Dec.
117
38
15
4
Atmos. balance -59
19
42
Year
724
208
308
44
Atmos. balance -160
352
152
OAK
Jan. & Feb.
Precip. 156
Intercep. loss 13
Soil evap. 9
Atmos. balance -134
Advection out. 9
Runoff 122
Mar. & Apr.
78
8
26
Atmos. balance -44
26
47
May & Jun.
117
20
Transpiration 100
16
Atmos. balance +19
116
26
Jul. & Aug.
189
26
99
6
Atmos. balance -58
115
16
Sep. & Oct.
67
13
63
Atmos. balance +9
63
4
Nov. & Dec.
117
8
1
28
Atmos. balance -80
29
63
Year
724
87
263
85
Atmos. balance -289
348
278

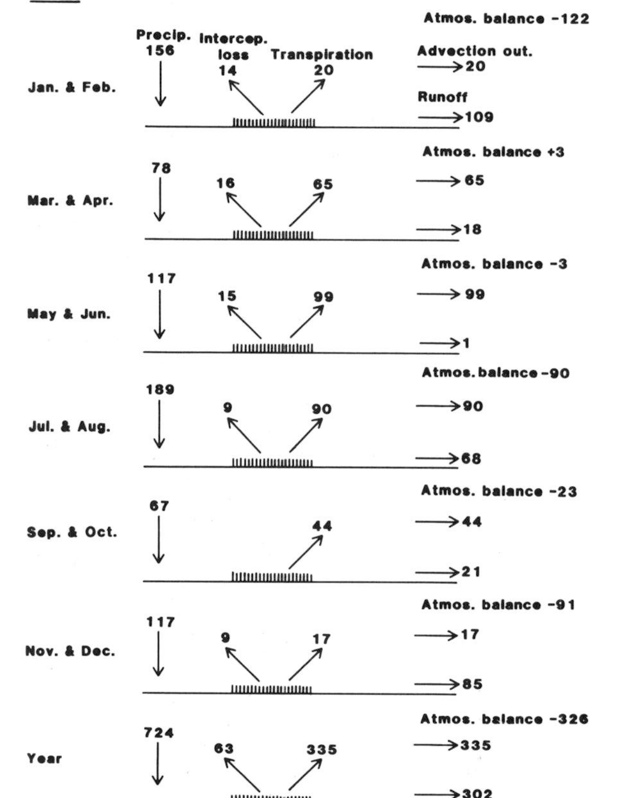

FIGURE 2.3 SIMULATED WATER BALANCE OF GRASS, PINE, OAK, AND WHEAT COVERED LANDSCAPES

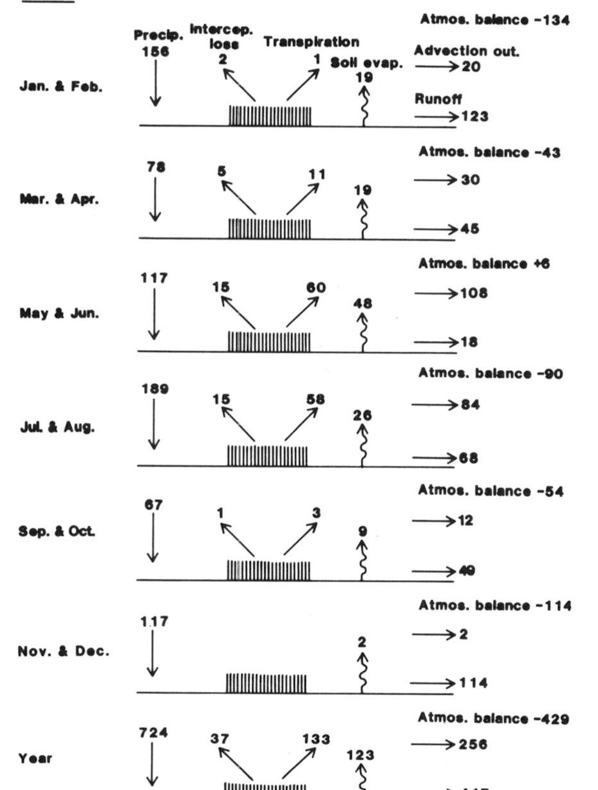

The same weather data were used in each case, but the multilayer crop model took into account the vegetation characteristics. Units are mm rainfall equivalents

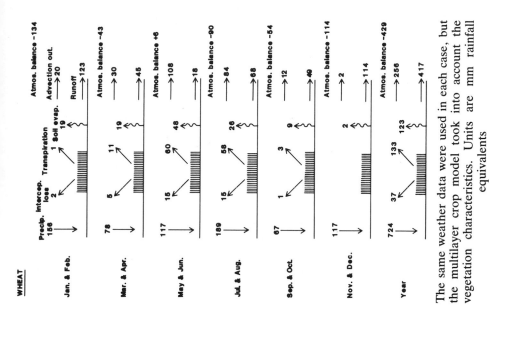

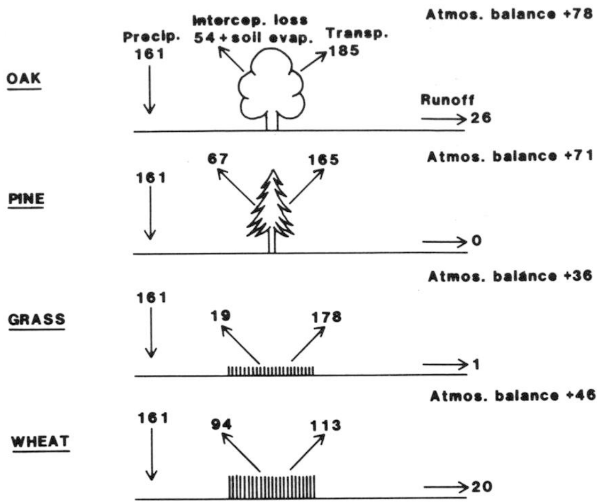

FIGURE 2.4 SIMULATED WATER BALANCE OF GRASS, PINE, OAK, AND WHEAT COVERED LANDSCAPES FOR DAYS 120–225
Units are mm rainfall equivalents

TABLE 2.1 SIMULATED WATER BALANCES FOLLOWING HEAVY RAIN AFTER A DRY PERIOD
Units are rainfall equivalents in mm

	Grass	Wheat	Oak	Pine
Soil moisture, day 225 (end of dry period)	67	64	24	26
Soil moisture, day 240 (after heavy rain)	127	126	124	125
Runoff, days 225–240	68.0	67.6	15.8	19.1

(After Lockwood and Sellers, 1983)

transpiration losses from oak forest are considerably increased under sunny warm conditions and will lead to a marked drying out of the soil.

These results may be of some palaeohydrological interest, particularly as Britain was once covered by deciduous forest. They suggest that the clearance of deciduous forest will result in changes in the hydrological regime even in the absence of any climatological changes. Almost certainly under deciduous forest surface runoff will have been much more concentrated in the late winter than now, with the summers and autumns lacking significant runoff. If the trees were deeply rooted, thus allowing extensive summer transpiration, and the growing season reasonably long, then it is likely that annual runoff will have been significantly reduced under deciduous forest compared with the present.

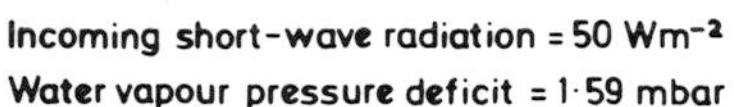

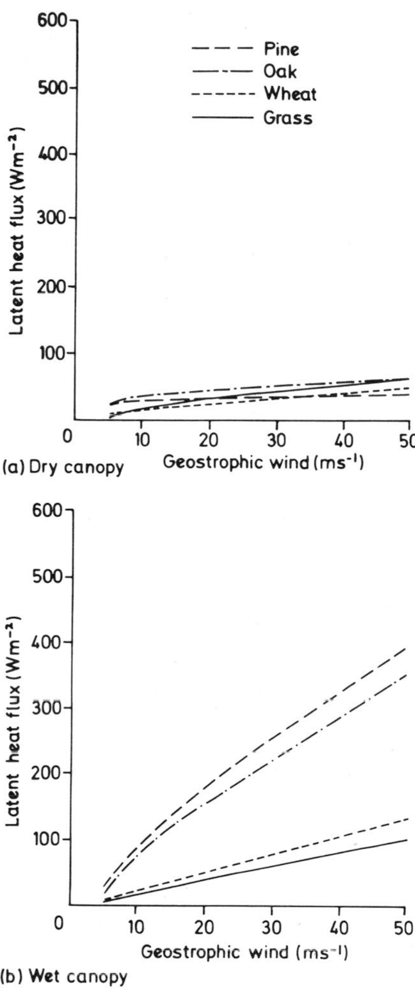

FIGURE 2.5 PREDICTED EVAPOTRANSPIRATION RATES FOR CANOPIES UNDER CONDITIONS OF LOW EVAPORATIVE POTENTIAL
(From Lockwood and Sellers, 1983)

Soil moisture and deserts

Idso *et al*. (1975) have shown that the albedo of drying bare soil, normalized for sun zenith angle effects, is a linear function of soil water content of a very thin surface layer (less than 0.2 cm thick) over a sizeable volumetric water content

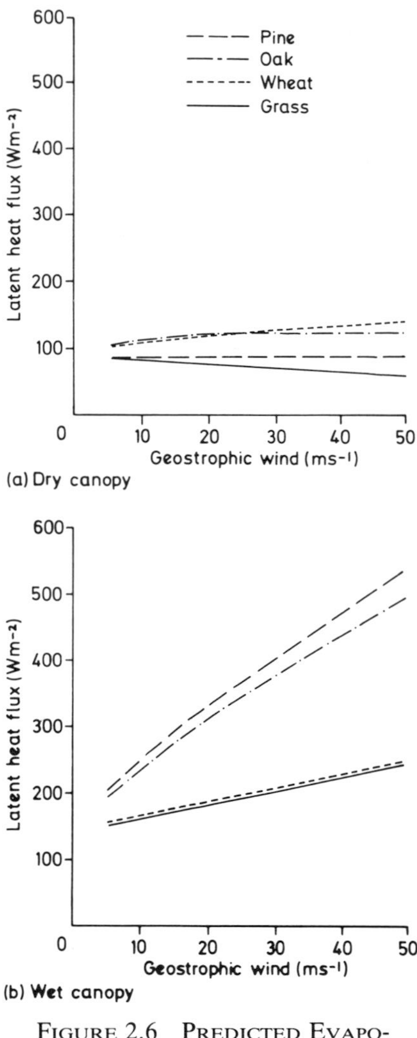

FIGURE 2.6 PREDICTED EVAPO-TRANSPIRATION RATES FOR CANOPIES UNDER CONDITIONS OF LOW VAPOUR PRESSURE DEFICIT
(From Lockwood and Sellers, 1982)

range. They also found that albedo is well correlated with the average soil water content of greater soil thicknesses. For Avondale loam soil at Phoenix, Arizona, U.S.A., they discovered that normalized albedo values varied from about 14 per cent for wet soil to just above 30 per cent for dry soils.

Charney (1975) has suggested that changes of albedo could be an important

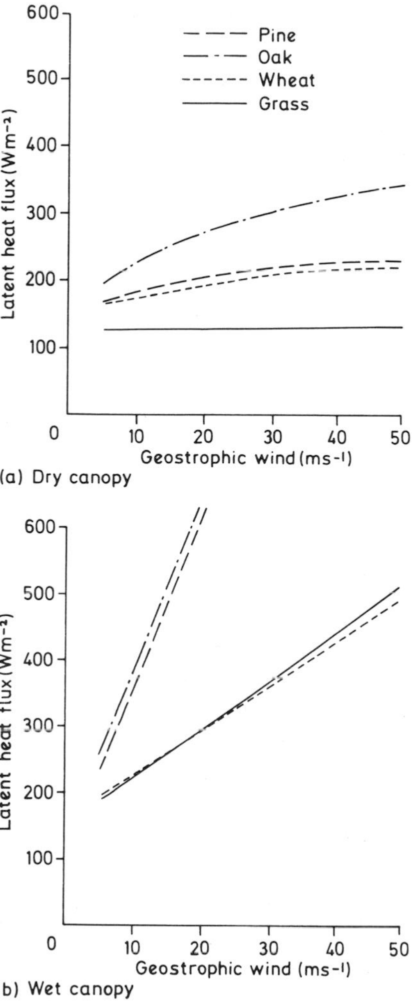

FIGURE 2.7 PREDICTED EVAPOTRANSPIRATION RATES FOR CANOPIES UNDER CONDITIONS OF MODERATE EVAPORATIVE POTENTIAL
(From Lockwood and Sellers, 1982)

factor in maintaining or creating deserts. He postulated that lack of rainfall leading to lack of vegetation will result in a higher albedo. The consequent radiation deficit requires sinking motion to maintain the heat balance and this leads to anticyclonic conditions, additional drying, and thus maintenance of the status quo. As indicated above, Charney's ideas depend on soil moisture, but in addition the amount of moisture in the soil may influence the weather more directly. This may occur because soil moisture effects evaporation and hence changes the proportion of net incoming radiation available as latent heat compared to that available as sensible heat. Walker and Rowntree (1977) recently reported two numerical experiments using a simple version of the Meteorological Office tropical atmosphere model. They used a simple zonally symmetrical land–sea distribution with land surface initially wet (W case) and then part wet part dry (D case). Albedo, radiation budget and all other aspects of the model were kept the same in order to isolate the response of the model to soil moisture. Evaporation was estimated in two phases. Firstly, actual evaporation, E_a, was assumed to be equal to the potential value, E_T, while the available water in the root layer of vegetation was more than half of its maximum value (taken as 10 cm). Then, for residual available water in the range $5 > \theta > 0$ cm, the ratio E_a/E_T was taken as $\theta/5$. The model area was divided into two parts to approximate to the real Africa, namely land north of 6°N and sea south of 6°N. In the D case it was assumed that there was no soil moisture in the band 14–32°N (approximately the Sahara) and for the remaining land 10 cm soil moisture was assumed. In the W case, 10 cm soil moisture was assumed for all land. The experiments were run for 10 model days (D case) and 20 model days (W case). The partitioning of energy between latent heat of evaporation and sensible heat flux was found to be determined by the dryness of the ground and by surface temperature. The experiments suggested positive feedback between the surface and the atmosphere: soil moisture in the W case aids rainfall which maintains the soil moisture; soil dryness in the D case inhibits rainfall, allowing persistence of the dry regime.

Snow and ice surfaces

General theoretical equations for snowmelt usually express snowmelt rate as a function of wind speed, humidity, rainfall, and solar radiation, as well as air temperature. For a monthly parameterization it is possible to neglect all terms except those concerning temperature and solar radiation. For a study of snowmelt during past periods it is particularly important to include solar radiation because it has varied during the ice ages due to changes in the earth's orbit (the Milankovitch mechanism). A suitable parameterization has been obtained by Pollard (1980) and tested on a variety of present-day data. According to Pollard (1980) monthly mean snowmelt may be estimated as follows:

$$\text{Monthly mean ablation} = \max\,[0; aQ + bT + c] \qquad (8)$$

where a, b, and c are constants based on crude surface energy balance estimates and actual glacial measurements; Q is the monthly mean solar radiation incident at the top of the atmosphere (W m^{-2}); and T is the monthly mean air temperature (°C). Pollard (1978) arrived at the following values for a, b, and c; $a = 0.32$ (g cm^{-2} month^{-1}), $b = 10.0$ (g cm^{-2} month^{-1})(°C)$^{-1}$, and $c = 47.0$ (g cm^{-2} month^{-1}). Therefore monthly mean ablation may be expressed in g cm^{-2} or to a good approximation cm depth rainfall equivalent. Ablation is defined as the reduction in mass of a vertical column of an ice/snow body due to the removal of H_2O out of the column by surface processes. It therefore includes not only melting and runoff but also evaporation. There will probably be significant discrepancies between equation (8) and particular examples of ablation, but such discrepancies should not be serious for the ice age problem as long as the summer months during which most ablation occurs are predicted reasonably well. Discrepancies will probably arise mainly because of seasonal variations of surface albedo. Equation (8) has been used in a zonal climate–ice sheet model by Pollard *et al.* (1980) and by Lockwood (1982).

Atmospheric–ocean interactions

Before considering ocean–atmosphere interactions it is useful to consider a few physical facts about the ocean. One of the most obvious is its high thermal inertia, which is 1200 times greater than that of the atmosphere. However, the stabilizing influence may not be as great as this simple calculation suggests, because of the different way the ocean is heated compared with the atmosphere. The latter is heated from below, particularly at low latitudes, and cooled from above, particularly at high latitudes. This distribution of heat sources and sinks maintains an atmosphere close to convective instability which readily overturns and mixes. In contrast both the heat sources and sinks occur at the surface of the ocean, and this results in a very stable vertical temperature structure in the tropical and equatorial ocean which inhibits vertical convection. Only in a few localized regions of the polar oceans is the whole water column close to convective instability. Furthermore, in the Arctic where there is strong cooling, convection is inhibited because of the presence of a large salinity gradient, or a halocline, at a depth of 200 m. The halocline is the result of the influx of fresh water from glaciers, runoff from the surrounding land masses, and precipitation, and this fresh water overlies the relatively warmer but more saline oceanic water. Because of the stability of the ocean the transfer of heat between the buoyant surface layers and the deeper ocean is strongly inhibited and for most of the time the deep layers have little direct influence on the atmosphere.

THE ASTRONOMICAL THEORY OF ICE AGES AND SIMPLE ENERGY BALANCE MODELS

Records of global ice volume for the past 700 K years, based on oxygen isotopic data from deep-sea cores and reflecting mainly the changing Northern

Hemisphere ice sheets, show a dominant cycle of roughly 100 K year period. The records also show smaller amplitude oscillations with spectral peaks at roughly 40 and 20 K year periods. It has been speculated that these variations in ice volume are the result of Milankovich (1941) insolation variations due to perturbations in the Earth's orbital parameters.

The astronomical theory of the ice ages has been investigated by Pollard, Ingersoll, and Lockwood (1980) using a simple climate model which includes ice sheets explicitly. The model has previously been outlined in Pollard (1978) and is shown in schematic form in Figure 2.8. There are two distinct parts corresponding to the two distinct timescales of the global seasonal weather and the long-term ice sheet response.

Following North (1975), the weather through one year over a spherical globe is described by a zonally averaged, one-level energy-balance equation for sea-level air temperature T:

$$\frac{\partial}{\partial t}[CT(x,t)] - \frac{\partial}{\partial x}\left[(1 - x^2)\frac{D}{R^2}\frac{\partial}{\partial x}(CT)\right] + [A + BT] = Q(x,t)(1-\alpha) + S. \quad (9)$$

Here x is sin (latitude) and t is time. All dependent variables are defined as ~1 month running-means, so daily correlations are effectively assumed constant. Boundary conditions are $(1 - x^2)^{1/2}\,\partial T/\partial x = 0$ at the poles $x = \pm 1$. The variables and their values for the 'standard' model are:

C = 4.6×10^7 J m^{-2} °C^{-1}, a constant seasonal heat capacity of the atmosphere–land–ocean system (equivalent to a layer of liquid water 11 m thick).

D = 0.501×10^6 m^2 s^{-1}, a linear diffusion coefficient acting over the whole thickness of the layer represented by C.

R = 6.36×10^6 m, radius of the Earth.

A = 207 W m^{-2}, B = 1.9 W m^{-2} °C^{-1}, net infrared radiation coefficients.

Q = zonal mean insolation at the top of the atmosphere, computed for any given era from orbital elements using a solar constant of 1360 W m^{-2}.

α = $r\alpha_c + (1 - r)\alpha_f$, earth–atmosphere albedo. $\alpha_c = 0.62$ represents areas covered by seasonal snow or ice sheet, and $\alpha_f = 0.31 + 0.08[(3x^2 - 1)/2]$ represents areas free of seasonal snow and ice sheet.

$r = 1$ north of 75°N to represent perennial Arctic Ocean sea-ice, and $r = 1$ south of 70°S to represent a fixed Antarctic ice sheet. At all other latitudes $r = 0.6$ when covered by seasonal snow or ice sheet, and $r = 0$ when free of seasonal snow and ice sheet.

S = 1.27 [J m^{-2} s^{-1}] per [g cm^{-2} month^{-1}], representing latent heat of

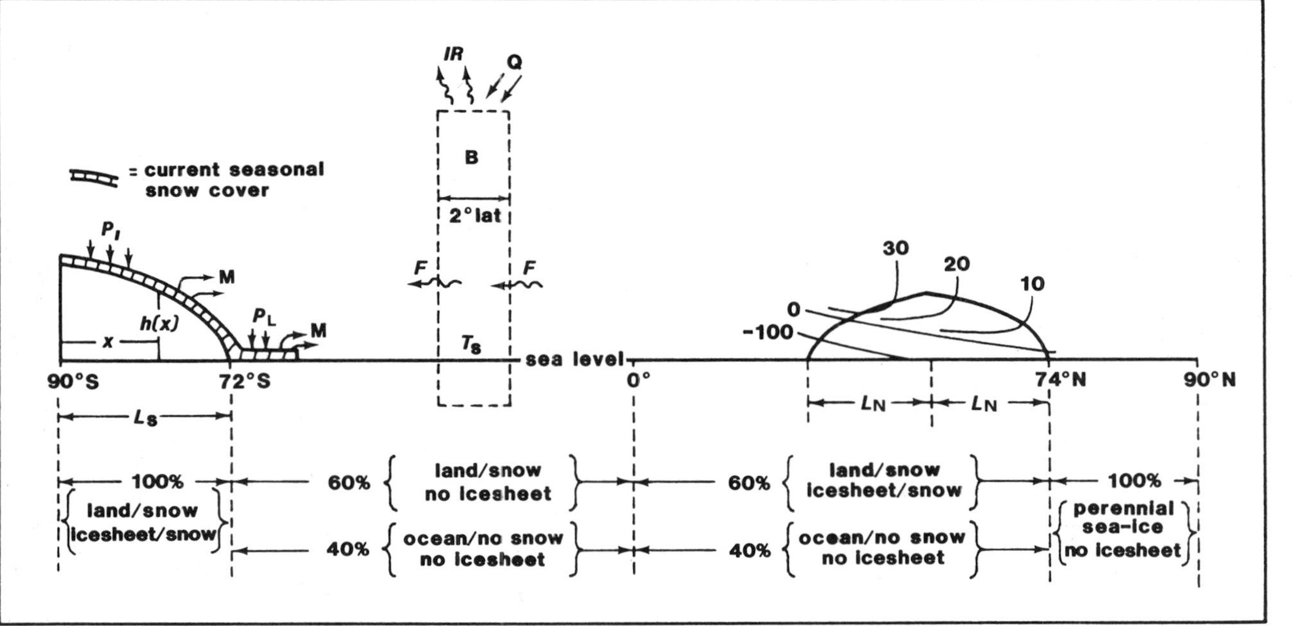

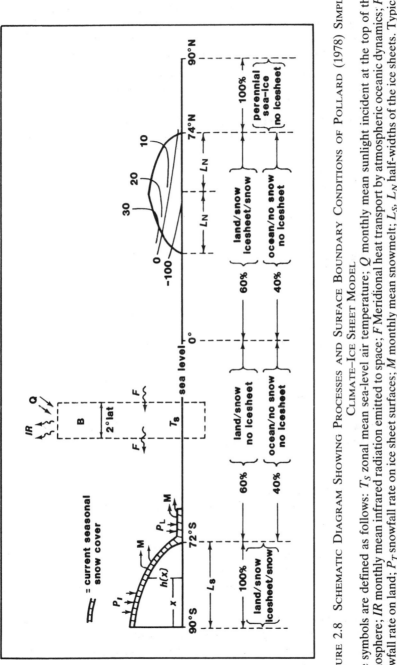

FIGURE 2.8 SCHEMATIC DIAGRAM SHOWING PROCESSES AND SURFACE BOUNDARY CONDITIONS OF POLLARD (1978) SIMPLE CLIMATE–ICE SHEET MODEL

The symbols are defined as follows: T_S zonal mean sea-level air temperature; Q monthly mean sunlight incident at the top of the atmosphere; *IR* monthly mean infrared radiation emitted to space; F Meridional heat transport by atmospheric oceanic dynamics; P_L snowfall rate on land; P_T snowfall rate on ice sheet surfaces; M monthly mean snowmelt; L_S, L_N half-widths of the ice sheets. Typical pattern of net accumulation minus ablation is shown for northern ice sheet; units are g cm^{-2} month^{-1}. Reprinted by permission from *Nature* Vol. 272, pp 233. Copyright © 1978a Macmillan Journals Limited

fusion released or required at each latitude by the varying amount of seasonal snow cover. The annual mean of S at each latitude is zero.

Most of these parameterizations (discussed, for instance, in Coakley, 1979) have found general use in many annual mean energy balance models and are based on annual mean data.

Seasonally varying snow cover on land and ice sheet surfaces is modelled diagnostically by parameterizing monthly snowmelt and snowfall as functions of the current air temperature T and insolation Q. For snowmelt equation (8) was used, and snowfall was assumed if the temperature was below 0°C.

Ice sheets are incorporated into the weather model following Weertman's (1976) simple treatment. Ice sheet flow under its own weight is approximated to be perfectly plastic, which constrains the model ice sheet profiles to always remain parabolic:

$$h(s) = [\lambda(L - |s|)]^{1/2}, \tag{10}$$

where h is the elevation of the ice sheet surface above sea-level, L is the half-width and s is the latitudinal distance from the ice sheet centre (with s taken positive towards the equator). The ice sheet base is assumed to be isostatically depressed to depths 0.5 $h(s)$ below sea-level. λ is a constant proportional to the yield stress of ice; for our standard model we use $\lambda = 10$ m, corresponding to a yield stress of ~0.7 bars.

The model ice sheet, representing the Laurentide and Scandinavian ice sheets of past eras, is constrained to extend equatorward with its northern tip fixed at 75°N (corresponding to the Artic Ocean shoreline). Where the margins of the real northern hemispheric ice sheets reached continental coastlines, further advance was prevented by rapid iceshelf and iceberg calving into the ocean, but their equatorial extent and overall volume were probably limited more by ablation on their southern flanks. Therefore, as in Weertman (1976), the long-term variation in the model ice sheet size is controlled by the net accumulation (snowfall) minus ablation (mostly snowmelt) on its southern half only (i.e. $s > 0$). Also, since its profile is constrained by equation (10), any change in size is determined simply by the total ice volume added to or removed from the entire southern half.

Model ice age curves were generated for the last several 100 K years by computing the seasonal climate as above once every 2 K years, with insolation calculated from actual Earth orbit perturbations. The change in ice sheet size for each 2 K year time step depends only on the net annual snow budget integrated over the whole ice sheet surface. In these model runs, the equatorward tip of the northern hemisphere's ice sheet oscillates through ~7° in latitude, correctly simulating the phases and approximate amplitude of the higher frequency components (~43 K year and 22 K year) of the deep-sea core data (Hays *et al.*, 1976). However, the model failed to simulate the dominant glacial–interglacial cycles (~100 to 120 K years) of this data.

Oerlemans (1980, 1981) has carried out experiments with a northern hemisphere ice sheet model that shows that the 100 K year cycle and its sawtooth shape may be explained by ice sheet/bedrock dynamics alone. According to Oerlemans this cycle seems to be an internally generated feature and is not forced by variations in the eccentricity of the Earth's orbit. There is a tendency to restore isostatic equilibrium under ice sheets, and in the simplest form this may be formulated by:

$$\frac{\partial h}{\partial t} = -w(H^* + 2h) \tag{11}$$

where h is the height of the bedrock with respect to its equilibrium value (the case of no ice sheet)
H^* is the elevation of the ice surface above sea level, and the time scale for adjustment is $\frac{1}{2}w$

In the steady state $h = \frac{1}{2}H^*$, which corresponds to an isostatic balance if the rock density is three times that of ice. In the numerical experiments carried out by Oerlemans (1981), the bedrock sinks or rises to restore the isostatic balance, and this mechanism has a timescale of $t = \frac{1}{2}w$. Now Oerlemans found that in the case of no bedrock adjustment ($t = \infty$) the ice sheet grows to a steady state in about 40 K years. If isostatic adjustment with a long time scale is included (t = 30 K years), the picture changes drastically. The ice sheet grows during about 50 K years, but then the bedrock has sunk so much that a large part of the ice surface comes below the equilibrium line which causes the ice sheet to decay rapidly. For smaller timescales for bedrock sinking, this feature occurs earlier.

Birchfield, Weertman, and Lunde (1982) have examined the response of a combined global zonal-averaged energy balance model and a continental ice sheet model to insolation anomalies produced by Earth orbital perturbations. They found that the presence of a high-latitude plateau significantly increases the sensitivity of the climate to the insolation perturbations. The sensitivity is maximised when the elevation of the summer snow line is near the elevation of the plateau, as appears to be the case with Baffin Island today. Both Birchfield, Weertman, and Lunde (1981) and Oerlemans (1980, 1981) have obtained encouraging agreement with some features of the glacial cycle by using a simple ice sheet model with a realistic time lag in the response of the bedrock to the ice load. Pollard (1982) has extended their basic model, first by including topography to represent high ground in the north. Improved results can then be obtained but only with unrealistic parameter values and for some aspects of the record. Pollard obtained further improvements by crudely

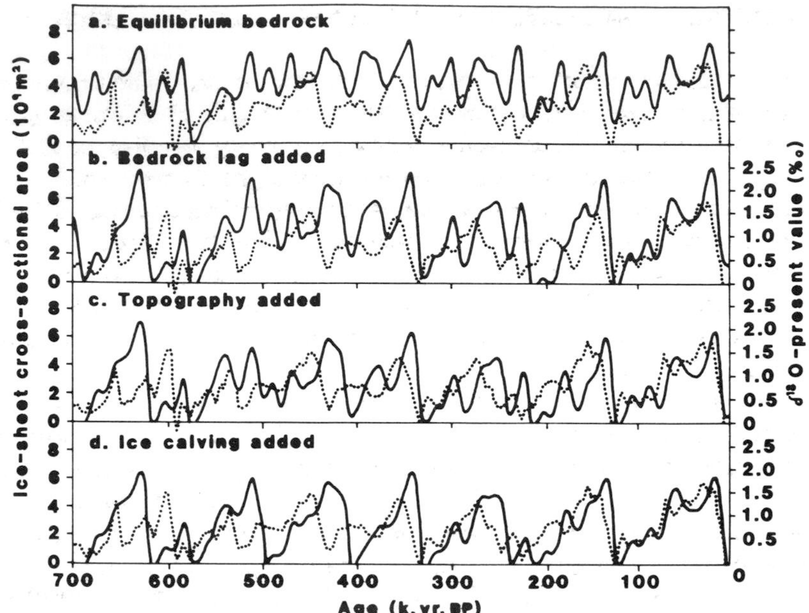

FIGURE 2.9 TOTAL CROSS-SECTIONAL AREA OF ICE VERSUS TIME FOR VARIOUS MODEL VERSIONS (solid curves)
The corresponding approximate ice volume can be obtained by multiplying by a typical east–west ice sheet dimension (~ 3000 km). The dotted curve in each panel is an oxygen isotope deep-sea core record minus its present value. (After Pollard, 1982). Reprinted by permission from *Nature* Vol. 296 pp 335. Copyright © 1982 Macmillan Journals Limited

parameterizing possible calving at the equatorward ice sheet tip during deglaciation by proglacial lakes and/or marine incursions. The resulting ice volume curves agree fairly well with the observed records and their power spectra over the past 700 K years (Figure 2.9).

Prell and Start (1982) have shown that two distinct climatic regimes occurred within the Quaternary. In both modes, spectra of indices show significant concentration of variance in frequency bands identified with orbital variation. However, the partitioning of climatic variance among the various orbital frequencies is distinctly different between the two modes. For example, the late Quaternary mode has significant concentrations of variance at periods around 100 K and 23 K years, whereas the early Quaternary mode spectra have more power at 41 K and 19 K years. So there are still major problems in the numerical modelling of ice sheet variations.

GENERAL CIRCULATION MODELS AND CLIMATIC CHANGE

General circulation models have the ability to portray many of the non-linear processes which serve to regulate atmospheric (and hence climatic) changes. In recent years they have been used extensively to study the climatic changes resulting from variations in one particular aspect of the climatic system. Such studies cover variations in the solar constant (Wetherald and Manabe, 1975), changes in atmospheric CO_2 content (Manabe and Stouffer, 1980), changes in vegetation cover (Potter *et al.*, 1981), changes in soil moisture content (Shukla and Mintz, 1982), etc. The climatic system is extremely complex and progress can only be made in the understanding of the system by considering variations in one particular aspect of it.

In the context of palaeohydrology interest focuses on those GCMs that model changes in the hydrological cycle. This involves not only producing realistic mathematical models of the atmosphere but also of surface/atmosphere interactions. Some models of surface/atmosphere interactions were discussed earlier in this chapter, and of particular interest are changes caused in the hydrological cycle by variations in surface vegetation. Simply altering ground cover affects surface albedo and runoff, changes the ratio of sensible to latent heat transport, and greatly modifies the surface winds. These variations in turn cause soil moisture, temperature, and erosion rates to change.

Examples of recent studies of vegetation change on regional climate are those by Potter *et al.* (1981) and Shukla and Mintz (1982). Using a statistical dynamical climate model, Potter *et al.* (1981) have computed the combined impact of desertification of the Sahara and deforestation of the tropical rain forest. The model is similar in many ways to GCMs in that it uses the basic conservation equations to calculate zonal mean fields of pressure, temperature, wind and humidity. For the experiment, two model equilibrium states were achieved: a control case with standard surface conditions corresponding to a pristine environment; and a perturbed case corresponding to the suggestion of Sagan *et al.* (1979) concerning albedo changes on tropical deforestation and desertification (see Table 2.2). The surface bulk transfer coefficients, the thermal and hydrologic characteristics were not altered; the only surface parameter modified for those areas perturbed was the surface albedo which, after modification, was held constant. The model computed a surface cooling of 0.6 K for the northern hemisphere, mostly concentrated at high latitudes.

CLIMAP (1976) have made a reconstruction of global sea surface temperature, ice extent, ice elevation, and continental albedo for the northern hemisphere summer at 18 K years B.P. Given the boundary conditions determined by CLIMAP for 18 K years B.P., both Gates (1976a, b) and Manabe and Hahn (1977) have used global general circulation models to

TABLE 2.2 ALBEDO CHANGES ON TROPICAL DEFORESTATION AND DESERTIFICATION

Process	Land type change	Centred on latitude bands	Surface albedo change
Desertification	Savanna to desert	20°N	0.16 → 0.35
Tropical deforestation	Forest to Savanna	10°S and equator	0.07 → 0.16

(After Sagan *et al.*, 1979)

simulate the climate of this time when the northern ice sheets were at their maximum extent. Both models estimate significantly lower precipitation rates for July and August over the tropical and subtropical continents than are observed today. Because tropical Africa and southern Asia receive a significant fraction of their precipitation during the summer monsoon this probably indicates drier annual conditions. Unfortunately, the two general circulation models yield different detailed results for Ice Age precipitation. There is some evidence (Lockwood, 1979) that lowland Britain was also drier during the Ice Age than at present. At this time much of lowland British precipitation probably came as showers during the summer with the winters being dry.

General circulation models have not been applied widely to the problem of simulating past climates. This is partly because they are extremely expensive in computer time, and also without a good knowledge of the boundary conditions at the time, which we do not have at present, the predictions will be of little value. For instance, atmospheric CO_2 content is known to have varied widely in the past (Oeschger *et al.*, 1980), and this will have had some influence on global climate. In the long-term significant advances will be made in the study of past climates by using GCMs with detailed contemporary boundary conditions, and this will probably be the direction of future research.

REFERENCES

Birchfield, G. E., Weertman, J., and Lunde, A. T. 1981. A paleoclimatic model of northern hemisphere ice sheets. *Quaternary Research*, **15**, 126–142.

Birchfield, G. E., Weertman, J., and Lunde, A. T. 1982. A model study of the role of high-latitude topography in the climatic response to orbital insolation anomalies. *Journal Atmospheric Sciences*, **39**, 71–86.

Charney, J. 1975. Dynamics of deserts and drought in the Sahel. *Quarterly Journal Royal Meteorological Society*, **101**, 193–202.

CLIMAP Project Members. 1976. The surface of ice-age earth. *Science*, **191**, 1131–1137.

Coakley, Jr., J. A. 1979. A study of climate sensitivity using a simple energy balance model. *Journal Atmospheric Sciences*, **36**, 260–269.

Gates, W. C. 1976a. Modelling the ice-age climate. *Science*, **191**, 1138–1144.

Gates, W. C. 1976b. The numerical simulation of ice-age climate with a global general circulation model. *Journal Atmospheric Sciences*, **33**, 1844–1873.

Hasselmann, K. 1981. Construction and verification of stochastic climate models. In Berger, A. (Ed.), *Climatic Variations and Variability: Facts and Theories*, Reidel, Dordrecht, 481–497.

Hays, J. D., Imbrie, J., and Shackleton, N. J. 1976. Variations in the Earth's orbit: pacemaker of the ice ages. *Science*, **194**, 1121–1132.

Idso, S. B., Jackson, R. D., Reginato, R. J., Kimball, B. A., and Nakayama, F. S. 1975. The dependence of bare soil albedo on soil water content. *Journal Applied Meteorology*, **14**, 109–113.

Lockwood, J. G. 1979. Water balance of Britain, 50,000 yr B.P. to the present day. *Quaternary Research*, **12**, 297–310.

Lockwood, J. G. 1982. Snow and ice balance in Britain at the present time, and during the last glacial maximum and late glacial periods. *Journal Climatology*, **2**, 209–231.

Lockwood, J. G., and Sellers, P. J. 1983. Some simulation model results of the effect of vegetation change on the near-surface hydroclimate. In Street-Perrot, A., Beran, M. and Ratcliffe. R. (Eds.), *Variations in Global Water Budget*, Reidel, Dordrecht, in press.

Lockwood, J. G., and Sellers, P. J. 1983. Comparisons of interception loss from tropical and temperate vegetation canopies. *Journal Applied Meteorology*, **21**, 1405–1412.

Manabe, S., and Hahn, D. G. 1977. Simulation of the tropical climate of an ice-age. *Journal Geophysical Research*, **82**, 3889–3911.

Manabe, S., and Stouffer, R. J. 1980. Sensitivity of a global climate model to an increase of CO_2 concentration in the atmosphere. *Journal Geophysical Research*, **85**, 5529–5554.

Milankovitch, M. 1941. *Canon of Insolation and the Ice Age Problem*, Royal Serbian Academy Special Publication 133, Belgrade, translated by Israel Program for Scientific Translation, Jerusalem, 1969.

Monteith, J. L. 1981. Evaporation and surface temperature. *Quarterly Journal Royal Meteorological Society*, **107**, 1–27.

North, G. R. 1975. Theory of energy–balance climate models. *Journal Atmospheric Sciences*, **32**, 2033–2043.

Oerlemans, J. 1980. Model experiments on the 100,000 yr. glacial cycle. *Nature*, **287**, 430–432.

Oerlemans, J. 1981. Some basic experiments with a vertically integrated ice sheet model. *Tellus*, **33**, 1–11.

Oeschger, H., Siegenthaler, U., and Heimann, M. 1980. The carbon cycle and its perturbation by man. In Bach, W., Panksath, J., and Williams, J. (Eds.), *Interactions of Energy and Climate*, Reidel, Dordrecht, 107–127.

Paulson, C. A. 1970. The mathematical representation of wind speed and temperature profiles in the unstable atmospheric surface layer. *Journal Applied Meteorology*, **9**, 857–861.

Pollard, D. 1978. An investigation of the astronomical theory of ice ages using a simple climate–ice sheet model. *Nature*, **272**, 233–235.

Pollard, D. 1980. A simple parameterization for ice sheet ablation rate. *Tellus*, **32**, 384–388.

Pollard, D. 1982. A simple ice sheet model yields realistic 100 K yr. glacial cycles. *Nature*, **296**, 334–338.

Pollard, D., Ingersoll, A. P., and Lockwood, J. G. 1980. Response of a zonal climate–ice sheet model to the orbital perturbations during the Quaternary ice ages. *Tellus*, **32**, 301–319.

Potter, G. L., Ellsaesser, H. W., MacCracken, M. C., and Ellis, J. S. 1981. Albedo change by man: test of climatic effects. *Nature*, **291**, 47–49.

Prell, W. C., and Start, G. G. 1982. Isotopic and carbonate evidence for two climatic modes in the Quaternary. *E.O.S.*, **63**, 1297.

Sagan, C., Toon, O. B., and Pollack, J. B. 1979. Anthropogenic albedo changes and the earth's climate. *Science*, **206**, 1363–1368.

Sellers, P. J., and Lockwood, J. G. 1981a. A computer simulation of the effects of differing crop types on the water balance of small catchments over long time periods. *Quarterly Journal Royal Meteorological Society*, **107**, 395–414.

Sellers, P. J., and Lockwood, J. G. 1981b. A numerical simulation of the effects of changing vegetation type on surface hydroclimatology. *Climatic Change*, **3**, 121–136.

Shukla, J., and Mintz, Y. 1982. Influence of land-surface evapotranspiration on the earth's climate. *Science*, **215**, 1498–1501.

Thom, A. S. 1972. Momentum, mass and heat exchange of vegetation. *Quarterly Journal Royal Meteorological Society*, **98**, 124–134.

Thom, A. S. 1975. Momentum, mass and heat exchange of plant communities. In Monteith, J. C. (Ed.), *Vegetation and the atmosphere. Vol. 1*, Academic Press, London, 57–109.

Thom, A. S., and Oliver, H. R. 1977. On Penman's equation for estimating regional evaporation. *Quarterly Journal Royal Meteorological Society*, **103**, 345–357.

Walker, J., and Rowntree, P. R. 1977. The effect of soil moisture on circulation and rainfall in a tropical model. *Quarterly Journal Royal Meteorological Society*, **103**, 29–46.

Weertman, J. 1976. Milankovitch solar radiation variations and ice age sheet sizes. *Nature*, **261**, 17–20.

Wetherald, R. T., and Manabe, S. 1975. The effects of changing the solar constant on the climate of a general circulation model. *Journal Atmospheric Sciences*, **32**, 2044–2059.

Background to Palaeohydrology
Edited by K. J. Gregory

3

Discharge: empirical observations and statistical models of change

J. B. THORNES

Bedford College, University of London

Discharge data and analyses rarely appear in palaeohydrological studies for two simple reasons. Firstly, the data for discharge is poor prior to the present century, rare in the last century, and practically non-existent before that (Potter, 1978). Secondly, palaeoflow reconstruction techniques as outlined in the rest of this book and in Starkel and Thornes (1981) do not permit sufficiently accurate reconstruction of palaeoflows to justify the application of sophisticated or even unsophisticated techniques of data analysis. The most we can expect for periods prior to the present millenium is the crude identification of stage or discharge value which are ordered in time. The essential reason for the palaeohydrologists interest in data records lies in the fact that most assumptions underpinning the subject have to be made on the basis of discharge and other variables such as precipitation, evaporation, drainage basin morphology, and channel and bank sediments. Changes in these relationships, intrinsic or extrinsic, are usually adduced to explain actual or perceived changes in discharge or conversely to infer, through discharge, changes in climatic regime. The investigation of past changes in runoff is one of the ways in which the relationships are established. They are also established, of course, by controlled laboratory and field experiments, and from theory but it is only by the examination of discharge records that the models of interactions between the variables may be validated.

PALAEOFLOWS AND DISCHARGE RECORDS

Contemporary applied problems in hydrology may be enhanced by studies of palaeoflow. For example, historic flows may be used to impose sensible constraints on the variability of synthetic series based on recent records and can be used for water resources management and for the design of hydraulic structures. Because of the existence of threshold effects it is important to have the variance of synthetic series sensibly circumscribed.

The establishment of theory used in palaeohydrology has generally proceeded along three lines as far as palaeoflow records are concerned. Firstly, in earlier studies, historic flow records were noted to have changes within them and these changes were then explained in terms of other changes affecting the catchment, usually changes in precipitation or human activity. Secondly, studies of the direct relationship between discharge and other variables have been investigated empirically, as in hydraulic and planform geometry, or theoretically as in studies of transport competence and sediment rating curves. Thirdly, there have been investigations of known historical changes within catchments and their impact on the discharge regime. Examples of all three types of study are found in Gregory (1977).

In this chapter we are concerned essentially with temporal changes in flow regime, highlighting first the temporal structure of hydrological data and then going on to discuss the types of data available, how to deal with missing data, and methods of detecting hydrological change. Spatial variability is as important as temporal change in palaeohydrology. Too often perceived climatic changes based on comparative analysis of sediments and morphology in different catchments are actually differences in hydrologic response due to the inherent spatial variability within catchments. This point has been well developed and explained by Schumm (1978).

DATA SOURCES AND STRUCTURE

The immediate source of past flow records is the data kept by national or regional water authorities. In most countries these are relatively short though in a few cases (e.g. the Nile) they may extend over several hundred years. Almost invariably they are much shorter than the longest available rainfall records. In Britain, for example, there was no statutory requirement to collect runoff records until the River Boards Act of 1948, though daily flows for the Thames at Teddington extend back to 1883 and for the Lee at Fields Weir back to 1851. Even with these relatively modern records great care has to be taken to avoid the impact of human interference. With contemporary water management this may occur on a massive scale and includes groundwater pumping, river water abstraction, effluent discharges, the export of water into and out of the catchment, evaporative loss by irrigation, and reservoir operation, to mention only the more obvious (Beran and Gustard, 1977). This problem is particularly acute where extensive groundwater abstractions are occurring, as from the Triassic Sandstones in the English River Severn, or where irrigation plays an important role as in semiarid zone rivers such as the Mojave or the Rio Segura in Spain.

In the nineteenth century lock levels provide a useful data source. Lock gauge boards were normally read at specified times of the day with extra readings at high levels (Potter, 1978). These have been extensively used to extend records back sometimes quite early into the nineteenth century. An extensive record of the Severn at Evesham for example back to 1855 is based on this source. In using lock records care has to be taken to ensure that the lock bypasses remained unaltered during the period of observation. Another comparable source for many countries in Europe are records maintained on reservoir levels and releases. These can be used in conjunction with estimates of evaporative losses to give 'naturalized' inflows. Before the time of formal records reliance has to be placed on individual accounts in newspapers and chronicles. These are dominantly related to the effects of floods on lives and property and it is important to ascertain the qualifications and credibility of the reporter as well as the objectives of the report. These sources may provide information on some or all of the following: circumstances leading to the floods such as extreme rainfall; the time of arrival of the flood at a particular place; the peak height; the period of inundation; the rate of recession; flood damage and changes to the river channel. Such reports may be useful in the evaluation of recent as well as more distant flood events as the study by Howe, Slaymaker, and Harding (1967) on flooding of the Severn at Shrewsbury indicates. There exist several major compilations of flood events in the British Isles and these are reviewed by Rodda, Downing, and Law (1976). A detailed study of a single river is that of Lopez-Bermudez (1979) on the Segura River, Spain which provides a detailed assessment of flood reports and associated climatic phenomena as far as 47 BC and similar compilations are available for Italian and French rivers.

Great care has to be taken in using both recent and older data which estimate flood volumes because gauging is both difficult and notoriously imprecise under flood conditions. Overbank flow and bed scour are the main contributary causes but these are compounded by the fact that engineers often use formulae such as the Manning equation to estimate extreme flows or other techniques of even more doubtful value under the prevailing conditions. There are a wide variety of estimators for ungauged flow ranging from slope–area methods to references from the simple widths of flow. These rely on the hydraulic geometry of the channel remaining stable (Knox, 1979) which is a situation unlikely to be met with at either high latitudes or in arid and semiarid environments, in rivers with heavy bed loads or in channels undergoing rapid metamorphosis due to human activity.

Both palaeohydrologists and applied hydrologists have a direct interest in decomposing and identifying the underlying components of the behavioural pattern, i.e. the statistical structure of the time series. For the former it enables the determination of the location and nature of changes arising from different physical sources; for the latter it provides a suitable basis for forecasting future

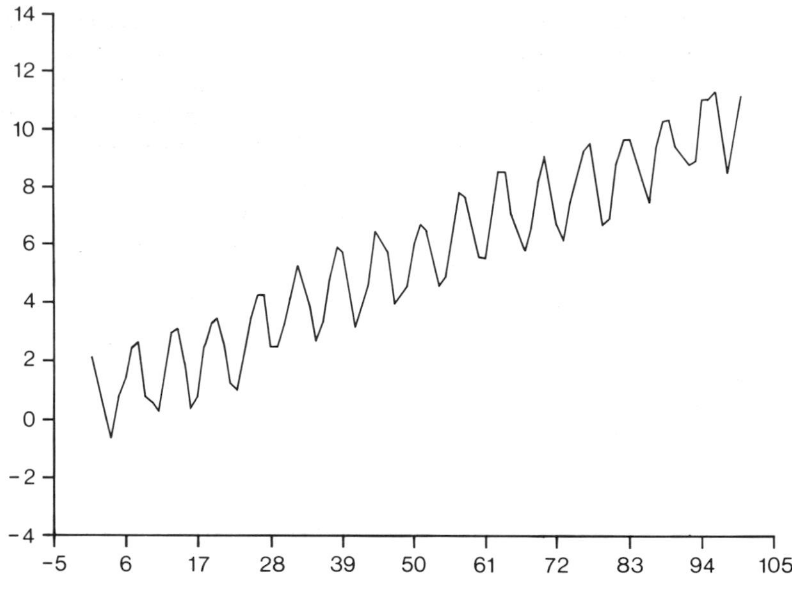

FIGURE 3.1 BASIC COMPONENTS OF TIME SERIES

values. Whereas the palaeohydrologist is more concerned with the long term pattern and tends to focus on the classical decomposition (Kisiel, 1969), in the last decade work on forecasting has emphasized continuously adaptive models and attempts to identify changes in structure on a more-or-less continuous basis (Culling, 1981).

For our purposes the classical decomposition is probably most useful. This envisages the data as made up of four separable components: namely long term trends, harmonic or cyclical components, persistence, and random noise (Figure 3.1). The discussion which follows is not a rigorous statistical treatment and for this the reader is referred to a standard text on time series (e.g. Kendall, 1975). Long term trends are those which might, for example, be identified by a plot of decadal running averages or by regression of five year mean values against time. This is often done for a 10 year or 30 year period though the plotting of simple running means tends to obscure the times when change occur in the records and residual mass curves (see below pp. 59) are often used as alternatives. Long term trends can usually be attributed to shifts in general climatic patterns directly, such as rainfall incidence or heat budget changes. The other source is human activity. The second component can be described by harmonic terms or by a plot of monthly averages over the entire period and reflects the seasonal component usually due to the annual march of climate. In urbanized or agriculturally dominated societies a diurnal cyclical component also arises from consumptive domestic and industrial water use and from irrigation respectively, and these effects have to removed before establishing the relationship between discharge and natural processes. The third component is persistence, which is identified by the tendency for successive observations to be well correlated with one another for relatively short time periods. Physically it may represent the drainage of the storage in a channel system following a flood (i.e. the recession limb of the hydrograph) or it may represent the characteristic flood-producing rainfall regime in a catchment. As such the persistence component may provide a sensitive indicator of the response characteristics of the catchment and changes in its relative role in the total variability may be used to identify changes induced by human activity. The final component is the set of uncorrelated events which are called random or white noise. This is what is left over after the other components have been accounted for and which has no identifiable pattern. Statistically speaking it comprises not only uncorrelated events but also non-systematic errors in data gathering. The general process of decomposition of a series into its main components comprises the identification, estimation and subtraction of each of the first three components successively leaving the residual random noise as the fourth component. The trends are normally removed by ordinary least squares regression analysis and the cyclical components by Fourier techniques involving identification using spectral analysis and estimation and removal by fitting harmonics, as outlined in Thornes and Brunsden (1977). A more detailed analysis permits the establishment of phase lags between paired seasonal series for different characteristics using cross-spectral analysis. For example, peak stage at Manaus lags systematically behind peak rainfall at the same locality. In determining lead–lag relationships between series using spectral estimation care must be taken to ensure that the variables themselves do not involve feedback (Granger, 1969) and that the two series realistically represent an input–output system (Dowling, 1974). Fortunately in the situations usually considered (e.g. Edwards and Thornes, 1973), this is often the case.

The persistence components are identified and fitted using autocorrelation functions and again estimated by least squares. Following Box and Jenkins (1970) there is now an extensive literature on the application of these techniques in hydrology and they can be extended for the estimation of seasonal components, though not without difficulty (Thornes, in press). The technique involves differencing the series at a length equal to the seasonal component, e.g. taking the first difference of flows twelve months apart to remove the annual cycle. The same technique can be ultimately extended to remove the long-term trend but is much less efficient than least-squares linear or polynomial regression. Its dominant value when compared with, say, using the monthly means is that it is parsimonious in parameters. This may not be a great advantage when trying to identify changes in the statistical structure of the series which is a major objective of palaeoflow analysis. It is worth noting that whereas annual precipitation series normally show no persistence (i.e. the observations are usually independent in time), this is not usually true of mean annual river flows because of interannual storage. For practical purposes the simple first order or second-order autoregressive persistence models seem adequate models for stationary annual runoff series (Yevjevich, 1977). This is an important result since it implies that the assumptions of regression analysis, used for detecting trends in annual series, are not usually met.

The discussion so far has been restricted to discharge volumes or rates but there is potentially an infinite number of parameters that could be used to identify change. These include, for example, mean daily flow rate, hydrograph characteristics such as time base or recession coefficients, the amplitude of seasonal variations, or the peak discharge of the annual flood. For each of these parameters there is a typical statistical distribution and a structure for serial values which is, in part at least, a function of the observational density. One of the problems in detecting hydrological response to known changes in other variables is to decide which of the many parameters is likely to be most sensitive, as the protracted discussion of the effects of vegetation removal indicates. This situation is rapidly improving with enhanced knowledge of hydrological processes (e.g. Eagleson, 1978). Regrettably, a certain hydrological equifinality means that there is no simple functional response to a change which could be identified from the hydrological record and it is this which makes the interpretation of past climate from hydrological data such a difficult and tenuous operation.

In practise, further back in time fewer parameters are available to choose from as the data become sparser and sparser. The most useful distinction is then between events which are statistically uncorrelated and those which are not; this distinction is important both to the issue of missing data and methods of detecting changes in hydrological data.

MISSING DATA

When data are statistically independent the probability of occurrence of a particular value is assumed to be independent of all other values. This is true of some annual data such as the annual flood series but almost all higher frequencies at least show persistence. The probability of independent discharge events is found from order statistics and traditionally is expressed by plots of exceedance probabilities against magnitude. A simple technique for detecting hydrological change is then to compare the distributions before and after some known effect (as in Figure 3.2 discussed below). The precision with which the curves can be described is a function of the number of data points and the degree to which assumptions as to the underlying distribution (e.g. Pearson Type III) are acceptable. When data is missing the distribution is censored. In bad cases some of the high values (Benson, 1950) or some of the low values (Leese, 1973) may be missing. In palaeoflow analysis the usual problem is

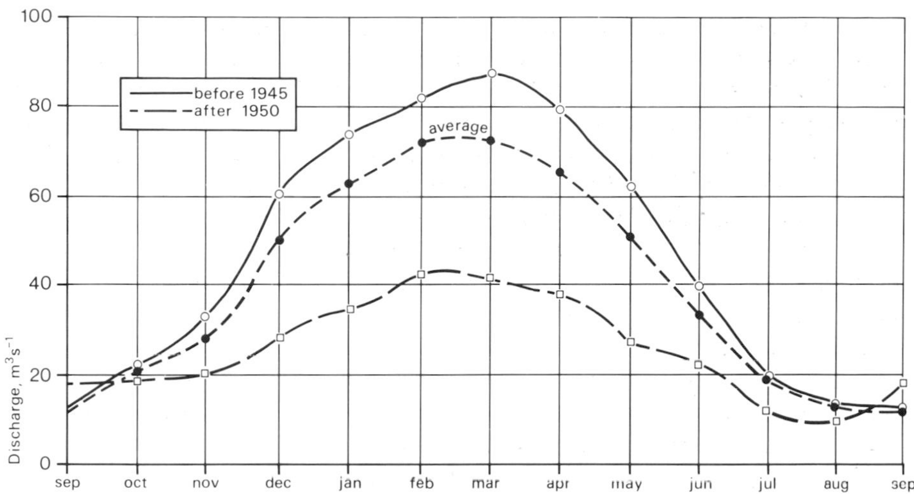

FIGURE 3.2 ANNUAL HYDROGRAPH OF RIO DUERO AT PENAFIEL
Illustrating changes due to dam construction

that several large values (historical floods) are separated by various lengths of missing record from a more continuous series. This problem has recently been studied by Zhang (1982) who derives the distribution for each rank in the ordered discharge statistic. The expectation of this statistic is, then, the exceedance probability of course and it is derived for the special case just outlined. As expected the net effect is to increase the exceedance probability of large flows, and vice-versa, but to provide an error band about the estimated probability as a function of the gap length.

In general, as with all cases of frequency analysis, a larger number of data points increases the precision of the estimates of the mean and variance. Gupta (1973) seeks an optimal sampling strategy which can provide for the minimum sample, the maximum amount of information about the mean and the variance. This is the inverse of the palaeoflow problem which, in this context, seeks to establish whether or not the existing data points are sufficient to establish adequately the mean and variance of a subset of the record which can then be compared with another subset. Mahloch (1974) uses multiple linear regression procedures to fill in gaps present in the data matrix (of variables, or stations against years) by regressing single rows with gaps against all other rows without gaps. As a row is filled it is added to the 'independent' variable system and another row with gaps is then evaluated. In palaeoflows such a technique might be used to investigate runs of exceptional values by comparing these with those produced by the Mahloch procedure for the same time points. This approach is somewhat analogous to the double-mass curve check for station consistency which also seeks to differentiate local and regional effects (Linsley, Kohler, and Paulhus, 1975).

For serially correlated data, if a harmonic structure is known or suspected, as in most runoff series, remarkably good reconstructions can be achieved from small numbers of data points if assumptions about trends and variance can be made. As the autocorrelation of a series increases the tolerable loss of information also increases (Thornes, 1982) so that monthly data usually provide a good reconstruction of Box–Jenkins models. Further study is needed of well defined long term series to evaluate the effects of missing information on palaeoflow reconstructions.

DETECTING CHANGE IN RECORDS

A wide variety of techniques have been explored to detect hydrological change. Generally three types of situation arise in palaeoflow studies: firstly, empirical searches for changes in both independent and serially correlated univariate data series; secondly, comparisons made before and after known changes in exogenous variables to evaluate the hydrological response; thirdly, attempts to establish relationships between variables where both are thought to be changing. These cases are also of interest to the theoretical and to the applied hydrologist because many statistical models assume that the data is stationary (roughly speaking that mean and variance are constant through time). Where this is not the case the non-stationary elements have to be identified and removed. Sometimes the removal of the non-stationary elements can induce further non-stationarity, for example when removing harmonic terms by fitting harmonics (Bullard, Yevjevich, and Kottegoda, 1976) or by seasonal differencing (Thornes and Clark, 1976). Here we illustrate but a few of the range of techniques.

In empirical searches of univariate series, such as the annual runoff series, it is common practise first to create averages of blocks of data and to visually inspect the data for changes (Figure 3.4). Sometimes running means are constructed. Both of these techniques tend to obscure the actual time at which change is initiated, particularly where this is relatively sharp. A more satisfactory technique is to plot residual mass curves which are graphs of cumulative percentage deviation from the mean. The quantity plotted is standardized to be zero at the beginning and end of the period and the graph rises during periods with average discharge greater than the mean for the entire period and vice-versa. An example of this type of study is Kraus' (1955) investigation of secular changes in tropical rainfall. The technique can be applied to monthly as well as to annual data and for a wide range of parameters. Another essentially non-parametric type of test involves the frequency of up-and down-crossings of the mean value and could be used to indicate runs of high or low flows in historical records and the average lengths between events above a particular threshold value (Nordin and Rosbjerg, 1970). This might be particularly useful for the study of lock gauge board crossings where the absolute flood heights are not recorded.

Regression analysis is widely used in empirical searches. At the simplest level the criterion of change is the significance of departure of a regression coefficient from zero. When it is assumed that the individual values are independent, the method is relatively tolerant of missing information. Successive fitting of linear or non-linear curves for subsets of the data may help to identify points of change. Thornes (1970) uses this technique by fitting linear regressions and allocating points to one or other of the curves in such a way as to minimize the joint error sum of squares for the combined data set. More recently Brown, Durbin, and Evans (1975) have addressed the same problem in an econometric context, and derived a more general model for detecting change. The process essentially consists of creating a regression by successively adding values at later and later time points and calculating the cumulative sums of departures of the residuals and the cumulative variance. Tests are provided for the significance of departure of the cumulative residuals. The generality derives in part from the fact that the regression is multivariate and a typical hydrological case might involve discharge through time being regressed against rainfall and evaporation estimates. The philosophy is somewhat analogous to the process of state variable estimation called Kalman filtering. Here the structural parameters of the model are allowed to change through time highlighting a particular point by change in the parameters (as opposed to the actual values). The models being 'updated' are usually of the autoregressive–moving average type.

Hydrological examples abound (e.g. Young and Beck, 1974) and Bennett (1976) has used the technique to detect changes in the behaviour of channel width series. Culling (1981) presents a theoretical discussion of the problem. The technique should be explored more fully in the palaeohydrological context but, as with trend surface fitting, there comes a point where the law of diminishing returns comes in with continuous adaptation to the curve or surface. Priestly (1976) has expressed the paradox perfectly:

> '...once the parameters are allowed to become time-dependent, the accuracy of the estimators is related in a fundamental way to the maximum rate at which the parameters are allowed to change.'

Much less work has been carried out on the patterns of change of seasonal components though there have been several attempts to apply Priestley's idea of 'evolving spectra' (Priestly, 1965). The somewhat simpler but perhaps more useful complex demodulation technique for identifying shifts in amplitude and phase have been deployed by Anderson (1975) to hydrological series and by Alexander (1979) to stream channel width series. Hirsch, Slack, and Smith (1982) have developed a test of trend which can tolerate strong seasonality in the data and compared the situations in which it performs better than ordinary linear regression. This 'seasonal regression' test is based on Kendall's widely used tau-test and is complimented by a slope estimator which can be used with a strong seasonal component.

The response of a system to a particular known change is often considered in terms of the change in the frequency plot (see Figure 3.2) or as a pair of successive fits of means or of linear regression curves. Unfortunately the ordinary student's t-test cannot be used to check for changes in the mean level of a time series since it is only valid if observations before and after the change vary about the mean normally, with a constant but not necessarily equal variance and also vary independently. These conditions are not usually satisfied since hydrologic series are usually serially dependent, often non-stationary, and frequently seasonal. The response of a system to a known particular change is a central problem in control theory and is widely treated in the literature. Geomorphologists and palaeohydrologists will find several treatments but a particularly useful one is Bennett and Chorley's *Environmental Systems* (1978) which essentially follows the Box–Jenkins and adaptive forecasting approaches. In hydrology the problem is met in two main contexts. Firstly, in direct applications of systems theory for modelling hydrological response (the hydrograph) to a unit input of change (a rainfall or a succession of rainfalls) by obtaining the unit impulse response function between the two. Secondly, the model of a univariate series can be modelled as a function of a series of pulsed random noise inputs (Box and Jenkins, 1970) so the methodology is well known. A development of the Box–Jenkins technique specifically for studying imposed change and called intervention analysis, has been applied by Hipel *et al.* (1976) to analyse whether or not a particular natural or man-induced change causes a significant change in the mean level of the series. Their example involved estimating the man-induced intervention on the mean level of the time series of the Nile flows during the period 1870–1945 as a result of the construction of the Aswan Dam, completed in 1920. The analysis indicates a decrease in the mean annual discharge of more than 20 per cent after 1902.

In more general terms it has to be accepted that the relationship between input and response is usually exceedingly complex due to time lagged responses, non-linearity and the stochastic nature of the processes involved. Time lagged response arises mainly from the existence of storage capacity within the system, or because of 'spatial-lag' where the response effects are translated across space by the processes involved and this actually takes time. These effects can sometimes be dealt with fairly simply by developing lag-dependent transfer functions between input (or cause) and output (or effect) either using parsimonious transfer function models (e.g. Lai, 1977) or spectral transfer functions, when the problems outlined earlier are usually encountered. Unfortunately palaeoflow data are rarely of sufficient quality to allow these techniques to be efficiently deployed. The second major problem is non-linearity which arises from the fact that the response of the system to an output is itself a function of the state of the system. Thus a unit input of rainfall produces less runoff from a dry than from a wet catchment. This problem is especially acute when the system is close to some threshold or bifurcation point when a small change carries it into a completely different class of behaviour. Non-linearity not only implies multiple steady states but also conditional stability or instability, unlike linear systems which tend to be either stable or unstable (Brunsden and Thornes, 1980). The third general problem in detecting changes from the historical record is the inherent variability in the input, the processes and the variables used for identifying response. Although the simple models outlined earlier assume a simple univariate response because of the inherent variability of the materials and spaces (e.g. hydraulic conductivity, Anderson, 1982), any historical case is to be regarded as only one particular realization of a complex stochastic process. Some of these realizations might be easily detected whereas others might not. As the dynamics of the process become better understood it should be easier to unravel the real world complexities which generate historical series. In the meanwhile palaeohydrology can assist in this process by dealing where possible with the responses to known changes in the exogenous variables. Any increase in information about processes which are *not* hydrologically governed (such as independent assessements of climatic change rather than inferences from channel sediments and morphology) and their effect or lack of effect on discharge will carry palaeohydrological studies some way in this direction.

TWO EXAMPLES

The purpose of the following examples is to illustrate the points made in the previous paragraph and to illustrate some of the simpler techniques outlined which, despite their limitations, may be useful in palaeoflow analysis.

The first example is a simple case of a univariate series subject to a well defined change at a particular date and hence amenable to relatively straightforward analysis. It is based on a data analysis by M. Michel (1979) of changes in runoff for the Rio Duero in north central Spain. The river drains the Iberian Mountains and the broad upland of the Altiplanice del Duero and has a strong seasonal component resulting mainly from the strong snowmelt which occurs in spring and the very high summer evaporation rates (Figure 3.2). Heavy abstractions occur for irrigation particularly in the summer months. The total area of the catchment above Peñafiel is 11,837 km^2 and between 1945 and 1950 about 20 per cent of the catchment was brought under the control of three reservoirs at Cuerda del Pozo (176 Hm^3), Linares del Arroyo (58 Hm^3), and Burgomillo (13.75 Hm^3). These dams were constructed to store peak flows for irrigation use. The exceedance probabilities are distributed as Pearson

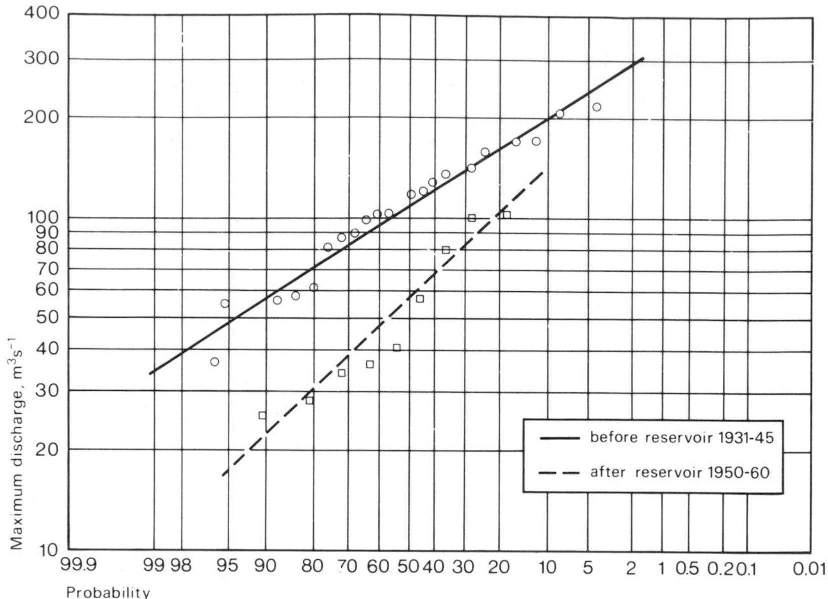

FIGURE 3.3 FLOW FREQUENCY CURVES: DUERO AT PENAFIEL

Type III and a plot of probabilities against discharge for the before and after years is given in Figure 3.3. They indicate a very appreciable reduction in the exceedance probability for a peak flood of a given magnitude and this is matched by a drop in the mean annual flow from 48.94 m^3 sec^{-1} to 26.33 m^3 sec^{-1} after completion of the reservoir. Investigation of a further unregulated catchment , that of the Rio Lobo, indicated no significant change in the peak flows during the same period indicating that the reduction in flow at Peñafiel was entirely attributable to artificial regulation and not a result of changes in rainfall (or snow) incidence over the catchment of the upper Duero.

The second example is taken from the Severn catchment. Rodda, Sheckley, and Tan (1978) carried out an analysis of the rainfall of England and Wales during the last 50 years and detected a decline in winter rainfalls (October--March) and a 10 year periodicity. The latter was fitted using an ARIMA model of the Box–Jenkins seasonal differenced type. By comparison the flows of the Severn at Bewdley indicated a slight increase in the median flows for the period 1851–1975 with generally a lower range of flows, the latter being less conspicuous in the Severn than in the Thames data, which were also examined. The authors concluded that land-use changes or artificial regulation of flows were probably responsible for this apparent discrepancy but noted that the rainfall for England and Wales as a whole might be different from that

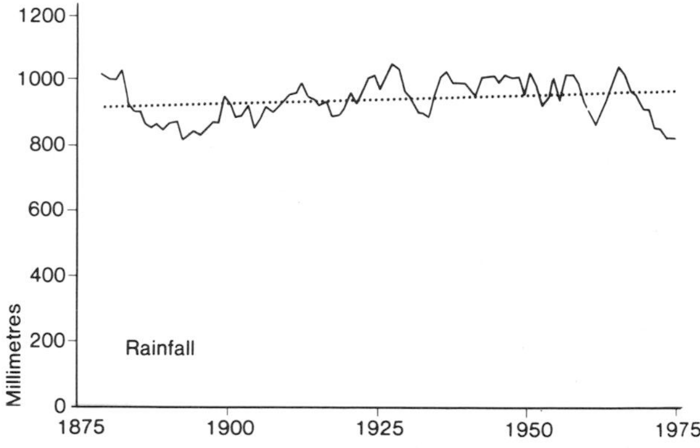

FIGURE 3.4 RAINFALL SERIES SEVERN CATCHMENT

of the Severn or Thames catchments. Together with Mr. J. Tucker, I have examined a rainfall series provided (together with runoff series) by Mr. T. Kitson of the Severn Trent River Authority for the years 1875–1975 (summed monthly totals). We have also investigated the Craddock (1976) series which show the deviations of annual rainfall for a 100 year period obtained from a group of statistics around Hereford and Ross-on-Wye. The data are composite in the sense that the results from stations had to be grouped. Although the Craddock series show up significant trends there is a slight, but significant positive trend in the annual series for the Severn catchment as a whole (Figure 3.4). Neither series show marked periodicity when subjected to autocorrelation or spectral analysis (but *cf.* Gray, 1976) other than a faint suggestion of a six year cycle. The overall increase in the annual totals could be related to the slight increase in discharge between the two subsections evaluated by Rodda *et al.* (1978) and would favour their suggestion that the national rainfall series do not adequately reflect what is taking place in the catchment.

Figures 3.5 and 3.6 are the 'naturalized' flows for Bewdley and Vyrnwy. The former is in the upland part of the catchment, the latter in the middle part of the catchment. Both series show statistically significant increases in annual flow, but that for Vyrnwy is much stronger than Bewdley (Figures 3.5 and 3.6). These results have been compared with an analysis of land-use based on Ministry of Agriculture returns since 1866. These are available from the Public Records Office at Kew and the data have been evaluated as a research source by Coppock (1958). In general there was a diminution in arable acreage from 1866–1939 followed by a steady rise in arable in the period 1940–1946 as mirrored by the county of Shropshire (Figure 3.7). We have taken a weighted acreage according to the percentage of each county within the catchment. In detail, an examination of parishes has revealed a strong contrast betwen upland and lowland parts of each county. Upland counties and parishes show virtually no change up to 1940 whereas in lowland counties the turn over to grass is very

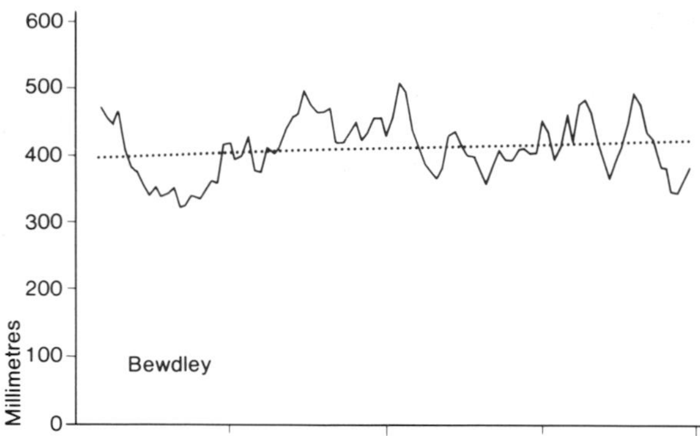

FIGURE 3.5 FLOWS AT BEWDLEY
Expressed as millimetres of depth over catchment, annual totals

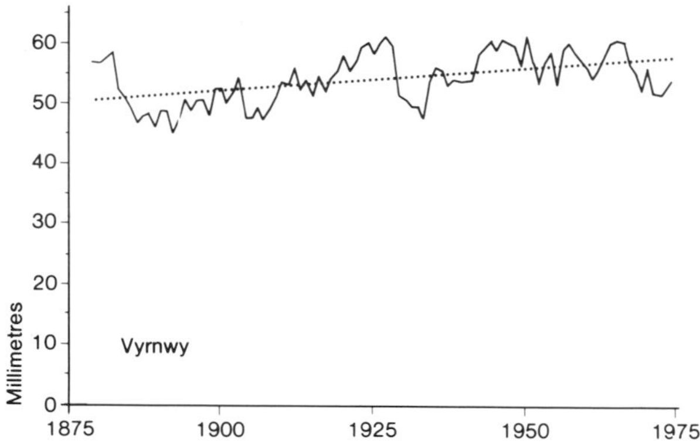

FIGURE 3.6 NATURALIZED FLOWS AT VYRNWY
Expressed as millimetres of depth over catchment, annual totals

strong. The result is that while Vyrnwy, dominated by counties with no change, had a strong increase in the naturalized annual flows, Bewdley, dominated by flows generated in the lowland counties, had very little change over the entire period. This paradox appears to be at variance with the idea that rural land-use change is responsible for the runoff patterns.

This example illustrates that although we can detect changes relatively easily, explaining them is significantly more difficult! One complex issue is the relationship of land-use change to the dynamic part of the channel. In a more detailed plot analysis of changes over the period 1930–1950, represented by two land use surveys, we found for example that changes from grass to arable occurred on average about 400 m from the nearest channel, whereas the much more controversial switches from heath to forest occurred closer to the channel

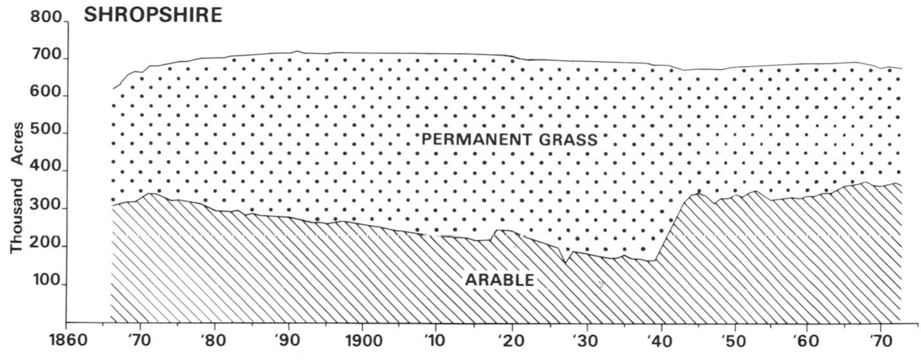

FIGURE 3.7 LAND-USE CHANGES: SHROPSHIRE

than any other type of change — an average of only 170 m. From the complete sample of plots it is clear that though the shift between arable and grass is the greatest by area, it tends to occupy regions in which the drainage density is lowest. This aspect of land use change in relation to hydrological sensitivity needs more careful investigation.

REFERENCES

Anderson, M. G. 1982. Modelling hillslope water status during drainage. *Transactions Institute of British Geographers*, **7(3)**, 337–354.

Anderson, M. G. 1975. Demodulation of stream flow series. *Journal of Hydrology*, **16**, 225–240.

Alexander, D. 1976. The role of profile disturbances in channel morphology. *Occasional Papers*, 29 Department of Geography, University College.

Bennett, R. J. 1976, Adaptive adjustments of channel geometry. *Earth Surface Processes*, **1**, 131–150.

Bennett, R. J., and Chorley, R. J. 1978. *Environmental Systems: Philosophy, Analysis and Control*, Methuen, 624pp.

Benson, M. A. 1950. Use of historical data in flood-frequency analysis. *Eos Transactions, American Geophysical Union*, **31(3)**, 419–429.

Beran, M. A., and Gustard, A. 1977. A study into the low-flow characteristics of British rivers. *Journal of Hydrology*, **35**, 147–157.

Box G. E. P. and Jenkins, G. 1970. *Time Series Analysis: Forecasting and Control*, Holden-Day, San Francisco.

Brown, R. L. J., Durbin, J., and Evans, J. M. 1975. Techniques for testing the constancy of regression relationships over time. *Journal Royal Statistical Society, Series B*, 149–163.

Brunsden, D., and Thornes, J. B. 1980. Landscape sensitivity and changes. *Institute of British Geographers, Transactions*, **4(4)**, 463–485.

Bullard, K. L. Yevjevich, V., and Kottegoda, N. 1976. Effects of misestimating harmonics in periodic hydrologic parameters. *Hydrology Papers 88*, Colorado State University, Fr. Collins, Colorado.

Coppock, J. T. 1958. The Agricultural Returns as a source for local history. *The Amateur Historian*, **4(2)**, 49–55.

Craddock, J. M. 1976. Annual rainfall in England and Wales since 1725. *Quarterly Journal Royal Meteorological Society*, **102**, 823–840.

Culling, W. H. 1981. Stochastic Processes. In Wrigley, N. and Bennett, R. J. (Eds.) *Quantitative Geography*, Routledge and Kegan Paul, London, 202–211.

Cunnane, C. 1978. Unbiased plotting position — a review. *Journal of Hydrology*, **37**, 205–222.

Dowling, J. M. 1974. A note on the use of spectral analysis to detect leads and lags in annual cycles of water quality. *Water Resources Research*, **10(2)**, 343–344.

Eagleson, P. S. 1978. Climate, soil and vegetation. *Water Resources Research*, **14(5)**, 705–776.

Edwards, A. M. C., and Thornes, J. B. 1973. Annual cycle of river water quality; A time series approach. *Water Resources Research*, **9(5)**, 1286–1295.

Granger, S. W. J. 1969. Investigating causal relations by econometric models and cross-spectral methods. *Econometrica*, **37**, 424–438.

Gray, B. M. 1976. Medium term fluctuations in south-eastern England. *Quarterly Journal Royal Meteorological Society*, **102**, 627–638.

Gregory, K. J. (Ed.) 1977. *River Channel Changes*, John Wiley, Chichester, 450pp.

Gupta, V. L. 1973. Information content of time-variant data. *Journal of Hydrological Division, American Society of Civil Engineers*, **99**, HY3, 383–393.

Hipel, K. A., Lennox, W. C., Unny, T. E., and McLeod, A. I. 1976. Intervention analysis in water resources. *Water Resources Research*, **11(6)**, 855–861.

Hirsch, R. M., Slack, J. R., and Smith, R. A. 1982. Techniques of trend analysis for monthly water quality data. *Water Resources Research*, **18(1)**, 107–121.

Howe, G. M., Slaymaker, H. O., and Harding, D. M. 1967. Some aspects of the flood hydrology of the upper catchments of the Severn and Wye. *Transactions Institute of British Geographers*, **41**, 33–58.

Knox, J. C. 1979. Geomorphic evidence of frequent and extreme floods. *Improved Hydrological Forecasting Why and How?*, American Society of Civil Engineers, 220–238.

Kraus, E. B. 1955. Secular changes of tropical rainfall regimes. *Quarterly Journal Royal Meteorological Society*, **82**, 108–210.

Lai, P. W. 1977. Stochastic–dynamic models for some environmental systems: transfer function approach. *London School of Economics Discussion Paper 61*, 43pp.

Leese, M. N. 1973. Use of censored data in the estimation of Gumbel distribution parameters for annual maximum flood. *Water Resources Research*, **9(6)**, 1534–1542.

Linsley, R. K., Kohler, M. A., and Paulhus, J. L. H. 1975. *Hydrology for Engineers*, 2nd Edition, McGraw-Hill.

Lopez-Bermudez, F. 1979. *Inundaciones catastroficas, precipitacioues torrenciales y erosion en la provincia de Murcia*, Papeles del Departmento de Geografia, VIII Universidad de Murcia, 50–91.

Mahloch, J. L. 1974. Multivariate techniques for water quality analysis. *Journal Environmental Engineering Division, American Society of Civil Engineers*, **100**, EE5, 1119–1131.

Michel, M. 1979. *The effect of reservoir implementation on the discharge of the River Duero in central Spain*. Unpublished Dissertation for B.Sc degree, Geography Department, King's College, London.

Nordin, C. F., and Rosbjerg, D. M. 1970. Application of crossing theory in hydrology. *Bulletin International Association Scientific Hydrology* XVI (3), 27–43.

Potter, H. R. 1978. The use of historic records for the augmentation of hydrological data. *Institute of Hydrology Report 46*, 58pp.

Priestley, M. B. 1965. Evolutionary spectra and non-stationary processes. *Journal Royal Statistical Society, Series B*, **27**, 204–237.

Priestley, M. B. 1976. Discussion of 'Techniques for testing the constancy of regression relationships over time'. by R. L. J. Brown, J. Durbin and J. M. Evans. *Journal Royal Statistical Society, Series B*, 164.

Rodda, J. C., Downing, R. A., and Law, F. M. 1976. *Systematic Hydrology*, Newnes-Butterworth, 399pp.

Rodda, J. C., Sheckley, A. V., and Tan, P. 1978. Water resources and climatic change. *Journal Institution of Water Engineers and Scientists*, **32**, 76–83.

Rodrigues-Iturbe, I. 1968. A modern statistical study of monthly levels of the Orinoco River. *Bulletin International Association Scientific Hydrology*, **13(4)**, 25–40.

Schumm, S. A. 1977. *The Fluvial System*, Wiley, New York.

Starkel, L., and Thornes, J. B. 1981. *Palaeohydrology of River Basins*, British Geomorphological Research Group, Technical Bulletin 28, 107pp.

Thornes, J. B. 1970. Observations on the hydraulic geometry of stream channels on the Xingu–Araguia watershed, Mato Grosso, Brazil. *Geographical Journal*, **1970**, 136–152.

Thornes, J. B. 1982. Problems in the identification of stability and structure from temporal data series. In Thorn, C. E. (Ed.), *Space and Time in Geomorphology*, George Allen and Unwin, 327–353.

Thornes, J. B. (in press). Models, missing data and management in water quality series. Proceedings 2nd Soviet–British Geographical Seminar, *Geoforum*.

Thornes, J. B., and Clarke, M. W. 1976. *The Stour raw data series and associated seasonal Box-Jenkins fitting problems*. Working Paper 5, Non-sequential water quality project, London School of Economics, 43pp.

Yevjedvich, V. 1963. Fluctuations of wet and dry years, part 1 research data assembly and mathematical models *Hydrology Paper*, **1**, Colorado state University, Ft. Collins, Colorado 55p.

Young, P. C., and Beck, B. 1974. The modelling and control of water quality in river systems. *Automatica*, **10**, 455–468.

Zhang, Y. 1982. Plotting positions of annual flood extremes considering extraordinary values. *Water Resources Research*, **18(4)**, 859–864.

Background to Palaeohydrology
Edited by K. J. Gregory

4

Patterns of sediment yield

D. E. Walling and B. W. Webb

Department of Geography, University of Exeter, UK.

THE CONTEXT

Sediment yield constitutes an important hydrological variable in any attempt to reconstruct the palaeohydrology of a drainage basin (*cf.* Schumm, 1977a). Such data provide a valuable index of the rate of fluvial denudation in the upstream area, and sediment transport rates are in turn important in influencing downstream channel and valley development. However, whereas the simple water balance equation and various physically-based modelling strategies make it possible to assemble meaningful estimates of the response of the discharge regime of a drainage basin to changes in inputs or catchment condition, much less is known about the factors controlling temporal variations in sediment yield. In the case of bed material load, appropriate sediment transport formulae may be coupled with flow data to produce estimates of sediment movement (e.g. Graf, 1971; ASCE, 1975). Reconstruction of past transport rates could therefore be based on information regarding the nature of the channel bed material and the discharge regime, provided various hydraulic parameters can be satisfactorily postdicted. However, there is no simple equivalent for estimating the non-capacity wash load component, which reflects supply from the slopes of the watershed. In most rivers this component will account for the major proportion of the total sediment yield. Some progress has been made in coupling erosion and sediment routing models in recent years (e.g. Smith, 1976; Li *et al.*, 1976; Alonso *et al.*, 1978; Beasley *et al.*, 1980) but lack of detailed data and computational complexities have precluded the widescale testing and application of this approach to simulating sediment yields.

In the absence of an effective theoretical base for predicting temporal variations in the pattern of sediment yield from a drainage basin, an empirical approach has frequently been adopted. The ergodic hypothesis of space-time substitution has been utilized to conjecture the response of a drainage basin to changes in climate or vegetation cover (e.g. Schumm, 1968). An empirical approach of this nature relies heavily on the availability of a wide range of sediment yield data at a variety of temporal and spatial scales, and data limitations have to date provided a significant limitation to progress. For example, the study of Fournier (1960) which attempted to develop global relationships between sediment yield and climate was based on data from only 96 rivers. The associated construction of a world map of sediment yields involved the extrapolation of this limited data set over the whole surface of the globe. The situation in North America stands somewhat in contrast, in that extensive national sediment monitoring programmes have permitted more detailed evaluation of climatic relationships (e.g. Langbein and Schumm, 1958; Dendy and Bolton, 1976), calculation of total sediment flux to the oceans (e.g. Curtis *et al.*, 1973) and local and regional analyses of the influence of individual physiographic controls on sediment yields (e.g. Inter-Agency Sedimentation Committee, 1947, 1965, 1976). However, the range of global conditions in North America is restricted and the applicability of these findings to other areas is consequently limited.

The International Hydrological Decade (1965–74) and the succeeding International Hydrological Programme have done much to promote sediment measuring activities in participating countries, and the global coverage of measuring stations has expanded greatly in recent years. Unfortunately, however, there is inevitably a significant time lag between the initiation of a monitoring programme and the generation and publication of representative sediment load data, and much of the potential of this expansion in providing an improved picture of the global behaviour of suspended sediment yields remains to be realized. Nevertheless, progress is evident and this chapter attempts to review current knowledge of spatial and temporal patterns of sediment yield in the context of its potential for palaeohydrological reconstruction, and to highlight uncertainties requiring further investigation and research. Attention will be given to the magnitude of suspended sediment yields, the relationship between sediment yields and climate, the influence of changing catchment condition on sediment yields, the linkage between upstream erosion and downstream sediment yield and the potential of lake sediment studies.

THE MAGNITUDE OF SUSPENDED SEDIMENT YIELDS

Recent increases in the availability of sediment yield data have provided a welcome extension to the documentation of the global range of sediment yield values. This increasing database is not, however, without its problems and Walling and Webb (1981) have emphasized the many uncertainties involved in assessing the reliability of individual data sources. The potential importance of these uncertainties are well illustrated by the controversy surrounding the sediment load of the Cleddau River on the west coast of South Island, New Zealand (Adams, 1980; Griffiths, 1979). The flow and sediment data for this river were collected by the New Zealand Ministry of Works and the two workers have independently used the same basic data to estimate the mean annual load by means of rating curves. Their estimates of 13 300t km^{-2} yr^{-1} (Griffiths, 1979) and 275t km^{-2} yr^{-1} (Adams, 1980) are different by nearly two orders of magnitude. The reliability of any estimate of sediment yield will depend on the adequacy of the sampling equipment, the sampling procedures, the frequency of sampling, the accuracy of the laboratory analysis and the flow data, and the method used to calculate the load from the individual samples. Moreover, additional uncertainty is introduced by the question of the length of record required to produce a representative estimate of the long-term mean. These various sources of error are difficult to assess without detailed evaluation of the measuring programmes employed by individual agencies, and this is normally impossible. *Caveat emptor* is therefore a very appropriate warning for any attempt to use sediment yield data from a variety of sources, since any patterns and relationships that exist could be more apparent than real. Furthermore, it must be clearly recognized that such data relate to the suspended sediment yield of the rivers concerned. Lack of information on bed load transport for most measuring sites precludes any attempt to consider total sediment yields.

Global minima for suspended sediment yield lie well below 2t km^{-2} yr^{-1}. For example Douglas (1973) cites yields of 1.3t km^{-2} yr^{-1} for the Brindabella catchment (26.1 km^2) and 1.7t km^{-2} yr^{-1} for the Queanbeyan River (172 km^2) in the Southern Tablelands and Highlands of New South Wales, Australia, and values of < 1.0t km^{-2} yr^{-1} have been reported for several rivers in Poland (Branski, 1975). Although improved availability and global coverage of sediment yield data can do little to modify our view of minimum levels of sediment yield, they have significantly changed our perception of the upper bound in recent years. In their reviews of global sediment transport rates Strakhov (1967) refers to a maximum of 2 000t km^{-2} yr^{-1} for the Sulak River in the USSR, and Gregory and Walling (1973) cite a maximum value of 8040t km^{-2} yr^{-1} for the Ching River in the People's Republic of China. Values in excess of 10 000t km^{-2} yr^{-1} have, however, now been reported for several rivers (Table 4.1). The highest value listed in Table 4.1 is for a small unnamed river in the northeast of Taiwan included on a map produced by Li (1976). In view of a degree of uncertainty surrounding this value, a leading contender for the global maximum value must be the Dali River in N. Shaanxi with a yield of 25 600t km^{-2} yr^{-1} from 96.1 km^2 and of 16 300t km^{-2} yr^{-1} from 3 893 km^2 documented by Mou and Meng (1980). This river is a tributary of the middle reaches of the Yellow River in the People's Republic of China and drains the gullied loess region.

TABLE 4.1 SOME RECENTLY REPORTED VALUES OF SEDIMENT YIELD IN EXCESS OF 10000t km^{-2} yr^{-1}

Country	River	Drainage area (km^2)	Mean annual sediment yield (t km^{-2} yr^{-1})	Source
People's Republic of China	Dali	96.1	25600	Mou and Meng (1980)
	Dali	187	21700	
	Dali	3893	16300	
Kenya	Perkerra	1310	19520	Dunne (1974)
Taiwan	(unknown)	(unknown)	31700	Li (1976)
Java	Cilutung	600	12000	Hardjowitjitro (1981)
	Cikeruh	250	11200	
New Guinea	Aure	4360	11126	Pickup *et al.* (1981)
North Island New Zealand	Waiapu	1378	19970	Griffiths (1982)
	Waingaromia	175	17340	
	Hikuwai	307	13890	
South Island New Zealand	Hokitika	352	17070	Griffiths (1981)
	Cleddau	155	13300	
	Haast	1020	12736	

Any attempt to account for the incidence of the high values of sediment yield referenced in Table 4.1, must take account of several contributing factors reflecting the erodibility of the terrain and the erosivity of the hydrometeorological regime. In the case of the tributaries of the Yellow River, the presence of highly erodible loess soils, the lack of vegetation cover and the semiarid climate are major controlling factors (Walling, 1981). The semiarid climate is again an important causal factor in the Kenyan example, but here severe disturbance of the catchment by agricultural activity is also of major significance. For Java and New Guinea, the steep relief, high rainfall totals, and agricultural activity are important and in South Island New Zealand the steep relief and very high rainfall (up to 9000 mm yr^{-1}) may again be invoked, although tectonic instability also plays a major role.

In the past there have been several attempts to combine the limited sediment yield data available at the time with notions concerning the influence of relief, tectonic stability, climate, geology and other factors to produce global maps of sediment yield. The work of Fournier (1960) and the Russian scientist, Lopatin, reported in Strakhov (1967), are two such studies that have been frequently referenced. Often it is not fully recognized that the maps are based on a very small number of actual observations of sediment yield (60 in the case of Lopatin and 96 for Fournier), and that they reflect very considerable subjective interpolation and extrapolation. The dangers associated with a limited database and subjective extension are clearly demonstrated by a comparison of the maps of these two workers (Figure 4.1). In terms of general levels, the sediment yields depicted on Fournier's map are frequently an order of magnitude greater than those shown by Strakhov. Furthermore, there are significant contrasts in the overall patterns depicted by the two maps.

Figure 4.2 presents a recent attempt by the authors to produce a world map of sediment yields which takes into account the improved data availability. It is based on data from nearly 1500 measuring stations and aims to represent the sediment yields of basins of 1000–10000 km^2. Small basins will frequently exhibit much higher levels. Although the database employed to construct Figure 4.2 represents a very considerable expansion on that available to the previous workers there are still large areas of the globe where sediment yield data are currently lacking. This is particularly so in the case of South America for which very few values are available. Subjective extrapolation and interpolation have still been necessary to map many areas, and it must be stressed that the yields reflect contemporary river transport and make no attempt to distinguish baseline levels from actual values which reflect the impact of human activity in a drainage basin. The map should not be viewed as a definitive representation of the magnitude and pattern of global sediment yields. Rather, it represents one stage in a continuing process of updating and improvement, but should prove more meaningful than other maps currently available.

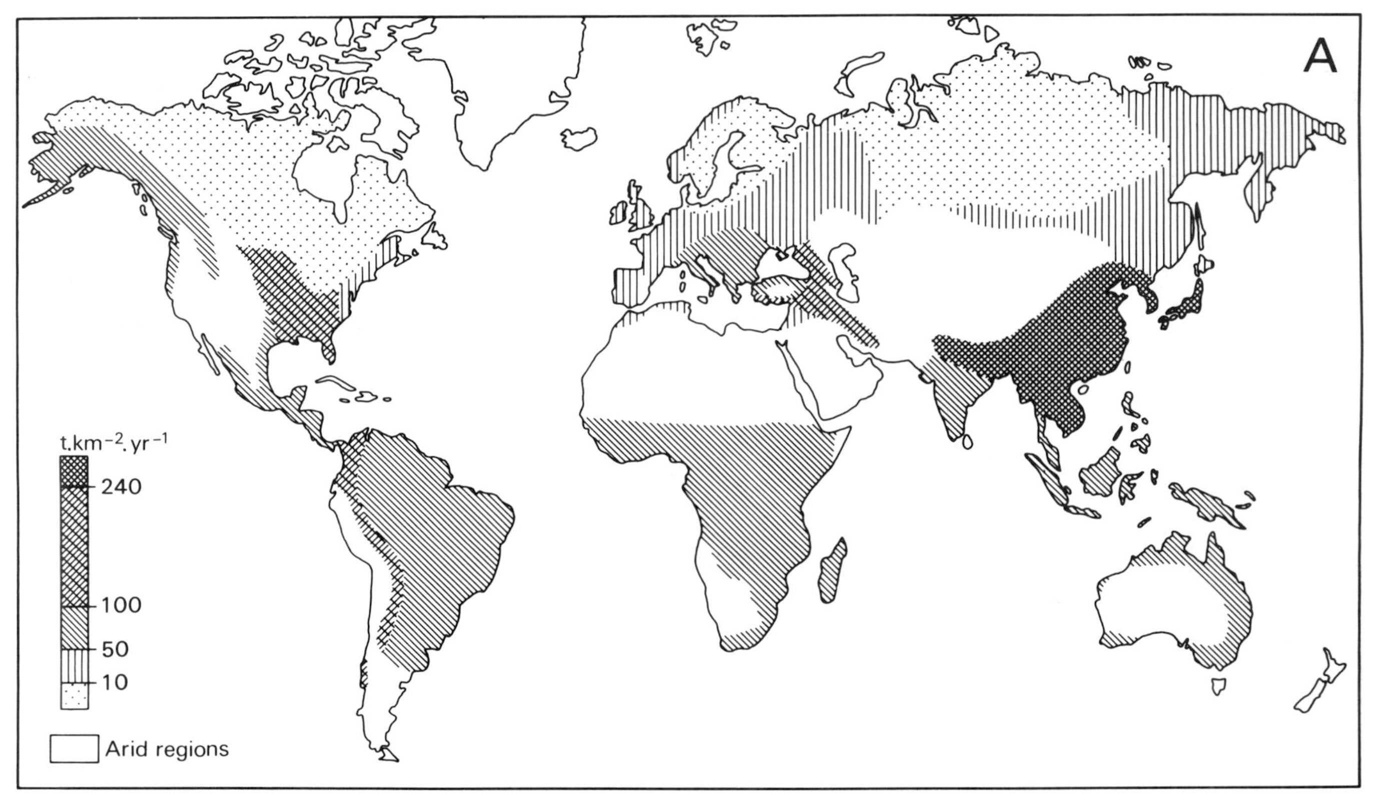
A
t.km^{-2}.yr^{-1}
240
100
50
10
Arid regions

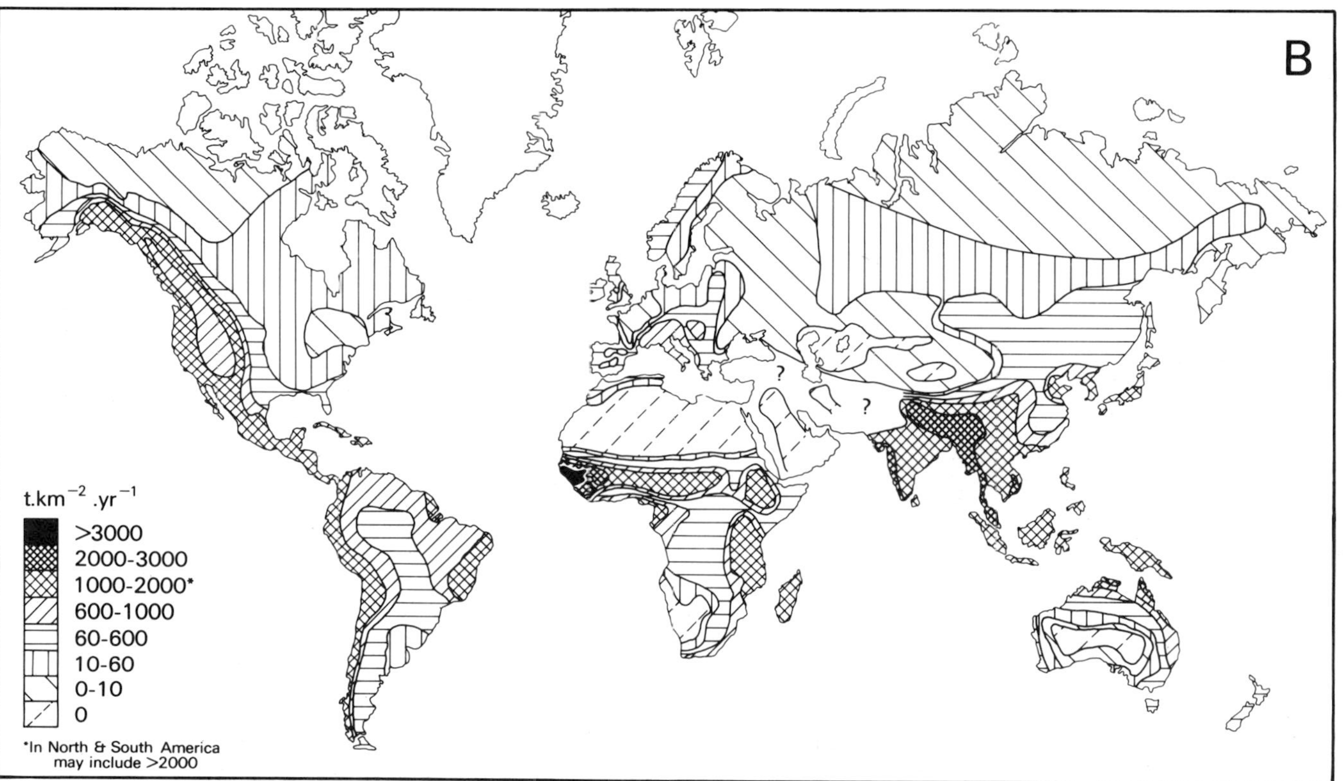

FIGURE 4.1 GLOBAL PATTERNS OF SEDIMENT YIELD ACCORDING TO STRAKHOV (1967) (A) AND FOURNIER (1960) (B)

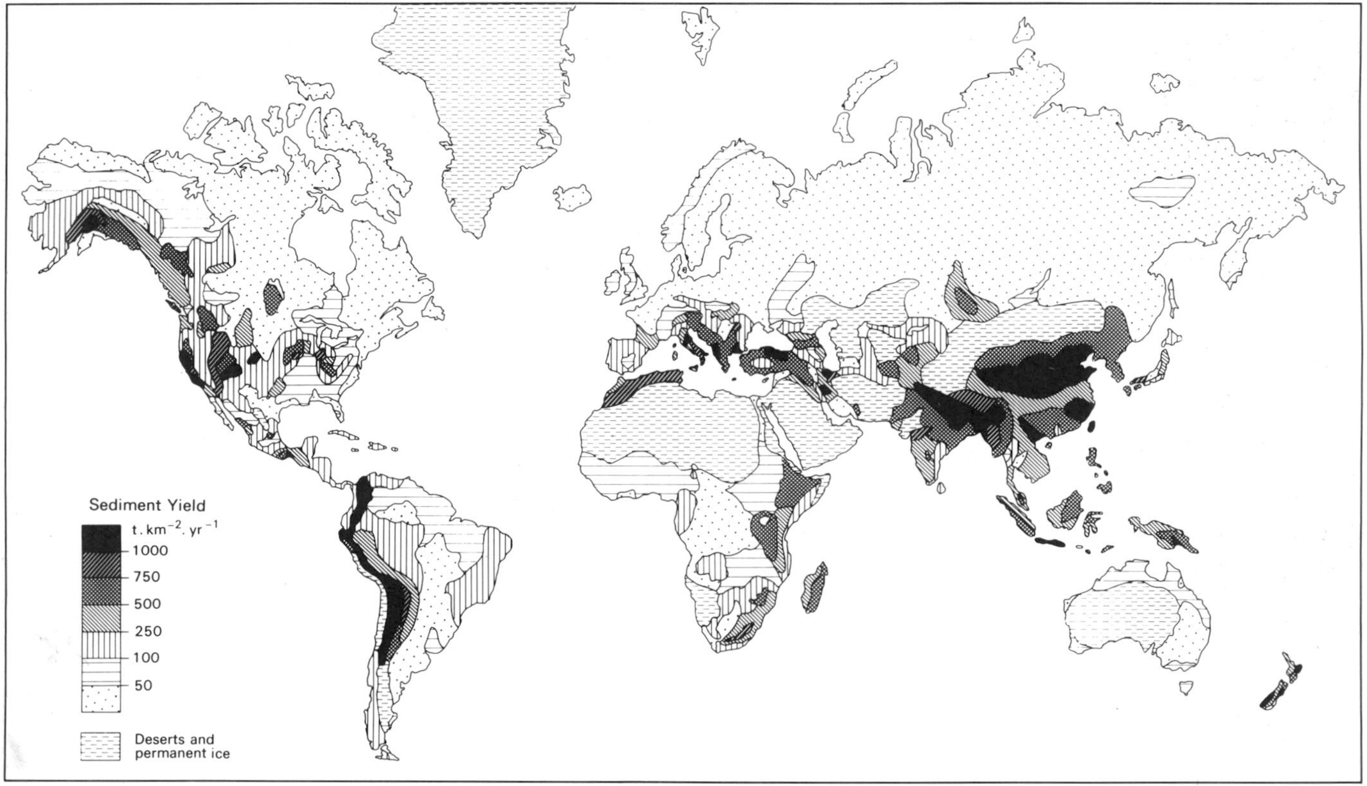

FIGURE 4.2 A REVISED MAP OF GLOBAL SEDIMENT YIELDS

To briefly review the patterns defined on Figure 4.2. it can be seen that maximum sediment yields are associated with the loess areas of China and the Cenozoic mountain areas around the Pacific margin. High values are also to be found in other mountain areas and in regions with mediterranean and semiarid climates and in the seasonally humid tropics. Based on this pattern, the current rate of sediment transport from the land surface of the globe to the oceans can be estimated at approximately 15×10^9t yr^{-1}, providing a global average sediment yield of approximately 150t km^{-2} yr^{-1}.

CLIMATE AND SEDIMENT YIELD

Relationships between climate and sediment yield have been sought by many workers, both as an end in themselves and as a means of postulating the influence of a changing climate on sediment yields through space–time substitution. It is the latter context which is of relevance to palaeohydrological reconstruction, and the approach pioneered by Schumm (1975) in applying the relationship between annual sediment yield and effective precipitation, proposed by Langbein and Schumm (1958), has now been widely documented and utilized. This relationship, depicted in Figure 4.3 has been convincingly explained in terms of the interaction of erosive energy and vegetation density. Maximum sediment yields occur at an annual effective precipitation of approximately 300 mm (i.e. semiarid areas). In areas with effective precipitation in excess of 300 mm, vegetation growth is promoted and increases surface protection. In drier areas, the available energy for erosion and transport is limited. This rational explanation has been termed the 'Langbein–Schumm rule' by a number of workers and the influence of climatic change on sediment yields can be readily estimated by evaluating the increase or decrease in annual effective precipitation. Effective precipitation is here defined as the annual precipitation required to generate the given annual runoff at a standardized annual mean temperature of 50°F and is not, as some workers (e.g. Golubev, 1982) have assumed, equivalent to precipitation minus evapotranspirational losses.

Although convincing and attractive in its simplicity, it is perhaps less well recognized that the Langbein–Schumm rule is not fully substantiated by existing sediment yield data and that its derivation introduces a degree of uncertainty into its acceptability. In the latter context, it should be emphasized that the curve depicted in Figure 4.3 was derived using data from 94 stream sampling stations by applying a group average technique. In this, the average sediment yields for a series of effective precipitation classes were calculated and the final curve was drawn through these points (*cf*. Figure 4.3). To this extent the curve is a statistical abstraction of the raw data and it has been suggested that removal or addition of a small number of key stations could substantially alter the form of the curve. In terms of its general applicability, it

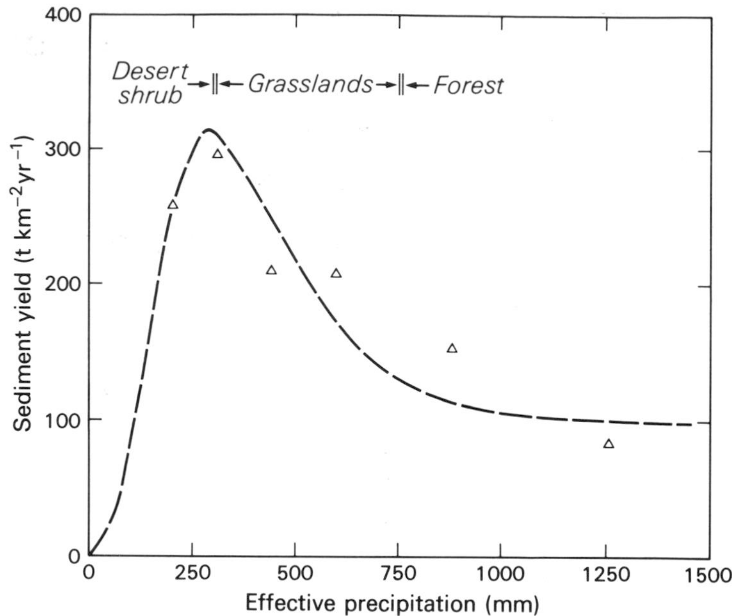

FIGURE 4.3 THE RELATIONSHIP BETWEEN SEDIMENT YIELD AND ANNUAL PRECIPITATION PROPOSED BY LANGBEIN AND SCHUMM (1958)

must also be accepted that the curve was derived from data for rivers in the USA and that the range of climatic conditions represented is therefore limited. Even within the USA, some regions such as the coastal areas of northern California, which exhibit high sediment yields at values of annual precipitation in excess of 1 000 mm (e.g. Janda and Nolan, 1979), do not conform to the pattern suggested by the curve.

Figure 4.4 attempts to demonstrate the extent to which the Langbein–Schumm rule is substantiated by global sediment yield data by considering relationships between annual sediment yield and annual precipitation and runoff produced by other workers. In the case of the USA, the generalized relationship proposed by Stoddart (1969) based on the data presented by Judson and Ritter (1964) and the more detailed analysis reported by Dendy and Bolton (1976) may be considered. The former relationship simply considers the average suspended sediment yield and annual runoff for the seven major drainage regions of the country, whilst the latter involved the use of the group average technique to generalize the relationship evidenced by information on reservoir siltation from over 500 locations. Both show evidence of the Langbein–Schumm rule, if it is assumed that sediment yields represented in the Judson and Ritter data would tend to zero as runoff decreased beyond the minimum value depicted. With the Dendy and Bolton curve, peak

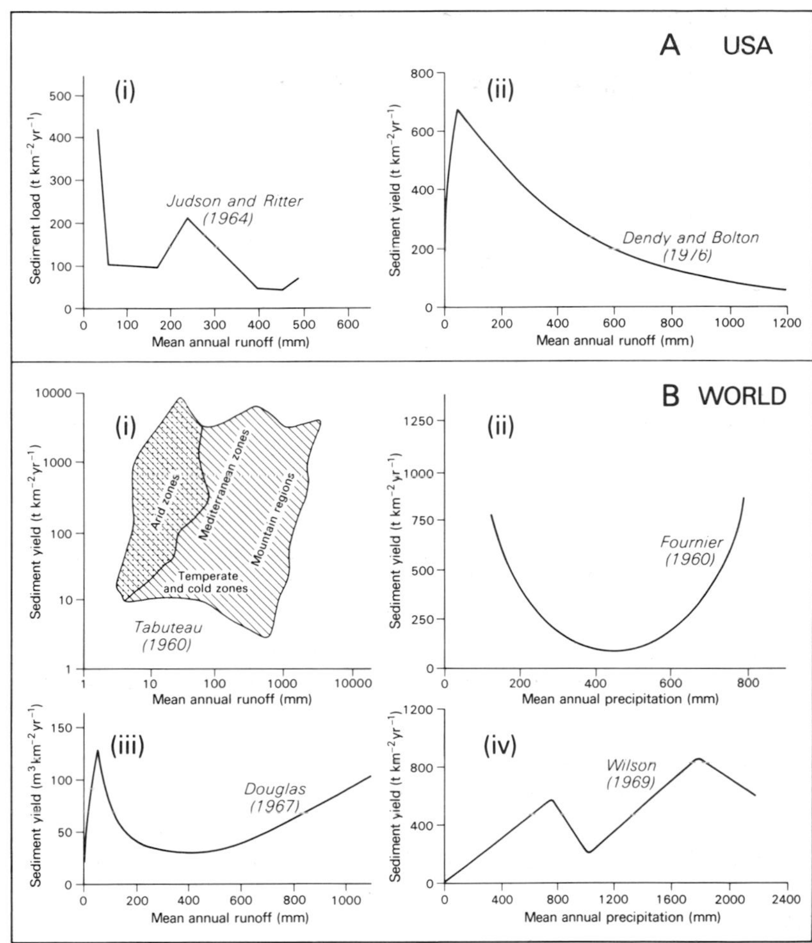

FIGURE 4.4 RELATIONSHIPS BETWEEN SEDIMENT YIELD AND ANNUAL PRECIPITATION AND RUNOFF ESTABLISHED FOR THE USA AND THE WORLD

sediment yields occur at between 25 and 75 mm of annual runoff. Converted to values of effective rainfall at 50°F, these would represent a range of annual effective precipitation between 450 and 500 mm, which is somewhat higher than the value of 300 mm at which peak sediment yields occur on the Langbein and Schumm curve.

With the four examples of relationships based on global data, evidence to confirm the Langbein–Schumm rule is less convincing. Thus, whilst the left-hand limbs of the relationships proposed by Fournier (1960) and Douglas (1967) may be equated with the Langbein and Schumm curve, both graphs show a tendency for sediment yields to increase again when annual precipitation and runoff exceed 1 000 mm and 500 mm respectively. Sediment yields at maximum values of precipitation and runoff are as high as those found at the peak reflecting semiarid conditions. The relationship proposed by Wilson (1969) is somewhat different in that it evidences two peaks at 750 and 1750 mm annujl precipitation. These have been related to sub-humid and tropical conditions and cast further doubt on the dominance of semiarid areas suggested by the Langbein–Schumm rule. Further diversity is introduced by the logarithmic plot of sediment yield versus annual runoff produced by Tabuteau (1960), which shows that a wide range of yields occurs in all regions, and that high sediment yields are not restricted to a particular level of annual runoff. In addition, this plot emphasizes the degree of simplification introduced by the use of a group averaging technique in the derivation of the curves presented by Fournier, Douglas, and Wilson.

That the curves based on data from the USA evidence some degree of conformity to the Langbein–Schumm rule, whereas those based on global data show little evidence of that trend, clearly merits comment. It could be suggested that the trend evident in the Langbein–Schumm rule is simply the spurious product of the group averaging technique applied to the particular range of annual precipitation and runoff and associated values of sediment yield encountered in the USA. This would, however, ignore the sound physical explanation underlying the Langbein–Schumm rule. It is perhaps more realistic to suggest that the Langbein–Schumm rule is meaningful in areas with continental climates, such as the USA, whereas it is less applicable in the case of seasonal climates (*cf*. Schumm, 1977a, pp. 29–30).

As a further test of the applicability of the Langbein–Schumm rule, Figure 4.5 presents a raw data plot of mean annual sediment yield versus mean annual precipitation for 1 246 measuring stations produced by Walling and Kleo (1979). The global database employed was both more extensive and more spatially representative than those used by previous workers. Although, some spatial inequality exists, all continents are presented (Table 4.2).A logarithmic scale has been used to facilitate plotting of the wide range of sediment yields encountered. No clear pattern emerges and a wide range of sediment yield values exists across the range of mean annual precipitation represented. Other factors such as relief (e.g. Indian rivers), the presence of highly erodible loess (Chinese rivers), and contrasts in the magnitude of human impact are important controls on the magnitude of sediment yield and Figure 4.5 suggests that these factors are of greater importance than annual precipitation in controlling global patterns of sediment yield.

Although Figure 4.5 shows little evidence of a clearly defined relationship between annual sediment yield and annual precipitation, an attempt has been made to follow the procedure used by previous workers seeking to define the relationship between these two variables. Figure 4.6A presents a subjectively fitted curve drawn through group-averaged values of sediment yield. In this instance, only drainage basins with an area less than 10 000 km^2 have been

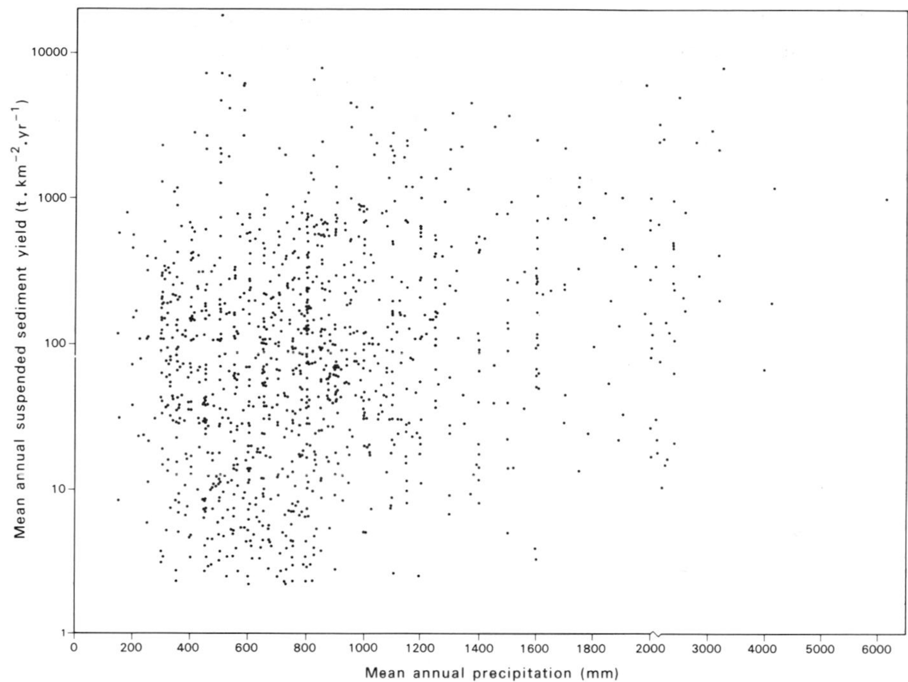

FIGURE 4.5 PLOT OF MEAN ANNUAL SEDIMENT YIELD VERSUS MEAN ANNUAL PRECIPITATION FOR THE GLOBAL DATABASE

TABLE 4.2 DISTRIBUTION OF SEDIMENT MEASURING STATIONS REPRESENTED IN THE GLOBAL DATABASE

Continent	Number
North America	400
Central and South America	169
Europe	396
Asia	243
Africa	13
Australasia	25
TOTAL	1246

(Based on Walling and Kleo, 1979)

included, in order to reduce the vast range of basin size represented in the original data compilation. The resultant curve based on data from 812 measuring stations differs substantially from its direct equivalents in Figures 4.3 and 4.4, although it evidences elements of their individual forms. For example, the first peak at 450 mm annual precipitation could reflect the simple relationship proposed by Langbein and Schumm (1958). Furthermore, the two initial peaks are similar in certain respects to the curve produced by Wilson (1969), but in this case the peaks are at 450 and 1 350 mm annual precipitation rather than at his maxima of 750 and 1 750 mm, which in fact correspond with the two troughs on Figure 4.6A.

An equivalent curve has been derived to represent the apparent relationship between sediment yield and mean annual runoff (Figure 4.6B). This is similar in certain respects to that produced by Douglas (1967), although the peak in semiarid areas is less pronounced, maximum yields occur in areas of high annual runoff, and the curve extends over a considerably greater range of runoff values. The precise form of the two curves presented in Figure 4.6 may represent little more than the spurious product of statistical aggregation, nevertheless, they must be viewed as being equally valid as those portrayed in Figures 4.3 and 4.4. With this in mind, there may be little justification in ascribing the three peaks shown in Figure 4.6A to specific climatic conditions, but it is suggested that they represent firstly, continental semiarid conditions, and, in the remaining two cases, the effects of seasonal precipitation regimes associated with areas of high rainfall mediterranean climate (*ca.* 1 250 mm precipitation) and of tropical monsoon conditions (> 2 500 mm precipitation). The correspondence of high sediment yields with zones of seasonal precipitation regime could be viewed as a direct development of the Langbein-Schumm rule, because the seasonal regime would restrict the growth of the vegetation cover, whereas the total erosive energy would be largely independent of the seasonality.

The considerable degree of scatter apparent in Figure 4.5 can be related to the wide variety of factors, in addition to mean annual precipitation, which control sediment yields. Further investigation of the global relationship between climate and sediment yield could consider alternative indices of climate or attempt to control for the effects of such factors as geology, relief, land use, or catchment size by using multivariate analysis or subsets of the main database. More detailed analysis of this nature would be particularly sensitive to the problems of data accuracy outlined previously and would require meaningful information on a variety of other physiographic variables. Such analysis will be undertaken when an improved and extended database has been assembled.

In the interim, it is perhaps instructive to consider the relationships between sediment yield and annual precipitation or runoff evidenced by small data sets from individual regions. With these, variability due to different measurement

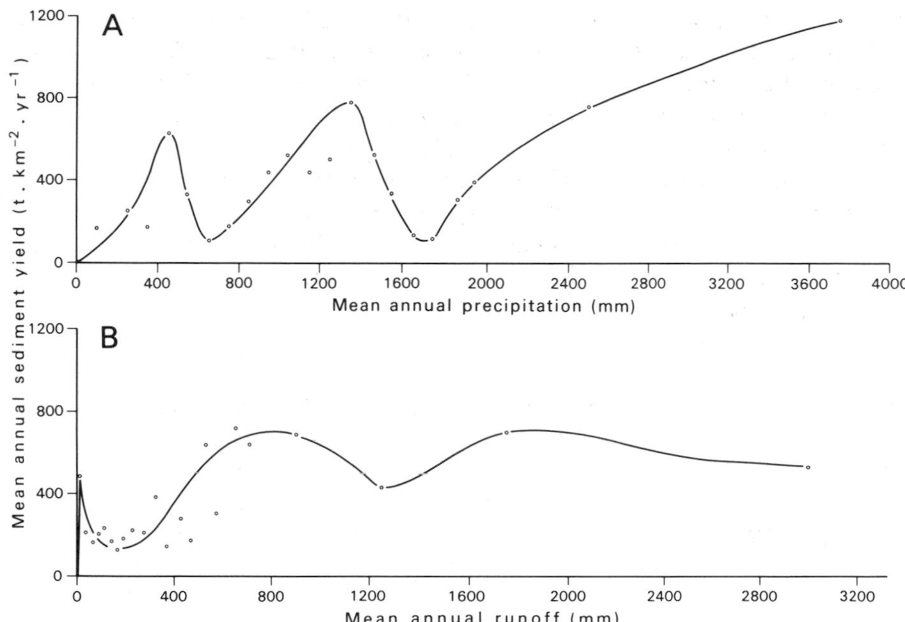

FIGURE 4.6 GENERALIZED RELATIONSHIPS BETWEEN MEAN ANNUAL SUSPENDED SEDIMENT YIELD AND MEAN ANNUAL PRECIPITATION (A) AND RUNOFF (B) PRODUCED BY THE GROUP AVERAGE TECHNIQUE

and load calculation techniques is likely to be minimized and a more restricted range of physiographic conditions will be embraced. Figure 4.7 presents plots of the relationship between annual sediment yield and annual precipitation or runoff for six countries or regions. In all cases, the relationship is positive and there is no evidence of the decrease in sediment yield at higher precipitation levels implicit in the Langbein–Schumm rule. The data for Rajasthan are derived from measurements in small nadis or reservoirs (Sharma and Chatterji, 1982) and are interesting in that they cross the threshold of annual effective precipitation where a downturn in sediment yield might be expected. Sharma and Chatterji (1982) have grouped the data according to both underlying lithology, examples of which are shown here, and to a general physiographic zonation, and in neither case is a downturn apparent. It is difficult to isolate the effects of land use on these relationships and it must be accepted that human activity may have reduced the cover density to a level where increased precipitation does not produce a decrease in sediment yield.

The data presented by Dunne (1979) for Kenya and depicted in Figure 4.7 reduce the uncertainty surrounding the effects of land use by representing relationships between sediment yield and annual runoff for individual land use classes. In all cases the relationship is positive, and even in the case of the drainage basins with a complete forest cover, there is no evidence of a downturn in sediment yield at increased runoff levels. Similar continuing upward trends of sediment yield with increasing precipitation and runoff are evidenced by data from Morocco and Eastern Australia (Figure 4.7).

Figure 4.7 also provides two examples, from Northern California and South Island New Zealand, of the relationship between sediment yield and annual precipitation in areas of high precipitation. The positive trend is again clearly apparent and provides some confirmation of the tendency for sediment yields to increase at maximum levels of precipitation suggested by the work of Douglas in Figure 4.4 and by the authors in Figure 4.6. This tendency can be related to the importance of mass movement as a sediment source in these mountain environments. In this case, increased precipitation results in reduced slope stability and increased sediment yield and cannot be interpreted in terms of the Langbein–Schumm rule which implicitly involves sheet erosion as the dominant source.

This brief review of some current evidence concerning the relationship between climate and sediment yield emphasizes that no simple relationship exists. It has been demonstrated that the pattern suggested by the Langbein–Schumm rule is not clearly apparent in a variety of data sets. This could in turn reflect the restriction of the rule to the continental climatic conditions of the USA or the effects of other physiographic variables in obscuring the influence of moisture availability. The positive trend of the relationships between annual sediment yield and runoff evident in Figure 4.7 suggests that the increase in erosivity associated with increased precipitation or runoff levels is not offset by increased protection by the vegetation cover, although the effects of human activity and land use practices in reducing this cover must also be considered. No simple or more meaningful alternative to the Langbein–Schumm rule can currently be proposed for use in palaeohydrologic reconstruction, but it is suggested that its limitations should be more fully and explicitly recognized.

THE INFLUENCE OF CHANGING CATCHMENT CONDITION ON SEDIMENT YIELD

In the context of palaeohydrological reconstruction, the influence of changing catchment condition is closely related to that of changing climate. The two are closely linked over longer time scales, although vegetation succession and development may also play an important role in the linkages involved. In the shorter term, the impact of human activity in modifying catchment condition and therefore sediment yields must be considered. In some areas of the world this activity spans a considerable period of time (*ca.* 5000 years), although in others it is of more recent incidence, and in general the intensity of the associated changes increases towards the present. It is the latter aspect which is

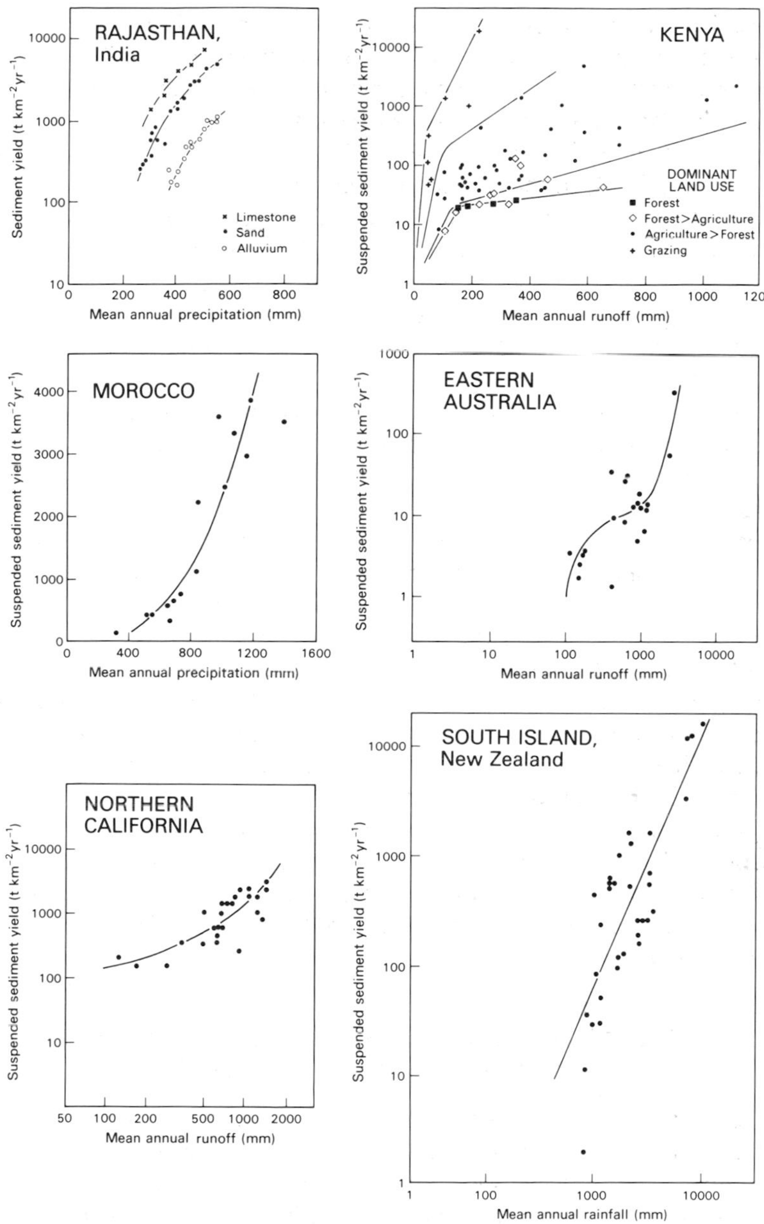

FIGURE 4.7 RELATIONSHIPS BETWEEN ANNUAL SEDIMENT YIELD AND MEAN ANNUAL PRECIPITATION AND RUNOFF FOR SPECIFIC REGIONS
Based on the work of Sharma and Chatterji (1982), Dunne (1979) Heusch and Milliès-Lacroix (1971), Douglas (1973), Janda and Nolan (1979), and Griffiths (1981)

most readily illuminated by study of current patterns of sediment yield. Again, space-time substitution can be used to postulate the changes resulting from an alteration in vegetation cover or land use. A classic example of this approach is provided by the curve produced by Wolman (1967) which attempts to show the temporal pattern of sediment yield from the Piedmont of the Eastern United States consequent upon various phases of land development including urbanization.

An analysis of contemporary erosion and sediment yield data can therefore usefully address the question of the potential magnitude of changes in sediment yield produced by land use changes. A valuable starting point is the various parameters employed in the Universal Soil Loss Equation (USLE) (Wischmeier and Smith, 1965, 1978) which attempts to summarize existing erosion plot data from the USA. In this equation the cover and management factor, *C*, expresses the influence of land use practice. An arbitrary value of 1.0 is taken to represent the clean-tilled fallow or bare soil situation which would exhibit maximum soil erosion and other land use conditions are expressed relative to this. Wischmeier (1975) presents estimates of the *C* factor for undisturbed areas of 0.003 for permanent pasture and rangeland with a good canopy and ground cover, and of 0.001 for woodland with a dense tree canopy and litter layer. These generalized factors suggest that land use activities could accelerate soil erosion by 300 to 1 000 fold. Roose (1977) presents a similar analysis of soil erosion in West Africa and suggests values of 0.001 for a forest or dense shrub cover and 0.01 for well developed savanna, providing a potential increase of soil loss of between 100 and 1 000 fold. In practice, typical agricultural activity would be characterized by a *C* value < 1.0 and following Golubev (1980) it could be proposed that the change from virgin grassland to cultivated land produces an increase in soil loss of one order of magnitude (i.e. 10 fold) whereas the conversion of natural forest produces an increase of two orders of magnitude (i.e. 100 fold).

These values represent major generalizations and it must be recognized that soil degradation could itself cause detrimental changes in soil character (factor *K* in the USLE), which has been assumed constant in the foregoing statements, and therefore further increase soil loss. Other factors may also not remain constant, since increased erosion may initiate a change in drainage density and therefore a change in the topographic factors *L* and *S* in the USLE. Furthermore, there are problems in equating changes in soil loss, which are essentially associated with sheet and rill erosion, with changes in downstream sediment yield which will reflect both the intervening processes of sediment delivery and additional supply of sediment from channel and gully erosion.

In an attempt to provide data more specifically related to sediment yields, Table 4.3 presents a selection of results from catchment studies where it has been possible to document the increase in sediment yield consequent upon a particular land use activity. In some instances the results have been obtained

TABLE 4.3 INCREASES IN SUSPENDED SEDIMENT YIELD CAUSED BY LAND USE CHANGE

Region	Land use change	Increase in sediment yield	Source
Rajasthan, India	Overgrazing	× 4 – 18	Sharma and Chatterji (1982)
Utah, USA	Overgrazing of rangeland	× 10 – 100	Noble (1965)
Oklahoma, USA	Overgrazing and cultivation	× 50 – 100	Rhoades *et al*. (1975)
	Cultivation	× 5 – 32	
Texas, USA	Forest clearance and cultivation	× 340	Chang *et al*. (1982)
N. California, USA	Conversion of steep forest to grassland	× 5 – 25	Anderson (1975)
Mississippi, USA	Forest clearance and cultivation	× 10 – 100	Ursic and Dendy (1965)
S. Brazil	Forest clearance and cultivation	× 4500	Bordas and Canali (1980)
Westland, New Zealand	Clearfelling	× 8	O'Loughlin *et al*. (1980)
Oregon, USA	Clearcutting forest	× 39	Fredriksen (1970)

from 'paired' catchment experiments, where the response of the disturbed basin has been compared with that of a neighbouring basin (*cf.* Walling, 1979) and in others they have been based on observations of changes through time in the sediment yield from a catchment. These data again suggest that the suspended sediment loads of many rivers may have increased by an order of magnitude or more as a result of agricultural and other human activity and provide a useful guideline for any attempt to reconstruct the sediment yield behaviour of a watershed . If this is attempted, it must be accepted that many forms of disturbance such as forest clearance will be relatively short-lived in their effect although 'recovery times' will vary according to environmental conditions (*cf.* Walling and Kleo, 1979). Furthermore, the possibility of rapid depletion of available sediment, as described by Rooseboom and Harmse (1979) in the case of the Orange River in South Africa and of critical threshold effects (*cf.* Schumm, 1977b), must also be entertained.

To briefly consider the significance of the above findings to longer term palaeohydrologic reconstruction, reference may be made to the work of Schumm (1968) who proposed a hypothetical series of curves illustrating the relation between precipitation and *relative* sediment yield during geologic time. Leaving aside discussion of the precise form of these curves, which has already been touched upon in the previous section, it is significant to note that Schumm suggests that peak sediment yields may have been of the order of four times higher during the Precambrian and early Palaeozoic when vegetation was absent and that they have subsequently decreased as the vegetation cover improved. The data contained in Table 4.3 suggest that the factor of four may be an underestimate and that sediment yields during these epochs, and indeed in areas newly exposed by retreating ice during the Quaternary, may have been an order of magnitude higher than those occurring under a well-developed vegetation cover.

LINKING UPSTREAM EROSION AND DOWNSTREAM SEDIMENT YIELD

In some exercises of palaeohydrological reconstruction it may prove attractive to relate downstream sediment yields to erosion rates operating within the basin. In this manner estimates of erosion rates could be converted to values of sediment yield or conversely evidence for past sediment yields could be extrapolated to provide estimates of erosion rates within the basin. However, any such attempt will face the considerable uncertainty surrounding the operation of sediment delivery or conveyance processes. It is well known that only a proportion, and perhaps a very small part, of the material representing the gross erosion within a watershed will be delivered to the catchment outlet. Some will be deposited further down the slope or at the foot of the slope, some will be deposited on the valley floors and floodplains and some will be stored in

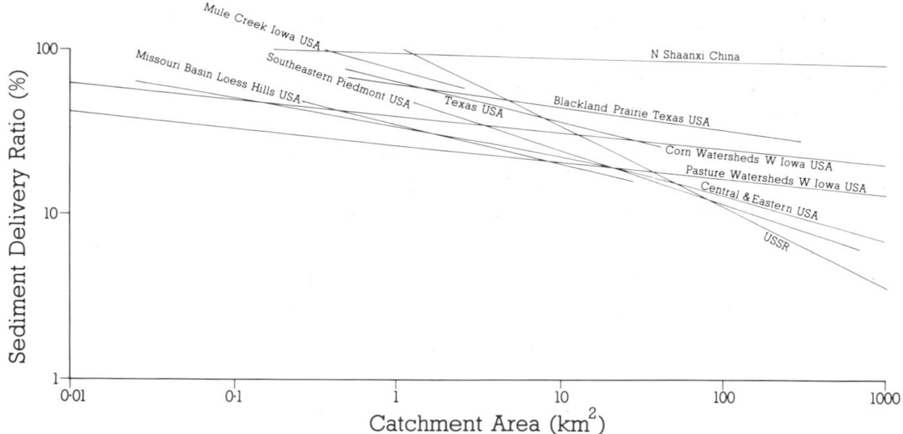

FIGURE 4.8 RELATIONSHIPS BETWEEN THE SEDIMENT DELIVERY RATIO AND CATCHMENT AREA PROPOSED BY VARIOUS WORKERS
Based on data presented in Mou and Meng (1980), Sokolovskii (1968), Piest *et al.* (1975), Renfro (1975), ASCE (1975), Williams (1977), and Roehl (1962)

the stream and river channels. The sediment delivery ratio concept (e.g. Robinson, 1977) defines this proportion through the sediment delivery ratio which is the ratio of the sediment yield at the catchment outlet to the gross erosion or sediment mobilization *within* the catchment. Values of the delivery ratio are frequently as low as 0.05 or 5 per cent. It is therefore necessary to define values of this ratio in order to establish the relationship between sediment yield and upstream rates of erosion, but few guidelines exist as to its estimation. This lack of guidance partly reflects the black box nature of the ratio and the large number of factors which could influence its magnitude, but more importantly it reflects the difficulty of establishing *actual* values of the ratio in the first place, in order to provide the dependent variable for developing prediction techniques. In most cases the ratio is established by comparing measured sediment yield with an estimate of sheet and rill erosion produced using a soil loss estimation procedure such as the USLE and augmented by a subjective assessment of the rate of gully and channel erosion to define the gross erosion.

Most of the existing studies which have attempted to relate the delivery ratio to catchment characteristics have employed the simple inverse relationship with catchment area. This is justified in terms of the increased opportunity for sediment deposition as basin size increases, which in turn is related to the increased travel distances and the increased incidence of lower angle slopes and valley bottomland. Figure 4.8 presents the various relationships between delivery ratio and catchment area that have been proposed in recent years. Considerable diversity exists in the precise form of the relationship, with catchments of 100 km^2 evidencing delivery ratios as high as *ca*. 100 per cent in the gullied loess plateau of the middle reaches of the Yellow River and as low as 10 per cent in the southeastern Piedmont of the USA. Some consistency has been established in the form of the relationship for catchments in Central and Eastern USA and the US Department of Agriculture Soil Conservation Service has proposed a generalized relationship for these areas (Renfro, 1975). Elsewhere, however, the estimation of a meaningful delivery ratio remains a major problem. More work is required to define the various processes and sinks involved in the delivery process and the sediment budget approach described by Dietrich and Dunne (1978), Lehre (1979), and Swanson *et al.* (1982) would seem to offer considerable potential. In this essentially conceptual approach, the various sediment sources within a basin are defined, and the sediment generated from these sources is routed to and through the channel system by considering the various sinks.

Figure 4.9 depicts diagrammatic representations of such budgets for three basins where investigators have attempted to define tentatively the source–yield linkages. In all cases the proportion of eroded sediment delivered to the basin outlet is relatively small, ranging from 53 per cent to 5.5 per cent, but considerable differences exist between the catchments represented in the precise form of the budget and in the location and importance of the various sinks. Application of this conceptual approach to other basins and parameterization of the various processes and pathways involved could lead to improved methods of defining the relationship between upstream erosion and downstream sediment yield.

A further problem associated with this relationship or linkage are the temporal discontinuities which may be involved. In this context, sediment eroded from one location in the watershed may be temporally stored and subsequently remobilized several times before reaching the catchment outlet. Sediment yield at the catchment outlet may therefore reflect the recent history of erosion and sediment delivery rather than contemporary erosion. This situation is clearly illustrated by the work of Trimble (1974, 1976, 1981) in Georgia and the Driftless area of Wisconsin, USA. Severe upland soil erosion occurred within these areas during the late 19th and early 20th centuries, but most of this sediment was deposited in the river valleys and only about 5 per cent was delivered to catchment outlets. Soil conservation measures implemented in the 1930's and thereafter severely reduced upland erosion but failed to produce a significant reduction in sediment yields (*cf*. Meade and Trimble, 1974) because sediment stored in the valley deposits was remobilized. The significance of this situation to the source–yield linkage is shown by the sediment budget for Coon Creek, Wisconsin, for the periods 1853–1938 and 1938–1975 depicted in Figure 4.9. The average sediment yields for the two periods were almost identical at 26 and 27 $\times$ 10^3 m^3 yr^{-1}, even though upland erosion had been reduced by about 25 per cent during the latter

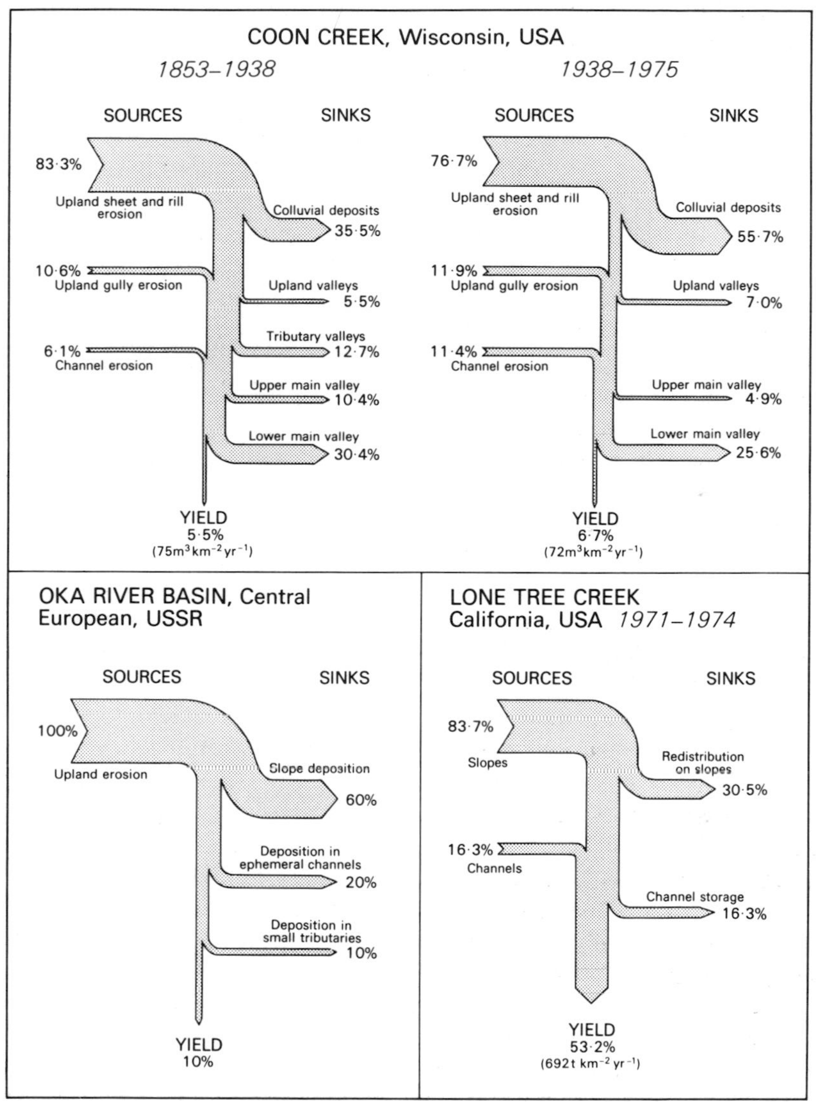

FIGURE 4.9 SEDIMENT BUDGETS FOR COON CREEK, WISCONSIN (360 km^2), 1853–1938 AND 1938–1975; LONE TREE CREEK, CALIFORNIA (1.74 km^2); AND THE OKA RIVER, USSR

Based on data presented by Trimble (1981), Lehre (1982), and Zaslavsky (1979)

period. The relationships between upland erosion and downstream sediment yield are thus very different due to the changes in storage that occurred.

THE PROSPECT

Current increases in national sediment measurement activities are to be welcomed as paving the way towards an improved understanding of spatial variations in sediment yield at the local, regional, and global levels. Application of the ergodic hypothesis of space–time substitution will in turn provide improved potential for palaeohydrological reconstruction. Similarly, improvements in our understanding of erosion and delivery processes produced by the development of the sediment budget approach must increase the potential for linking the reconstruction of sediment yields to erosion rates within the basin. Nevertheless, reliance upon space–time substitution necessarily imposes limitations on the potential for such reconstruction. Alternative approaches must be sought. The development of integrated erosion–transport–deposition models at the basin scale as advocated by Fleming (1975, 1981) would appear to offer considerable potential for evaluating the effects of changes in hydrological inputs and catchment condition, but their effective implementation must await an improved understanding of the many processes involved and the incorporation of physically-based functions that obviate calibration problems. Further difficulties will arise in bridging the gap between the essentially *hydrological* modelling of erosion and sediment yield at the level of the storm event and the simulation of longer term *geomorphological* aspects of catchment response, such as the temporal discontinuities of sediment delivery discussed above.

An important problem that faces attempts to reconstruct the sediment yield response of a drainage basin, is the paucity of long-term records of sediment transport. The availability of records covering, for example, the last 100 years would be of very considerable value in attempting to understand temporal variation in sediment yields and their response to climatic fluctuations and changes in catchment condition. Record lengths will inevitably increase as time proceeds, but opportunities to document particular changes in sediment response may not recur. In this context it is advocated that more attention should be directed to the potential of lake sedimentation studies in reconstructing temporal patterns of sediment yield. Oldfield (1977) and O'Sullivan (1979) have stressed the value of lake sediment studies in illuminating the environmental history of the upstream drainage basin and it is clear that much can bc lcarnt about rates of sediment yield. A classic example of the potential of such work is the study undertaken by Davis (1976) in Frains Lake, Michigan, USA. This lake is located at the outlet of a 0.18 km^2 drainage basin and a detailed exercise of sediment coring, and core analysis and dating enabled the pattern of inorganic sediment influx in terms of volume to be reconstructed over the past 200 years. Volume influx rates can be readily converted to estimates of sediment yield and Figure 4.10A reproduces the generalized history of sediment yield from the Frains Lake drainage basin over this period proposed by Davis (1976). The graph shows low rates of sediment yield in pre-settlement times, rising by a factor of up to 70 with the onset of settlement and agricultural clearance after 1830, and stabilizing after 1900 at a rate about 10 times the pre-settlement rate. Interestingly, this picture is very similar to that produced by Wolman (1967) for the Piedmont of Maryland, although his reconstruction was based on a study of contemporary sediment yields from basins under different land use and involved space–time substitution. Other less detailed core analysis for Frains Lake indicated that pre-settlement sediment yields had remained relatively constant throughout the whole of the Holocene at about 9t km^{-2} yr^{-1}.

The work by Davis at Frains Lake necessitated analysis of a considerable number of cores, in order to permit calculation of the *volumes* of inorganic sediment involved. Volume data are required to produce estimates of sediment yield from the catchment area. In many lake sediment studies data are available from only a single core or at best from a very small number of cores, and changes in sediment yield over the period represented by the core must be inferred from changes in the rate of sediment accumulation at that point. This invokes the assumption that the pattern of sedimentation over the lake floor has remained constant through time. There is, however, considerable evidence to suggest that this assumption is frequently unjustified, since sediment focusing may cause changes in the relationship between the rate of sediment accumulation at a point and the overall sediment influx to the lake. Evidence for changes in sediment yield through time based on analysis of single cores must therefore be treated with caution, but it is nevertheless of interest to consider the evidence presented by four such studies (Figure 4.10B), particularly since they cover a longer time span than the curve for Frains Lake. In this case the vertical axes of the plots refer to sediment accumulation rates rather than sediment yield.

The results from Mirror Lake, New Hampshire, USA and Lake Trummen, Sweden, based respectively on the work of Likens and Davis (1975) and Digerfeldt (1972), are of note in that both suggest that sediment yields were relatively high during the late glacial period. This can be tentatively related to the lack of vegetation cover and high availability of sediment within a landscape recently exposed by glacial retreat. Evidence from the environment of Mirror Lake suggests that a forest became established about 9000 B.P. and this is reflected in the stabilizing of sediment accumulation rates at that time. The data from Lake Trummen also point to an increase in sediment accumulation rates towards the present which doubtless relates to the influence of human activity within the surrounding drainage basin. The evidence from Braeroddach Loch, Scotland is not entirely consistent with the previous two

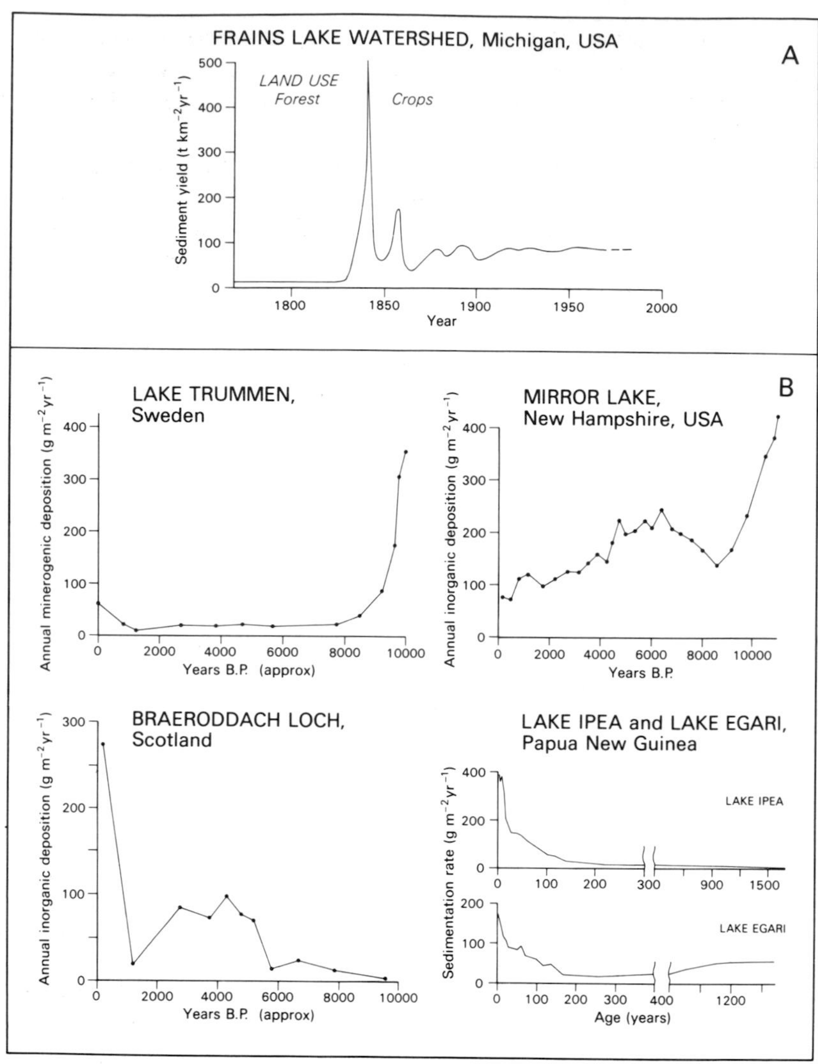

FIGURE 4.10 EVIDENCE FROM LAKE SEDIMENT STUDIES
Illustrating the reconstruction of (A) sediment yields to Frains Lake, Michigan, proposed by Davis (1975) and of (B) sedimentation rates in Mirror Lake, New Hampshire; Lake Trummen, Sweden; Braeroddach Loch, Scotland and two small lakes in Papua New Guinea based on data presented by Likens and Davis (1975), Digerfeldt (1972), Edwards and Rowntree (1980), and Oldfield *et al.* (1980)

cases in that although there is evidence of increased rates of sedimentation in the past 400 years, again related to human activity, there is no indication of high rates of accumulation during the immediate post-glacial. The core data for Papua New Guinea are particularly interesting in again showing the impact of human activity on sedimentation rates during the past 300 years which Oldfield *et al.* (1980) have related tentatively to the intensification of land use resulting from the introduction of the sweet potato and more definitively to the post-1950 impact of 'western' peoples.

More detailed studies of these and other lakes that would permit the estimation of volumes of sediment influx and therefore sediment yield would clearly be of very great value in the context of palaeohydrological reconstruction. It is to be hoped that the potential for rapid inter-core correlation offered by magnetic measurements (e.g. Bloemendal *et al.*, 1979) will facilitate such studies. Interpretational problems will still exist in terms of defining the trap efficiency of the lake body, dating the sediment cores, and distinguishing allochthonous and autochthonous sources of minerogenic particulate material, but these problems must be weighed against the many uncertainties surrounding attempts at space–time substitution. Much could be learnt by closer cooperation of limnologists and fluvial geomorphologists, for the work of the former understandably frequently fails to take full advantage of the potential for palaeohydrological reconstruction. Furthermore, interpretation of lake sedimentation rates in terms of erosional activity in the drainage basin can clearly benefit from a geomorphological perspective. For example, it is worthwhile to consider again the case of Coon Creek, Wisconsin, detailed in Figure 4.9 and to conjecture that this flowed into a lake. A record of almost uniform rates of sediment influx during the period 1853–1973 provided by this lake would conceal very great changes in erosional behaviour within the watershed.

REFERENCES

Adams, J. 1980. High sediment yields from major rivers of the Western Southern Alps, New Zealand. *Nature*, **287**, 88–89.

Alonso, C. U., De Coursey, D. G., Prasad, S. N., and Bowie, A. J. 1978. Field test of a distributed sediment yield model. In *Verification of Mathematical and Physical Models in Hydraulic Engineering*, ASCE, New York, 671–678.

Anderson, H. W. 1975. Sedimentation and turbidity hazards in wildlands. In *Watershed Management, Proceedings of a Symposium of the Irrigation and Drainage Division*, ASCE, New York, 347–376.

ASCE, 1975. *Sedimentation Engineering*, ASCE, New York, 745pp.

Beasley, D. B., Huggins, L. F., and Monke, E. J. 1980. Planning for water quality using the Answers approach. In *Hydrologic Transport Modeling Symposium*, American Society of Agricultural Engineers, S. Joseph, Michigan, USA, 21–30.

Bloemendal, J., Oldfield, F., and Thompson, R. 1979. Magnetic measurements used to assess sediment influx at Llyn Goddionduon. *Nature*, **280**, 50–53.

Bordas, M. P., and Canali, G. E. 1980. The influence of land use and topography on the hydrological and sedimentological behaviour of basins in the basaltic region of South Brazil. In *The Influence of Man on the Hydrological Regime with special Reference to Representative and Experimental Basins*, International Association of Hydrological Sciences Publication no. 130, 55–60.

Branski, J. 1975. Ocena denudacji dorzecza Wisley na podstawie wynikow pomiarow rumowiska unoszonego. *Prace Instytutu Meteorologii i Gospodarki Wodnej*, **6**, 1–58.
Chang, M., Roth, F. A., and Hunt, E. V. 1982. Sediment production under various forest-site conditions. In Walling, D. E. (Ed.) *Recent Developments in the Explanation and Prediction of Erosion and Sediment Yield*, International Association of Hydrological Sciences Publication no. 137, 13–22.
Curtis, W. F., Culbertson, J. K., and Chase, E. B. 1973. Fluvial-sediment discharge to the oceans from the conterminous United States. *US Geological Survey Circular 670.*
Davis, M. B. 1976. Erosion rates and land use history in Southern Michigan. *Environmental Conservation*, **3**, 139–148.
Dendy, F. E., and Bolton, G. C. 1976. Sediment yield–runoff–drainage area relationships in the United States. *Journal of Soil and Water Conservation*, **32**, 264–266.
Dietrich, W. E., and Dunne, T. 1978. Sediment budget for a small catchment in mountainous terrain. *Zeitschrift für Geomorphologie*, Supplementband 29, 191–206.
Digerfeldt, G. 1972. The post-glacial development of Lake Trummen. *Folia Limnologica Scandinavica*, **16**, 104pp.
Douglas, I. 1967. Man, vegetation and the sediment yield of rivers. *Nature*, **215**, 925–928.
Douglas, I. 1973. *Rates of Denudation in Selected Small Catchments in Eastern Australia*, University of Hull, Occasional Papers in Geography, no. 21, 127pp.
Dunne, T. 1974. Suspended sediment data for the rivers of Kenya. Unpublished Report.
Dunne, T. 1979. Sediment yield and land use in tropical catchments. *Journal of Hydrology*, **42**, 281–300.
Edwards, K. J., and Rowntree, K. M. 1980. Radiocarbon and palaeoenvironmental evidence for changing rates of erosion at a Flandrian stage site in Scotland. In Cullingford, R. A., Davidson, D. A., and Lewin, J. (Eds.) *Timescales in Geomorphology*, Wiley, Chichester, 207–223.
Fleming, G. 1975. Sediment erosion–transport–deposition simulation: state of the art. In *Present and Prospective Technology for Predicting Sediment Yields and Sources*, US Dept. Agriculture Publication ARS–S–40, 274–285.
Fleming, G. 1981. The sediment problem related to engineering. In Tingsanchali, T., and Eggers, H. (Eds.), *Proceedings of the South-east Asian Regional Symposium on Problems of Soil Erosion and Sedimentation*, Asian Institute of Technology, Bangkok, 3–14.
Fournier, F. 1960. *Climat et Erosion*, P.U.F., Paris, 201pp.
Fredriksen, R. L. 1970. Erosion and sedimentation following road construction and timber harvest on unstable soils in three small Western Oregon watersheds. *US Forest Service Research Paper, PNW 104*, 15pp.
Golubev, G. N. 1980. Agriculture and water erosion of soils: a global outlook. *International Institute for Applied Systems Analysis Working Paper*, 80–129, 21pp.
Golubev, G. N. 1982. Soil erosion and agriculture in the world: an assessment and hydrological implications. In Walling, D. E. (Ed.), *Recent Developments in the Explanation and Prediction of Erosion and Sediment Yield*, International Association of Hydrological Sciences Publication no. 137, 261–268.
Graf, W. H. 1971. *Hydraulics of Sediment Transport*, McGraw-Hill, New York, 544pp.
Gregory, K. J. and Walling, D. E. 1973. *Drainage Basin Form and Process*, Arnold, London, 458pp.
Griffiths, G. A. 1979. High sediment yields from major rivers of the western Southern Alps, New Zealand. *Nature*, **282**, 61–63.

Griffiths, G. A. 1981. Some suspended sediment yields from South Island catchments, New Zealand. *Water Resources Bulletin*, **17**, 662–671.

Griffiths, G. A. 1982. Spatial and temporal variability in suspended sediment yields of North Island basins, New Zealand. *Water Resources Bulletin*, **18**, 575–584.

Hardjowitjitro, H. 1981. Soil erosion as a result of upland traditional cultivation in Java Island. In Tingsanchali, T., and Eggers, H. (Eds.), *Proceedings of the South-east Asian Regional Symposium on Problems of Soil Erosion and Sedimentation*, Asian Institute of Technology, Bangkok, 173–179.

Heusch, B. and Milliès-Lacroix, A. 1971. Une méthode pour estimer l'écoulement et l'érosion dans un bassin. Application au Maghreb. *Mines et Geologie* (Rabat) no. 33.

Inter-Agency Sedimentation Committee, 1947. *Proceedings of the Federal Inter-Agency Sedimentation Conference 1947*, US Water Resources Council, Washington DC, USA.

Inter-Agency Sedimentation Committee, 1965. *Proceedings of the Federal Inter-Agency Sedimentation Conference, 1963*, US Dept. Agriculture Misc. Pub. 970, 933pp.

Inter-Agency Sedimentation Committee, 1976. *Proceedings of the Third Federal Inter-Agency Sedimentation Conference*, US Water Resources Council, Washington DC, USA.

Janda, R. J., and Nolan, K. M. 1979. Stream sediment discharge in Northwestern California. In *Guidebook for a Field Trip to observe Natural and Management-related Erosion in Franciscan Terrane of Northern California*, US Geological Survey, Menlo Park, California, IV–I–IV–27.

Judson, S., and Ritter, D. F. 1964. Rates of regional denudation in the United States. *Journal of Geophysical Research*, **69**, 3395–3401.

Langbein, W. B., and Schumm, S. A. 1958. Yield of sediment in relation to mean annual precipitation. *Transactions American Geophysical Union*, **39**, 1076–1084.

Lehre, A. K. 1979. Sediment budget of a small California Coast Range drainage basin. In Davies, T. R. H., and Pearce, A. J. (Eds.), *Erosion and Sediment Transport in Pacific Rim Steeplands*, International Association of Hydrological Sciences Publication no. 132, 123–139.

Lehre, A. K. 1982. Sediment budget of a small coast range drainage basin in North–Central California. In Swanson, F. J. *et al.* (Eds.), *Sediment Budgets and Routing in Forested Drainage Basins*, US Forest Service General Technical Report PNW–141, 67–77.

Li, R. M., Stevens, M. A., and Simons, D. B. 1976. Water and sediment routing from small watersheds. In *Proceedings of the Third Federal Inter-Agency Sedimentation Conference*, US Water Resources Council, Washington DC, USA, 1–193–1–204.

Li, Y-H. 1976. Denudation of Taiwan Island since the Pliocene epoch. *Geology*, **4**, 105–107.

Likens, G. E., and Davis, M. B. 1975. Postglacial history of Mirror Lake and its watershed in New Hampshire, USA: an initial report. *Verhalt International Verein. Limnologie*, **19**, 982–993.

Meade, R. H., and Trimble, S.W. 1974. Changes in sediment loads in rivers of the Atlantic drainage of the United States since 1900. In *Effects of Man on the Interface of the Hydrological Cycle with the Physical Environment*, International Association of Hydrological Sciences Publication no. 113, 99–104.

Mou, J. and Meng, Q. 1980. *Sediment delivery ratio as used in the computation of the watershed sediment yield*, Biejing, China.

Noble, E. L. 1965. Sediment reduction through watershed rehabilitation. In *Proceedings of the Federal Inter-Agency Sedimentation Conference 1963*, US Dept. Agriculture Misc. Pub. 970, 114–123.

O'Loughlin, C. L., Rowe, L. K., and Pearce, A. J. 1980. Sediment yield and water quality responses to clearfelling of evergreen mixed forests in western New Zealand. In *The Influence of Man on the Hydrological Regime with Special Reference to Representative and Experimental Basins*, International Association of Hydrological Sciences Publication no. 130, 285–292.

O'Sullivan, P. E. 1979. The ecosystem watershed concept in the environmental sciences — a review. *International Journal of Environmental Studies*, **13**, 273–281.

Oldfield, F. 1977. Lakes and their drainage basin as units of sediment-based ecological study. *Progress in Physical Geography*, **1**, 460–504.

Oldfield, F., Appleby, P. G., and Thompson, R. 1980. Palaeoecological studies of lakes in the Highlands of Papua New Guinea. I. The chronology of sedimentation. *Journal of Ecology*, **68**, 457–477.

Pickup, G., Higgins R. J., and Warner, R. F. 1981. Erosion and sediment yield in Fly River drainage basins. In Davies, T. R. H. and Pearce, A. J. (Eds.), *Erosion and Sediment Transport in Pacific Rim Steeplands*, International Association of Hydrological Sciences Publication no. 132, 438–456.

Piest, R. F., Kramer, L. A., and Heinemann, H. G. 1975. Sediment movement from loessial watersheds. In *Present and Prospective Technology for Predicting Sediment Yields and Sources*, US Dept. Agriculture Publication ARS–S–40, 130–141.

Qian, N. and Dai, D. 1980. The problems of river sedimentation and the present status of its research in China. In *Proceedings of the International Symposium on River Sedimentation*, Guanghua Press, Beijing, 3–39.

Renfro, G. W. 1975. Use of erosion equations and sediment delivery ratios for predicting sediment yield. In *Present and Prospective Technology for Predicting Sediment Yields and Sources*, US Dept. of Agriculture Publication ARS–S–40, 33–45.

Rhoades, E. D., Welch, N. H., and Coleman, G. A. 1975. Sediment-yield characteristics from unit source watersheds. In *Present and Prospective Technology for Predicting Sediment Yields and Sources*, US Dept. Agriculture Publication ARS–S–40, 125–129.

Robinson, A. R. 1977. Relationships between soil erosion and sediment delivery. In *Erosion and Sediment Transport in Inland Waters*, Proceedings of the Paris Symposium, International Association of Hydrological Sciences Publication no. 144, 159–167.

Roehl, J. W. 1962. Sediment source areas, delivery ratios and influencing morphological factors. *International Association of Hydrological Sciences Publication* no. 59, 202–213.

Roose, E. J. 1977. Use of the universal soil loss equation to predict erosion in West Africa. In *Soil Erosion Prediction and Control*, Soil Conservation Society of America, Ankeny, Iowa, USA, 60–74.

Rooseboom, A., and Harmse, H. J. von M. 1979. Changes in the sediment load of the Orange River during the period 1929–1969. In *The Hydrology of Areas of Low Precipitation*, International Association of Hydrological Sciences Publication no. 128, 459–470.

Schumm, S. A. 1965. Quaternary palaeohydrology. In Wright, H. E., and Frey, D. G. (Eds.), *The Quaternary of the United States*, Princeton University Press, Princeton, 783–794.

Schumm, S. A. 1968. Speculation concerning palaeohydrologic controls of terrestrial sedimentation. *Geological Society of America Bulletin*, **79**, 1573–1588.

Schumm, S. A. 1977a. *The Fluvial System*, Wiley, Chichester, 356pp.

Schumm, S. A. 1977b. An experimental study of geomorphic thresholds. *Final Report, Colorado State University, Fort Collins, Colorado, USA*.

Sharma, K. D., and Chatterji, P. C. 1982. Sedimentation in Nadis in the Indian arid zone. *Hydrological Sciences Journal*, **27**, 345–352.

Smith, R. E. 1976. Simulating erosion dynamics with a deterministic distributed watershed model. In *Proceedings of the Third Federal Inter-Agency Sedimentation Conference*, US Water Resources Council, Wasington DC, USA, 1–163–1–173.

Sokolovskii, D. L. 1968. *Rechnoi Stok. Osnovy Teorii i Metodiki Raschetov*, Gidrometeorologischeskoe Izdatel'stvo, Leningrad, USSR.

Stoddart, D. R. 1969. World erosion and sedimentation. In Chorley, R. J. (Ed.), *Water Earth and Man*, Methuen, London 43–64.

Strakhov, N. M. 1967. *Principles of Lithogenesis*, Vol.I, Oliver and Boyd, Edinburgh, 245pp.

Swanson, F. J., Janda, R. J., Dunne, T., and Swanston, D. N. 1982. *Sediment Budgets and Routing in Forested Drainage Basins*, US Forest Service General Technical Report PNW–141, 165pp.

Tabuteau, M. M. 1960. Etude graphique pour les conséquences hydro-érosives du climat mediterrané. *Association Géographique Française Bulletin*, **295**, 130–142.

Trimble, S. W. 1974. *Man-induced Soil Erosion on the Southern Piedmont*, Soil Conservation Society of America, Ankeny, Iowa, USA.

Trimble, S. W. 1976. Sedimentation in Coon Creek Valley, Wisconsin. In *Proceedings of the Third Federal Inter-Agency Sedimentation Conference*, US Water Resources Council, Washington DC, USA, 5–100–5–112.

Trimble, S. W. 1981. Changes in sediment storage in the Coon Creek Basin, Driftless Area, Wisconsin, 1853 to 1975. *Science*, **214**, 181–183.

Ursic, S. J., and Dendy, F. E. 1965. Sediment Yields from small watersheds under various land uses and forest covers. In *Proceedings of the Federal Inter-Agency Sedimentation Conference 1963*, US Dept. Agriculture Misc. Pub. 970, 47–52.

Walling, D. E. 1979. Hydrological processes. In Gregory, K. J., and Walling, D. E. (Eds.), *Man and Environmental Processes*, Butterworth, London, 57–81.

Walling, D. E. 1981. Yellow River which never runs clear. *Geographical Magazine*, **53**, 568–575.

Walling, D. E., and Kleo, A. H. A. 1979. Sediment yields of rivers in areas of low precipitation: a global view. In *The Hydrology of Areas of Low Precipitation*, International Association of Hydrological Sciences Publication no. 128, 479–493.

Walling, D. E., and Webb, B. W. 1981. The reliability of suspended sediment load data. In *Erosion and Sediment Transport Measurement*, Proceedings of the Florence Symposium, June 1981, IAHS Publication no. 133, 177–194.

Williams, J. R. 1977. Sediment delivery ratios determined with sediment and runoff models. In *Erosion and Sediment Transport in Inland Waters*, Proceedings of the Paris Symposium, International Association of Hydrological Sciences Publication no. 122, 168–179.

Wilson, L. 1969. Les relations entre les processus géomorphologiques et le climat modern comme méthode de paléoclimatologie. *Revue de Géographie Physique et de Géologie Dynamique*, Sér. 2, **11**, 303–314.

Wischmeier, W. H. 1975. Estimating the soil loss equation's cover and management factor for undisturbed areas. In *Present and Prospective Technology for Predicting Sediment Yields and Sources*, US Dept. of Agriculture Publication ARS–S–40, 118–124.

Wischmeier, W. H., and Smith, D. D. 1965. *Predicting Rainfall Erosion East of the Rocky Mountains*, US Dept, Agriculture Handbook no. 282.

Wischmeier, W. H., and Smith, D. D. 1978. *Predicting Rainfall Erosion Losses — Guidebook to Conservation Planning*, US Dept. Agriculture Handbook no. 537.

Wolman, M. G. 1967. A cycle of sedimentation and erosion in urban river channels. *Geografiska Annaler*, **49A**, 383–395.
Zaslavsky, M. N. 1979. *Erozia Pochv* (Soil Erosion), Mysl Publishing House, Moscow.

Background to Palaeohydrology
Edited by K. J. Gregory

5

Palaeovelocity and palaeodischarge determination for coarse gravel deposits

JUDITH K. MAIZELS

Department of Geography, University of Aberdeen

Coarse fluvial sedimentary sequences have been extensively investigated as a means of environmental reconstruction. Particle size, sorting, fabric and shape measures, and lithofacies analyses have been widely used in the determination of selected palaeoflow parameters including flow regime, flow direction, stream competence, and rates of sediment sorting, abrasion, and transport. Channel form and the sedimentary characteristics of channel infills have also been used to determine estimates of palaeodischarges in finer grained or high sinuosity channels. All the methods of palaeoflow determination are based on direct analogies either with hydraulic relations between sediment and flow parameters in present-day streams or flumes, or with geomorphic relations between channel morphology, channel sediments and discharge measures observed in present-day rivers. These two approaches to palaeodischarge analysis in turn reflect the contrasting conditions which are represented by the estimated palaeodischarge values (Miall, 1978). The hydraulic or micro-approach (Jopling, 1971), provides estimates of short-term or instantaneous velocity or discharge conditions, as a result of local adjustments of hydraulic geometry around a mean or equilibrium channel condition. The geomorphic or macro-approach indicates the development of longer term equilibrium or quasi-equilibrium conditions between catchment, channel and flow parameters (e.g. Langbein and Leopold, 1964).

The development of these two approaches has produced a large number of possible methods for palaeovelocity and palaeodischarge determination. However, many of these methods are based on empirical data and experimental work which have well-defined limits of sediment and flow conditions, such that their applicability must in turn be severely limited to selected and specific environments. The micro-approach appears the more appropriate for coarse gravel deposits, but for certain coarse bed environments, for example those resulting from massive floods or from flow with very large bed roughness, there may still be no appropriate methods of determining palaeoflow velocities. Any discussion of the possible methods of palaeovelocity and palaeodischarge determination must assess the applicability of these various methods to particular sedimentary environments, the assumptions underlying each of the methods, the nature of the field data required for these methods, and the significance that the resulting estimates may have in the geomorphic interpretation of palaeovelocity and palaeodischarge values.

Palaeovelocity and palaeodischarge determinations generally require an initial input of selected field data. The field measurements may include the energy gradient or bed slope, an estimate of resistance to flow (often taken as some measure of particle size, representing grain resistance), and cross-sectional form or area such as the width, depth, hydraulic radius, or cross-sectional area. While measures of particle size and, usually, bedslope, may be determined directly in the field, flow depth (as opposed to channel depth) is best estimated from equations describing boundary shear conditions, where flow depth is taken as an index of the length of the wetted perimeter (i.e. the hydraulic radius) presenting the boundary shear which acts to retard flow velocities. Resistance to flow can best be estimated by relating measures of bed particle size to known velocity equations, both of empirical and theoretical derivation, for equivalent flow conditions. Palaeodischarge may also be estimated from either theoretical or empirical relations. Hence palaeovelocity and palaeodischarge determination requires a combination of field measurement and analysis of force mechanics and friction controls affecting flow velocities. This paper examines firstly the main assumptions underlying the application of selected flow equations to palaeohydraulic analysis; and secondly proceeds to describe some of the methods of estimating flow resistance and palaeovelocity. The final part of the paper deals with methods of palaeodischarge determination, and concludes with some recommendations for palaeohydraulic analysis of coarse gravel deposits.

ASSUMPTIONS OF PALAEOHYDRAULIC ANALYSIS OF COARSE GRAVEL DEPOSITS

The assumptions identified as underlying the methods of palaeohydraulic analysis of coarse gravel deposits are basically those relating to certain conditions of open-channel flow (see Chow, 1959; Henderson, 1966; Raudkivi, 1976).

Steady, uniform flow

The most fundamental assumption in applying flow equations to palaeohy draulic analysis is that of steady, uniform flow. Steady, uniform flow means that no changes in velocity, or at least in the form of the vertical velocity profile, take place through time or through a downstream reach of the stream. It is assumed that no acceleration or deceleration of flow takes place at the given channel cross-section, and that the cross-section area of the channel remains relatively rigid. This basic assumption has important implications for palaeohydraulic analysis in the following four respects. Firstly, it means that the bedslope of the channel is parallel with the energy gradient of the stream, and therefore that field measurements of bedslope can be taken to represent the total energy slope in the former stream. Secondly, it means that one may accept the continuity equation of

$$Q = w.d.v \tag{1}$$

for palaeodischarge determination (all notation explained in Appendix 1, pp. 132–3). This equation is basic to the concept of flow equilibrium, such that a change in one independent variable will be matched by mutual adjustment of the other variables in order to maintain dynamic equilibrium around a mean value (Hey, 1978). In unsteady, non-uniform flow, however, this relation may not be maintained, as a result of local flow acceleration or deceleration, changes in water density, and movement of flood waves (Henderson, 1966). Thirdly, since the channel is assumed to have a rigid boundary, the flow equations are applicable to flow up to the condition of bed motion, but not beyond. Once the bed itself becomes mobile or 'live', shear resistance to flow is markedly reduced, and alternative flow equations need to be adopted (e.g. Church and Gilbert, 1975). Finally, the assumption of steady, uniform flow allows one to assume also that palaeovelocity estimates relate to representative flow conditions rather than to rapid, instantaneous, local changes in velocity. This seems perhaps the most questionable of the four implications of steady, uniform flow for palaeohydraulic analysis, since coarse gravels can generate largescale or macro turbulence and eddies that promote particle lifting and motion (e.g. Hartung and Scheuerlein, 1967; Baker, 1973a, b; Bradley and Mears, 1980).

Fully turbulent flow over rough boundary

Flow equations derived for laminar or transitional flow depend particularly on water viscosity as a control on resistance, but viscosity is of little significance in fully turbulent flow. Under these conditions, the characteristic logarithmic vertical velocity profile developed by Prandtl and von Karman (Chow, 1959; Leopold, Wolman, and Miller, 1964; Henderson, 1966; Simons, 1969) may be accepted:

$$\frac{V}{V^*} = \frac{C_1}{\kappa} \log \left(\frac{d}{K_s}\right) + C_2 \tag{2}$$

This equation is developed for rough boundary flow, that is where the bed particles protrude through the laminar sublayer close to the channel bed. The values of the constants c_1 and c_2 have been determined experimentally to approximate 2.303 and 8.5 respectively for fully turbulent flow over a rough bed. Von Karman's constant is taken to be equal to 0.4 for clear water. Acceptance of the Prandtl–Karman velocity equation is crucial in the development of certain friction factors for palaeohydraulic analysis of coarse gravel deposits.

Small relative bed roughness

A dominant control on velocity distribution is clearly the relative roughness of the bed, as expressed by the ratio d/D, or d/K_s in equation (2). The flow equations available for palaeohydraulic analysis of coarse gravel deposits relate to channels exhibiting relatively low relative roughness values, i.e. those where D/d lies between about 0.067 and 0.3 or 0.4 (Kellerhals, 1967; Day, 1978), and where flow depth is between 2.5 and 15 times the height of the bed roughness elements. Where relative flow depth is much greater, smooth boundary conditions begin to dominate the vertical velocity profile, while where relative flow depth is smaller, no known velocity law has been developed (Day, 1978), and largescale bed roughness conditions start to become dominant. Velocity profiles then become completely disrupted, other sources of flow resistance are introduced in association with non-uniform flow conditions, and wakes and eddies are generated around individual clasts (e.g. Ashida and Bayazit, 1973; Kellerhals, 1973; Day, 1977, 1978; Bathurst, 1978; Bathurst et al. 1981; Thompson and Campbell, 1979).

Uniform distribution of representative resistance elements

Although flow resistance may be derived from three different sources namely from skin or grain resistance, from form resistance including bedforms, bends, etc., and from local spill resistance such as around boulders (Leopold, Wolman, and Miller, 1964), most flow equations for coarse bed streams assume that resistance is dominated by the size of grains on the channel bed (but see assumption pp. 109 below). Where flow is assumed to be steady and uniform, form and spill resistance should be negligible, and hence skin resistance may indeed represent the dominant source of resistance. However, several important problems arise in trying to meet this assumption.

Determination of representative grain size

It is generally assumed that the largest size fraction on the bed has primary control on grain resistance, and hence on eddy intensity and particle entrainment characteristics (e.g. Limerinos, 1970; Church and Gilbert, 1975; Church and Kellerhals, 1978; Day, 1978). Most traditional flow resistance equations are based on measures of the 50th and 84th percentile (D_{50} and D_{84} respectively) of the whole bed particle size distribution, the latter simply representing two standard deviations above the median grain size. Limerinos (1970) concludes that although the use of D_{84} is convenient, there is no hydraulic justification for its use. Similarly, Hey (1979) regards D_{84} as having relatively low sampling errors compared with higher percentiles, but recommends using 3.5 D_{84} in the development of the friction factor (see below). With the recognition that the few largest clasts on the bed can adequately represent the dominant grain roughness control on flow resistance, the traditional measures of D_{50} and D_{84} are gradually being replaced in more recent studies with alternative grain size measures. Henderson (1966) and Lane and Carlson (1954) suggest the use of D_{75}, since this is meant to represent the size at which the bed tends to become armoured. However, several more recent studies have used percentiles between D_{84} and D_{99}. Fahnestock and Bradley (1973), Baker and Ritter (1975), Church and Kellerhals (1978), and Day (1978) use D_{90}; Burkham and Dawdy (1976) recommend using D_{95}; while Parker and Peterson (1980) recommend adopting Kamphuis' approach of adopting 2 D_{90} in the development of the friction factor, compared with Hey's (1979) 3.5 D_{84} (above) which is intended to be equivalent to D_{99}. Similarly, several workers have measured simply the largest single clast (D_x) (e.g. Malde, 1968; Butler, 1977), or the largest five (D_{5x}) (e.g. Church 1978), the largest 10 (D_{10x}) (e.g. McDonald and Banerjee, 1971; Scott and Gravlee, 1968 in Baker, 1973a; Maizels, 1980, 1982, 1983), or the largest 25 (D_{25x}) clasts (e.g. Clague, 1975) in a deposit. It appears that D_{10x} is approximately equivalent to D_{95}, some value between D_{25x} and D_{50x} is more equivalent to D_{84}, while D_{99} is equivalent to the largest one or two clasts (Figure 5.1). Neither Limerinos (1970) nor Bray (1979), however, found any significant difference in the friction factors as determined from different percentile measures.

Normally, the intermediate clast diameter (*b*-axis) is measured, but Limerinos (1970), in particular, claims that the minimum diameter (*c*-axis) is a more representative indication of relative roughness height since it indicates height of projection of clast into the flow. However, he found no significant difference in determination of the friction factor using different axial measures. In addition, the position of the clast on the bed, its degree of imbrication, embedding in underlying sands, or relative height above adjacent clasts, will all affect the actual projection height. Nevertheless, critical relative roughness ratios may be altered by choice of axial measurement. Further research is

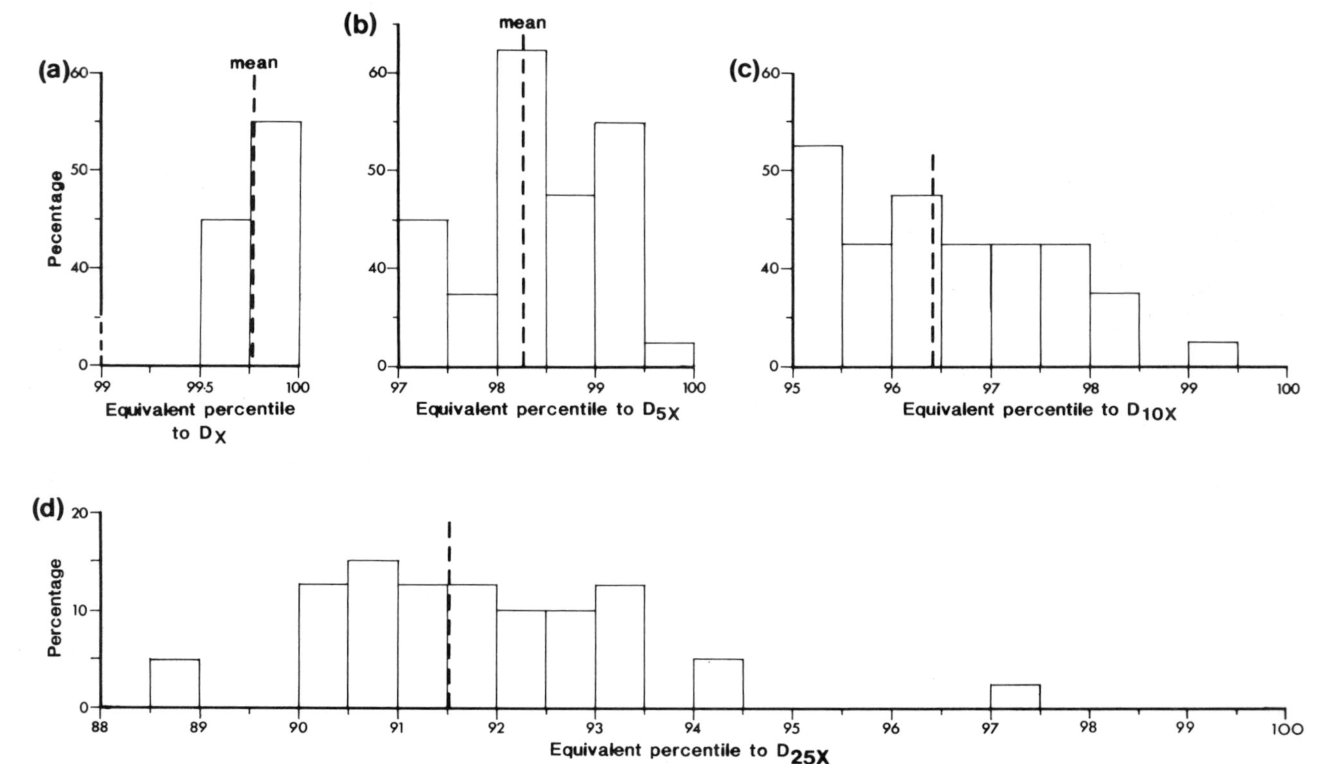
(a)
Pecentage
mean
Equivalent percentile to D_X
(b)
mean
Equivalent percentile to D_{5X}
(c)
Equivalent percentile to D_{10X}
(d)
Percentage
Equivalent percentile to D_{25X}

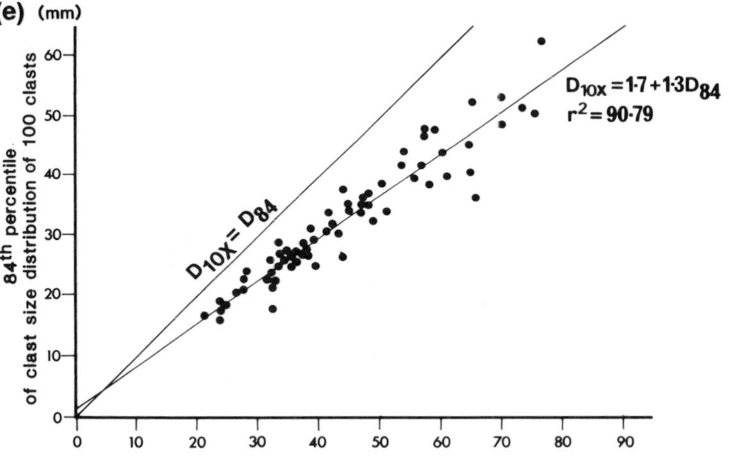

FIGURE 5.1 PARTICLE SIZE EQUIVALENTS BASED ON PERCENTILE MEASURES AND MEASURES OF LARGEST PARTICLES (b-AXIS)
(Example of 40 samples each of 100 clasts from gneissose and granite bed material in two upland rivers, N.E. Scotland) (a) Single largest clast (Dx). (b) Five largest clasts (D_{5x}). (c) Ten largest clasts (D_{10x}). (d) Twentyfive largest clasts (D_{25x}). (e) Relation between D_{84} and D_{10x}

required into the development of a truly representative grain roughness measure. The development of an armoured bed will clearly affect ease of particle motion, but this also means that sampling of clasts should be restricted to a single armoured layer, rather than including any underlying finer-grained layer that may have remained protected from flow (e.g. Kellerhals, 1967).

Uniform distribution of grain roughness elements

Most flow equations have been developed on the basis of a uniform distribution of grain roughness elements. Particles are hence assumed to exhibit no systematic spacing (as may be associated with bedform development), such that initiation of motion may be regarded as a random probabilistic or stochastic process (ASCE, 1966; Church, 1978). However, it is also widely recognized that particles are differentially distributed, and moved, on the channel bed according to their degree of packing, imbrication and armouring (e.g. Howard, 1980). In addition, bed and banks present different degrees of grain resistance (e.g. see Hey, 1979). Church (Church and Gilbert, 1975; Church, 1978) differentiates between different types of boundary condition that will affect particle motion, namely, a 'normal' boundary with random packing of grains, an 'overloose' boundary exhibiting open work, matrix-poor packing, and an 'underloose' boundary where particles are imbricated. Hence, differences in grain packing may be accommodated in certain palaeohydraulic analyses.

Uniform particle shape

The stochastic element of grain movement also assumes that particles are of equal sphericities. However, it is well known that more spherical clasts are more readily entrained because they are less likely to be imbricated, but that once entrained 'roller' type clasts move greater distances (e.g. Nevin, 1946; Lane and Carlson, 1954; Helley, 1969). Flow equations do not at present account for differences in clast shape.

Particle size limits

The lower particle size limits for which the flow equations are applicable are reasonably well defined. From experimental work and data from natural streams, measures of stream competence can only be functionally related to particle size where particles exceed 5–8 mm in diameter. This minimum size represents the lower boundary of so-called 'coarse' gravel deposits. At lower sizes, viscous forces begin to become effective rather than inertial forces, and palaeovelocity or force functions can no longer be predicted directly from clast diameter; these methods are only applicable at high Reynolds numbers (i.e. $>$

ca. 10^4). The upper size limit is, by contrast, very poorly defined. Palaeoflow equations have been applied to boulders exceeding *ca.* 10 m in diameter (Novak, 1973; see also Birkeland, 1968; Malde, 1968; Baker 1973a), but the validity of applying standard flow equations to such large bed materials needs further testing.

Absence of bedforms

The flow equations adopted for most palaeohydraulic analysis of coarse gravel deposits assume skin resistance is dominant, and that form resistance may be neglected. Until recently, it was believed that coarse gravels rarely exhibit bedforms of any kind, while bar deposits have been shown to present little additional resistance to flow during high stages (Parker and Peterson, 1980). However, recent evidence has demonstrated a wide range of coarse-grained bedforms, including the giant current ripples of the Lake Missoula flood, *ca.* 4 m in height (Baker 1973a, b); the 'gigantic bars' of the Melon Gravel deposited during the Lake Bonneville flood (Malde, 1968), up to 2 km in length, and *ca.* 50 m in height; and coarse-grained antidune deposits and associated 'transverse ribs' (Fahnestock, 1963; Galay, 1967; Hand *et al.*, 1969; McDonald and Banerjee, 1971; Boothroyd, 1972; Boothroyd and Ashley, 1975; Baker, 1977; Shaw and Kellerhals, 1977; Koster, 1978; Shaw and Healy, 1980), generally 0.2 to 2.5 m high, comprising imbricated pebbles, cobbles and boulders. Hence bedforms may well be present in coarse gravel deposits, but are only likely to affect flow velocities during low stage with shallow depths and high relative roughness. Shaw and Kellerhals (1977) and Koster (1978) have used the wavelength of transverse ribs to estimate palaeoflow velocities based on a relation developed by Kennedy (1961, 1963 in Shaw and Kellerhals, 1977; Koster, 1978).

Low suspended sediment concentration

Most flow equations assume clear water conditions during sediment entrainment and deposition. In many Pleistocene proglacial and periglacial gravel bed streams, suspended sediment concentrations are likely to have been very high (e.g. Malde, 1968; Novak, 1973; *cf.* Fahnestock, 1963; Fahnestock and Bradley, 1973; Østrem, 1975) resulting in greatly increased specific fluid weights and fluid densities. Komar (1970) demonstrated an increase in the specific weight of water from 1.00 to 1.18 g cm^{-3} from clear water to 9 per cent sediment concentration. Similarly, Boothroyd and Ashley (1975) used a water density value of 1.0013 g cm^{-3} for suspended sediment concentration of 2 g/l in water of 4°C (i.e. higher viscosity conditions); and Koster (1978) claimed one should use a density value of 1.005 g cm^{-3} when estimating flow velocities of sediment-laden waters. An increase in suspended sediment concentration is also likely to reduce the value of von Karman's coefficient (see equation (2), pp. 104) (Jopling, 1966; Itakura and Kishi, 1980; Wang, 1981). This in turn will be reflected in a steeper vertical velocity gradient because of a reduction in the amplitude of velocity fluctuations above the channel bed (i.e. decreased eddy diffusion), such that mean friction factors may be reduced by up to 28 per cent, and mean flow velocities thereby increased (Vanoni and Nomicos, 1960; Yalin and Finlayson, 1972). This in turn means, of course, that the peak competent velocity for bedload sediment transport may be reduced in heavily laden streams (e.g. Novak, 1973; Bradley and Mears, 1980).

Maximum availability of fluvial sediment

The final major assumption relates to the hydraulic interpretation of palaeovelocity and palaeodischarge determinations. Maximum sediment availability is assumed, such that all sizes of sediment are available for transport; hence, the maximum particle sizes present reflect stream competence rather than the lack of any larger particles in the source region (Nevin, 1946; Church, 1978; Koster, 1978). If larger sizes are not available to the stream then estimates of palaeovelocities will represent minimum values (Nevin, 1946). In most glacial and periglacial environments this assumption may prove acceptable (e.g. Maizels, 1976), but there may be exceptions, as may have occurred in some of the Upper Thames terraces (Clarke and Dixon, 1981).

It is further assumed that all particles sampled are of fluvial origin, in that they have been transported to the site of deposition by the appropriate competent stream flows, and are not derived from agents of non-fluvial deposition such as ice-rafting (e.g. Krumbein and Lieblein, 1956), bank slumping, avalanche or other valleyside debris inputs, or biogenic activity (Fahnestock, 1963; Church, 1978), nor from the results of non-uniform flow such as secondary currents, shifting channel position and 'waves' of sediment motion (Church and Kellerhals, 1978).

PALAEOVELOCITY DETERMINATION

The term 'velocity' can assume a number of different meanings, and the type of palaeovelocity being predicted (or post-dicted) will depend on the method of determination that is employed (Bradley and Mears, 1980). Thus one may wish to predict mean velocity (most flow equations), bed velocity (e.g. Helley, 1969; Bradley and Mears 1980), local or maximum instantaneous velocity (e.g. Bridge, 1978), critical erosion or critical transport velocity (e.g. Hjulstrom, 1935; Sundborg, 1956), 'competent' velocity (e.g. Menard, 1950; Jopling 1966), or the complete vertical velocity profile. Although these different measures of velocity may all be closely interrelated (Novak, 1973), many workers have demonstrated that one cannot expect to predict any velocity value with great accuracy (e.g. Sundborg, 1956; Jopling, 1966; Baker and Ritter, 1975; Bradley and Mears, 1980), largely because one cannot ensure that all the assumptions underlying the method have been fully met. Stream velocity can be highly variable at a single cross-section, not only as a result of changes in water volume transfers, but also as a result of random fluctuations in the magnitude and direction of flow, of changing degrees of local turbulence, and of the changing balance between the forces propelling and resisting flow. Hence, methods of estimating some value of flow velocity must take account of the controls on turbulence and, in particular, of these two main opposing forces operating in fully turbulent, rough bed streams, since flow velocity represents an end result of the balance between these forces.

The tractive force concept

The tractive force concept was introduced by Du Boys as a measure of stream competence. It represents the condition on the stream bed when the propulsive force of flow, F_u, where

$$F_u = -\gamma A \Delta h \tag{3}$$

is equalled by the resisting force, F_d, where

$$F_d = \tau_o \Delta x P \tag{4}$$

This resisting force acts as a shearing force (τ_o) parallel to the direction of flow, over a given wetted perimeter (P) and length of reach (Δx) which provide the boundary surface for shear. When these forces are equal, an equilibrium flow condition of zero net force, F_n, exists, which is represented by Du Boys' static equilibrium equation for boundary shear in steady flow:

$$F_n = \gamma A \Delta h - \tau_o P \Delta x = 0 \tag{5}$$

$$\therefore \tau_o = \gamma RS \tag{6}$$

S represents the bedslope of the channel, since even for non-uniform flow, this value will be closer than is the water surface slope to the energy slope (O'Brien and Rindlaub, 1934). Hence, Henderson (1966) has demonstrated that this equation may be applied to non-uniform steady flow conditions as long as the type of slope measured is indicated. In wide, shallow flows (i.e. width–depth ratios $\geqslant$ *ca* 15; Hey, 1979) which are characteristic of many coarse grained channels, the hydraulic radius, R, may be replaced by maximum flow depth, d, to give

$$\tau_o = \gamma d S \tag{7}$$

The use of the hydraulic radius in Du Boys' equation (6) is based on the assumption that shear stresses are uniformly distributed through the flow: however, since this is known not to be the case (e.g. Shih and Grigg, 1967) no additional theoretical error is contributed by its substitution by d. Nevertheless, if flow depth values considerably exceed equivalent hydraulic radius values, O'Brien and Rindlaub (1934) claim that shear forces may be overestimated by up to 20 per cent.

'Critical' flow conditions and the critical tractive force

'Critical' flow conditions in the context of coarse gravel movement represent the values assumed by the whole range of flow parameters at the moment (or moments) when bed particles begin to be entrained, i.e. these are the conditions of flow when the critical threshold for particle entrainment has been reached. Hence, critical conditions of flow parameters such as mean velocity, bed velocity, width, depth, discharge, slope, tractive force, bed and form roughness, suspended sediment concentration, fluid density, viscosity, temperature, and so on, may be identified as those occurring at the threshold condition for bed particle movement. As Jopling (1966) points out, it is much easier to determine this critical threshold velocity than the true velocity, for most palaeoflow analysis is based on direct functional relationships that have been developed between the critical tractive force or critical flow velocity (i.e. stream competence) and a measure of particle size. Under these conditions, estimates of palaeovelocity and palaeodischarge, as well as of other palaeoflow parameters, must relate to critical flow conditions. Hence one can distinguish between, for example, the true mean flow velocity which is not necessarily related to movement of bed particles, and the critical mean flow velocity, which is a direct function of the size of bed particles that have been entrained. Since one is measuring the largest bed particles, this critical flow condition must in turn represent a maximum flow event at that particular sedimentary horizon, i.e. it represents the maximum competence to entrain the particles at each level in the sedimentary record.

The critical flow condition is best described by Shields' criterion (1936) which represents the critical shear stress due to sediment particles on the bed. It represents the condition of tractive force at which bed particles begin to move, rather than the condition at which all particles are in motion (ASCE, 1966). Shields derived two dimensionless parameters representing two opposing forces: firstly, the drag force of the fluid on the grains, ϕ, where

$$\phi = \tau_c/\{(\gamma_s - \gamma)D\} \tag{8}$$

and secondly, the resistance of the particles on the fluid, ψ, which is the particle Reynolds number, where

$$\frac{1}{\psi} = V^* D/\nu \tag{9}$$

The critical shear velocity, V^*, is a theoretical velocity value which expresses the relationship between Du Boys' tractive force equation, equation (6) and the fluid density at critical flow conditions:

$$V^* = (\tau_c/\rho_c)^{1/2} \tag{10}$$

i.e. it represents the velocity that would give rise to the critical shear stress, τ_c. The relationship between ϕ and $1/\psi$ yields Shields' entrainment function which is often considered to be constant for coarse bed particles at high Reynolds numbers, such that ϕ in equation (8) assumes a value of 0.056 (Shields, 1936; Graf, 1971), This allows one to predict the value of the critical tractive force (from equation (8)):

$$\tau_c = 0.056\,(\gamma_s - \gamma)\,D \tag{11}$$

Where γ_s and γ are taken as 2650 and 1000 kg m^{-3} respectively,

$$\tau_c = 0.092\,D \tag{12}$$

While for streams with *ca.* 10 per cent suspended sediment concentration (i.e. $\gamma \approx 1250$ kg m^{-3})

$$\tau_c = 0.078\,D \tag{13}$$

i.e. a lower critical tractive force is required where suspended sediment concentrations are increased.

Shields function forms an important basis for palaeohydraulic analysis, particularly in the estimation of shear stress and flow depths, and hence certain limitations of this approach should be recognized. Recent work has demonstrated that Shields' coefficient ϕ is not necessarily constant for large bed particles. Shields' function was originally derived from flume experiments in fully developed turbulent flow with artificially flattened non-cohesive sand beds. Since the initiation of particle motion is believed to represent a random stochastic process, based on velocity fluctuations and eddy forces operating on a mixture of particle sizes on the bed, Shields' function of 0.056 must represent the mean value of a broad zone, or confidence band, relating ϕ and $1/\psi$. In addition, Shields (1936) admits that for large particle Reynolds numbers, the plotted relation has only been extrapolated. Church (Church and Gilbert, 1975; Church, 1978) has demonstrated that the value of ϕ may indeed vary from as low as 0.0 to as much as 0.1, depending on sediment sorting, particle shape, and particularly on the degree of looseness of packing of the boundary (see *Bankfull flow depth from thickness of sedimentary deposits* above) (and see also Peterson and Mohanty, 1960; Helley, 1969; Gessler, 1971; Baker 1973a;

Cheetham, 1976, 1978; Bradley and Mears, 1980; Bray, 1981; Griffiths, 1981). Baker and Ritter (1975) also demonstrate that where particle packing is denser, or where bed imbrication, armouring or differential protection of smaller particles by larger ones occur, particles will require much higher tractive forces than those predicted by Shields' function. By contrast, where largescale or macroturbulence occurs, as is possible in very deep, high gradient streams, particles may be moved by lift forces in addition to shear forces, and hence Shields' function may overestimate the actual tractive forces involved in particle motion.

Estimates of critical flow depth

Flow depth may be estimated in several different ways.

Maximum flow depth from geomorphological evidence of high water levels

Major flood events such as those produced by the drainage of ice-dammed lakes or jökulhlaups may leave evidence of wash limits, scour marks, lichen trim lines etc., on adjacent slopes, and a number of workers have used such indices to estimate the maximum overall water levels for given flood events (e.g. Birkeland, 1968; Malde, 1968; Baker, 1973a, b; Gregory, 1976; Costa, 1978; Petts, 1980).

Bankfull flow depth from thickness of sedimentary deposits

Morton and Donaldson (1978) estimated bankfull channel depth from the thickness of coarse channel infill sediments, while a number of other workers have estimated bankfull depth from the thickness of either (fine-grained) epsilon cross-stratfraction units in fluvial sedimentary sequences (e.g. Moody-Stuart, 1966; Elliott, 1976; Bridge, 1978; Rust, 1978; Allen and Mange-Rajetzky, 1982), or thickness of channel or lateral bar deposits (e.g. Cotter, 1971; Padgett and Ehrlich, 1976; Miall, 1977).

Bankfull channel depth from field measurement of exposed palaeochannels

The depth of exposed palaeochannels is relatively easy to measure in the field so long as banks are fairly well defined (e.g. Baker and Ritter, 1975; Maizels, 1982), but this depth value is not necessarily related to the flow event that entrained the sediment particles on the channel bed. Actual flow depth may have exceeded channel depth (i.e. during overbank or flood flow), equalled channel depth (i.e. represents bankfull flow), or been lower than channel depth (i.e. low stage inchannel flow). In addition, in many coarse gravel rivers, especially braided rivers, banks are difficult to determine (e.g. Miall, 1977), and this reduces the accuracy of any field measurements.

Critical flow depth from hydraulic equations

This approach should provide the most accurate estimate of flow depth since it relates to the condition of sediment entrainment of the bed particles being examined (e.g. Koster, 1978). Critical flow depth may be predicted from equations (7) and (12) such that

$$d_c = \frac{\tau_c}{\gamma S} \tag{14}$$

$$= 0.092\, D\, S^{-1} \text{ for clear water} \tag{15}$$

and

$$= 0.078\, D\, S^{-1} \text{ for sediment-laden water} \tag{16}$$

or, in general terms,

$$d_c = C_1\, DS^{-1} \tag{17}$$

where

$$C_1 = \frac{\phi(\gamma_s - \gamma)}{\gamma} \tag{18}$$

Comparison of measured channel depths and estimated critical flow depths using equation (15) for a series of 69 palaeochannels on an abandoned West Greenland sandur indicate that the ratio of critical flow depth, d_c, to measured channel depth, d, where

$$d_r = \frac{d_c}{d} \tag{19}$$

is highly variable. In some deposits, d_c values may largely exceed values of d; for the Watson *D* terrace palaeochannels in West Greenland, for example, 98.5 per cent of d_c values exceed d values (Figure 5.2a), while for the T3 palaeochannels at Narssarssuaq in Southwest Greenland (Figure 5.2b), only 73.2 per cent of d_c values exceed d values. Hence, over 25 per cent of flows in the latter deposit were below bankfull stage during entrainment of the largest sediment particles measured on the channel bed. Nevertheless, in neither deposit were more than 10 per cent of critical flow depths approximately equal to bankfull depth (i.e. $\pm$ 0.1 m).

Friction factors and velocity equations

The 'representative' particle size of bed material is used at two stages of palaeovelocity determination. As demonstrated above, a measure of particle size is used in the determination of the critical tractive force for motion, and thence as a means of calculating the critical flow depth. Particle size is also used, however, to estimate the degree of skin resistance presented by particles

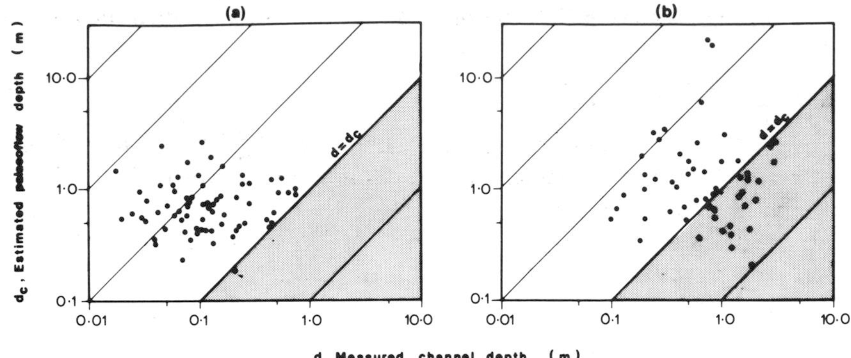

FIGURE 5.2 RELATION BETWEEN MEASURED CHANNEL DEPTH AND ESTIMATED CRITICAL FLOW DEPTH FOR TWO SETS OF ABANDONED SANDUR CHANNELS IN WEST GREENLAND
(a) Watson 'D' terrace, Søndre Strømfjord (Sample No. = 69). (b) Narssarssuaq T3 terrace (Sample No. = 57)

on the channel boundary, acting to retard flow velocities. In the latter calculations, critical flow velocity becomes an inverse function of particle size, since the larger particles will present a greater resisting force to flow.

The relation between grain size and critical flow velocities for particle entrainment was graphically demonstrated by Hjulström's (1935) classic competency diagram. However, this diagram was based on homogeneous sediment sizes and surface velocities in water depths exceeding 1 m (Jopling, 1966). In addition, the plotted function was determined experimentally for sands only, and was just extrapolated to finer and coarser sediments. Sundborg (1956) provided a modified function for finer, more cohesive sediments, while Novak (1973) has clearly demonstrated that for coarser sediments (*ca.* 50 mm to 10 m diameter) the Hjulström curve underpredicts mean entrainment velocities by up to *ca.* 10 times. It would seem a more rational approach to be able to predict critical velocities according to changing flow depth and boundary conditions for mixtures of sediment sizes, and this is most successfully achieved by estimation of skin resistance presented by the channel.

Three well-known methods of estimating skin resistance have been adopted by hydraulic engineers working in coarse bed, fully turbulent, low relative roughness streams. These methods are based initially on the well-established hydraulic relation for turbulent flow which states that resistance to flow is equal to the ratio of the tractive force to the square of the velocity (Leopold, Wolman, and Miller, 1964):

$$\text{Resistance} = \tau/V^2 \tag{20}$$

In order to estimate the extent to which resistance has affected mean flow velocity, one can substitute Du Boys boundary shear stress function (equation

(7)) into this last equation (20) to obtain an expression for critical mean flow velocity

$$\bar{V}_c = C\sqrt{R_c S} \approx C\sqrt{d_c S} \qquad (21)$$

where C is Chezy's friction coefficient, and = $(8\ g/f)^{1/2}$ (see later). Chezy recognized that mean flow velocity is a function of the product of flow depth, slope and some resistance factor.

Chezy's approach has largely been superseded by the widely used Manning equation, because of the relative ease of estimating the resistance, or bed roughness coefficient, n:

$$\bar{V}_c = \frac{R^{2/3} S^{1/2}}{n} \approx \frac{d_c^{2/3} S^{1/2}}{n} \qquad (22)$$

Although n may be estimated reasonably easily from empirical published values for a wide range of channel forms and stream flow conditions (e.g. Chow, 1959), their accuracy is open to considerable doubt. The empirical values of n are totally inappropriate for application to palaeoflow analysis where three-dimensional channel form may be obscured, where local bed roughness can be highly variable, and where no present day analogues may actually exist (e.g. catastrophic draining of ice-dammed lakes; Kjartansson 1967; Birkeland, 1968; Malde, 1968; Baker 1973a, b). Computational methods of determining the value of n have been proposed by Strickler (1923) and Limerinos (1970). These are more accurate than empirical estimates for determining n, but are based only on skin resistance (Strickler approach) or skin resistance and relative roughness (Limerinos approach). Strickler (1923) derived a simple functional relationship between n and the representative particle diameter such that

$$n = 0.039\, D_{50}{}^{1/6} \qquad (23)$$

This has since been shown to be a direct function of the Darcy–Weisbach friction factor, f (Williamson, 1951 in Henderson, 1966) (see below). Limerinos (1970) has tested a number of different methods of calculating n, and concluded that n was best determined by the use of a logarithmic function relating the resistance coefficient to a measure of relative roughness:

$$n = \frac{0.113 d^{1/6}}{1.16 + 2.0 \log(d/D_{84})} \qquad (24)$$

Either of the predictive equations for n (equations (23) and (24)) may be substituted into the Manning formula to provide the palaeohydrologist with a simple means of determining critical mean flow velocity from inputs only of the 'representative' particle diameter, D, and the bedslope, S. The Manning

–Strickler approach may be defined as

$$\bar{V}_c = \frac{d_c^{2/3} S^{1/2}}{0.039\, D^{1/6}} \tag{25}$$

and from equation (23)

$$\bar{V}_c = \frac{5.24\ D^{1/2}}{S^{1/6}} \tag{26a}$$

(where Shields coefficient = 0.056 and for clear water flow), and

$$\bar{V}_c = \frac{5.63\ D^{1/2}}{S^{1/6}} \text{ (where } \phi = 0.07 \text{ and } \gamma = 1.018 \text{ g cm}^{-3}) \tag{26b}$$

This formula, however, is not dimensionally correct. The Manning–Limerinos approach provides the equation

$$\bar{V}_c = \frac{d_c^{2/3}\ S^{1/2}}{\left\{\dfrac{0.113\ d_c^{1/6}}{1.16 + 2.0 \log (d_c/D)}\right\}} \tag{27}$$

$$= 8.85\ C_1^{1/2}\ D^{1/2}\left[1.16 + 2.0 \log \left(\frac{C_1}{S}\right)\right] \tag{28}$$

The Darcy–Weisbach friction factor has recently attracted more popularity in the palaeohydraulic literature since it is dimensionally correct, is theoretically related to the von Karman–Prandtl 'universal velocity law' (see '*Critical flow conditions and the critical tractive force*, pp. 112–4), and may be used in combination with a channel form parameter in prediction of flow velocity (e.g. Hey, 1979). The Darcy–Weisbach friction factor, f, for fully developed, uniform flow over a rough bed may be defined as

$$\frac{1}{\sqrt{f}} = C \log \left(\frac{aR}{K_s}\right) \tag{29}$$

Where $f = V^*/V^2$, and the coefficient C has been determined from $C = 2.30/(\kappa \sqrt{8})$, where κ is the von Karman coefficient (Silberman *et al.*, 1963; Leopold, Wolman, and Miller, 1964; Hey, 1979). The coefficient 'a' is a function of 'c', of channel cross-sectional shape, and the spacing of bed roughness elements (Silberman *et al.*, 1963). Hey (1979), using Keulegan's approach, demonstrates that the value of this coefficient 'a' may be determined directly from an index of channel cross-sectional shape, and ranges between

11.09 for very wide channels to 13.46 for circular pipes; Hey (1979) provides a simple graphical means of determining the value of 'a'. This approach is only feasible in palaeohydraulic analysis, therefore, where both flow depth and width have been determined. Hence equation (29) is really more suitable for estimating resistance in confined channel flow conditions. For conditions of intermediate or uncertain flow width, f may be estimated from relative roughness alone:

$$f = 0.113 \left(\frac{d}{D} \right)^{1/3} \tag{30}$$

(Henderson, 1966). However, neither of these methods of estimating flow resistance take account of the shape, packing and spacing of roughness elements along the channel boundary. Hey (1979), however, recommends a procedure for determining separately bed and bank resistance to flow, but this is unlikely to be possible for palaeohydraulic analysis unless the original bank form and sediments are preserved. Similarly, Knight and McDonald (1979) suggest the use of a modified friction factor to take account of variable size, shape and spacing of roughness elements on the channel boundary. At present no completely satisfactory methods of estimating the total resistance to flow are available.

Mean critical flow velocity may be estimated from the Colebrook–White equation based on the Darcy–Weisbach friction factor:

$$\bar{V}_c = \left[\frac{(8gdS)}{f} \right]^{1/2} \tag{31}$$

$$= 8.86\ C_1^{1/2}\ D^{1/2} \left[C_2 \log \left(\frac{aC_1}{3.5S} \right) \right] \tag{32}$$

For normal boundary conditions (i.e. $\phi = 0.056$) and clear water flow, (i.e. $\gamma =$ 1 000 kg m^3), critical mean velocity may be estimated from:

$$\bar{V}_c = 5.46\ D^{1/2} \left[\log \frac{0.0264a}{S} \right] \tag{33}$$

while for underloose boundary conditions (e.g. imbricated sediments, $\phi =$ 0.07, Church, 1978) and sediment-laden flow (i.e. $\gamma =$ 1018 kg m^3), the equivalent velocity value may be estimated from:

$$\bar{V}_c = 7.70\ D^{1/2} \left[\log \frac{0.0294a}{S} \right] \tag{34}$$

(These different approaches are summarized in Table 5.1.)

A number of workers have also adopted the use of regime-type equations in estimates of mean flow velocity, and have recently applied them to gravel reaches (e.g. Kellerhals, 1967; Bray, 1978). The regime approach assumes, however, that the stream is in equilibrium, such that it undergoes no degradation or aggradation of its bed. This assumption is unlikely to be met in coarse bed fluvial sediment sequences in the stratigraphic record.

The five main methods of palaeovelocity determination discussed in this section are summarized in Table 5.1, where general equations are given (Table 5.1A), together with specific equations related either to clear water flow in 'normal' boundary channels (I in Table 5.1B) or to sediment-laden flow (i.e. 10 per cent suspended sediment concentration) in 'underloose' or imbricated boundary channels (II in Table 5.1B). Some results of palaeovelocity estimates using the five basic equations for these two flow conditions are presented for a series of 69 palaeooutwash channels examined on an abandoned terrace surface in West Greenland (Table 5.2C). The Manning–Limerinos approach (Method 3) provides some of the lowest estimates of palaeovelocities for both clear water flow (1.62 ± 0.42 m s^{-1}) and sediment-laden flow (1.80 ± 0.45 m s^{-1}), while the highest estimates result from the Chezy and Manning–Strickler approaches (Methods 1 and 2) (maxima of 3.27 ± 0.62 m s^{-1} and 3.98 ± 0.60 m s^{-1} respectively.) Methods based on estimates of Manning's n and D_{84} appear to predict higher palaeovelocities than those based on prediction of the Darcy–Weisbach f factor (Figures 5.3 and 5.4). Hence the five approaches have provided estimates that differ by up to 100 per cent for a given flow condition, while increases in suspended sediment concentration and degree of bed imbrication, have increased palaeovelocity estimates by up to *ca.* 45 per cent (Method 5). These results emphasize firstly, the high variability and inaccuracy associated with palaeovelocity determination, and secondly the need to determine an 'envelope' of likely values based on the application of a wide range of approaches.

PALAEODISCHARGE DETERMINATION

Determination of palaeodischarge from coarse gravel deposits is most reliably achieved by application of the flow continuity principle, expressed as equation (1). As in the case of palaeovelocity determinations, however, the term 'discharge' may be defined in a number of different ways. Flood discharges of different recurrence intervals, ranging from the mean annual flood, for example, to the 1 000-year flood may be responsible for deposition of particular gravel deposits; alternatively, the sediments may reflect deposition during a maximum instantaneous or short-lived flood event of unknown frequency and unrelated for example, to longer term climatic change (e.g. Baker and Penteado-Orellana, 1977; Hey, 1978; Rose *et al.*, 1980). The discharge event

Table 5.1 Selected Methods of Palaeohydraulic Analysis of Coarse Gravel Deposits

A. Selected methods of palaeohydraulic analysis of coarse gravel deposits

Method 1 Chezy equation	Method 2 Manning – Strickler equation	Method 3 Manning – Limerinos equation	Method 4 Darcy – Weisbach – Henderson equation	Method 5 Darcy – Weisbach – Hey equation
Calculation of critical palaeoflow velocity				
$\bar{V}_c = C_3\sqrt{d_c s}$ (21) $= 78{\cdot}41\ C_1^{1/6}S^{1/3}D^{1/2}$ (44)	$\bar{V}_c = \frac{d_c^{2/3}S^{1/2}}{0{\cdot}039D^{1/6}}$ (25) $= \frac{25{\cdot}64C_1^{2/3}D^{1/2}}{S^{1/6}}$ (50)	$\bar{V}_c = \frac{d_c^{2/3}S^{1/2}}{\left\{\frac{0{\cdot}113d_c^{1/6}}{1{\cdot}16 + 2{\cdot}0 \log d_c/D}\right\}}$ (53) $= 8{\cdot}85C_1^{1/2}D^{1/2}\left[1{\cdot}16 + 2{\cdot}0 \log\left(\frac{C_1}{S}\right)\right]$ (28)	$\bar{V}_c = \left(\frac{8gd_cS}{f}\right)^{1/2}$ (31) $= 26{\cdot}35C_1^{1/3}D^{1/2}S^{1/6}$ (59) $\left[\text{n.b. } f = 0{\cdot}113\left(\frac{d_c}{D}\right)^{1/3}\right]$ (30)	$\bar{V}_c = \left(\frac{8gd_cS}{f}\right)^{1/2}$ (31) $= 8{\cdot}86C_2D^{1/2}\left[C_1 \log\left(\frac{aC_1}{3{\cdot}5S}\right)\right]$ (32) $\left[\text{n.b. } \frac{1}{\sqrt{f}} = C_2 \log\left(\frac{ad_c}{3{\cdot}5D}\right)\right]$ (65)
Calculation of critical palaeodischarge (specific discharge)				
$q_c = \frac{78{\cdot}41C_1^{7/6}D^{3/2}}{S^{2/3}}$ (45)	$q_c = \frac{25{\cdot}64C_1^{5/3}D^{3/2}}{S^{7/6}}$ (37)	$q_c = \frac{\left\{8{\cdot}85C_1^{3/2}D^{3/2}\left[1{\cdot}16 + 2{\cdot}0 \log\left(\frac{C_1}{S}\right)\right]\right\}}{S}$ (54)	$q_c = \frac{26{\cdot}35C_1^{4/3}D^{3/2}}{S^{5/6}}$ (60)	$q_c = \frac{8{\cdot}86C_1^{3/2}D^{3/2}}{S} \cdot \left[C_2 \log\left(\frac{aC_1}{3{\cdot}5S}\right)\right]$ (66)

B. Summarized for I. clear water over 'normal' boundary; II. sediment-laden flow (10% conc.) over 'underloose' boundary

	Method 1	Method 2	Method 3	Method 4	Method 5
	Calculation of critical palaeoflow velocity				
I	$\bar{V}_c = 52{\cdot}72S^{1/3}D^{1/2}$ (46)	$\bar{V}_c = \frac{5{\cdot}24D^{1/2}}{S^{1/6}}$ (26a)	$\bar{V}_c = 2{\cdot}69\ D^{1/6}\left[1{\cdot}16 + 2{\cdot}0 \log\left(\frac{0{\cdot}0924}{S}\right)\right]$ (55)	$\bar{V}_c = 11{\cdot}91\ D^{1/2}S^{1/6}$ (61)	$\bar{V}_c = 5{\cdot}46\ D^{1/2}\left(\log \frac{0{\cdot}0264a}{S}\right)$ (33)
II	$\bar{V}_c = 53{\cdot}68S^{1/3}D^{1/2}$ (47)	$\bar{V}_c = \frac{5{\cdot}63D^{1/2}}{S^{1/6}}$ (26b)	$\bar{V}_c = 2{\cdot}84\ D^{1/6}\left[1{\cdot}16 + 2{\cdot}0 \log\left(\frac{0{\cdot}1029}{S}\right)\right]$ (56)	$\bar{V}_c = 12{\cdot}35\ D^{1/2}S^{1/6}$ (62)	$\bar{V}_c = 7{\cdot}70\ D^{1/2}\left(\log \frac{0{\cdot}0294a}{S}\right)$ (34)
	Calculation of critical specific discharge				
I	$q_c = \frac{4{\cdot}87D^{3/2}}{S^{2/3}}$ (48)	$q_c = \frac{0{\cdot}48D^{3/2}}{S^{7/6}}$ (51)	$q_c = \frac{0{\cdot}25\ D^{3/2}\left[1{\cdot}16 + 2{\cdot}0 \log\left(\frac{0{\cdot}0924}{S}\right)\right]}{S}$ (57)	$q_c = \frac{1{\cdot}1\ D^{3/2}}{S^{5/6}}$ (63)	$q_c = \frac{0{\cdot}51\ D^{3/2}}{S}\left(\log \frac{0{\cdot}0264a}{S}\right)$ (67)
II	$q_c = \frac{5{\cdot}52D^{3/2}}{S^{2/3}}$ (49)	$q_c = \frac{0{\cdot}58D^{3/2}}{S^{7/6}}$ (52)	$q_c = \frac{0{\cdot}29\ D^{3/2}\left[1{\cdot}16 + 2{\cdot}0 \log\left(\frac{0{\cdot}1029}{S}\right)\right]}{S}$ (58)	$q_c = \frac{1{\cdot}27\ D^{3/2}}{S^{5/6}}$ (64)	$q_c = \frac{0{\cdot}79\ D^{3/2}}{S}\left(\log \frac{0{\cdot}0294a}{S}\right)$ (68)

Equation numbers given in brackets after equation

TABLE 5.2 ESTIMATES OF PALAEOFLOW CONDITIONS FOR ABANDONED SANDUR CHANNELS, SØNDRE STRØMFJORD, WEST GREENLAND

A. INPUT VARIABLES	
No. of palaeochannels	69
Channel width (m)	10.82 ± 8.31
Channel depth (m)	0.18 ± 0.16
Channel slope (m m^{-1})	0.0135 ± 0.0056
Clast size (D_{10x}; m)	0.0937 ± 0.0356

B. ESTIMATED CHANNEL BOUNDARY CONDITIONS

	I Clear water/ 'Normal' boundary		II Sediment-laden water/ 'Underloose' boundary	
	$\bar{x}$	σ	$\bar{x}$	σ
Critical flow depth (m)	0.75	0.44	0.83	0.49
d/d_c ratio	0.28	0.27	0.25	0.24
Channel w/d ratio	89.46	121.47	89.46	121.47
Flow w/d_c ratio	19.14	21.67	17.18	19.46
Channel relative roughness d/D	1.80	1.32	1.80	1.32
Flow relative roughness d_c/D	8.57	4.99	9.54	5.56

C. PALAEOVELOCITY ESTIMATES (using Methods 1–5 in Table 5.1)				
Method 1 (ms^{-1})	2.93	0.45	3.98	0.60
Method 2	3.27	0.62	3.52	0.67
Method 3	1.62	0.42	1.80	0.45
Method 4	1.73	0.41	1.79	0.42
Method 5	2.25	0.47	3.28	0.68
D. SPECIFIC PALAEODISCHARGE ESTIMATES				
Method 1 ($m^2 s^{-1}$)	2.45	1.09	3.15	1.39
Method 2	2.64	2.03	3.16	2.43
Method 3	1.42	1.59	2.33	1.92
Method 4	1.31	0.82	1.55	0.95
Method 5	1.87	1.56	3.02	2.50
E. MEAN CHANNEL PALAEODISCHARGE ESTIMATES				
Method 1 ($m^3 s^{-1}$)	26.54	27.52	34.08	35.34
Method 2	29.76	37.52	35.61	44.89
Method 3	21.37	27.46	25.92	33.24
Method 4	15.41	19.03	17.79	21.97
Method 5	20.66	26.62	33.50	43.09

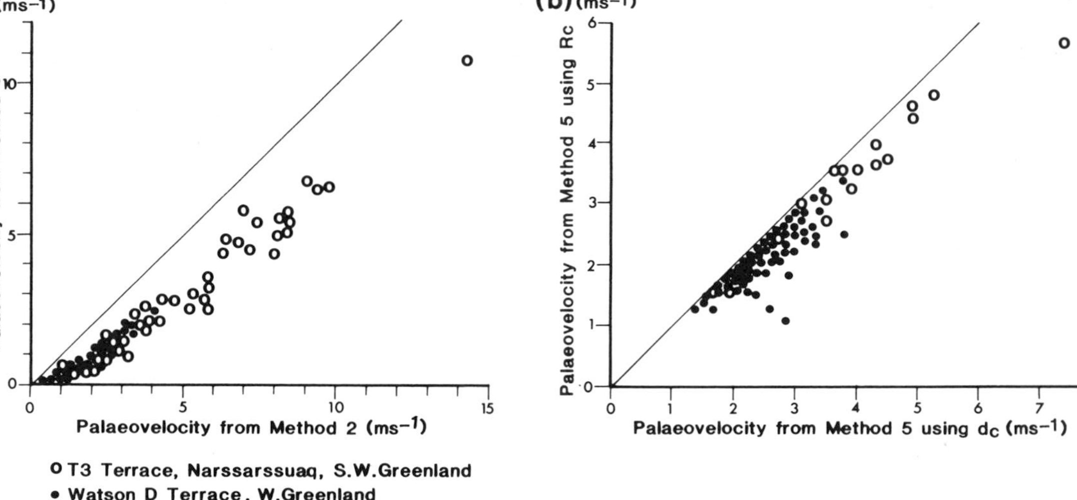
(a) (ms-1)
Palaeovelocity from Method 5
0
5
10
15
Palaeovelocity from Method 2 (ms-1)
O T3 Terrace, Narssarssuaq, S.W.Greenland
• Watson D Terrace, W.Greenland
(b) (ms-1)
Palaeovelocity from Method 5 using Rc
0
1
2
3
4
5
6
7
8
Palaeovelocity from Method 5 using dc (ms-1)

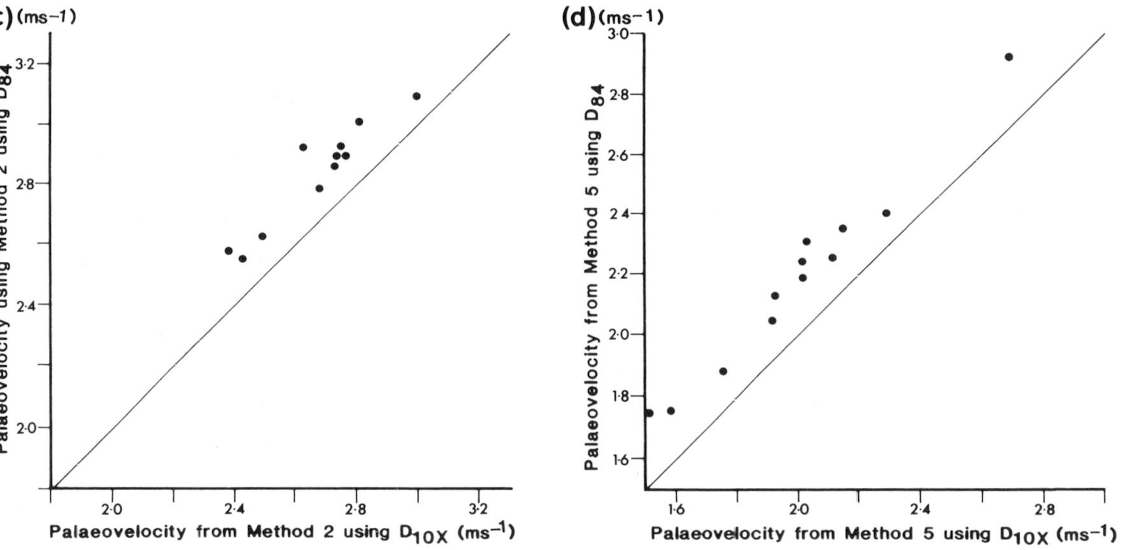

FIGURE 5.3 PALAEOVELOCITY DETERMINATIONS BASED ON DIFFERENT INPUT VARIABLES AND FRICTION FACTORS

(a) Comparison of palaeovelocities based on Manning–Strickler approach (Method 2) and Darcy–Weisbach–Hey approach (Method 5). (Abandoned sandur channels, West Greenland). (b) Comparison of palaeovelocities based on input of critical flow depth or hydraulic radius in Darcy–Weisbach–Hey approach (Method 5). (Abandoned sandur channels, West Greenland). (c) Comparison of Manning–Strickler palaeovelocity estimates (Method 2) based on inputs of D_{10x} and D_{84}. (North Esk sandur terrace, N.E. Scotland. (d) Comparison of Darcy–Weisbach–Hey palaeovelocity estimates (Method 5) based on inputs of D_{10x} and D_{84}. (North Esk sandur terrace, N.E. Scotland)

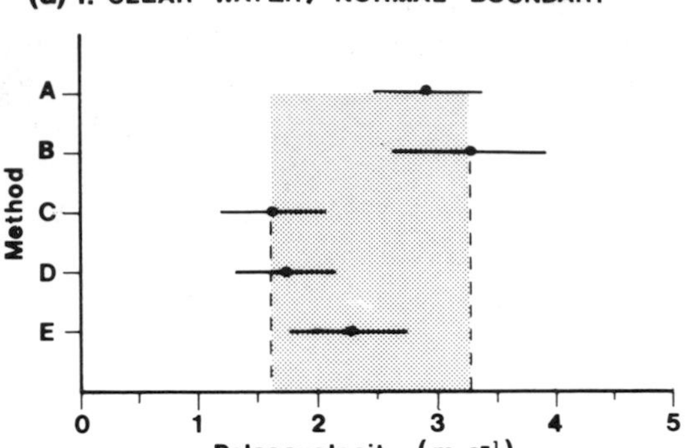

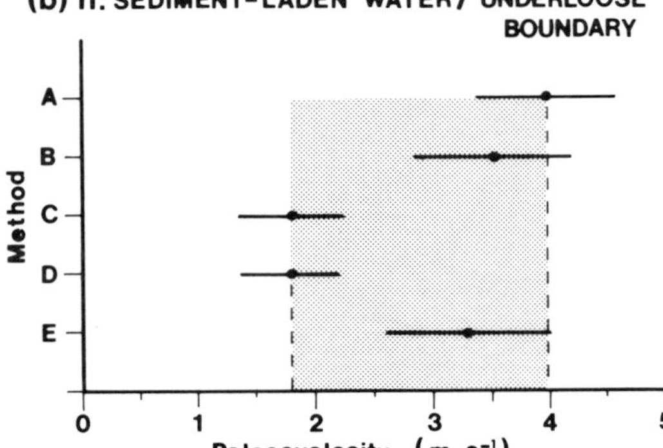

FIGURE 5.4 COMPARISON OF FIVE METHODS OF PALAEOVELOCITY DETERMINATION FOR 69 ABANDONED SANDUR CHANNELS, SØNDRE STRØMFJORD, WEST GREENLAND

(Methods A–E correspond to Methods 1–5 in Table 5.1;
(a) For clear water/'normal' boundary conditions
(b) For sediment-laden water/'underloose' boundary conditions

being predicted largely depends on the nature of the input variables of flow width, depth and velocity, with the main distinction being related to whether in-channel or overbank flow occurred during entrainment and deposition of the sediments being examined. Where overbank flow is known to have occurred (i.e. $d_r > 1$), palaeodischarge estimates must take account of two important considerations. Firstly, at least two different flow environments are present, with channels generally exhibiting higher velocities, greater depths, lower relative roughness and higher Reynolds numbers than the overbank, or bar top, flow environment (e.g. Simons, 1969; Myers, 1982). In addition, water, sediment (including bank slump material), and momentum will be lost (and perhaps gained farther downstream) from the main channel to the overbank environment, while main channel flow velocities themselves are likely to be retarded by proximity to reduced bartop flow velocities. Hence the two flow environments do not operate independently, but no simple methods of accommodating these transfer processes are yet available for palaeodischarge analysis. Secondly, some estimate of total flow width is required to determine total discharge, although specific discharge per unit width of flow can be obtained from the product of depth and velocity measures. This may be obtained where palaeochannels are exposed at the surface (e.g. Maizels, 1982a, b), where lateral wash limits etc. are present (see pp. 114) or, if one assumes stable or equilibrium flow conditions, where width may be predicted using regime-type equations (e.g. Henderson, 1966; Kellerhals, 1967; Hey, 1978; Ferguson, 1981). However, this latter approach is not recommended for palaeochannel analysis. Where overbank flow has occurred, the total flow width must be divided into two components: channel width, w_c, and overbank or bartop flow width, w_b (e.g. Riggs, 1978). w_c and w_b may be fairly readily determined where a single channel and flood plain deposit can be identified (Figure 5.5a) but in dealing with a braided channel environment this approach can only be adopted where all, or much of, the original channel system is exposed at the surface (Figure 5.5b). It is extremely unwise to predict flow width from a 'guesstimate' of the 'likely' width–depth ratio (e.g. Cheetham, 1976). The total number and mean width of palaeochannels may then be either determined from the whole surface, or predicted from remnants of the original surface (e.g. Maizels, 1982, 1983). Estimates of relative roughness, friction factors, critical mean depth, velocity and discharge need to be made separately for individual channels and for each area of overbank flow, and total discharge then determined according to the total width of each type of flow or mean number of (predicted) channels present across the surface during the given discharge event. Total discharge may be determined either from the product of mean discharge per channel (or bar) and the number of channels (or bars):

$$Q_T = (\bar{Q}_c \times n_c) + (\bar{Q}_b \times n_b) \tag{35}$$

or from the product of specific discharge per unit width and total flow width for each type of flow environment.:

$$Q_T = (q_c \cdot w_c) + (q_b \cdot w_b) \tag{36}$$

For channel and overbank deposits exposed in sedimentary sequences, Leeder (1978) suggests a method of calculating the fractional areas of each type of deposit as a function of the time and residence intervals associated with valley infill information. This approach might help establish the likely total discharges for a given stratigraphic alluvial horizon.

The value of Q_c, the total critical discharge, may be estimated either directly from substitution of w, d_c and $\bar{V}_c$ in the continuity equation, or by combining

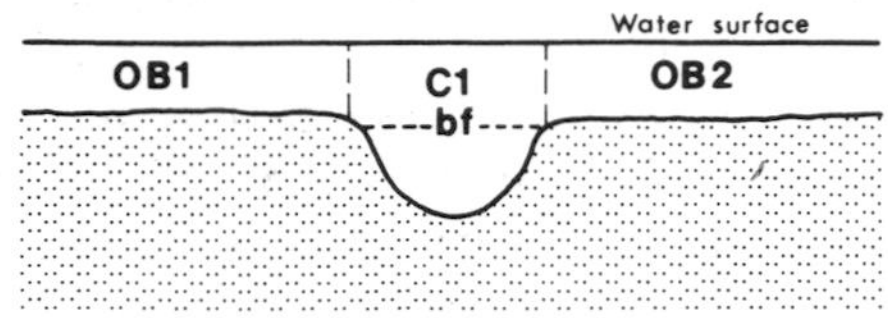

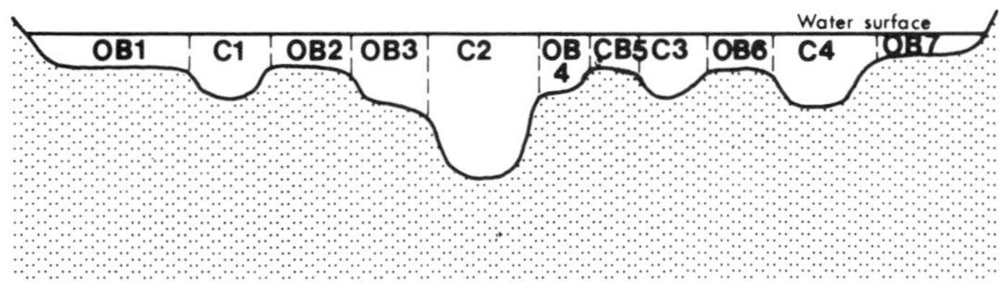

FIGURE 5.5 DIFFERENTIATION OF CHANNEL AND OVERBANK FLOW DOMAINS FOR PALAEODISCHARGE DETERMINATION
(a) Simple channel and overbank flow situation; (b) Complex channel and overbank flow situation: OB = overbank flow domain; C = channel flow domain

earlier equations (Table 5.1) into simplified forms requiring inputs only of particle diameter and slope. This approach has been adopted by Henderson (1966), Clague (1975) and Maizels (1983). Hence the combination of equations (16) and (23), for example, provides an expression for the critical specific discharge, q_c:

$$q_c = \frac{25.64\ C_1^{5/3}\ D^{3/2}}{S^{7/6}} \tag{37}$$

and for total critical discharge, Q_c:

$$Q_c = \frac{25.64\ C_1^{5/3}\ D^{3/2}}{S^{7/6}}\ w = q_c w \tag{38}$$

(Manning–Strickler approach). Alternative formulae based on Limerinos, Darcy–Weisbach and Hey friction coefficients are listed in Table 5.1. Five alternative approaches to palaeodischarge determination have been proposed in the literature (see summary in Ethridge and Schumm, 1978). These methods include:

1. The application of *empirical discharge–form relations* (e.g. Leopold and Wolman, 1957; Dury, 1964, 1965, 1976; Schumm, 1960, 1965, 1968a, b, 1972) where mean annual flood, mean annual discharge or bankfull discharge may be estimated from channel width, depth, meander wavelength and sediment characteristics. Although very widely used in palaeodischarge analysis (e.g. Moody–Stuart, 1966; Cotter, 1971; Leeder, 1973; Elliott, 1976; Padgett and Ehrlich, 1976; Baker, 1978; Jackson, 1978; Morton and Donaldson, 1978; Rust, 1978; Petts, 1980) these relations are not applicable to non-temperate streams, to gravel bed streams with high sediment transport rates, or to overbank flood flows (Bridge 1978; Allen and Mange-Rajetzky, 1982).
2. *Empirical drainage area–discharge relations* have been established for a large number of basins, especially in humid temperate alluvial environments (e.g. Dury, 1965). However, such relations are clearly not tenable for gravel bed streams, particularly those related to periglacial or glacial environments. Clague (1975) has extended analysis of area–discharge relations to include the relation of peak meltwater discharge to glacier area in glaciated catchments (and see Maizels, 1978), such that

$$Q_{pk} = 2.1\,A^{0.99} \tag{39}$$

3. A third approach is through modifications of Leopold and Wolman's (1957) *discriminant function for braided and meandering streams*, based on analysis of channel slope and bankfull discharge. Cheetham (1976, 1978) calculated the regression equation for Leopold and Wolman's braided river data only, to derive a predictive equation for bankfull discharge:

$$Q_{bf} = 0.000585\,S^{-2.01} \tag{40}$$

   This equation has been used in several palaeohydrologic studies of coarse gravel deposits (e.g. Cheetham, 1976, 1978; Clarke and Dixon, 1981), but only represents maximum in-channel flow amounts, and assumes that no overbank flow occurred.
4. Regime-type equations may be applied to equilibrium or 'graded' alluvial channels (i.e. where no aggradation or degradation of the channel bed has occurred), where width, depth, velocity, slope and bed roughness are known to be functionally related to bankfull discharge (e.g. Leopold and Maddock, 1953; Simons and Albertson, 1960; Henderson, 1966; Charlton *et al.*, 1978 in Ferguson, 1981; Hey, 1978). For example, Charlton *et al.* (1978) in Ferguson (1981) obtained such relationships for 23 gravel-bed rivers in England and Wales, from which bankfull discharges could be predicted

$$Q_{bf} = 0.053\,w^{2.22} \tag{41}$$

$$Q_{bf} = 18.69\,d^{2.50} \tag{42}$$

$$Q_{bf} = 2.73\, V^{6.67} \tag{43}$$

However, as noted above, the assumption of stable channel development in a fluvial sedimentary sequence is unlikely to be acceptable (e.g. Burkham, 1981).

5. A final approach is through the application of present-day relationships established by Schumm (1965, 1968b) between temperature, precipitation, sediment yield, sediment concentration, and runoff amounts. However, since these *environmental conditions* associated with the palaeoriver system are generally unknown, this approach is not recommended.

None of these five alternative approaches is very useful for analysis of coarse bed palaeochannel deposits. Instead, the main methods recommended for palaeodischarge determination are those based on the associated methods of palaeovelocity determination, as summarised in Table 5. These methods are presented both in general terms (Table 5.1A), and for clear water flow in 'normal' boundary channels and for sediment-laden flow in 'underloose' boundary channels (Table 5.1B). The application of these equations to determination of the specific discharge (discharge per unit width) and total channel discharge to the 69 palaeooutwash channels on the abandoned sandur terrace in West Greenland indicate that minimum discharge amounts are predicted by the Darcy–Weisbach–Henderson approach (Method 4), maximum discharge amounts by the Chezy and Manning–Strickler approaches (Methods 1 and 2) and intermediate discharges by the remaining two approaches (Table 2, D and E). Again palaeodischarges predicted by each approach differ by up to 100 per cent, while those predicted for sediment-laden flow exceed those of clear water flow by up to 62 per cent (Method 5). Hence, the Darcy–Weisbach–Hay approach appears most sensitive to changes in suspended sediment concentration and boundary conditions. These results relate only to in-channel flow conditions although it can be seen that mean flow depth exceeds mean channel depth by *ca* 0.57 m (Table 5.2A, B). Estimates of total discharge of in-channel flow (using Method 2) and overbank or bar top flow (using Method 5) have been made for a series of palaeooutwash deposits in the proglacial Austerdalen Valley in southern Norway (Figure 5.6) where deposits of different age have been approximately dated using lichenometry for the 1750–1870 AD period. These estimates indicate that bar top total palaeodischarges generally exceeded those of channel flow discharges, largely because of high flow depths across the steeply graded bar tops. The overall decline in total maximum discharge exhibited by these deposits for the 120-year period following the Little Ice Age maximum appears to be associated with a period of glacier retreat.

CONCLUSION

The development of palaeohydraulic analytical techniques is still at an early and experimental stage. The application of palaeovelocity and palaeodis-

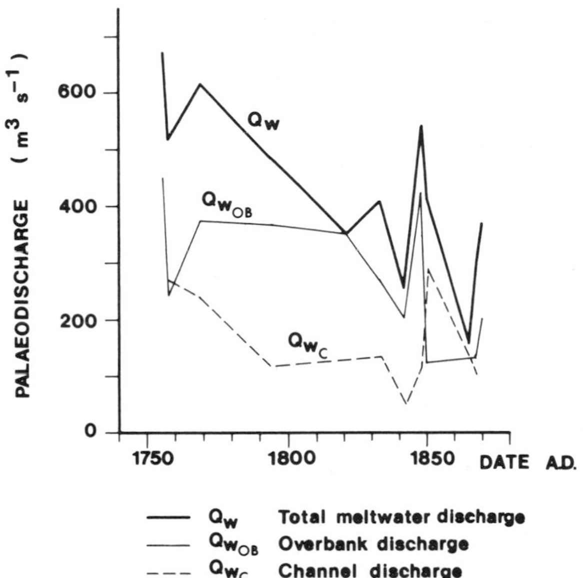

FIGURE 5.6 ESTIMATES OF CHANNEL, BAR TOP AND TOTAL FLOW DISCHARGE FOR 11 PALAEO-OUTWASH DEPOSITS IN AUSTERDALEN, SOUTHERN NORWAY, DATING FROM C. 1756 TO 1870 AD

equations for coarse gravel deposits is still based on limited empirical data and theoretical analysis of flow, sediment entrainment and depositional conditions within a rather restricted range of coarse-grained river environments. The equations proposed so far should only be applied after careful assessment of the underlying assumptions that first need to be met, and of the nature of the measured field data to be input to such equations. None of the methods suggested can be expected to provide very accurate results, but they should at least indicate the order of magnitude of palaeovelocities and palaeodischarge amounts.

Where possible, several methods of palaeovelocity or palaeodischarge determination should be adopted, each based on different assumptions and involving different independent controls on palaeoflow estimates (e.g. Bradley and Mears, 1980). In addition, palaeoflow estimates should be related to associated confidence limits, thereby providing an 'envelope' of likely palaeovelocity and palaeodischarge values for a given deposit (e.g. Table 5.2). Of the five main methods of palaeovelocity determination discussed in this paper (summarized in Table 5.1) it appears that the Manning–Limerinos approach and the Darcy–Weisbach–Hey approach provide values tending toward the centre or lower range of the 'envelope'. The Manning–Limerinos approach is recommended for coarse-bed gravels related to unknown or uncertain flow width, while the Darcy–Weisbach approach appears more appropriate for conditions of known and well-defined flow width or in-channel flow. These approaches allow the estimation of a 'first approximation' of palaeoflow conditions, but there are still many flow relations, resistance effects, fluid, sediment and channel boundary relations that are not fully understood and have not yet been accounted for in methods of palaeohydraulic analysis of coarse gravel deposits.

APPENDIX 1 KEY TO NOTATION

A	Cross-sectional area of channel (Equations (3), (5)); glacier area (Equation (39))
a	long axis of particle; coefficient of channel cross-sectional shape (in Hey palaeovelocity equations)
b	Intermediate axis of particle
C	Chezy friction coefficient
C_1	Palaeovelocity coefficient related to Shields' function $= \phi(\gamma_s - \gamma)$
C_2	Coefficient related to Darcy–Weisbach–Hey method of palaeovelocity determination $= 2.30/(\kappa\sqrt{8})$
C_3	Coefficient in Chezy palaeovelocity equation $= \left(\frac{8g}{f}\right)^{1/2}$
c	Short axis of particle; coefficient in palaeovelocity equation; as subscript refers to critical flow conditions
D	Size of bed roughness element
D_{84}	84th percentile of size distribution (*b*-axis)
D_{10x}	Mean diameter of 10 largest bed particles (*b*-axis)
d	Channel depth
d_c	Critical flow depth
d_r	Ratio of critical flow depth to channel depth $= d_c/d$
f	Darcy–Weisbach friction factor
F_d	Propulsive force of flow
F_n	Net force of flow
F_u	Resisting force of flow
g	Acceleration due to gravity
Δh	Difference in elevation between upstream and downstream limits of stream reach
K_s	Height of 'representative' bed roughness element
n	Manning's roughness coefficient
n_b	No. of bars (in equation (35))

n_c	No. of channels (in equation (35))
P	wetted perimeter
Q	Total flow discharge
Q_b	Total bartop or overbank discharge
Q_{bf}	Bankfull discharge
Q_c	Total channel discharge (in equation (38)); Total critical discharge
Q_{pk}	Peak flow discharge (maximum instantaneous discharge)
Q_T	Total discharge (in equations (35) and (36))
q	Specific discharge (per unit width of flow)
q_c	Specific in-channel discharge
q_b	Specific bartop or overbank discharge
R	Hydraulic radius
S	Channel gradient
V	Flow velocity
$\bar{V}$	Mean flow velocity
$\bar{V}_c$	Critical mean flow velocity
V^*	Shear velocity
V^*_c	Critical shear velocity
w	Width
w_c	Channel width
w_b	Bartop or overbank flow width
γ	Specific weight of water
γ_s	Specific weight of sediment
K	von Karman's coefficient
ν	Kinematic viscosity
ρ	Mass density of water
τ_o	Tractive force
τ_c	Critical tractive force
φ	Shields bedload entrainment function
$\frac{1}{\psi}$	Particle Reynolds number ψ

REFERENCES

Allen, J. R. L., and Mange-Rajetzky, M. 1982. Sediment dispersal and palaeohydraulics of Oligocene rivers in the eastern Ebro Basin. *Journal of Sedimentary Petrology*, **29**, 705–716.

American Society of Civil Engineers. 1966. Sediment transport mechanics: initiation of motion. *Proceedings American Society Civil Engineers, Journal Hydraulics Division*, HY2, 291–314.

Ashida, K., and Bayazit, M. 1973. Initiation of motion and roughness of flows in steep channels. *Proceedings 15th Congress International Association of Hydraulic Research*, **1**, A58.1–A58.10.

Baker, V. R. 1973a. Palaeohydrology and sedimentology of Lake Missoula flooding in eastern Washington. *Geological Society of America Special Paper 144*, 79 pp.

Baker, V. R. 1973b. Erosional forms and processes for the catastrophic Pleistocene Missoula floods in eastern Washington. In Morisawa, M. E. (Ed.), *Fluvial Geomorphology*, State University of New York, Binghampton, 123–148.

Baker, V. R. 1977. Stream-channel response to floods with examples from Central Texas. *Bulletin Geological Society of America*, **88**, 1057–1071.

Baker, V. R. 1978. Adjustment of fluvial systems to climate and source terrain in tropical and sub-tropical environments. In Miall, A. D. (Ed.), *Fluvial Sedimentology*, 211–230.

Baker, V. R., and Penteado-Orellana, M. M. 1977. Adjustment to Quaternary climatic change by the Colorado River in Central Texas. *Journal of Geology*, **85**, 395–422.

Baker, V. R., and Ritter, D. F. 1975. Competence of rivers to transport coarse bed material. *Bulletin Geological Society of America*, **86**, 975–978.

Bathurst, J. C. 1978. Flow resistance of large-scale roughness. *Proceedings American Society Civil Engineers, Journal Hydraulics Division*, **104**, HY12, 1587–1603.

Bathurst, J. C., Li, R-M, and Simons, D. B. 1981. Resistance equation for large-scale roughness. *Proceedings American Society Civil Engineers, Journal Hydraulics Division*, **107**, HY12, 1592–1613.

Birkeland, P. W. 1968. Mean velocities and boulder transport during Tahoe-age floods of the Truckee River, California — Nevada, *Bulletin Geological Society of America*, **79**, 137–142.

Boothroyd, J. C. 1972. Coarse-grained sedimentation on a braided outwash fan, northeast Gulf of Alaska. *Office of Naval Research Technical Report*, No. 6–CRD, Univ. S. Carolina, 127p.

Boothroyd, J. C. and Ashley, G. 1975. Process, bar morphology, and sedimentary structures on braided outwash fans, northeastern Gulf of Alaska. In Jopling, A. V., and McDonald, B. C. (Eds.), *Glaciofluvial and Glaciolacustrine Sedimentation Society Economic Palaeontologists and Mineralogists Special Publication*, **23**, 193–222.

Bradley, H. C. and Mears, A. I. 1980. Calculations of flows needed to transport coarse fraction of Boulder Creek alluvium at Boulder, Colorado. *Bulletin Geological Society of America*, **91**, 1056–1090.

Bray, D. I. 1979. Estimating average velocity in gravel-bed rivers. *Proceedings American Society Civil Engineers, Journal Hydraulics Division*, **105**, HY9, 1103–1122.

Bray, D. I. 1981. Estimating average velocity in gravel-bed rivers. Discussion. *Proceedings American Society Civil Engineers, Journal Hydraulics Division*, **107**, HY4, 515–517.

Bridge, J. S. 1978. Palaeohydraulic interpretation using mathematical models of contemporary flow and sedimentation in meandering channels. In Miall, A. D. (Ed.), *Fluvial Sedimentation*, 723–742.

Burkham, D. E. 1981. Uncertainties resulting from changes in river form. *Proceedings American Society Civil Engineers, Journal Hydraulics Division*, **107**, HY5, 593–610.

Burkham, D. E., and Dawdy, D. R. 1976. Resistance equations for alluvial-channel flow. *Proceedings American Society Civil Engineers, Journal Hydraulics Division*, **102**, HY10, 1479–1489.

Butler, P. R. 1977. Movement of cobbles in a gravel-bed stream during a flood season. *Bulletin Geological Society of America*, **88**, 1072–1074.

Chow, V. T. 1959, *Open-channel Hydraulics*, McCraw-Hill, 680 pp.

Cheetham, G. H. 1976. Palaeohydrological investigations of river terrace gravels. In Davidson, D. A., and Shackley, M. O. (Eds.), *Geomorphology, Earth Science and the Past*, Westview Press Inc., Boulder, Co., 335–344.

Cheetham, G. H. 1978. Late Quaternary palaeohydrology: the Kennet Valley case-study. In Jones, D. K. C. (Ed.), *The Shaping of Southern England*, Academic Press, 203–223.

Church, M. 1978. Palaeohydrological reconstructions from a Holocene valley fill. In Miall, A. D. (Ed.), *Fluvial Sedimentology*, 743–772.

Church, M., and Gilbert, R. 1975. Proglacial fluvial and lacustrine environments. In Jopling, A. V., and McDonald, B. C. (Eds.), *Glaciofluvial and Glaciolacustrine Sedimentation, Society Economic Palaeontologists and Mineralogists Special Publication*, No. 23, 22–100.

Church, M., and Kellerhals, R. 1978. On the statistics of grain size variation along a gravel river. *Canadian Journal of Earth Sciences*, **15**, 1151–1160.

Clague, J. J. 1975. Sedimentology and paleohydrology of Late Wisconsinan outwash, Rocky Mt. Trench, Southeastern British Columbia. In Jopling, A. V., and McDonald, B. C. (Eds.), *Glaciofluvial and Glaciolacustrine Sedimentation,* Society Economic Palaeontologists and Mineralogists Special Publication, No. 23, 223–23?

Clarke, M., and Dixon, A. J. 1981. The Pleistocene braided river deposits in the Blackwater Valley area of Berkshire and Hampshire, England. *Proceedings Geologists Association,* **92**, 139–157.

Costa, J. E. 1978. Colorado Big Thompson flood: geologic evidence of a rare hydrologic event. *Geology*, **6**, 617–620.

Cotter, E. 1971. Paleoflow characteristics of a Late Cretaceous river in Utah from analysis of sedimentary structures in the Ferron Sandstone. *Journal of Sedimentary Petrology*, **41**, 129–138.

Day, T. J. 1977. Resistance equation for alluvial-channel flow. Discussion. *Proceedings American Society Civil Engineers, Journal Hydraulics Division*, **103**, HY5, 582–584.

Day, T. J. 1978. Aspects of flow resistance in steep channels having coarse particulate beds. In Davidson-Arnott, R., and Nickling, W. (Eds.), *Research in Fluvial Geomorphology*, Geo Abstracts Ltd., 45–58.

Dury, G. H. 1964. Principles of underfit streams. *U.S. Geological Survey Professional Paper 452A*, 67 pp.

Dury, G. H. 1965. Theoretical implications of underfit streams. *U.S. Geological Survey Professional Paper* 452C.

Dury, G. H. 1976. Discharge prediction, present and former, from channel dimensions. *Journal Hydrology*, **20**, 219–245.

Elliott, T. 1976. The morphology, magnitude and regime of a Carboniferous fluvial-distributary channel. *Journal of Sedimentary Petrology*, **46**, 70–76.

Ethridge, F. G., and Schumm, S. A. 1978. Reconstructing Paleochannel morphologic and flow characteristics: methodology, limitations and assessment. In Miall, A. D. (Ed.), *Fluvial Sedimentology*, 703–721.

Fahnestock, R. K. 1963. Morphology and hydrology of a glacial stream — White River, Mount Rainier, Washington. *U.S. Geological Survey Professional Paper 422A*, 70 pp.

Fahnestock, R. K., and Bradley, W. C. 1973. Knik and Matanuska Rivers, Alaska: a contrast in braiding. In Morisawa, M. E. (Ed.), *Fluvial Geomorphology*, 221–250.

Ferguson, R. I. 1981. Channel forms and channel changes. In Lewin, J. (Ed.), *British Rivers*, 90–125.

Galay, V. J. 1967. Observed forms of bed roughness in an unstable gravel river. *Proceedings 12th Congress International Association Hydraulic Research*, Ft. Collins, Colorado, **4**, A11.1–A11.10.

Gessler, J. 1971. Aggradation and degradation. In Shen. H. W. (Ed.). *River Mechanics* vol. 1, Fort Collins, Colorado, 8.1–8.23.

Graf, W. H. 1971. *Hydraulics of Sediment Transport*, McGraw-Hill, 509 pp.

Gregory, K. J. 1976. Lichens and the determination of river channel capacity. *Earth Surface Processes*, **1**, 273–285.

Griffiths, G. A. 1981. Flow resistance in coarse gravel bed rivers. *Proceedings American Society Civil Engineers, Journal Hydraulics Division*, **107**, HY7, 899–918.

Hand, B. M., Wessel, J. M. and Hayes, M. O. 1969. Antidunes in the Mount Toby Conglomerate (Triassic) Massachusetts. *Journal of Sedimentary Petrology* 39, 1310–1316.

Hartung, F. and Scheuerlein, H. 1967. Macroturbulent flow in steep open channels with high natural roughness. *Proceedings 12th Congress International Association Hydraulics Research*, Ft. Collins, Colorado, **4**, A1.1–A1.8.

Helley, E. J. 1969. Field measurement of the initiation of large bed particle motion in Blue Creek near Klamath, California. *U.S. Geological Survey Professional Paper 562G*, 19pp.

Henderson, F. M. 1966. *Open channel flow*, MacMillan, 522p.

Hey, R. D. 1978. Determinate hydraulic geometry of river channels. *Proceedings American Society Civil Engineers, Journal Hydraulics Division*, **104**, HY6,

Hey, R. D. 1979. Flow resistance in gravel-bed rivers. *Proceedings American Society Civil Engineers, Journal Hydraulics Division*, **105**, HY4, 365–379.

Hjulström, F. 1935. Studies in the morphological activity of rivers as illustrated by the River Fyris. *Geological Institute University Uppsala, Bulletin*, **25**, 221–528.

Howard, A. D. 1980. Thresholds in river regimes. In Coates, D. R., and Vitek, J. D. (Eds.), *Thresholds in Geomorphology*, 227–258.

Itakuri, T. and Kishi, T. 1980. Open channel flow with suspended sediments. *Proceedings American Society Civil Engineers, Journal Hydraulics Division*, **106**, HY8 1325–1343.

Jackson, R. G. 1978. Preliminary evaluation of lithofacies models for meandering alluvial streams. In Miall, A. D. (Ed.), *Fluvial Sedimentology*, 543–576.

Jopling, A. V. 1966. Some principles and techniques used in reconstructing the hydraulic parameters of a paleoflow regime. *Journal of Sedimentary Petrology*, **36**, 5–49.

Jopling, A. V. 1971. Some techniques used in the hydraulic interpretation of fluvial and fluvioglacial deposits. In Yatsu, E. *et al.* (Eds.), *Research Methods in Geomorphology*, Proc. 1st Guelph Symp. on Geomorphology 1969, 93–116.

Kellerhals, R. 1967. Stable channels with gravel-paved beds. *Proceedings American Society Civil Engineers, Journal Waterways Harbours Division*, **93**, WW1, 63–84.

Kjartansson, G. 1967. The Steinholtshlaup, central south Iceland on January 15th 1967. *Jökull*, **17**, 249–262.

Knight, D. W. and MacDonald, J. A. 1979. Open channel flow with varying bed roughness. *Proceedings American Society Civil Engineers, Journal Hydraulics Division*, **105**, HY9, 1167–1183.

Komar, P. D. 1970. The competence of turbidity current flow. *Bulletin Geological Society America*, **81**, 1555–1562.

Koster, E. H. 1978. Transverse ribs: their characteristics, origin and paleohydraulic significance. In Miall, A. D. (Ed.), *Fluvial sedimentology*, 161–186.

Krumbein, W. C. and Lieblein, J. 1956. Geological application of extreme-value methods to interpretation of cobbles and boulders in gravel deposits. *Transactions American Geophysical Union*, **37**, 313–319.

Lane, E. W. and Carlson, E. J. 1954. Some observations on the effect of particle shape on the movement of coarse sediments. *Transactions American Geophysical Union*, **35**, 453–462.

Langbein, W. B. and Leopold, L. B. 1964. Quasi-equilibrium states in channel morphology. *American Journal of Science,* **262**, 782–794.

Leeder, M. R. 1973. Fluviatile fining-upwards cycles and the magnitude of palaeochannels. *Geological Magazine*, **110**, 265–276.

Leeder, M. R. 1978. A quantitative stratigraphic model for alluvium, with special reference to channel deposit density and interconnectedness. In Miall, A. D. (Ed.), *Fluvial Sedimentology*, 587–596.

Leopold, L. B., and Maddock, R. Jr. 1953. The hydraulic geometry of stream channels and some physiographic implications. *U.S. Geological Survey Professional Paper 252*, 57 pp.

Leopold, L. B., and Wolman, M. G. 1957. River channel patterns — braided, meandering and straight. *U.S. Geological Survey Professional Paper 262B*, 85 p.

Leopold, L. B., Wolman, M. G., and Miller, O. P. 1964. *Fluvial Processes in Geomorphology*, Freeman.

Limerinos, J. T. 1970. Determination of the Manning coefficient from measured bed roughness in natural channels. *U.S. Geological Survey Water-Supply Paper 1898B*, 47 pp.

Maizels, J. K. 1976. *A Comparison of Pleistocene and Present-day Proglacial Environments, with Particular Reference to Morphology and Sedimentology*, Unpublished Ph.D. thesis, Univ. London, 675 pp.

Maizels, J. K. 1978. Débit des eaux de font, charges sédimentaires et taux d'érosion dans le massif du Mont Blanc. *Revue Géographie Alpine*, **66**, 65–91.

Maizels, J. K. 1980. Paleohydrology of some Holocene sandur deposits, Søndre Strømfjord and Nassarssuaq, West Greenland. In *Final Report, Aberdeen University West Greenland Expedition* 1979, 89–127.

Maizels, J. K. 1982. Channel changes, palaeohydrology and deglaciation, evidence from some Late glacial sandur deposits of north east Scotland. *Quaternary Studies Poland*, No. 4, in press.

Maizels, J. K. 1983. Proglacial channel systems: change and thresholds for change over long, intermediate and short timescales. In Collinson, J., and Lewin, J. (Eds.), *Modern and ancient fluvial systems*, Special Publication *International Association Sedimentologists*.

Malde, H. E. 1968. The catastrophic Late Pleistocene Bonneville Flood in the Snake River Plain, Idaho. *U.S. Geological Survey Professional Paper 596*, 52 pp.

McDonald, B. C., and Banerjee, I. 1971. Sediments and bed forms on a braided outwash plain. *Canadian Journal Earth Sciences*, **8**, 1282–1301.

Menard, H. W. 1950. Sediment movement in relation to current velocity. *Journal of Sedimentary Petrology*, **20**, 148–160.

Miall, A. D. 1977. A review of the braided-river depositional environment. *Earth Science Reviews*, **13**, 1–62.

Miall, A. D. 1978. Fluvial sedimentology: an historical review. In Miall, A. D. (Ed.), *Fluvial sedimentology*, 1–47.

Moody-Stuart, M. 1966. High- and low-sinuosity stream deposits, with examples from the Devonian of Spitsbergen. *Journal of Sedimentary Petrology*, **36**, 1102–1117.

Morton, R. A., and Donaldson, A. C. 1978. The Guadaloupe River and delta of Texas — a modern analogue for some ancient fluvial–deltaic systems. In Miall, A. D. (Ed.), *Fluvial Sedimentology*, 773–787.

Myers, W. R. C. 1982. Flow resistance in wide rectangular channels. *Proceedings American Society Civil Engineers, Journal Hydraulics Division*, **108**, HY4, 471–482.

Nevin, C. 1946. Competency of moving water to transport debris. *Bulletin Geological Society America*, **57**, 651–674.

Novak, I. D. 1973. Predicting coarse sediment transport: the Hjulstrom curve revisited. In Morisawa, M. E. (Ed.), *Fluvial Geomorphology*, 13–25.

O'Brien, M. P., and Rindlaub, B. D. 1934. The transportation of bedload by streams. *Transactions American Geophysical Union*, **15**, 593–602.

Østrem, G. 1975. Sediment transport in glacial meltwater streams. In Jopling, A. V., and McDonald, B. C. (Eds.), *Glaciofluvial and Glaciolacustrine Sedimentation, Society Economic Palaeontologists and Mineralogists Special Publication No. 23*, 101–122.

Padgett, G. V., and Ehrlich, R. 1976. Paleohydrologic analysis of a Late Carboniferous fluvial system, southern Morocco. *Bulletin Geological Society America*, **87**, 1101–1104.

Parker, G., and Peterson, A. W. 1980. Bar resistance of gravel-bed streams. *Proceedings American Society Civil Engineers, Journal Hydraulics Division*, **106**, HY10, 1559–1575.

Peterson, D. F. and Mohanty, P. K. 1960. Flume studies of flow in steep, rough channels. *Proceedings American Society Civil Engineers, Journal Hydraulics Division*, **86**, 55–76.

Petts, G. E. 1980. *Indentification and interpretation of river channel changes*, Loughborough University Tech., Dept. Geog. Occ. Paper 2, Research Series Geom. I, 57 pp.

Raudkivi, A. J. 1976. *Loose Boundary Hydraulics*, Pergamon, 2nd Edn. 331 pp.

Riggs, H. C. 1978. Streamflow characteristics from channel size. *Proceedings American Society Civil Engineers, Journal Hydraulics Division*, **104**, 87–96.

Rose, J., Turner, C., Coope, G. R., and Bryan, M. D. 1980. Channel changes in a lowland river catchment over the last 13,000 years. In Cullingford, R. A., Davidson, D. A., and Lewin, J. (Eds.), *Timescales in Geomorphology*, J. Wiley, Chichester, 159–175.

Rust, B. R. 1978. The interpretation of ancient alluvial successions in the light of modern investigations. In Davidson-Arnott, R., and Nickling, W. (Eds.), *Research in Fluvial Geomorphology*, Proc. 5th Guelph Symp. Geom. 1977, Geo Abstracts Ltd. 67–105.

Schumm, S. A. 1960. The shape of alluvial channels in relation to sediment type. *U.S. Geological Survey Professional Paper 352B*, 17–30.

Schumm, S. A. 1965. Quaternary palaeohydrology. In Wright, H. E., and Frey, D. G. (Eds.), *The Quaternary of the United States*, Princeton Univ. Press, 783–794.

Schumm, S. A. 1968a. River adjustment to altered hydrologic regimen — Murrumbidgee River and palaeochannels. *U.S. Geological Survey Professional Paper 598*.

Schumm, S. A. 1968b. Speculations concerning palaeohydrologic controls of terrestrial sedimentation. *Bulletin Geological Society America*, **79**, 1573–1588.

Schumm, S. A. 1972. Fluvial palaeochannels. In Rigby, J. K., and Hamblin, W. K. *(Eds.), Recognition of ancient sedimentary environments, Society of Economic Palaeontologists and Mineralogists Special Publication*, **16**, 98–107.

Shaw, J., and Healy, T. R. 1980. Morphology of the Onyx River system, McMurdo Sound Region, Antarctica. *New Zealand Journal Geology and Geophysics*, **23**, 223–238.

Shaw, J., and Kellerhals, R. 1977. Palaeohydraulic interpretation of antidune bedforms with applications to antidunes in gravel. *Journal of Sedimentary Petrology*, **47**, 257–266.

Shields, A. 1936. Anwendung der Ähnlichkeitsmechanic und der Turbulenzforschung auf die Geschiebebewegung. Preussische Versuchsanstalt für Wasserbau und Schiffbau (Berlin). Mitteil. No. 26, 26 pp. (Transl. by W. P. Ott and J. C. Van Uchelen. Application of similarity principles and turbulence research to bedload movement.) U.S. Department Agriculture, Soil Conservation Service, Cooperative Lab., Calif. Inst. Technol., 70 pp.

Shih, C. C., and Grigg, N. S. 1967. A reconsideration of the hydraulic radius as a geometric quantity in open channel hydraulics. *Proceedings 12th Congress International Association Hydraulic Research* Ft. Collins, Colorado, **1**, A36.1–A36.9.

Silberman, E., Carter, R. W., Einstein, H. A., Hinds, J., and Powell, R. W. 1963. Friction factors in open channels. Progress Report. Task Force on Friction Factors in open channels of the Committee on Hydromechanics of the Hydr. Div. *Proceedings American Society Civil Engineers, Journal Hydraulics Division*, **69**, HY2, 97–143.

Simons, D. B. 1969. Open channel flow. In Chorley, R. J. (Ed.), *Water, Earth and Man*, 297–318.

Simons, D. B., and Albertson, M. L. 1960. Uniform water conveyance channels in alluvial material. *Proceedings American Society Civil Engineers, Journal Hydraulics Division*, HY5, 33–71.

Strickler, A. 1923. Beiträge zur Frage der Geschwindigkeits — formel und der Rauhigkeitszamen für Ströme, Kanäle, und geschlossene Leitungen. *Mitteil. des Amtes für Wasserwirtschaft*, Bern. No. 16, 77 pp.

Sundborg, A. 1956, The River Klaralven: a study of fluvial processes. *Geografiska Annaler*, **38**, 127–316.

Thompson, S. M., and Campbell, P. L. 1979. Hydraulics of a large channel paved with boulders. *Journal Hydraulics Research*, **17**, 341–354.

Vanoni, V. A., and Nomicos, G. N. 1960. Resistance properties of sediment-laden streams. *Transactions American Society Civil Engineers*, **125**, Paper No. 3055, 1140–1167.

Wang, S. 1981. Variation of Karman constant in sediment-laden flow. *Proceedings American Society Civil Engineers, Journal Hydraulics Division*, **107**, HY4, 407–417.

Yalin, M. S., and Finlayson, G. D. 1972. On the velocity distribution of the flow carrying sediment in suspension. In Shen, H. W. (Ed.), *Sedimentation*, 8.1–8.18.

Background to Palaeohydrology
Edited by K. J. Gregory

6

The role of magnetic studies in palaeohydrology

F. OLDFIELD

Department of Geography, University of Liverpool

The aim of the present chapter is to review the role of magnetic measurements in palaeohydrological and related studies. Since the techniques and applications of magnetic measurement are relatively new and rapidly developing it takes the form of a progress report and it considers both case studies in various stages of completion and future prospects in areas where the techniques have not yet been evaluated. Its main purpose is to make readers aware of the value of magnetic measurements, and interested in extending their application to new contexts and problems. Whenever possible the text has been written to complement the contributions of Thompson (1980) and Dearing (1982) to recently published IGCP volumes. Some emphasis is given to recent developments in instrumentation since these have, over the last few years, brought the techniques out of the specialized palaeomagnetic laboratory into a much wider range of academic and real-world environments.

MAGNETIC ORDER

Not only do iron oxides form ~ 2 per cent of the Earth's crust but they are among the most sensitive and ubiquitous indicators of environmental conditions and processes. This applies whether the processes themselves are of 'natural' occurrence, as for example in the soil profile or sediment record, or of directly anthropogenic origin as in the case of particulate pollution resulting from fossil fuel combustion. Moreover, the end products of many iron-transforming processes, whether these be igneous activity, weathering, soil formation, or industrial combustion, often include magnetic oxides which, once formed, are persistent and conservative in a wide range of depositional contexts. Hence magnetic properties are often indicative of sources and of palaeoenvironmental processes. Many advances in the environmental sciences arise from the advent of techniques which make it possible to decipher, record, organize, and interpret newly-perceived forms of order in environmental systems. Recently, rapid and enhanced insight into magnetic order in rocks, soils, peat, and sediments has been made possible by the development of modern electronics and computing capability linked to sensitive measuring systems based on well-established but newly applied physical principles. When Mackereth (1971) confirmed the chronological significance of the geomagnetic secular variation record in the Windermere sediments, he used an Astatic Magnetometer requiring quite a large dedicated wooden building remote from road traffic and similar disturbing influences. Within just over a decade, we have acquired the capacity to do comparable measurements more accurately and to add a variety of other environmentally significant magnetic parameters, using portable battery driven equipment which can be carried to and operated anywhere in the world.

Palaeomagnetism and mineral magnetism

At the outset, it is essential to recognize a distinction between magnetic properties which record the nature of the earth's magnetic field at some period in the past (Natural Remanent Magnetization — NRM) and those magnetic properties which are entirely independent of the ancient field and a function of the magnetic mineralogy of the sample. These latter are measured by using artificially-generated fields in ways which illuminate the nature of the magnetic minerals present in a sample or core. We term the latter *mineral magnetic* properties in contradistinction to the former which are truly *palaeomagnetic*. The main concern of this chapter is with mineral magnetic properties though it is important to remember the potential importance of palaeomagnetic secular variation studies in establishing recent chronologies of sedimentation in lacustrine and marine environments.

Holocene palaeomagnetism and sediment chronology

Under favourable circumstances sediments can provide a basis for reconstructing changes in the Earth's magnetic field. This applies not only to slowly accumulating deep sea sediments which record polarity changes of major significance in the long term geological record (e.g. Foster and Opdyke, 1970), but also to more rapidly accumulating lake (Thompson and Turner, 1979), near-shore marine (Creer, 1974), and possibly cave (Noel, 1982), sediments which often make it possible to resolve the shorter term and lower amplitude secular variation changes which certainly characterize the last ten thousand years. In virtually all the cases studied so far, the palaeomagnetic record is preserved in the sediment through the deposition, orientation, and physical 'locking-in' of small grains of magnetite a few microns or less in size, yielded from the eroding drainage basin. The record they preserve of the orientation of the Earth's field shortly after deposition can be resolved into Inclination (*I*) and Declination (*D*) components which can then be compared with the Observatory and archaeomagnetic record for the last two millenia, and dated at favourable localities by ^{14}C assay for earlier times. This provides a dated record of *I* and *D* 'turning points' which can then be used as a 'master curve' for chronological purposes, wherever they can be identified in sediment sequences from the same region. Since the sedimentary record of NRM *intensity* (*J*) reflects mostly sedimentological rather than geomagnetic variations, it cannot be used as a dating tool in this way (though it can provide a basis for within-site core correlation and *relative* chronology). As a rough guide, present indications are that dated secular variation 'master curves' can be transferred from site to site over areas at least as large as western Europe. Favourable sites are ones with fine, undisturbed, and largely allochthonous minerogenic sediments derived from catchments reasonably rich in magnetic minerals either as primary oxides in, for example, basic igneous rocks, or as secondary oxides formed in the soil (Mullins, 1977). These requirements have been met by sites spanning a wide range of lithological variation from basalt-bound L. Neagh in N. Ireland to the lakes of the Jura in eastern France. Although large lakes have generally been shown to have a better palaeomagnetic record than very small ones, even sites of less than 100 ha surface area have provided useful results. The construction of a well supported and detailed independent 'calibration' chronology for each region is an essential prerequisite for using secular variation as a dating tool (Figure 6.1), the more so since suggestions that the secular variation record is harmonic (Creer *et al.*, 1972) and/or parallel on an intercontinental scale (Creer and Tucholka, 1982), though beguiling in terms of the relatively simple dynamo models they generate, are poorly supported by the empirical evidence so far available. Use of pneumatic Mackereth corers (Mackereth, 1958, 1969) in sediment sampling for Holocene palaeomagnetic study often permits detailed declination-based chronologies to be established for whole unopened cores within hours of retrieval. Reconstruction of the secular variation record for any point on the earth for the period 1600 AD to present (Thompson and Barraclough, in press) means that for this time interval, spanning the major impact of human activity at many sites, the palaeomagnetic record in the sediments may often be extremely useful (Oldfield *et al.*, 1980) especially in combination with short-lived radioisotope (e.g. ^{137}Cs and ^{210}Pb) measurements (e.g. Oldfield and Appleby, in press).

MINERAL MAGNETIC STUDIES

The value of mineral magnetic measurements in palaeohydrology lies in the

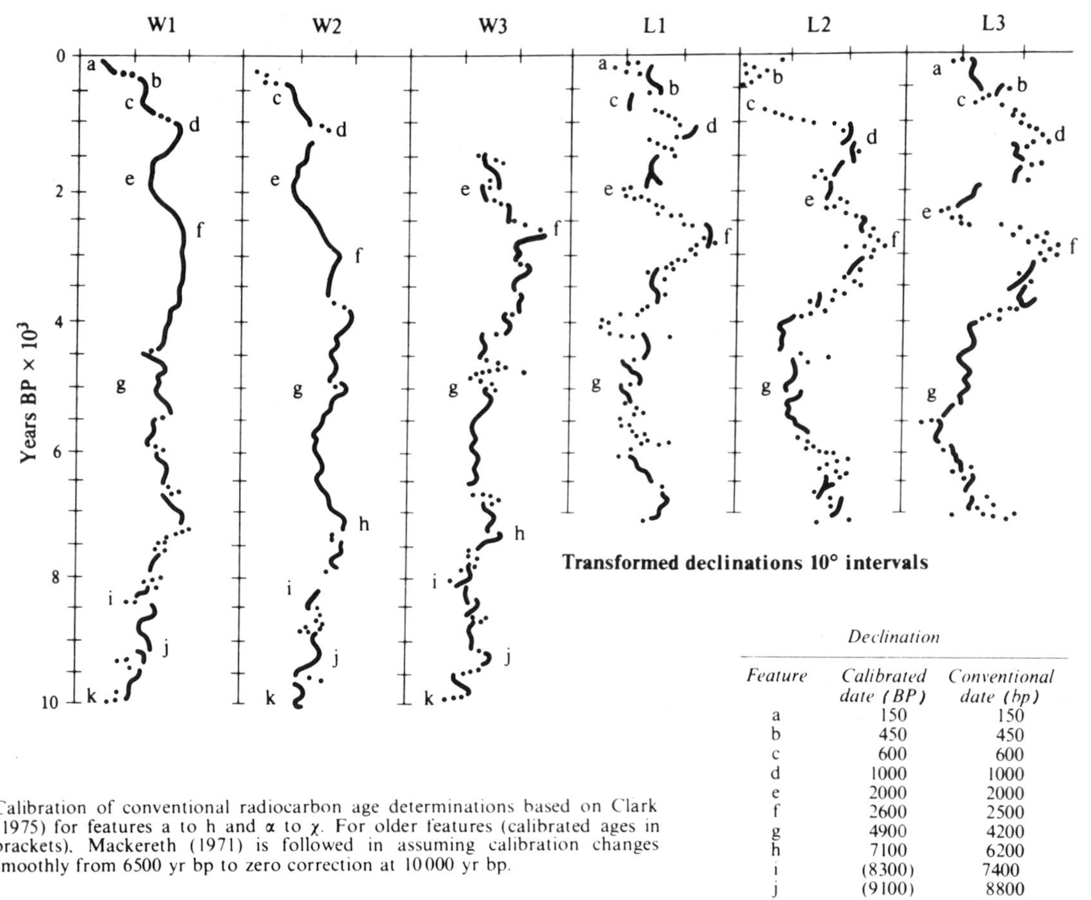

Calibration of conventional radiocarbon age determinations based on Clark (1975) for features a to h and α to χ. For older features (calibrated ages in brackets), Mackereth (1971) is followed in assuming calibration changes smoothly from 6500 yr bp to zero correction at 10 000 yr bp.

	Declination	
Feature	*Calibrated date (BP)*	*Conventional date (bp)*
a	150	150
b	450	450
c	600	600
d	1000	1000
e	2000	2000
f	2600	2500
g	4900	4200
h	7100	6200
i	(8300)	7400
j	(9100)	8800
k	(10000)	10000

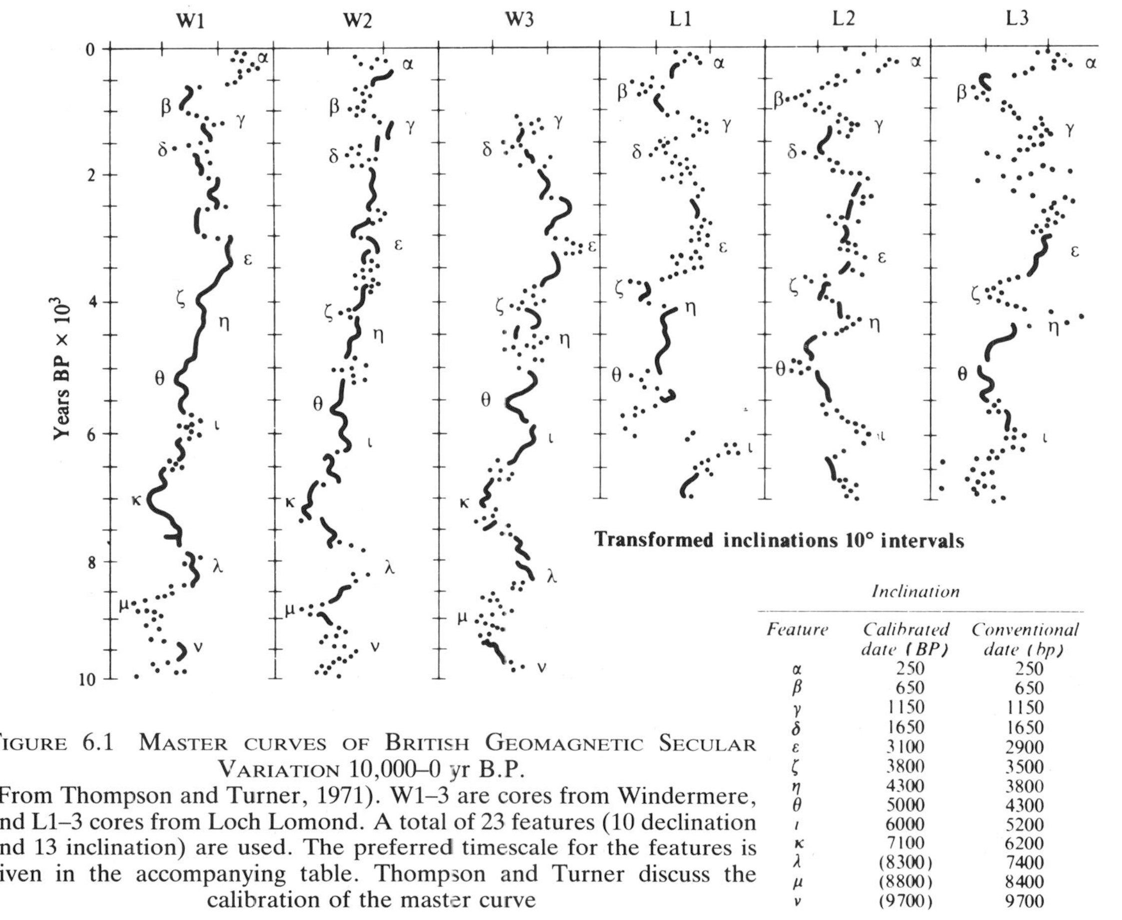

FIGURE 6.1 MASTER CURVES OF BRITISH GEOMAGNETIC SECULAR VARIATION 10,000–0 yr B.P.
(From Thompson and Turner, 1971). W1–3 are cores from Windermere, and L1–3 cores from Loch Lomond. A total of 23 features (10 declination and 13 inclination) are used. The preferred timescale for the features is given in the accompanying table. Thompson and Turner discuss the calibration of the master curve

	Inclination	
Feature	*Calibrated date (BP)*	*Conventional date (bp)*
α	250	250
β	650	650
γ	1150	1150
δ	1650	1650
ε	3100	2900
ζ	3800	3500
η	4300	3800
θ	5000	4300
ι	6000	5200
κ	7100	6200
λ	(8300)	7400
μ	(8800)	8400
ν	(9700)	9700

speed, ease, and non-destructive nature of the techniques, the 'environmentally diagnostic' nature of many of the magnetic parameters and the degree to which these parameters are conserved for long periods in many contexts.

For practical purposes we may divide instrumentation into systems designed to measure *magnetization in low but artificially created magnetic fields* and systems designed to measure some form of artificial *magnetic 'remanance'* — i.e. behaviour arising from having placed the sample in a higher magnetizing field from which it has been withdrawn before measurement.

The most commonly used 'in-field' measurements are of magnetic susceptibility. What is recorded is the effect of a sample, core, or surface on the magnetic flux generated by a coil or ferrite probe. The effect can be electronically detected and digitized. In the Bartington Instruments system, alternative sensors can be used to measure the magnetic susceptibility of surfaces, profiles, and sections, cores of soil or sediments held vertically or horizontally, and single samples of any size or type from 1 ml to 300 ml or more total volume, whether wet or dry. Readings take from 1 to 10 s depending on sample strength, and full sample characterization may take from 5 to 500 s. In weakly magnetic materials allowance has to be made for the diamagnetic (i.e. negative) susceptibility of sample holders and packing material (normally some form of carbon) and of the sample matrix where this is rich in organic material, calcium carbonate, or pure silica. Careful use of susceptibility measured at low and high frequencies on the sample often indicates the relative abundance of surface soil derived material (see below).

Mineral magnetic remanence measurements involve at least two pieces of equipment, one to magnetize and one to measure. So far, most work has been done using an electromagnet adjustable to varying field strengths up to ~ 1 Tesla (about 20 thousand times the Earth's field) to grow Isothermal Remanent Magnetizations (IRM's) in samples. The recently developed Molspin pulse magnetizer which uses capacitors and a coil to generate precisely controlled, uniform fields from 0.002 to 0.3 Tesla will fulfil almost all the requirements of mineral magnetic studies and replace the cumbersome and costly electromagnet for most purposes.

Virtually any sensitive magnetometer is suitable for IRM measurement though the best combination of sensitivity, convenience, economy, and portability is provided by the Minispin fluxgate system. This is portable and battery driven and together with the Bartington susceptibility system and a pulse magnetizer it provides the basis for a highly mobile mineral magnetic laboratory.

The following account and illustrations concentrate almost exclusively on studies which have been or can be completed using the instrumentation briefly noted above. Full definition of parameters lies outside the scope of this chapter and readers should refer to Thompson (1980) and Oldfield *et al.* (1978).

Core correlation and sediment influx

Dearing (1982) and Thompson (1980) have dealt with this aspect of mineral magnetic measurements in recently published IGCP volumes. Dearing's account deals largely with methods of correlation in lake sediments, and stresses the role of magnetic measurements in establishing synchroneity, in extending absolute chronologies from dated cores and in quantifying many aspects of past material flux within lake–watershed ecosystems. Thompson deals with both the chronological and correlative aspects of magnetic measurements in lake sediments. The present account concentrates on a single illustration which complements the case studies to which Dearing and Thompson refer and extends the approach to a much more weakly magnetic system than those previously studied.

Mirror Lake (Figure 6.2) is a granite-bound lake lying close to the Hubbard Brook catchment very intensively studied in recent times (Bormann and Likens, 1969). The fresh sediments of the lake are almost entirely composed of water, carbon and biogenic silica (all of which are diagmagnetic) and the impact of human activity in the catchment has been too slight to generate the major changes in magnetic mineral concentrations which make reliable core correlation possible using whole core or even single sample susceptibility measurements alone. Instead, IRM measurements have been used to identify correlatable horizons in the immediately pre- and post-colonial sediments. Figure 6.3 shows IRM *vs*. accumulated dry weight traces from 10 cores located on Figure 6.2. Intercore correlations are possible and they can be used along with more detailed magnetic characterization (Oldfield and Appleby, in press) to illuminate processes such as sediment focusing within the lake and the history and impact of anthropogenic disturbance in the catchment.

Recent studies of Lake Washington sediments (Oldfield *et al.*, in press; Oldfield and Appleby, in press) show how effective whole core susceptibility scans are for identifying anomalous, disturbed and discontinuous sedimentation in both short (50 cm) and long (~ 20 m) cores. This is of particular importance in morphometrically complex, steep-sided, or tectonically-disturbed basins since it can often save the investment of wasted time and effort involved in applying more time consuming techniques to unrewarding material.

Magnetically based core correlation has been applied successfully to East Mediterranean (Oldfield, Bloemendal, and Hunt, unpub.) and to suites of deep sea sediments from the Eastern Atlantic (Robinson, 1982). In both cases, as in long lake sequences (Oldfield *et al.*, 1978; Oldfield *et al.*, in press),

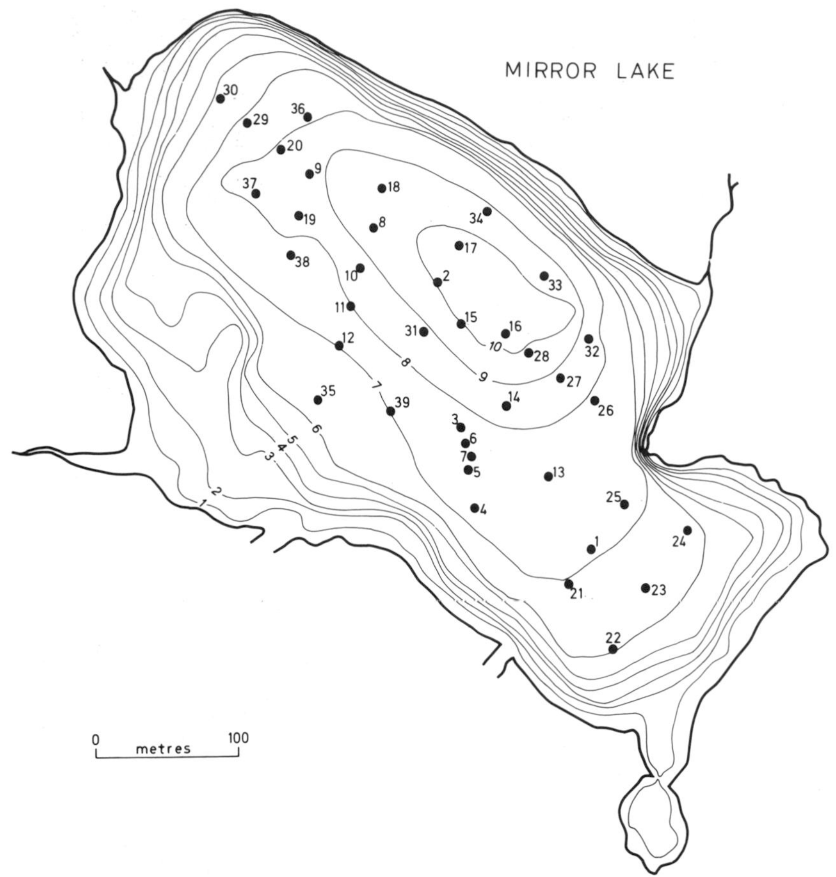

FIGURE 6.2 BATHYMETRIC MAP OF MIRROR LAKE, NEW HAMPSHIRE
Showing the location of 39 short cores taken in February 1982 and spanning the mud–water interface and the top 30–50 cm of sediment

the major changes often reflect climatic shifts which in turn control weathering and sedimentation regimes (*cf.*, Bloemendal, 1981; Robinson, 1982; Kent, 1982). Little attempt has been made to apply mineral magnetic measurements systematically to the correlation of alluvial or estuarine sequences. Results from the Rhode River, a tidal arm of Chesapeake Bay suggest that magnetic correlations over more than 10–20 m spatially, are only feasible where major source and rate shifts have taken place (see below). In this connection, the remarks of Dearing (1982) concerning the effects of sorting on magnetic properties should be borne in mind.

Magnetic differentiation: weathering and pedogenesis

The visual expression of soil formation is often dominated by the distribution

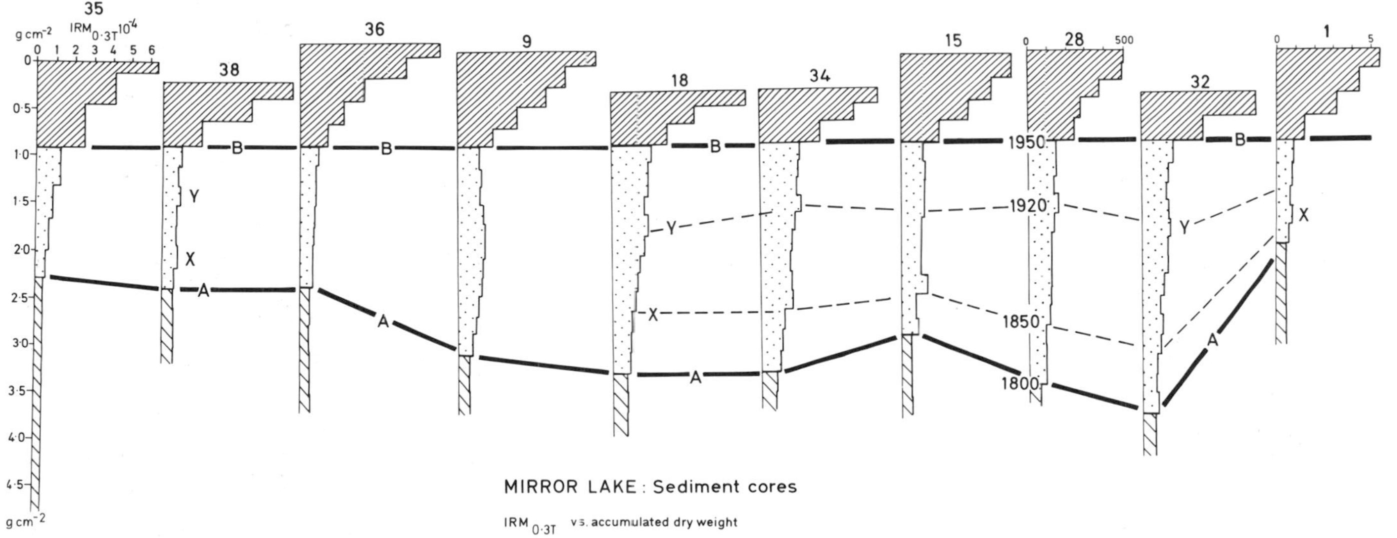

FIGURE 6.3 PLOTS OF IRM.$_{3T}$ *VS* ACCUMULATED DRY WEIGHT FOR TEN CORES FROM MIRROR LAKE LOCATED ON FIGURE 6.2 IRM.$_{3T}$ is measured by magnetizing the sample in a uniform field of 0.3 Tesla using a Molspin Pulse Magnetizier, then measuring the Isothermal Remanent Magnetization (IRM) so induced, on a Minispin slow-speed fluxgate Magnetometer. The major variations recorded here mostly reflect changes in the concentration of magnetic minerals washed into the lake from the catchment. The overall trend to increasing values towards the surface is thus consistent with a greater input of allochthonous minerogenic catchment derived particulates as a result of human activity. horizons A, B, and where identifiable X and Y, are thought to be synchronous. The very approximate timescale of sedimentation expressed as years AD alongside the trace from Core 28 is derived using the methods outlined in Appleby and Oldfield (1978). A fuller interpretation of the mineral magnetic record in these and other cores from the site is given in Oldfield and Appleby (in press)

and nature of iron compounds in the regolith. As a result, magnetic parameters which reflect the transformations and states of lithospherically-derived iron can often be used to characterize material from different depths and horizons in the profile. One of the best documented (though far from fully understood) processes is termed magnetic enhancement (Le Borgne, 1955; Mullins, 1977; Longworth *et al.*, 1979); and it gives rise to higher concentrations of fine grained ferrimagnetic minerals (e.g. magnetite) in surface layers. This can usually be identified in down profile susceptibility or IRM measurements (Figure 6.4). On most lithologies this 'enhancement' is accompanied by other mineral magnetic changes which reflect iron transformations in the near surface layers (Figure 6.5). These properties once established in the soil are remarkably persistent and they form the basis for sediment source identification described in the next section.

Over the full depth of a soil profile, the magnetic properties reflect many other processes than the enhancement mechanisms referred to above (Figure 6.5). Since palaeosols give rise to quite recognizable magnetic properties (Harvey *et al.*, 1981; Poutiers, 1975) it is at least possible that systematic mineral magnetic study of palaeosols will be of value in the reconstruction of past weathering and pedogenic regimes. Results so far suggest that magnetic properties are preserved in soils save where gleying has been active. This tends to convert both primary and secondary magnetic minerals to soluble forms, attacking the finest grains and reduced oxides first.

Sediment source identification

Initial studies confirmed that magnetic measurements could be used successfully to identify the dominant source of suspended sediments in the Jackmoor Brook catchment (Oldfield *et al.*, 1979; Walling *et al.*, 1979). Here the lithology and channel morphometry were extremely favourable. Subsequent studies based on stream, lake, and estuarine sediments show that the mineral magnetic approach once adapted to the lithologies under study, can allow similar differentiation between surface and erosion scar derived material in a wide range of contexts. The use of quadrature susceptibility (Mullins and Tite, 1973; Oldfield *et al.*, in press; Oldfield and Bartington, 1982) is especially promising since it often allows identification of soil-derived secondary magnetic minerals irrespective of underlying lithology (Figure 6.6.)

A magnetic approach to catchment studies

The ubiquity and sensitivity of magnetic minerals, the speed and versatility of measuring equipment and the persistence of magnetic linkages between source and sediment make the emerging methodology ideally suited to both process- and reconstruction-oriented catchment studies and more especially to that

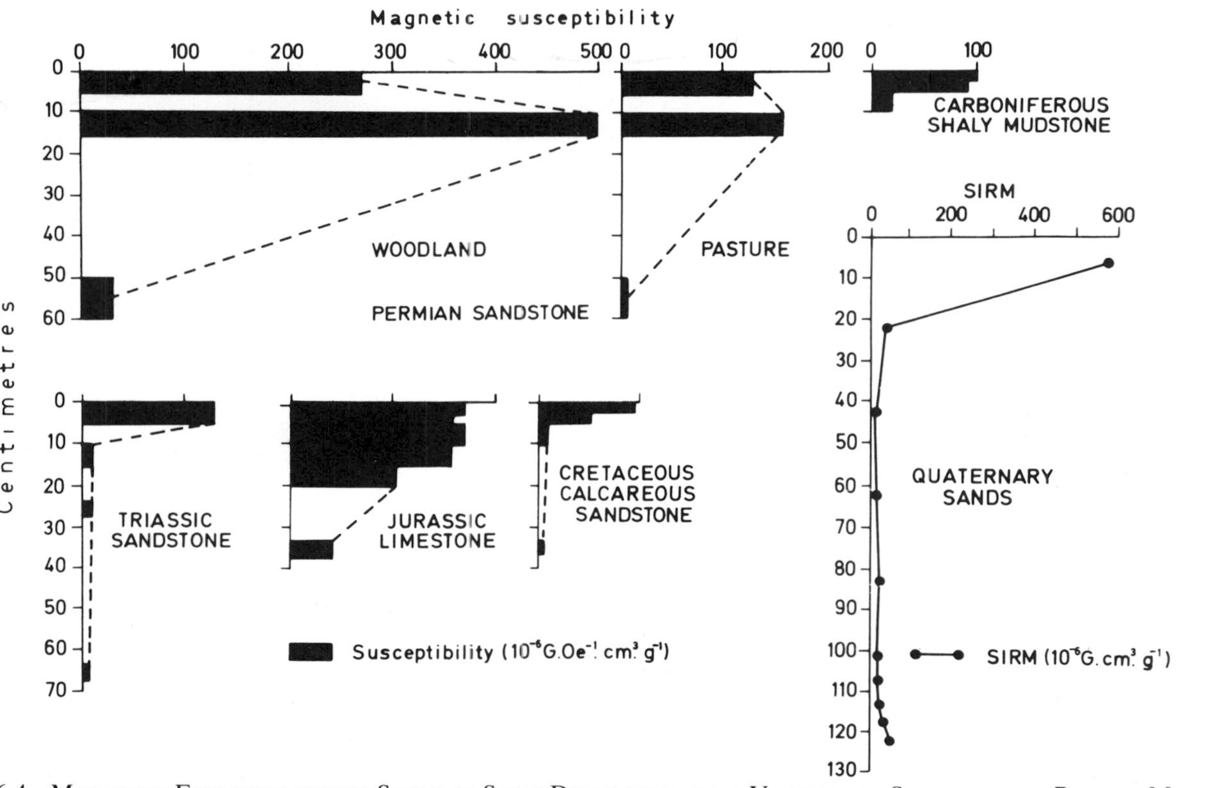

FIGURE 6.4 MAGNETIC ENHANCEMENT IN SURFACE SOILS DEVELOPED ON A VARIETY OF SEDIMENTARY PARENT MATERIALS
In the case of soil profile from the Landes only the top sample was sufficiently rich in magnetic minerals to give a measurable susceptibility. The enhancement is shown equally well in the SIRM values. SIRM is measured in the same way as $IRM_{.3T}$ (see caption to Figure 6.3) save that the remanance is grown in a higher 'saturating' field, (ca. 1.0 Tesla) using a conventional electromagnet

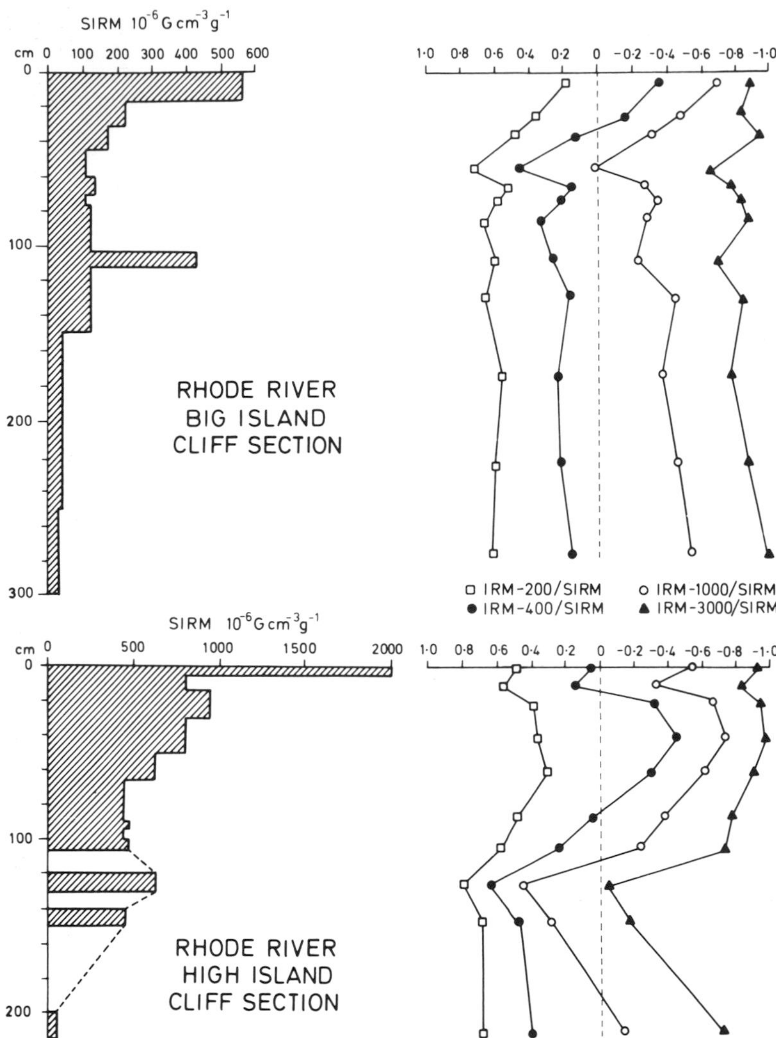

FIGURE 6.5 DOWNPROFILE VARIATIONS IN SIRM (SEE FIGURE 6 CAPTION) AND RELATED PARAMETERS FOR TWO DEEPLY WEATHERED SOIL PROFILES IN THE RHODE RIVER, CHESAPEAKE BAY AREA

The profiles show peak SIRM at the surface, as a result of enhancement (see text) and at 110 and 125 cm respectively as a result of iron enrichment in illuviated horizons. The normalized parameters plotted on the right of each profile are independent of concentration and reflect changes in magnetic mineralogy and grain size. They are measured by placing previously saturated and measured samples in a sequence of increasing reverse magnetic fields and remeasuring the IRM at each stage (*cf*. Oldfield *et al*., 1978, 1979). The horizons of higher *sub*surface IRM are associated with a relatively 'harder' magnetic mineral assemblage, and the relatively greater importance of an antiferromagnetic component (almost certainly goethite) can be inferred

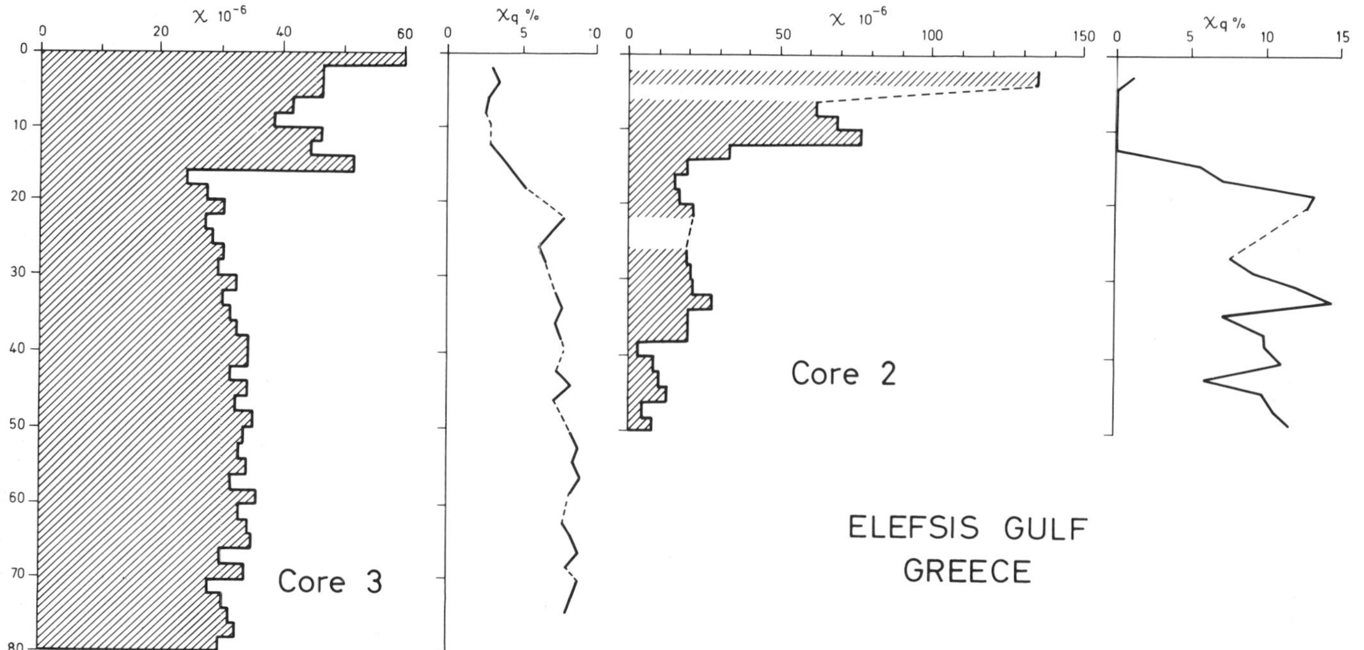

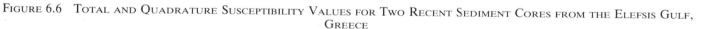

FIGURE 6.6 TOTAL AND QUADRATURE SUSCEPTIBILITY VALUES FOR TWO RECENT SEDIMENT CORES FROM THE ELEFSIS GULF, GREECE

Core 2 is close to and Core 3 more distant from a major iron and steel works (Scoullos *et al.*, 1979). The impact of particulate discharge from the industrial complex has given rise to increased susceptibility values in the top 15–18 cm of each core. Below this level there is a significant quadrature component (χ_q) confirming that input from eroding soils is the main source of magnetic oxides in the sediment. Above this as a result of the industrial input the χ_q component is unmeasurable in Core 3 and diluted to proportionally low values in Core 2. Thus quadrature susceptibility can be used to identify specifically soil-derived material

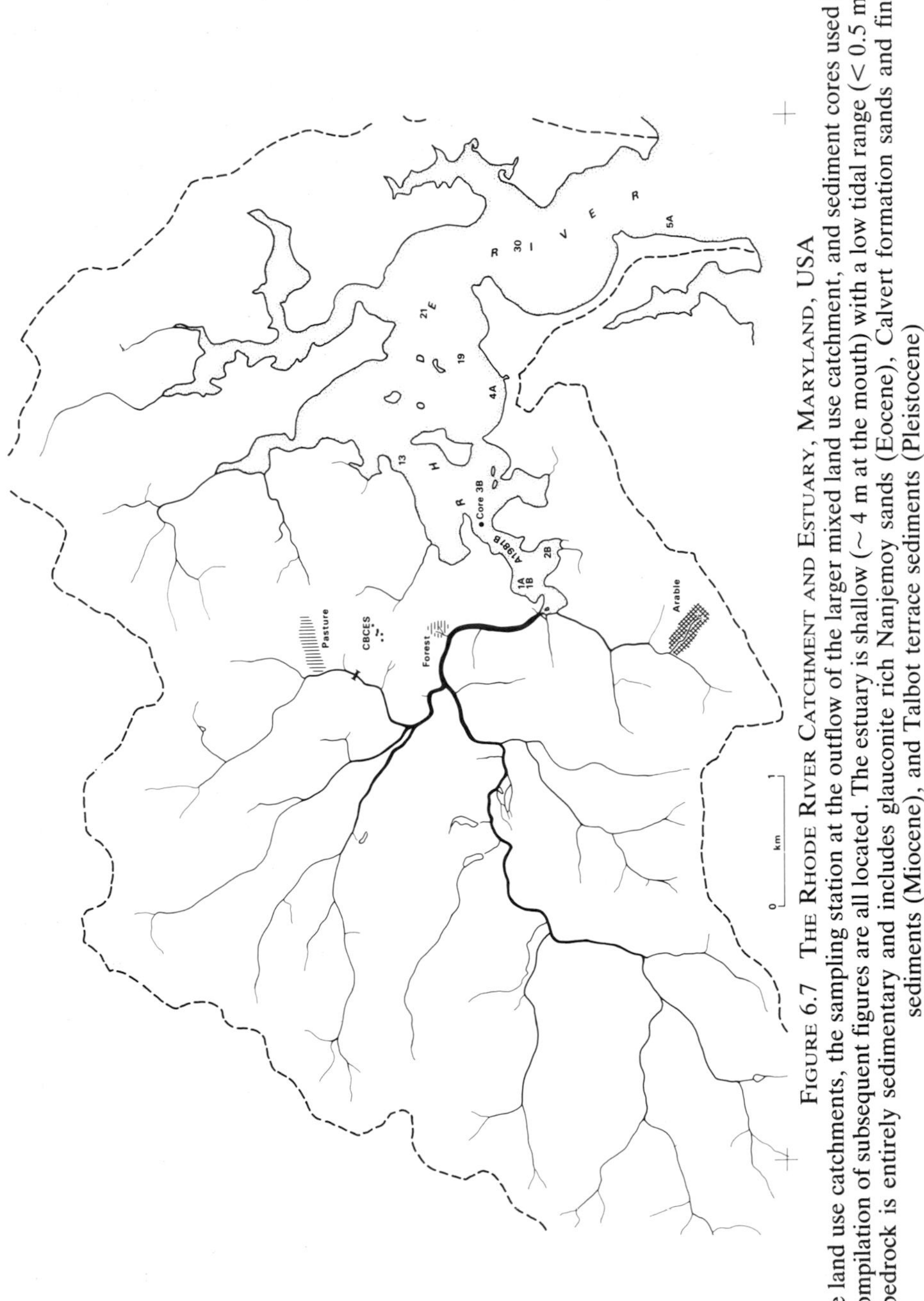

FIGURE 6.7 THE RHODE RIVER CATCHMENT AND ESTUARY, MARYLAND, USA

Single land use catchments, the sampling station at the outflow of the larger mixed land use catchment, and sediment cores used in the compilation of subsequent figures are all located. The estuary is shallow (~ 4 m at the mouth) with a low tidal range (< 0.5 m). The bedrock is entirely sedimentary and includes glauconite rich Nanjemoy sands (Eocene), Calvert formation sands and finer sediments (Miocene), and Talbot terrace sediments (Pleistocene)

integration of the two approaches so strongly advocated in recent times (see e.g. Oldfield, 1977). Preliminary results from the Rhode River are presented here to support and illustrate this contention.

Figure 6.7 locates the main instrumented reaches and subcatchments within the Rhode River watershed and the position of the main sediment cores used in the study. Mineral magnetic measurements began in 1980 and the project was designed as a methodological exploration as well as a location-specific case study. Initial work included catchment surface surveys using loop and probe sensors, in order to provide an empirical basis for site selection in soil and section sampling. Whole core susceptibility scans were carried out on some 20 cores collected by J. Donohue, then of the Smithsonian Institution, in order to assess rapidly any major spatial and temporal variations in recent (mostly post 1600 AD) sedimentation and to indicate cores of particular importance in further studies. Suspended sediment samples from integrated samplers at weir sites at the outlet from single land use and mixed catchments, and estuarine suspensates were also measured.

The main potential sediment sources within the system are surface soils and eroding tidal cliff sections in the Eocene Nanjemoy formation. Mutually independent magnetic parameters were used on both a bulk and a particle size specific basis to characterize material in all phases of the system from eroding soils and cliffs to 'mature' sediments. In this way it was hoped to dispose of any possibility that sediment-source linkages were spurious or coincidental. Preliminary results are shown in Figures 6.8 to 6.11.

Figure 6.8 shows some of the results of the preliminary catchment surveys and Figure 6.9 a sequence of whole core scans from the estuarine sediments. Figure 6.10 gives a more detailed magnetic characterization of the main potential sediment source types and Table 6.1 summarizes the main dimensions of contrast. Figure 6.11, like Figure 6.10, plots two magnetic parameters for several of the sediment cores and shows that deposition at marginal sites close to present day Nanjemoy cliffs has been dominated by sediment derived from these. In the more centrally located cores, however, whereas the older sediment is also characterized by magnetic parameters indicative of unweathered substrate, the more recent sediment has clearly been derived largely from eroding surface soils.

This shift is tentatively dated to the mid-19th century. It correlates with pollen analytical evidence for a major increase in the intensity of land use and tillage (Oldfield and Appleby, in press) and is clearly of great significance in any appraisal of the recent hydrological-sedimentological history of the system. The continuing studies of contemporary sedimentation, using the same magnetic parameters to characterize and quantify particulate flux, can be set within this historical context. It should also be noted that on the basis of pilot measurements, one of the most promising chronological tools for dating the sediments appears to be comparison of the record of post 1600 AD

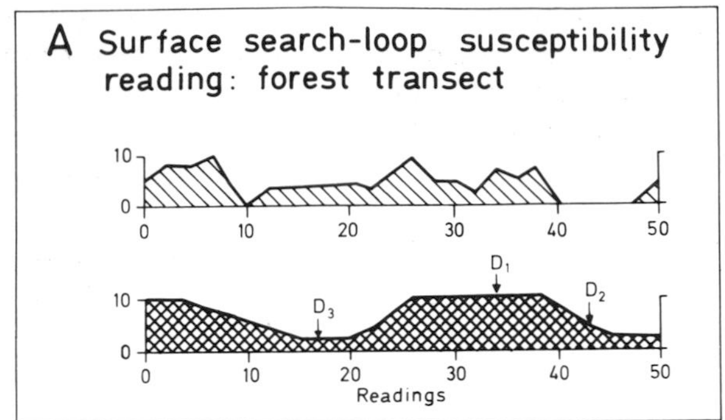

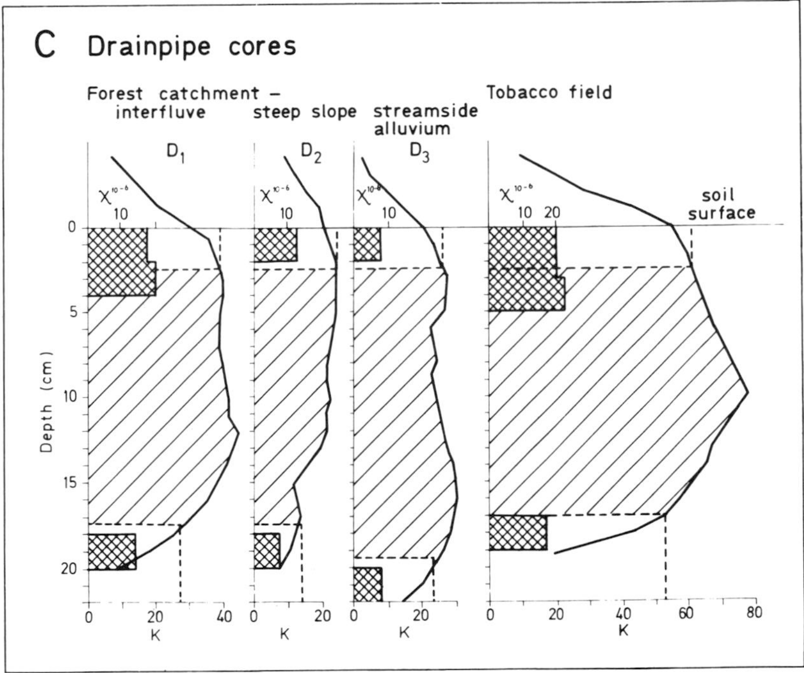

FIGURE 6.8 RHODE RIVER. INITIAL MAGNETIC MEASUREMENTS FROM THE FIRST STAGE OF THE CATCHMENT SURVEY

A Plots surface susceptibility readings along a transect in the Forest catchment (Figure 6.7). Readings are generally higher on slope crests than on the gleyed valley floors. B Shows frequency plots of surface readings from several catchment environments. The bimodal peak in the Pasture catchment (Figure 6.7) reflects varying degrees of enhancement on Calvert (higher) and Nanjemoy (lower) substrates. The within-channel muds have a range of readings closer to Nanjemoy surface material than to channel side erosion scars. C Volume susceptibility scans (see

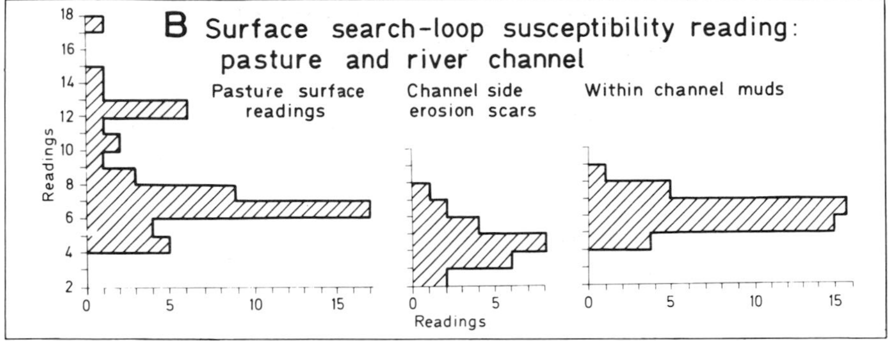

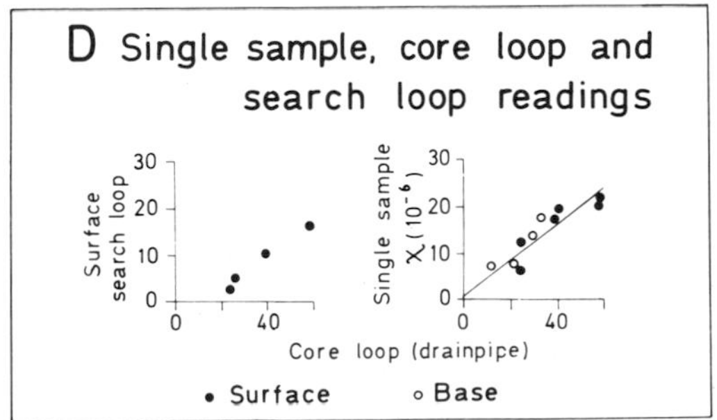

Rhode River catchment
Maryland, U.S.A.

Field survey magnetic susceptibility results

Figure 6.9 caption) of four plastic drainpipe soil cores rapidly confirms a fairly uniform concentration of magnetic minerals in the top 15–20 cm of the soil. Single sample susceptibility (χ) readings from the top and bottom of each are mass specific. Cores D_1, D_2 and D_3 are from the Forest–transect plotted in A. The fourth core is from a soil profile developed on Calvert sands on the edge of the cultivated catchment located on Figure 6.7. D Plots to show the relationship between the upper values obtained from the drainpipe core scans and both the single sample measurements from surface samples (left hand graph) and the surface search loop readings (right hand graph) for the area cored

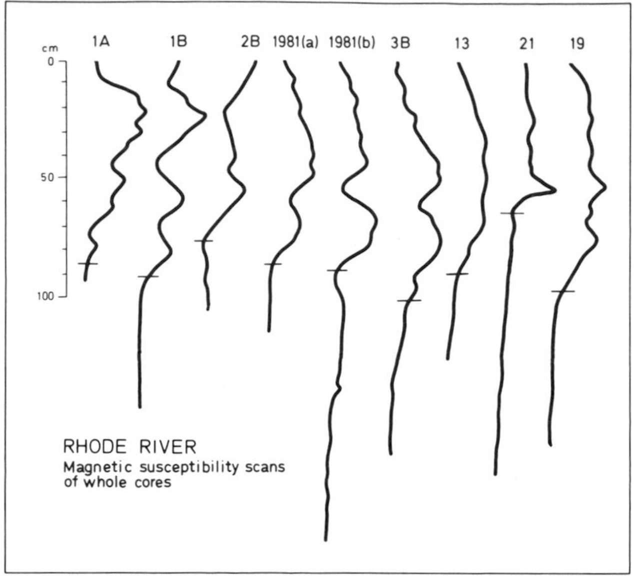

FIGURE 6.9 VOLUME SUSCEPTIBILITY SCANS OF NINE SEDIMENT CORES FROM 'CENTRAL' LOCATIONS IN THE RHODE RIVER ESTUARY

The traces record the varying volume concentration of magnetic minerals at each depth in the core. They are obtained by passing a 7 cm internal diameter coil over the unopened core tube and taking readings at 2 cm intervals. All the core scans show higher susceptibility values above the levels identified by the horizontal bars. These higher values are the result of the increased input of surface soil derived sediments during recent times of more intensive cultivation (Oldfield *et al.*, in press, and Oldfield and Appleby, in press)

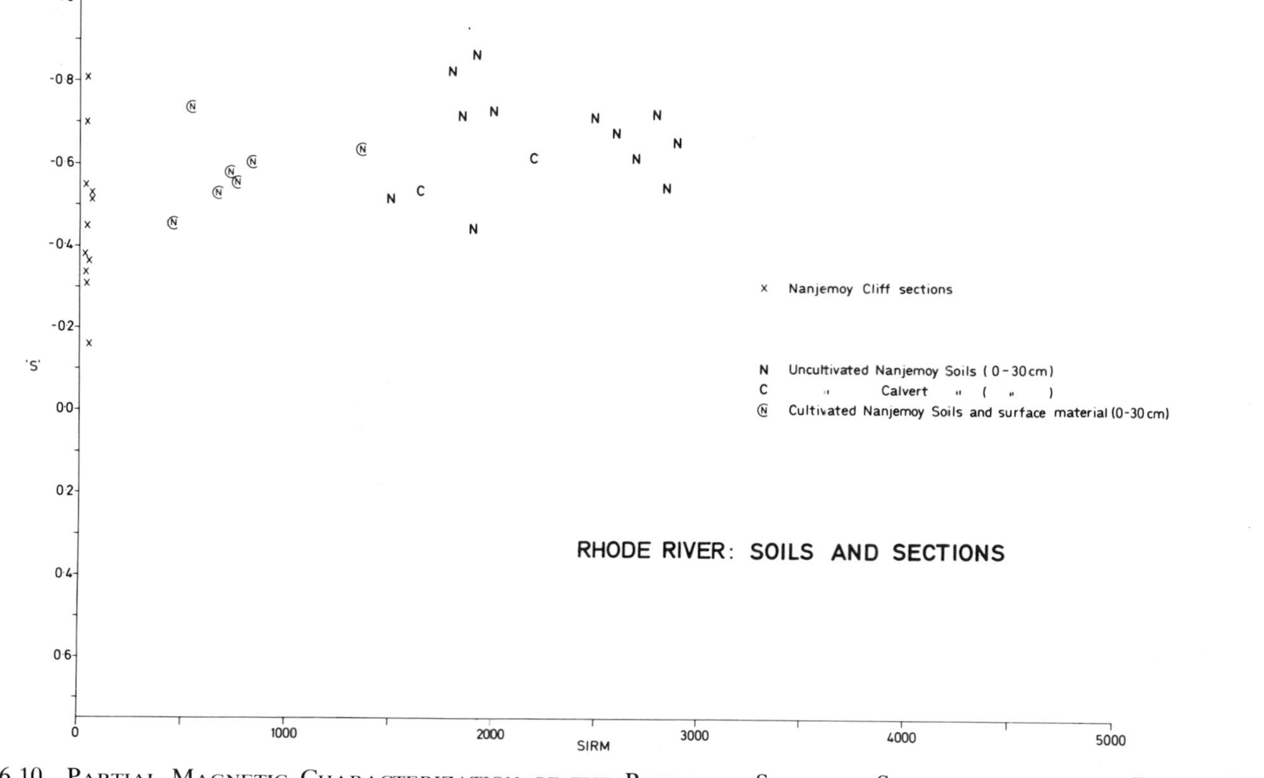

FIGURE 6.10 PARTIAL MAGNETIC CHARACTERIZATION OF THE POTENTIAL SEDIMENT SOURCES WITHIN THE RHODE RIVER CATCHMENT

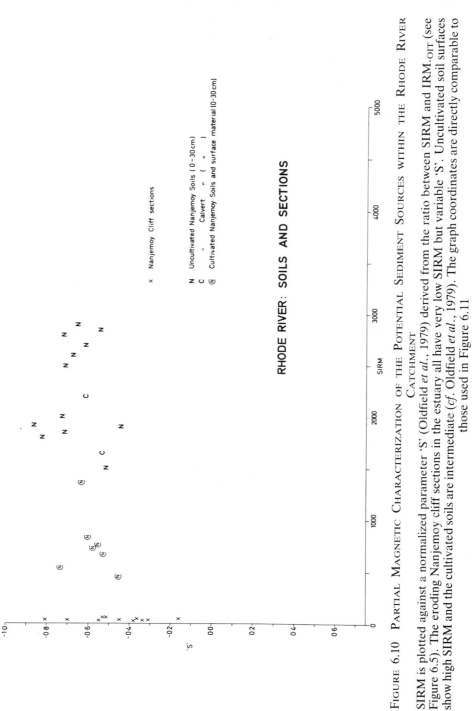

SIRM is plotted against a normalized parameter 'S' (Oldfield *et al.*, 1979) derived from the ratio between SIRM and $IRM_{\cdot OIT}$ (see Figure 6.5). The eroding Nanjemoy cliff sections in the estuary all have very low SIRM but variable 'S'. Uncultivated soil surfaces show high SIRM and the cultivated soils are intermediate (*cf.* Oldfield *et al.*, 1979). The graph coordinates are directly comparable to those used in Figure 6.11

TABLE 6.1 RANGE OF MINERAL MAGNETIC PARAMETERS FOR POTENTIAL SEDIMENT SOURCES FROM THE RHODE RIVER CATCHMENT

Potential sources	$\chi^{10^{-6}}$	$\chi_q\%$	$SIRM^{10^{-6}}$	SIRM/χ	$\frac{IRM_{-1000}}{SIRM}$
Surface soils	8–65	6–15	500–6000	40–90	−0.30 to −0.95
Weathered/illuviated substrates	< 8	0–3	< 700		+0.50 to −0.40
Unweathered substrates	1–11	< 1	< 100	< 15	−0.10 to −0.80

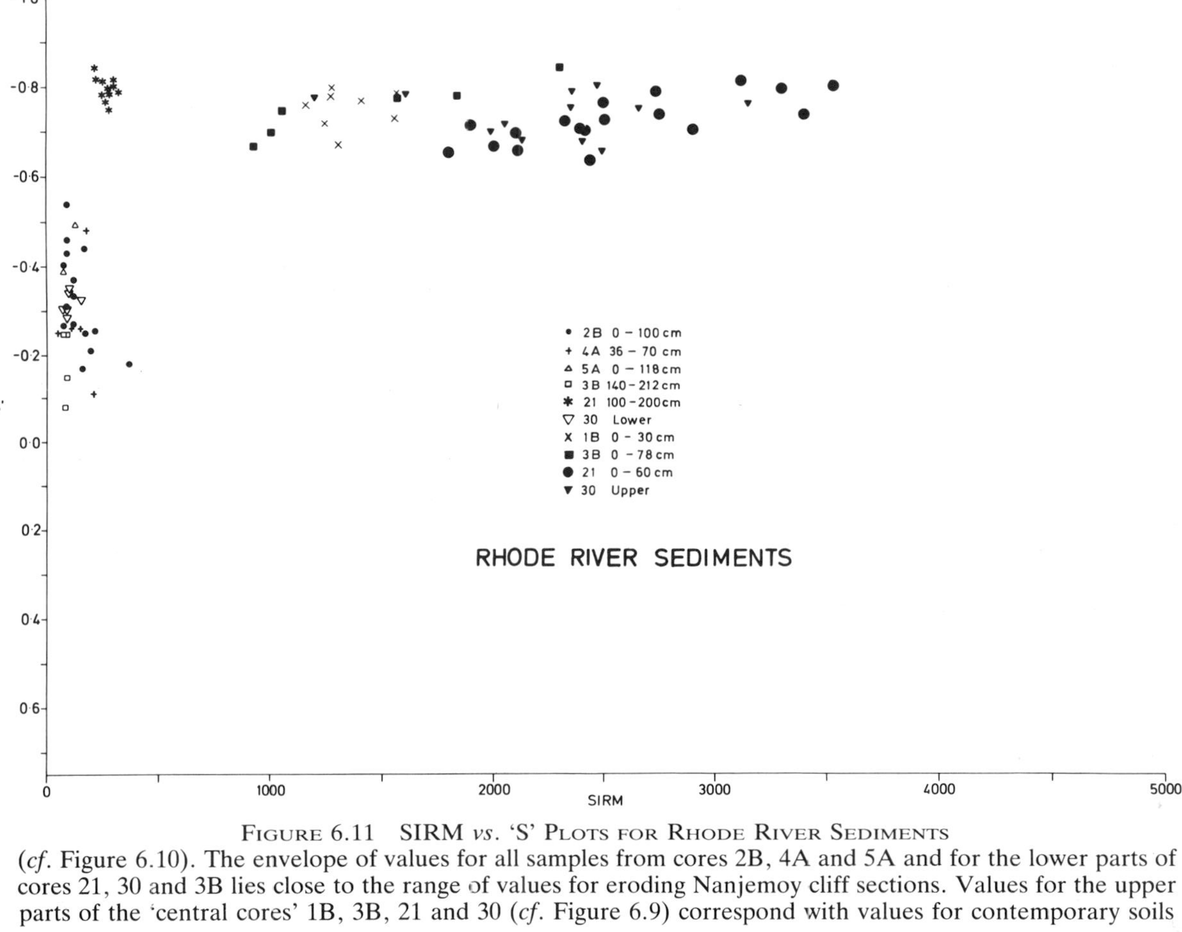

FIGURE 6.11 SIRM *vs*. 'S' PLOTS FOR RHODE RIVER SEDIMENTS
(*cf*. Figure 6.10). The envelope of values for all samples from cores 2B, 4A and 5A and for the lower parts of cores 21, 30 and 3B lies close to the range of values for eroding Nanjemoy cliff sections. Values for the upper parts of the 'central cores' 1B, 3B, 21 and 30 (*cf*. Figure 6.9) correspond with values for contemporary soils

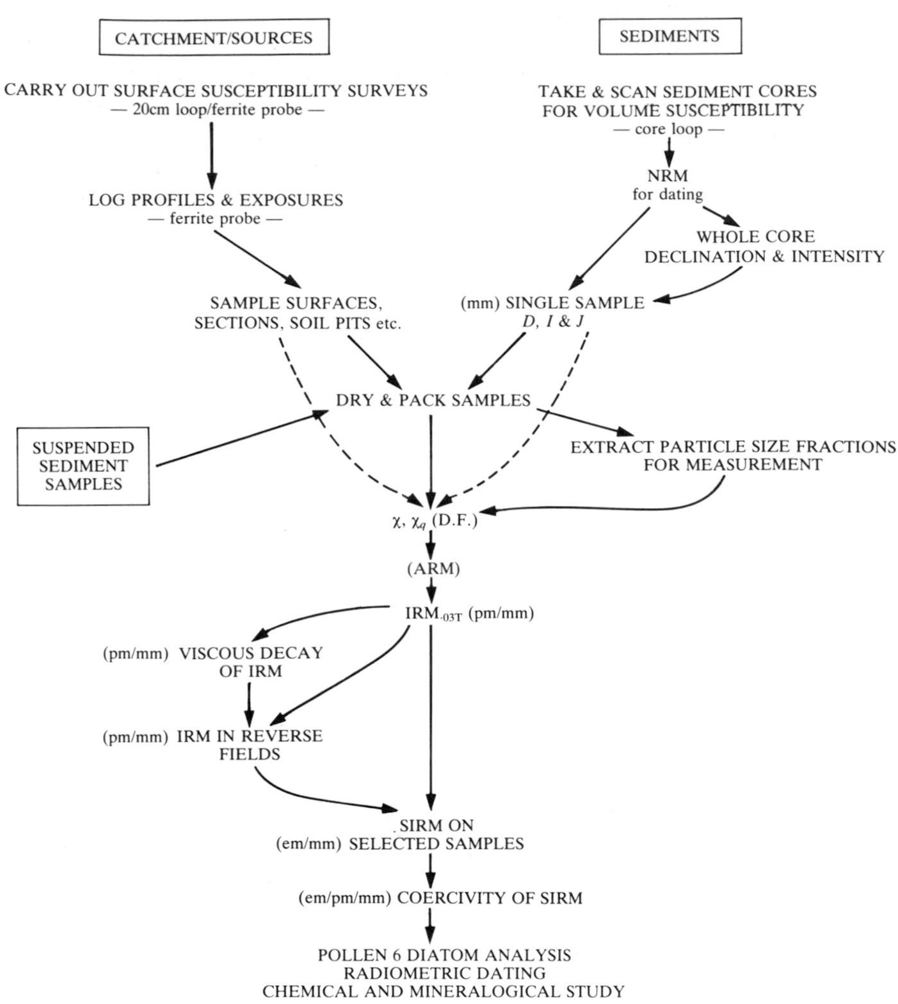

mm Minispin magnetometer
pm Molspin pulse magnetizer
em Electromagnet

20 cm loop	Alternative
core loop	Barlington
ferrite probe	Instruments
D.F. (dual frequency)	Sensors

ARM Anhysteretic Remanent Magnetization

FIGURE 6.12 THE MAGNETIC APPROACH TO CATCHMENT AND SEDIMENT STUDIES — SOME POSSIBLE MEASUREMENT SEQUENCES

geomagnetic secular variation which they contain with that reconstructed for the area by Thompson and Barraclough (in press).

The Rhode River case study points the way to a coherent magnetic methodology leading through from the survey and appraisal stages to detailed site specific measurements. Figure 6.12 shows an attempt to portray this in the form of an annotated flow diagram. It is capable of almost infinite modification in response to the problems and exigencies of particular locations.

PROSPECTS

The foregoing account is far from cxhaustivc and many possibilities remain to be explored systematically for the first time. Particularly interesting for the palaeohydrologist are those we may hope to see emerge from the application of mineral magnetic measurements to long sediment sequences from major lake basins in non-glaciated regions, to near-shore marine sediments in morphogenetically dynamic areas, to cave sediments, to alluvial fills and river terrace sequences, and to loess successions. These are in addition to the range of applications already introduced by case studies of the type used as illustrations in the present review. By now, we can confidently identify this emerging area of methodology as an important new element in the future course of palaeohydrological studies.

REFERENCES

Appleby, P. G., and Oldfield, F. 1978. The calculation of lead-210 dates assuming a constant rate of supply of unsupported ^{210}Pb to the sediment. *Catena,* **5**, 1–8.

Bloemendal, J., 1981. Palaeoenvironmental implications of the magnetic characterisations of sediments from D.S.D.P. site 514, S.E. Argentine Basin. *Initial Reports of the D.S.D.P.*

Bormann, F. H., and Likens, G. E. 1969. The watershed ecosystem concept and studies of nutrient cycles. In Van Dyne, G. (Ed.), *The Ecosystem Concept in Natural Resource Management*, 49–76. Academic Press.

Creer, K. M. 1974. Geomagnetic variations for the interval 7000 to 25000 yr BP as recorded in a core of sediment from Station 1474 of the Black Sea cruise of 'Atlantis II'. *Earth and Planetary Science Letters,* **23**, 34–42.

Creer, K. M., Thompson, R., Molyneux, L., and Mackereth, F. J. H. 1972. Geomagnetic secular variation recorded in the stable magnetic remanence of recent sediments. *Earth and Planetary Science Letters,* **14**, 115–127.

Creer, K. M., and Tucholka, P. 1982. Secular variation as recorded in Lake Sediments: a discussion of North American and European results. *Philosophical Transactions of the Royal Society, London,* Series A 590, in press.

Dearing, J. A. 1982. Core correlation and total sediment influx. *IGCP*. Volume, Section 10 16.

Foster, J. H. and Opdyke, N. D. 1970. Upper Miocene to Recent magnetic stratigraphy in Deep-sea sediments. *Journal of Geographical Research,* **75**, 4465–4473.

Harvey, A. M., Oldfield, F., Baron, A. F., and Pearson, G. W. 1981. *Earth Surface Processes.*

Kent, D. V. 1982. Apparent correlation of palaeomagnetic intensity and climatic records in deep-sea sediments. *Nature,* **299**, 538–539.

Le Borgne, E., 1955, Susceptibilite magnetique anomale due sol superficiel. *Annales de Geophysique,* **11**, 399–419.

Longworth, G., Becker, L. W. Thompson, R., Oldfield, F., Dearing, J. A., and Rummery, T. A. 1979. Mössbauer effect and magnetic studies of secondary iron oxides in soils. *Journal of Soil Science,* **30**. 93–110.

Mackereth, F. J. H. 1958. A portable core sampler for lake deposits. *Limnology and Oceanography,* **3**, 181–191.

Mackereth, F. J. H. 1969. A short core sampler for sub-aqueous deposits. *Limnology and Oceanography,* **14**, 145–151.

Mackereth, F. J. H. 1971. On the variation in direction of the horizontal component of remanent magnetization in lake sediment. *Earth and Planetary Science Letters,* **12**, 332–338.

Mullins, C. E., and Tite, M. S. 1973. Magnetic viscosity, quadrature susceptibility and frequency dependence of susceptibility in single-domain assemblies of magnetite and maghemite. *Journal of Geophysical Research,* **78**, 804–809.

Mullins, C. E. 1977. Magnetic susceptibility of the soil and its significance in soil science: a review. *Journal of Soil Science,* **30**, 93–110.

Noel, M. 1982. Palaeomagnetic studies of core sediments from Britain and Sarawak. U. K. Geophysical Assembly Abstracts. *Geophysical Journal of the Royal Astronomical Society.*

Oldfield, F. 1977. Lakes and their drainage basins as units of sediment-based ecological study. *Progress in Physical Geography,* **1**, 460–504.

Oldfield, F., Thompson, R., Dearing, J. A., and Garrett-Jones, S. E. 1978. Some magnetic properties of lake sediments and their possible links with erosion rates. *Polish Archives for Hydrobiology,* **25**, 321–334.

Oldfield, F., Rummery, T. A., Thompson, R., and Walling, D. E. 1979. Identification of suspended sediment sources by means of magnetic measurements. *Water Resources Research,* **15**, 211–218.

Oldfield, F., Appleby, P. G., and Thompson, R. 1980. Palaeoecological studies of three lakes in the Highlands of Papua Guinea. *Journal of Ecology,* **68**, 457–477.

Oldfield, F., and Bartington, G. W. 1982. Sediment source identification using high and low frequency susceptibility sensors. U.K. Geophysical Assembly Abstracts. *Geophysical Journal of the Royal Astronomical Society.*

Oldfield, F., and Appleby, P. G. In Press. The effects of catchment disturbance and sediment redistribution on ^{210}Pb dating. *Chemical Geology.*

Oldfield, F., Barnosky, C., Leopold, E. B., and Smith, J. P. In Press. Mineral magnetic studies of lake sediments — a brief review. *Proceedings of the 3rd International Symposium on Palaeolimnology, Developments in Hydrobiology*, Junk, The Hague.

Oldfield, F., Bloemendal, J., and Hunt A., Unpublished mineral magnetic data from E. Mediterranean sediment cores.

Poutiers, J. 1975. Sur les proprietes magnetiques de certain sediments continentals et marins; application. These de Doctorat. Univ. de Bordeaux.

Robinson, S. G. 1982. Two applications of mineral–magnetic techniques to deep-sea sediment studies. U.K. Geophysical Assembly Abstracts. *Geophysical Journal of the Royal Astronomical Society.*

Scoullos, M., Oldfield, F., and Thompson, R. 1979. Magnetic monitoring of marine particulate pollution in the Elefsis Gulf, Greece. *Marine Pollution Bulletin,* **10**, 288–291.

Thompson, R. 1980. Palaeomagnetic correlation and dating. In Berglund., B. (Ed.) *Lake Sediments*, IGCP Volume, 39–59.
Thompson, R. and Turner, G. M. 1979. British geomagnetic master curve 10,000–0yr B.P. for dating European sediments. *Geophysical Research Letters,* **6**, 249–252.
Thompson, R., and Barraclough, D. R. 1982. *Journal of Geomagnetism and Geoelectricity, Kyoto* (in press)
Walling, D. E., Peart, M., Oldfield, F., and Thompson, R. 1979. Identifying suspended sediment sources by magnetic measurements on filter paper residue. *Nature,* **281**, 110–113.

Background to Palaeohydrology
Edited by K. J. Gregory

7

Soils and hydrological change

MARK HAYWARD and IAN FENWICK

University of Reading, Department of Geography

Much important palaeoenvironmental information is extracted in various ways from floodplains and terraces. Surface and buried soils of such bodies of sediment have been used previously to infer environmental conditions penecontemporaneous with, and subsequent to, deposition (e.g. in Britain by Chartres, 1980, and Rose and Allen, 1977). Strictly palaeohydrological deductions are more difficult to make than broad climatic inferences, but as the alluvial record is a terrestrial repository of information on environmental change, we should examine that record thoroughly. Pedological aspects of floodplains offer some clues to palaeohydrology, and those aspects of alluvial soils of greatest interest would appear to be soil layering, texture, profile features, and mircomorphology. These are discussed in turn below but it is first necessary to examine the relationship between soil formation and hydrology.

MODELS OF HYDROLOGY AND SOIL FORMATION

Most of the classic models of soil formation (e.g. Jenny, 1941) have been cast in a total environmental context. Thus soil profile properties are seen to be dependent on climatic and vegetational inputs as well as on parent material and, in an ill-defined way, on 'relief'. However, precisely what aspects of the latter are important has never been adequately established. Undoubtedly, a major consideration is the influence which relief has on slope hydrology. In most models the soil is seen as a static entity in relation to the flow of water and sediment through the drainage basin. But when considering the relationships between hydrology and soil formation, one is obliged to consider water movements through the drainage basin, both as surface and subsurface flow. Furthermore, these movements will be closely related to transfers of soil and sediment which modify the landscape, and to the hydrological regime and hence they can initiate renewed, and often very different, soil formation.

To accommodate these realities we may turn to the nine-unit land–surface model (Figure 7.1), initially advanced by Dalrymple, Blong, and Conacher (1968) and refined by Conacher and Dalrymple (1977). Their assumption that the critical considerations are the lateral movements of surface materials and the lateral and vertical translocation of soil constituents offers a real chance of understanding the relation between soils and hydrology.

Lateral movements

Thus, while in unit 1 vertical pedogenic processes such as eluviation are dominant, they argue that in unit 2 lateral throughflow is paramount and carries solutes and suspended solids with it (Figure 7.1). However, it is only in the steeper units 3, 4, and 5 that sediment is also translated across the surface. This may be either as creep (unit 3), as falls and slides (unit 4), or, additionally in unit 5, as surface wash.

Unfortunately, even now, there is only a very modest amount of observational data to support some of these assertions. For instance, it is only recently that we have gained a fuller understanding of the patterns of water movement across slopes. It has become increasingly evident that overland flow plays little part in the disposal of water from catchments in the temperate zone. But through the work, notably, of Whipkey (1965), Weyman (1970), and Knapp (1974) it is now clear that lateral water transfer is concentrated in horizons of high permeability which overlie a less permeable interface. In particular, the highly porous O/A horizons often act as the principal transmitters in podzolic soils, especially on the upper slopes. However, in brown earths and generally on lower slopes, vertical percolation and transmission from upslope concentrate water in the coarsely weathered C horizon whence it provides most of the baseflow by lateral seepage (Weyman, 1970).

It has also become increasingly clear that water transmission across slope units is spatially concentrated. Bunting (1961) recognized that moisture movement was focused on seepage lines or percolines, related to concavities in the land surface. This point was established theoretically by Kirkby and Chorley (1967). However, still further concentration of transmitted moisture into subsurface pipes has been identified by many workers in recent years. For instance, Gilman and Newson (1980) have mapped an extensive pipe network in Nant Gerig, central Wales. Such networks appear to be common on units 3 and 5, especially where the soils are podzolic. By and large, the channels are up to 5 cm diameter and are found above impeding layers such as Bf or Bt horizons and immediately above any fine textured Ea horizons. There seems to be little doubt that a significant part of runoff is channelled through these systems, as shown by Gilman and Newson (1980).

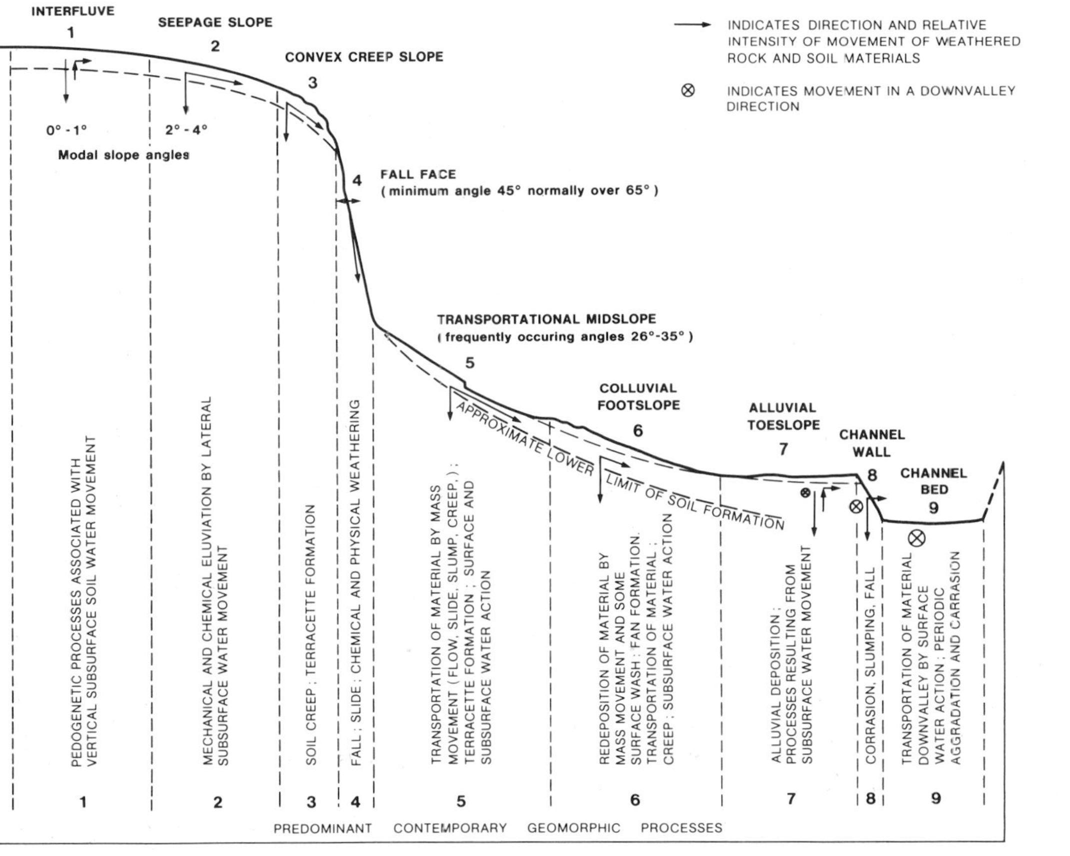

FIGURE 7.1 THE NINE-UNIT LANDFORM MODEL AS PROPOSED BY DALRYMPLE *et al* (1968)
Reproduced by permission of Gebrüder Borntraeger

That pipes are able to transmit both salts in solution and clays in suspension can easily be demonstrated by direct field sampling (see Conacher and Dalrymple, 1977, p. 21). However, what is not so clear is how such movements and, perhaps of more general relevance, lateral translocation via matrix flow, compare in effectiveness with the vertical redistribution of materials by infiltrating waters. Conacher (1975) has affirmed that salinization in parts of Western Australia can be attributed to lateral movements through the soil. This apart, little data are to hand which would establish the relative roles of vertical and lateral translocation.

Attempts have been made to approach the problem from a different standpoint by examining the variation in various soil attributes with topographic variables. Furley (1971) and Whitfield and Furley (1971) were able to show marked changes in soil properties with slope angle and distance from the interfluve on slopes on the Chalk and on Palaeozoic sediments in southern Scotland. On upper convex slopes (units 2 and 3) there appears to be a removal and transport of material across the slope but a lack of vertical movement. By contrast, the pattern in the depositional zone (units 5 and 6) is unclear, save that pH generally rises in response to the accumulation of bases leached from upslope. Adopting a 'balance-sheet' approach Huggett (1976) has argued that lateral movements of clay, silt, and the transition elements are apparent, especially in concavities and in the upper horizons i.e. those zones where throughflow would be expected to be most prevalent. Fanning, Hall, and Foss (1973) produced evidence from Maryland which pointed to the lateral movement of iron from the wetter soils of units 6 and 7 of a loamy catena.

PALAEOSOLS

Evidence of hydrological change in soils during the past 15 ka is often to be found by the examination of palaeosols, whether buried beneath more recent sediments or surviving as relict features in modern soils or sedimentary accumulations. The most informative sites are, therefore, likely to be in zones of sediment deposition such as (1) areas of periodic marine inundation, (2) colluvial footslopes (units 5 and 6), and (3) (alluvial) floodplains (unit 7).

Areas of marine inundation

The fen margins around the Wash in eastern England have revealed a palaeosolic catena buried beneath fen peats and marine clays. Freely draining members of the catena which are developed in Early to Middle Devensian fen margin sands and gravels, have been classed as iron or humus–iron podzols by Valentine and Dalrymple (1975). In that the 'sealing' peat has been dated to *ca*. 4100 a B.P. the soil assemblage suggests that this part of the landscape was characterized by very freely-draining conditions for part, if not most, of the Late Devensian and Early Flandrian stages.

Colluvial soils

Undoubtedly, under conditions of reduced soil strength associated with periods of high moisture content or reduced binding by extensive root systems, substantial downslope movements of soil have taken place. Colluvial soils which accumulate on footslopes have the potential to record aspects of slope hydrology (and hence palaeohydrology), having been in many cases the products of water movement across and near to the surface of the soil. The evidence for hydrological change comes from field observations, included subfossils, and the microscopic examination of thin sections. Recent advances in the study of slope soils owe much to the work of H. J. Mücher and co-workers (e.g. Kwaad and Mücher, 1979; Mücher, 1974; Mücher *et al.*, 1972, 1981; Mücher and Vreeken, 1981; Vreeken and Mücher, 1981).

It is not possible in the space available to review many studies of colluvial soils, but it should be stressed that much redeposited material will have a pedogenic origin, having been derived from the surface horizons of soil upslope. Nevertheless, it is possible for continued erosion higher on a slope to produce an inverted soil profile, with former subsurface, less altered material becoming superimposed on former topsoil (Valentine and Dalrymple, 1976).

The stratigraphy, morphology, and subfossils of colluvial deposits frequently give evidence of changes in local slope movements which are related to hydrological conditions. At their most simple these changes may be from stability to instability, and back to stability, as seen in dry valley sections in southern England. The periods of instability are marked by much more intensive slope processes, including sheet- and rillwash, and appear to have occurred during periods when the soil was unprotected, when the subfossil fauna and flora indicate a cold climate and a relatively sparse ground cover, or when human activities such as tree clearance or ploughing occurred (Evans, 1966; Evans and Valentine, 1974; Kerney, 1963; Vreeken and Mücher, 1981).

Some of these deposits have been shown to be of late glacial age but, in the case of Flandrian accumulations, if one takes a mean rate of 1–10 mm of soil creep/year as being not unreasonable in the temperate zone at the present time then it is clearly most unlikely that they could have been transported under present hydrological conditions. Movement may have been initiated by woodland clearance but, to continue, it seems that soil strength must have been much lower than at present as a result of higher moisture content. Either winter temperatures were lower thus producing a saturated layer during the melt period, or precipitation levels were substantially higher. In that other palaeoclimatic evidence does not indicate markedly lower temperatures during the Flandrian it seems that precipitation and soil moisture levels must have been much greater especially during the Iron Age.

Vreeken and Mücher (1981) demonstrate the importance of a thorough stratigraphic and sedimentological analysis of loess and its colluvial derivatives, and in the same deposit Mücher and Vreeken (1981) used micromorphology to interpret the origin of some of the material. The authors examined the microstructures in a complex of silt loams and compared them with features simulated in the laboratory. Their detailed study apparently has enabled distinctions to be made between the hillslope processes that redeposited the loess, e.g. rainwash with '*smooth turbulent*' flow, rainsplash, and very thin laminar flow. The ability to draw these distinctions has important implications for the reconstruction of slope palaeohydrology. Similar studies of lateral surficial movements may be combined with assessments of the relative role of vertical downward movement (translocation) in order to reconstruct the hydrology of colluvial soils.

Alluvial soils

Although colluvial materials contain a record of changed hydrological conditions, the most direct responses to variations in river regimes are to be found in soils of the alluvial tract. Direct sedimentary response may be witnessed in the soil layering and in textural discontinuities while the pedological imprint is contained in a variety of profile features and, especially, in the soil micromorphology.

Soil layering

Soil layering in alluvium cannot be studied without acknowledging that the parent material is a stratified deposit. Furthermore, during the accumulation of that deposit pedogenic processes are likely to have modified it to some degree, in addition to weathering of the constituents prior to transport. In many floodplains few depositional events will be big enough, or pedogenesis strong enough, to lead to a pronounced division into unaltered sediment and buried sola. Usually soil formation and sedimentation will overlap. At any one time pedogenesis may be severely restricted by the addition of new material, and conversely, deposits will often be incorporated rapidly by active soil fauna. Nevertheless, floodplain sites protected from channel erosion for a sufficiently long time, and undergoing slight vertical accretion, will develop distinct horizons. In sections through alluvium, then, one will encounter both sedimentary strata and pedogenic horizons, with many layers possessing properties of both (*cf*. Bridges, 1973; Bunting, 1967, p. 118). This is an important premise underlying any study of alluvial soils, and should enable one to move on quickly from the urge to employ any crude, quasiquantitative index such as used by Bilzi and Ciokosz (1977) and Meixner and Singer (1980).

Because the lower parts of alluvial soils will usually be permanently moist, and because the deposits obviously lie close to the river channel, changes in hydrological regime may be registered last of all in these soils. On the other hand, related changes in fluviatile sedimentation could be recorded quickly. In addition, the drainage properties of alluvial soils vary according to topography and position on the floodplain. Therefore, the soil profiles should be viewed in relation to sedimentary environments, and catenary sequences established where appropriate.

Soil texture

The texture of alluvial soils is a product of fluvial sedimentation, which is related to basin hydrology. However, the inferential leap from texture to hydrology is a wide one, except possibly in the case of ephemeral streams (*cf.* Vita-Finzi, 1971). In humid temperate environments the jump is not to be recommended. The texture of the alluvium is only partly due to the sedimentary processes themselves: sediment sources dictate the availability of particles of different sizes. Thus changes in the catchment boundary and in the provision of sediment may be recorded as texture changes. Furthermore, since texture is a sedimentological as well as a pedological property, and sedimentology has moved a long way from simplistic grain size interpretations of sedimentary environments, it would be foolhardy to isolate texture as a useful hydrological indicator. If grain size is to be used as a source of palaeoenvironmental information, it must be used in association with other data in a stratigraphic context.

Thus, for example, in the case of the River Kennet in southern England the palaeobiological evidence has confirmed Holyoak (1980) in the belief that the change in the alluvial materials from gravels to clays was associated with a substantial decrease in discharge (see Chapter 18). In other river systems this change which took place with the climatic amelioration at *ca.* 10 000 a B.P. resulted in rather different responses. One such example, the middle Severn, is discussed below.

Profile features

Evidence of alluvial soil formation is often discernible in cutbanks in the form of peds (Figure 7.2). A concentric sequence of scroll bars, or segments of floodplain of differing ages, may display varying degrees of ped development. Peds have several hydrological implications. Firstly, crumb-like and fine subangular blocky peds are common in the root zone, and will have a restricted development where the water table is permanently high. Secondly, subsoil peds often attributed to wetting and drying stresses (White, 1966), will also be

FIGURE 7.2 FINE SUBANGULAR BLOCKY PEDS AND SOME LARGER AGGREGATES IN THE SURFACE HORIZON NEAR WROXETER (site 16, Figure 7.7)
This shows some of the best developed soil structure among the Severn floodplain soils. The trowel is about 250 mm long

restricted in alluvial soils with a high water table. Thirdly, the alluvial soils of floodplains which become terraces may retain the catenary ped features of the former active surface, but burial may destroy the evidence of soil structure. Too little has been documented about changes on burial to make any generalizations. Pedality is strongly influenced by texture, and although the size of subsoil peds may give an indication of the moisture regime (Graaff, 1978), the influence of grain size should be borne in mind when comparing different soils.

Several chemical transformations are wrought in floodplain conditions on account of the high water-table. Foremost in these gley soils is the reduction of iron and manganese compounds when enveloped in water with minimal dissolved oxygen. Thus ferric oxides are reduced, dissolved and tend to be leached from the profiles or to move toward areas of higher oxidation potential. The dissolution of iron is encouraged by the formation of ferrous–organic complexes by chelating agents from fresh organic debris. Phosphates too tend to be mobilized.

Bloomfield (1973) has suggested that the grey colours in gleys are due to the presence of organic matter closely associated with small clay particles. He has also argued that blue colouration is attributable to vivianite (ferrous phosphate) while spots of black are most likely formed by ferrous sulphide. Both the latter compounds are readily oxidized so that the soils are likely to turn brown as soon as the water-table falls or the soils are exposed. However, the grey colour will persist even when the conditions which produced the gleying have long since changed.

But what fate awaits these dissolved materials? Schuylenborgh (1973), arguing from thermodynamic assumptions, and Brummer (1973) have indicated that, in groundwater gleys, iron will move toward the major cracks and pores around which iron oxide separations, even concretions, will form. However, Russell (1973) suggests that as the water-table rises at the beginning of the wet season, reducing conditions will be initiated along old root channels. Some of the ferrous iron produced on the walls of these channels will diffuse into the peds where it may be oxidized in the following summer (Figure 7.3).

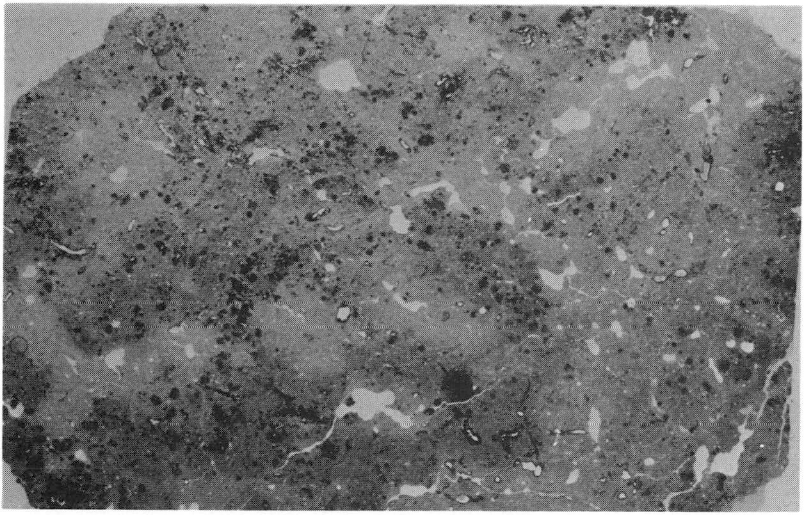

Figure 7.3 Ferruginous Mottles and Nodules at a Depth of 1.3 m in Alluvium Near Fitz (site 3, Figure 7.7)

Frame width about 35 mm. Note that most of the larger voids are associated with pale aureoles depleted of iron. Concentrations of ferric oxides and hydroxides occur in the matrix

Clearly, the direction of diffusion will depend on the direction of the moisture potential gradient. Thus iron concretions are likely to occur in zones of high pH or Eh in those areas to which iron has moved over the winter months. Such concretions of iron — either as segregations or concretions — are likely to survive changed hydrological circumstances, such as the lowering of the water-table, since they can only be remobilized by chelates.

In general, manganese compounds, which are much more easily reduced than ferric compounds, will concentrate in areas higher up the profile than their iron counterparts. Nevertheless, they relate to the position of the water-table so that fossil concentrations of manganese must be indicative of a former water-table position.

Recent research has supported detailed field observations by laboratory experiments (Veneman *et al.*, 1975; Vepraskas and Bouma, 1976). Observed mottle patterns could be matched with moisture regime and redox fluctuations. Short periods of saturation lead to the formation of manganiferous mottles, but long periods of anaerobic conditions lead to the removal of manganese from the profile. Between these two extremes manganese oxides are associated with iron oxides and hydroxides in many mottles. As with pedality, terracing of a floodplain may preserve the catenary relationships of hydromorphic features (although formerly reduced sediment could be oxidized), but burial might lead to the converse, with reduction of iron and manganese oxides.

However, it may be difficult in the fossil situation to differentiate between concentrations due to former groundwater and to drainage impedance. This problem may be resolvable if the experimental results of Blümel (1962) are of general application. He was able to show that iron oxide precipitations in groundwater gleys were generally as tubules or flakes whereas concretions resulted from pseudogleying i.e. gleying due to a slowly permeable horizon.

Micromorphology

Polished thin sections of soil viewed under a petrological microscope enable interpretations made in the field to be clarified or amplified, and frequently yield important new information. Soil mottle relationships to pores and matrix are more easily seen microscopically than in the field. In some circumstances it is possible to determine whether the features seen are still forming, or are relict, or whether there has been a change in the nature and intensity of hydromorphic processes. A detailed study examining these problems would tie in more readily than any of the other pedological methods to a study of basin hydrological changes. Fortunately soil mottles and concretions are more easily interpreted than many features seen in thin section. However, if a sequence of changes is being inferred, the micromorphological 'chronology' must be subservient to stratigraphy. Furthermore, the influence of variable materials must be borne in mind.

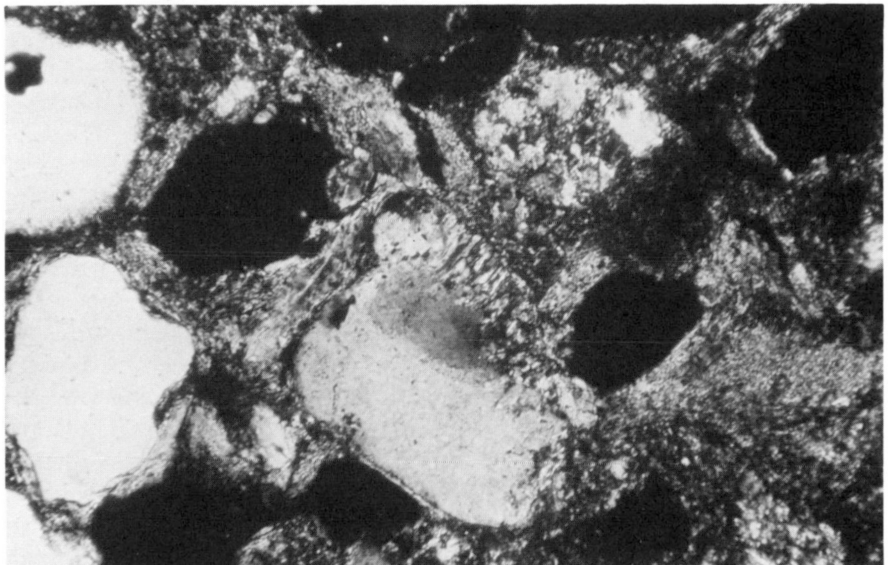

FIGURE 7.4 POSSIBLE TRANSLOCATED CLAY, UNDERDALE, EAST OF SHREWSBURY (site 9, Figure 7.7)
Well orientated clay interpretable as illuvial can be seen in some thin sections from floodplain soils. This relatively sandy material shows one form: well oriented clay is partly filling packing voids between quartz grains. Frame width 0.8 mm

Soil drainage may lead to translocatory phenomena in alluvial soils as much as in soils developing elsewhere (Figure 7.4). It is impossible to be specific about the general significance of coating materials in soils of floodplains. For example, it is possible that flood drawdown can lead to the translocation of particles of various sizes, while at the opposite end of the timescale, well developed alluvial soils may display eluvial and illuvial horizons while still within an active floodplain. Brammer (1970) has recorded coatings produced by flood events, and Brakenridge (1981) in a study of some floodplain trenches suggests that translocation of material may occur at any time after deposition. Micromorphological evidence may suggest at what stage any observed void or grain coatings formed, if consistent relationships to other features can be observed. The method used to establish the sequence of events is a sort of 'plane-stratigraphy' much as used to classify features of the lunar surface. Thus, a ferruginous mottle extending over a clay coat obviously 'post-dates' the coating, whereas a clay coat in an old root channel through an unstratified matrix must have formed well after deposition of the sediment.

This section has attempted to show how alluvial soils form part of the soil sediment system of the drainage basin and how they may be used as part of a palaeohydrological study. Since so many palaeohydrological data come from alluvial sediments it is natural that the soils formed in these deposits should receive some attention if they still bear the imprint of the floodplain soil-forming environment. In order to illustrate how alluvial soil data might be used, some results of studies of the Kennet and Severn floodplains are presented in the following section.

CASE STUDIES: KENNET AND SEVERN

The River Kennet

Few drainage basins have remained sufficiently stable for the land surface to take on the simple form of Figure 7.1. During the past 15 000 years incision and aggradation have both been widespread in the British Isles. Thus low terrace features have been formed only to be followed by substantial aggradation. Associated with these changes and reversals have been similar ones in stream discharge and in the calibre of sediment transported. In the case of the River Kennet (Berkshire, UK) Holyoak (1980) points to gravel deposition at *ca*. 4 m above present river level at *ca*. 27.4 ka BP. Subsequent to this the stream incised to approximately 3.5 m below present river level where it cut a channel in the underlying Tertiary sediments. By *ca* 10 000 a BP it had almost completed the gravel aggradation but thereafter the river was transformed into a solute-load stream, albeit with a little suspended sediment. Fen peat accumulations were soon initiated probably by *ca*. 9700 a B.P. and these developed for at least 1000 years. The subsequent record has been largely destroyed as a result of peat digging.

Less certainly, a 2 m gravel terrace, overlain by *ca*.1 m of silt loam was created at some time during the Late Devensian cold stage but subsequent to 27.4 ka B.P. With the abandonment of its 2 m level the R. Kennet has left a highly complex pattern of soils (Figure 7.5) on this surface of deep, silt-rich profiles separated by much shallower silt units over gravel. Examination of these areas reveals that the deeper silt soils are developed in overbank material infilling the former braided channels while those with a gravelly subsoil represent former bars (Figure 7.6). Clearly we have preserved here in the gravel surface a record of a traction load stream of variable discharge and Cheetham (1980) estimates that mean annual bankfull discharge was between 4 and 10 times that of the present. However, it may be that the palaeochannels cannot be interpreted so simply (see Bryant, this volume, Chapter 18).

Micromorphological analysis by Chartres (1975) of the silt-rich profiles on the 2 m terrace revealed that the soil is mostly undisturbed — presumably an indication that pedogenesis has been confined to the Flandrian. However, there are also signs that hydrological conditions have changed during the development of these soils. For instance, Chartres identified both iron and manganese nodules with diffuse margins in the uppermost horizon; these

FIGURE 7.5 PALAEOCHANNELS ON THE 2 m TERRACE OF THE R. KENNET NEAR THEALE, BERKSHIRE
The dark tones represent the channels and the paler areas gravel bars

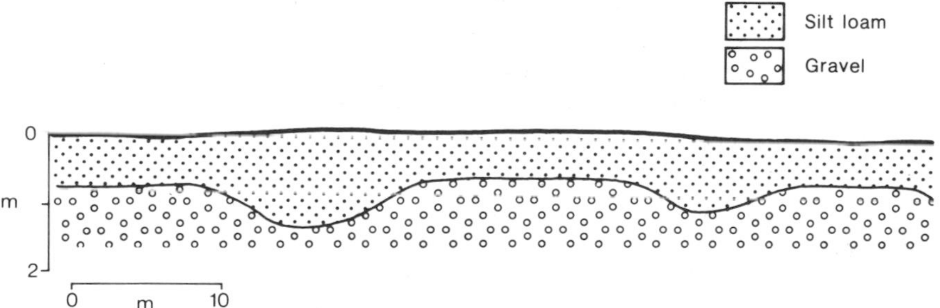

FIGURE 7.6 SECTION ACROSS A CHANNEL/BAR SYSTEM AS SHOWN IN FIGURE 7.5

suggest active growth. In some cases the nodules consisted of a core of manganese oxides surrounded by iron-rich material. By analogy with the results of Veneman *et al.* (1976) this would appear to indicate that the periods during which the horizon has remained anaerobic have tended to become more prolonged. It would be reasonable to suppose that this is due to the progressive decrease in the permeability of the illuvial Bt horizon as a result of the build-up of clay coatings in the major voids.

The River Severn floodplain between the Vyrnwy confluence and Buildwas

The landforms of the Severn floodplain in the Shropshire lowlands have been

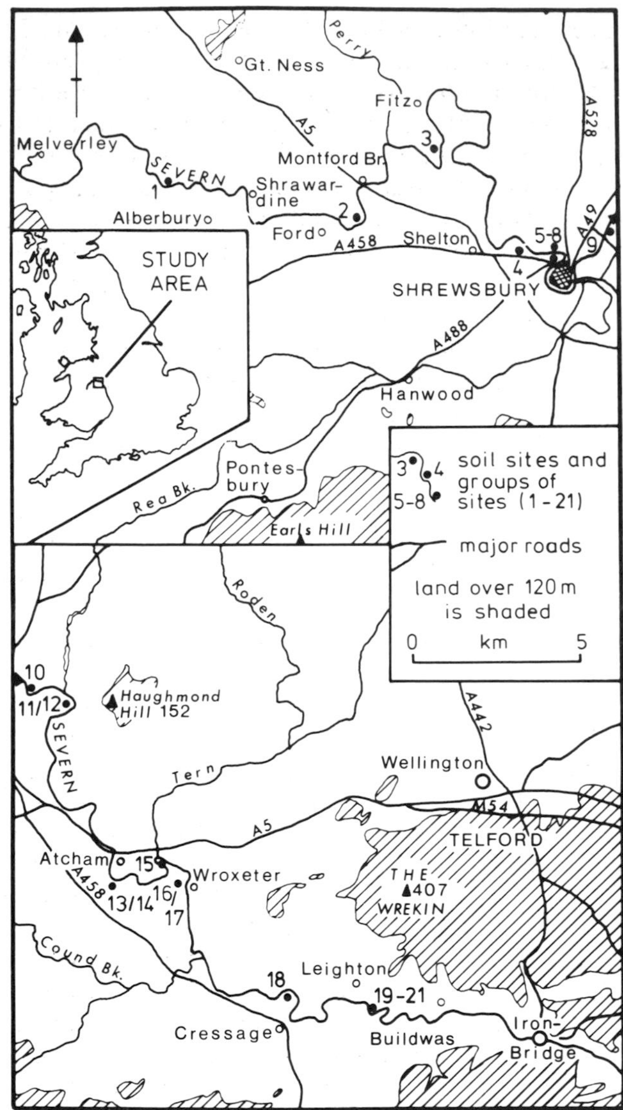

Figure 7.7 Location Map of the Area of the Severn Floodplain Referred to in the Text

mapped at a scale of 1:10 000 and analytical data have been obtained for twenty-one cutbank sections (Hayward, 1982) and the locations are shown on Figure 7.7. The Flandrian fill was found to have been laid down mostly as point bar-like deposits along a channel of low sinuosity. These floodplain ridges contain a good deal more silt and clay than the rare point bars forming at present. For much of its length in the study area the channel of the Severn, with a low gradient of only *ca.* 0.3 m km^{-1}, has migrated only slowly during the deposition of the present floodplain. Archaeological evidence from preserved fishweir sites (Pannett, 1981) and ridge-and-furrow ploughmarks indicates that channel shift in historical times has been very slight. Therefore, some of the floodplain must be quite old, although dating evidence is very poor.

A consideration of the stratigraphy and geomorphological setting of the sites enables a simple threefold classification of the sections and associated soils to be made. This grouping has been portrayed in Figure 7.8, which also schematizes some of the properties of the soils.

The classification has an important textural element: the supposed oldest sites expose the finest alluvium, whereas the youngest are underlain by the coarsest alluvium; sites of intermediate age lie between the extremes of silty clay and sand. A, B, and C of Figure 7.8 may be referred to as the 'oldest', 'intermediate', and 'youngest' types of alluvium. Recent channel and flood sediments are younger still. Typical sequences of horizons in the top 2 m are shown in Figure 7.8. When several exposures are recorded some boundaries will be more important than others, namely those between beds, members, and formations. Likewise, the major texture change shown in Figure 7.8, and recorded in the field at at least two widely separated sites, is thought to be an important break signifying more than channel shift or simple variability within a sedimentary environment. The silty clay beneath the boundary extends to the probable basal gravels of the floodplain sequence, about 4 m below the surface.

This change from silty clay to loams is thought to indicate a change in sediment load from one dominated by derived 'geological' materials provided by river cliffs and slopes in glacigenic sediment, to one in which 'pedological' materials, namely topsoil and other pre-weathered material, were important. In profiles similar to B of Figure 7.8 the pedological materials appear to be most important. This inference is supported by micromorphological observations and soil colour which suggest that clay–iron colloids are important to greater depths in the section than elsewhere. Such materials are shown as 'Brown Soil matrix' in the micromorphological sketches of Figure 7.8, which are shown for selected depths in the three soil types. Below the surface horizon at type-A and -B sites segregation of the iron into mottles is sometimes accompanied by patches, often around voids, of clay with some preferred orientation.

Alluvium corresponding to C of Figure 7.8 is found under many floodplain ridges adjoining and usually accordant with the modern channel. Most show little evidence of soil development. Soil properties reflect the alternation of sandy and relatively muddy sediments laid down on a point bar-like ridge.

The pattern of sediment changes may have resulted as follows. A possible early Flandrian phase of gravel transport in a low gradient river was followed by a change to a suspended load channel in which the original bar surfaces were covered. This change may have been brought about by the restriction of detritus

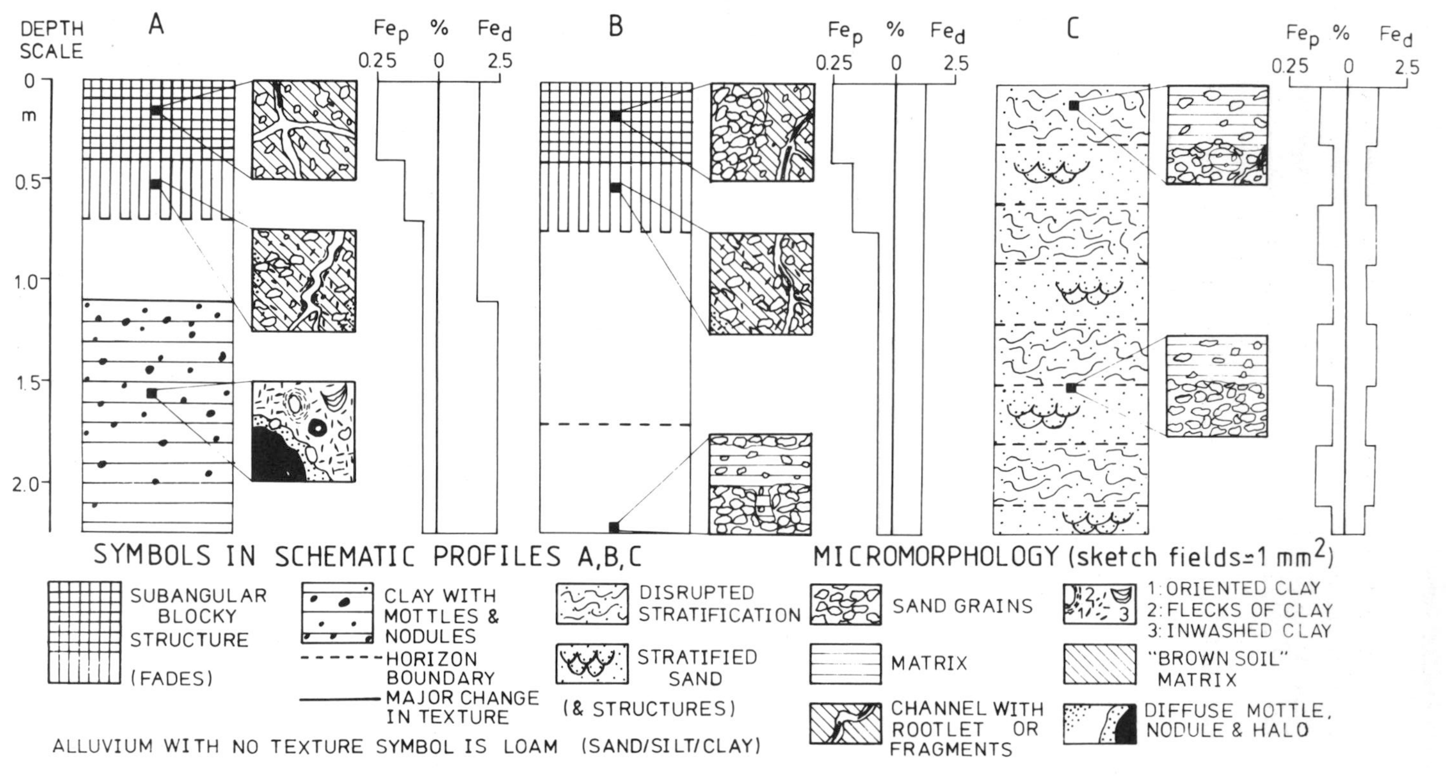

FIGURE 7.8 SUMMARY DIAGRAM OF SOME PROPERTIES OF SEVERN ALLUVIAL SOILS

Soil types A, B, and C are schematic profiles derived from at least two sites each. Trends of pyrophosphate- and dithionite-extractable iron (Fe_p and Fe_d respectively) are shown, together with general representations of texture and structure, and sketches of micromorphology

supply as slopes became stable and soil profiles deepened under forest. At the same time the runoff regime probably became more equable. No radiometric or other age estimates are at present available to date the change. By analogy with dated alluvium elsewhere, however, a date of about 8 000 to 9 000 ^{14}C a B.P. may be suggested. The earliest date on the tributary Avon is 8 480 ± 140 ^{14}C a B.P. (Shotton, 1978), but this determination was not from the base of the succession. Beales' core from Crose Mere in north Shropshire (Beales, 1980) has an important lithological change at 8 730 ± 200 ^{14}C a B.P., from mineral to organic detritus. In addition, Rose *et al.*, (1980) believe that the River Gipping has barely moved from its inherited Late Glacial course in about 9 000 years, and it is easy to concur with this view for much of the course of the Severn. Dating of the initiation of the fine grained Flandrian fills in this country remains meagre, but the 8–9 000 a B.P. date suffices for the present.

Stratigraphic evidence from Wroxeter (D. J. Pannett, pers. comm.), where the bed of the Roman road was excavated at a depth corresponding to the major boundary in A of Figure 7.8, suggests that the first phase of deposition may have lasted up to Roman times. There is good evidence from the work of Shotton (1978) and Beales (1980) for sedimentary and environmental changes at about this time. Shotton's organic-poor 'buff-red silty clay' is estimated to have started accumulating in the lower Severn and Avon valleys at about 2 600 ^{14}C a B.P. In Beales's closely dated core the effects of clearance are especially in evidence after 2 310 ± 85 ^{14}C a B.P. The intensive clearance, it is suggested, led to the erosion of topsoil which contributed to the loamy alluvium at type-A and -B sites. The simultaneous response of runoff is likely to have been a change to a steeper flood hydrograph than in the preceding phase.

The inferred following period of sandy ridge formation could have been induced or accentuated by clearance of the floodplain itself; indeed, Mediaeval riparian use and land drainage must have altered the floodplain environment profoundly. More rapid runoff and faster flood flows probably resulted, leading to reduced mud deposition, and consequently the dominance of sand in the floodplain ridges which formed at this time. This phase of sedimentation appears to be over (possibly due to a reduction in the input of sediment), and channel marginal sedimentation is slight. There may have been slight incision, bedrock and pre-floodplain sediment are exposed at places at low stage.

The sequence of changes outlined above, which is based on evidence from sampled sections, is constrained by the limitations of the exposures. The inferences have been based on stratigraphy more than pedology, but the soils found during the investigation have properties strongly determined by the texture of the sediments in which they are forming. To this can be added the fact that the type-A and -B soils have moderately to well developed topsoil and subsoil structures which extend to a depth of up to 1 m. This probably reflects favourable drainage in floodplain soils bordering a high capacity channel in which the water level is more than 4 m below the bank tops at low flows.

Similarly, mottling is seldom prominent. However, at two type-A sites abundant ferrimanganiferous concretions occur from *ca* 1.0 to in excess of 2 m depth (Figure 7.8); these may be perched somewhat above the level at which redox fluctuations presently could produce strong segregation of iron and manganese, while below the sediment appears to be dominated by areas depleted of iron (which suggests removal at some time by long-term saturation). Thus, the hydromorphic features at these sites may be relict, dating from before riparian improvements. Soils inferred to be younger are coarser textured and the absence of possible relict concretions does not necssarily mean that these soils post-date floodplain clearance.

Insufficient micromorphological data are available to establish a micromorphological sequence of development which can be related to hydrology. In thin sections from some older soils features attributable to pedogenic clay illuviation can be seen, and these show variable relationships to hydromorphic features: some coatings are transgressed by ferruginous mottles, whereas others occur in matrix depleted of iron. The clay translocation features may be indicative of free profile drainage, and thus be fairly recent. Stretching the timescale to its fullest extent would give more time for the considerable reorganization in the oldest soils to have taken place, however.

Finally, it is submitted that such pedological data as are relevant to hydrology in this study are strongly influenced by the textural variability among the studied sites, and that this variation is temporal as well as spatial. The latter is not strongly displayed in the studied sites because all were from relatively elevated, well-drained floodplain. On the other hand, the changes through time may well owe their instigation to Man's activities in the drainage basin and on the floodplain.

ACKNOWLEDGEMENT

M. Hayward's contribution was written at McMaster University, Hamilton, Ontario, Canada, during a visit made possible by a Britain–Canada Exchange Award granted by the Foundation for Canadian Studies in the UK.

REFERENCES

Beales, P. W. 1980. The Late Devensian and Flandrian vegetational history of Crose Mere, Shropshire. *New Phytologist,* **85**, 133–161.

Bilzi, A. F., and Ciokosz, E. J. 1977. A field morphology rating scale for evaluating pedological development. *Soil Science,* **124**, 45–48.

Bloomfield, C. 1973. Some chemical properties of hydromorphic soils. In Schlichting, E., and Schwertmann, U. (Eds.), *Pseudogley and Gley*, 7–14, Verlag Chemie, Weinheim.

Blume, H. P. 1973. Genese und Ökologie von Hangwasserböden. In Schlichting, E. and Schwertmann, U. (Eds.), *Pseudogley and Gley*, 187–194, Verlag Chemie, Weinheim.

Blümel, F. 1962. Shapes of ferric oxide hydrate precipitations in gleys and pseudogleys. *Zeitschrift für Pflanzenernährung, Düngung und Bodenkunde,* **98**, 258–264.

Brakenridge, G. R. 1981. Late Quaternary floodplain sedimentation along the Pomme de Terre River, southern Missouri. *Quaternary Research,* **13**, 62–76.

Brammer, H. 1970. Coatings in seasonally flooded soils. *Geoderma,* **6**, 5–16.

Brewer, R. 1964. *Fabric and Mineral Analysis of Soils*, Wiley, New York, 470pp.

Bridges, E. M. 1973. Some characteristics of alluvial soils in the Trent Valley, England. In Schlichting, E. and Schwertmann, U. (Eds.), *Pseudogley and Gley*, 247–254, Verlag Chemie, Weinheim.

Brummer, G. 1973. Redoxreaktionen als merkmalsprägende Prozesse hydromorpher Böden. In Schlichting, E., and Schwertmann, U. (Eds.), *Pseudogley and Gley*, 17–29. Verlag Chemie, Weinheim.

Bunting, B. T. 1961. The role of seepage moisture in soil formation, slope development and stream initiation. *American Journal of Science,* **259**, 503–518.

Bunting, B. T. 1967. *The Geography of Soil*, Hutchinson, London, 213pp.

Chartres, C. J. 1975. *Soil Development on the Terraces of the River Kennet*. Unpublished Ph.D. Thesis, University of Reading, 318pp.

Chartres, C. J. 1980. A soil sequence in the Kennet valley, central southern England. *Geoderma,* **15**, 143–156.

Cheetham, G. H. 1980. Late Quaternary palaeohydrology: the Kennet Valley case study. In Jones, D. K. C. (Ed.), *The Shaping of Southern England*, 203–223, Academic Press, London.

Conacher, A. J. 1975. Throughflow as a mechanism responsible for excessive soil salinisation in non-irrigated, previously arable lands in the West Australian wheatbelt: a field study. *Catena*, **2**, 31–67.

Conacher, A. J., and Dalrymple, J. B. 1977. The nine-unit landsurface model: an approach to pedogeomorphic research. *Geoderma,* **18**, 1–154.

Dalrymple, J. B., Blong, R. J., and Conacher, A. J. 1968. A hypothetical nine-unit land surface model. *Zeitschrift für Geomorphologie,* **12**, 60–76.

Evans, J. G. 1966. Late-Glacial and Post-Glacial subaerial deposits at Pitstone, Buckinghamshire. *Proceedings of the Geologists' Association,* **77**, 347–364.

Evans, J. G., and Valentine, K. W. G. 1974. Ecological changes induced by prehistoric Man at Pitstone, Buckinghamshire. *Journal of Archaeological Science,* **1**, 343–351.

Fanning, D. S., Hall, R. L., and Foss, J. E. 1973. Soil morphology, water tables, and iron relationships in soils of the Sassafras drainage catena in Maryland. In Schlichting, E. and Schwertmann, U. (Eds.), *Pseudogley and Gley*, 71–79, Verlag Chemie, Weinheim.

Fitzpatrick, E. A. 1980. *The Micromorphology of Soils*, Department of Soil Science, University of Aberdeen, 186pp.

Furley, P. A. 1971. Relationships between slope form and soil properties developed over Chalk parent materials. In Brunsden, D. (Ed.), *Slopes: Form and Process*, Institute of British Geographers Special Publication No. 3, 141–164.

Gilman, K., and Newson, M. D. 1980. *Soil Pipes and Pipeflow — a Hydrological Study in Upland Wales*, British Geomorphological Research Group, Research Monograph, Geo Abstracts, Norwich, 114pp.

Graaff, R. H. M. van der. 1978. Size of subsoil blocky peds in relation to textural parameters, depth and drainage. In Emerson, W. W., Bond, R. D., and Dexter, R. E. (Eds.), *Modification of Soil Structure* 87–96, Wiley, Chichester.

Hayward, M. 1982. *Floodplain Landforms, Sediments and Soil Formation: the River Severn, Shropshire*. Unpublished Ph.D. Thesis, University of Reading, 324pp.

Hodgson, J. M. 1974. *Soil Survey Field Handbook*. Technical Monograph No. 5. Soil Survey, Harpenden, 99pp.

Holyoak, D. T. 1980. *Late Pleistocene Sediments and Biostratigraphy of the Kennet Valley, England*. Unpublished Ph.D. Thesis, University of Reading, 512pp.

Huggett, R. J. 1976. Lateral translocation of soil plasma through a small valley basin in the Northam Great Wood, Herts. *Earth Surface Processes,* **1**, 99–109.

Jenny, H. 1941. *Factors of Soil Formation: A System of Quantitative Pedology*. McGraw-Hill, New York. 281pp.

Kerney, M. P. 1963. Late-glacial deposits on the Chalk of south-east England. *Philosophical Transactions of the Royal Society of London, B*, **246**, 203–254.

Kirkby, M. J., and Chorley, R. J. 1967. Throughflow, overland flow and erosion. *Bulletin of the International Association of Scientific Hydrology,* **12**, 5–21.

Knapp, B. J. 1974. Hillslope throughflow observation and the problem of modelling. In Gregory, K. J., and Walling, D. E. (Eds.), *Fluvial processes in instrumented watersheds*. Institute of British Geographers Special Publication No. 6, 23–31.

Kwaad, F. J. P. M., and Mücher, H. J. 1977. The evolution of soils and slope deposits in the Luxembourg Ardennes near Wiltz. *Geoderma,* **17**, 1–37.

Kwaad, F. J. P. M., and Mücher, H. J. 1979. The formation and evolution of colluvium on arable land in northern Luxembourg. *Geoderma,* **22**, 173–192.

Meixner, R. E., and Singer, M. J. 1981. Use of a field morphology rating system to evaluate soil formation and discontinuities. *Soil Sicence,* **131**, 114–123.

Mücher, H. J. 1974. Micromorphology of slope deposits: the necessity of a classification. In Rutherford, G. K. (Ed.), *Soil Microscopy*, Proceedings of the fourth International Working Meeting on Soil Micromorphology, 553–566, The Limestone Press, Kingston, Ont.

Mücher, H. J., and Vreeken, W. J. 1981. (Re)deposition of loess in southern Limbourg, the Netherlands. 2. Micromorphology of the Lower Silt Loam Complex and comparison with deposits produced under laboratory conditions. *Earth Surface Processes and Landforms,* **6**, 355–363.

Mücher, H. J., Carballas, T., Guitian Ojea, F., Jungerius, P. D., Kroonenberg, S. B., and Villar, M. C. 1972. Micromorphological analysis of effects of alternating phases of landscape stability and instability on two soil profiles in Galicia, N.W. Spain. *Geoderma,* **8**, 241–266.

Mücher, H. J., de Ploey, J., and Savat, J. 1981. Response of loess materials to simulated translocation by water. *Earth Surface Processes and Landforms,* **6**, 331–336.

Pannett, D. J. 1981. Fishweirs of the River Severn. In *The Evolution of Marshland Landscapes*, Oxford University Department of External Studies, 144–157.

Rose, J., and Allen, P. 1977. Middle Pleistocene stratigraphy of south-east Suffolk. *Journal of the Geological Society,* **133**, 83–102.

Rose, J., Turner, C., Coope, G. R., and Bryan, M. D. 1980. Channel changes in a lowland river catchment over the last 13,000 years. In Cullingford, R. A., Davidson, D. A., and Lewin, J. (Eds.), *Time-scales in geomorphology*, 159–176, Wiley, Chichester.

Russell, E. W. 1973. *Soil Conditions and Plant Growth*. 10th Edition, Longman, London, 849pp.

Schuylenborgh, J. van. 1973. Report on Topic 1.1: sesquioxide formation and transformation. In Schlichting, E., and Schwertmann, U. (Eds.), *Pseudogley and Gley*, 93–102, Verlag Chemie, Weinheim.

Shotton, F. W. 1978. Archaeological inferences from the study of alluvium in the lower Severn-Avon valleys. In Limbrey, S., and Evans, J. G. (Eds.), *The Effect of Man on the Landscape: the Lowland Zone*. Council for British Archaeology Research Report No. 21, 27–31.

Valentine, K. W. G. and Dalrymple, J. B. 1975. The identification, lateral variation and chronology of two buried channels at Woodhall Spa and West Runton, England. *Quaternary Research* **5**, 551–590.

Valentine, K. W. G., and Dalrymple, J. B. 1976. The identification of a buried paleosol developed in place at Pitstone, Bucks. *Journal of Soil Science,* **27**, 541–553.
Veneman, P. L. M., Vepraskas, M. J., and Bouma, J. 1976. The physical significance of soil mottling in a Wisconsin toposequence. *Geoderma,* **15**, 103–118.
Vepraskas, M. J., and Bouma, J. 1976. Model experiments on mottle formation simulating field conditions. *Geoderma,* **15**, 217–230.
Vita-Finzi, C. 1971. Heredity and environment in clastic sediments: silt/clay depletion. *Geological Society of America Bulletin,* **82**, 187–190.
Vreeken, W. J. and Mücher, H. J. 1981. (Re)deposition of loess in southern Limbourg, the Netherlands. 1. Field evidence for conditions of deposition of the Lower Silt Loam Complex. *Earth Surface Processes and Landforms,* **6**, 337–354.
Weyman, D. R. 1970. Throughflow on slopes and its relation to the stream hydrograph. *Bulletin of the International Association of Scientific Hydrology,* **15**, 25–33.
Whipkey, R. Z. 1965. Subsurface stormflow from forested slopes. *Bulletin of the International Association of Scientific Hydrology,* **10**, 74–85.
White, E. M. 1966. Subsoil structure genesis: theoretical consideration. *Soil Science,* **101**, 135–141.
Whitfield, W. A. D., and Furley, P. A. 1971. The relationship between soil patterns and slope form in the Ettrick Association, south-east Scotland. In Brunsden, D. (Ed.), *Slopes: Form and Process*, Institute of British Geographers Special Publication No. 3, 165–175.

Background to Palaeohydrology
Edited by K. J. Gregory

8

Archaeology and palaeohydrology

SUSAN LIMBREY

University of Birmingham

Studies in archaeology and in palaeohydrology must necessarily have a close relationship to each other since, for the prehistoric period and to a lesser but still significant extent in historic times, archaeological evidence forms an important element in our understanding of the impact of human activity on the hydrology of the landscape.

In the history of study of valley sediments, the recognition of archaeological material by people working in gravel and brickearth extraction and in drainage projects and construction has often drawn attention to sites whose deposits would not otherwise have been recorded. A closer watch on such sites has often been kept by both amateur and professional archaeologists than by those who might be interested in the deposits themselves, and though the quality of recording of stratigraphy and description of deposits may not always have been high, some chronological information may be obtainable, even from the minimal provenance data attached to objects long stored in museums. Where deposits are no longer accessible and for deposits containing no organic material datable by radiocarbon, such relics may be the only means of obtaining a date. In recent years, with the establishment of professional archaeological organizations throughout the country and their specific concern with areas of urban development, gravel extraction, and road construction which threaten aracheological remains, and with increasing awareness among archaeologists of the importance of palaeoenvironmental studies in association with their work, the amount of information of palaeohydrological significance to be derived in this field is greatly increased.

The nature of archaeological data and the limitations inherent in archaeological research preclude our ever being able to obtain a complete picture of the pattern of settlement and the mode of exploitation of the landscape for any particular period. What we can hope to do is to study spatial and temporal changes in hydrologically significant factors. This chapter provides a survey of recent views on settlement and land use, derived from ground and aerial survey and excavation and from palaeobotanical and zoological studies. Direct evidence of river behaviour and the condition of valley floors derived from archaeological excavation is presented for a number of sites, covering between them a significant part of the Holocene period. The chapter is restricted in its coverage in two ways: firstly, evidence from Scotland and from Ireland and is not included; secondly, no attention is given to the lower reaches of valleys where sea level changes introduce a variable factor of estuarine influence into the alluvial sequence. In Britain there are important archaeological zones in these situations, for example the Somerset Levels, the Fenland, and a number of towns, particularly London. Each has a great abundance and high quality of both archaeological and environmental evidence concerning the changing relationship of alluvial and maritime regimes and the response of inhabitants to changing environment; it is a very large field of study and requires separate treatment.

EXPLOITATION AND ORGANIZATION OF THE LANDSCAPE

The Mesolithic period

In considering the behaviour of pre-agricultural people we can only understand their impact on the hydrology of the landscape by correlating archaeological evidence for their presence and their mode of subsistence with palynological evidence suggesting anthropogenic disturbance of vegetation. Evidence that non-agricultural people in Britain did open up forest and burn areas of vegetation, causing soil disturbance sufficient to leave traces in lake sediments, and perhaps initiating retrogressive succession of vegetation and degradation of soils in the areas of more fragile ecosystem, is summarized and discussed by Simmons *et al.* (1981). Archaeological views of the period include suggestions of hunting groups following herds of red deer into the uplands in summer, where they used fire in clearing forest or to prevent its encroachment and to maximize forage and facilitate hunting. Management of the woodland to perpetuate the hazel crop as high forest encroached, or to establish such a crop around newly created clearings, is another option for which pollen evidence is cited. The majority of mesolithic sites consists of scatters of flint implements and knapping waste in plough soil, beneath the mor humus of heathland and moorland soils and beneath blanket peat, and such sites cannot be closely enough dated for us to estimate population densities. It has usually been assumed that the population of Britain was very low during the mesolithic period, the number of sites reflecting the transient nature of the camps of a mobile people, but the suggestion that closing forest put pressure on a large enough population for them to have to go to considerable effort to maintain food supplies as the forests closed around them does seem to be supported by the palynological evidence. They certainly had axes with which they could fell trees, and there is enough evidence to show that they were indeed responsible for widespread episodes of fire.

In areas of lowland forest on base-rich soils, the hydrological impact of mesolithic people is perhaps best described as slight, scattered, and transient, forest regeneration after clearance being complete and erosions scars healing, e.g. at Hockham Mere (Sims, 1978), but their impact where equilibrium in the soil–plant system was delicately balanced, in upland areas and on base-poor, coarse textured parent materials in the lowlands may have been critical. Failure of forest to regenerate in certain upland areas e.g. Dartmoor (Simmons, 1964) and in the lowland heaths, e.g. Iping Common (Keefe *et al.*, 1965), had a permanent hydrological consequence, even if it was only the earlier onset of a regime which would have come about eventually in the complete absence of man, and which would anyway have been hastened by the greater disturbance of subsequent populations. Its greater significance, however, is probably that it conditioned the landscape for the people who followed, and thus influenced their exploitation of it in ways which had repercussions in a wider sphere. This chain reaction is a factor which has to be taken into account throughout the succeeding periods.

The Neolithic period

The introduction of farming to Britain during the fourth millennium bc† must be seen as a process of considerable hydrological significance. The pattern of introduction and the temporal and spatial extent of overlap and transition between mesolithic and neolithic ways of life are not well established, and the arable/pastoral relationships in our early farming are still very uncertain. We know that neolithic people had domestic cattle, pigs, sheep, and a few goats, and that they grew emmer wheat, two-row and six-row barley, some flax and probably beans, and it seems most likely that mixed farming was generally practised. The possibility of mesolithic hunters taking to herding, maybe via poaching and then rustling, without adopting crop growing, would establish an alternative system which could have persisted for some time. Though some archaeologists argue for maintenance of livestock entirely on forest resources without there being any necessity to clear forest to create pasture and hay meadow, this would still have involved sufficient opening of the canopy to affect interception and evapotranspiration (Limbrey, 1978). Any trampling of the forest floor by cattle and rooting by pigs in excess of that of the wild population would have had an impact on infiltration, runoff and erosion.

†Note: dates or period indicated as bc, ad are based on uncalibrated radiocarbon dates; AD indicates calendar dates.

Smith (1981) discusses the episode of the elm decline, considering both climate and human impact, showing that it coincides with absolute reduction in total tree pollen and with other indicators of human activity, and summarizing evidence that any critical climatic change probably came rather later. Rowley-Conway (1982) and Rackham (1980), however, have attempted to quantify exploitation of elm foliage as fodder and arrive at figures for human population and numbers of cattle necessary to explain such a large reduction in pollen production by elm trees which are very large in comparison with archaeologically visible early neolithic populations. The ubiquitous and synchronous elm decline in the late fourth millennium bc follows widespread but small and transient clearance episodes whose attribution to the mesolithic population or to a pioneer phase of farming is very uncertain. Thereafter, clearances in which initial arable land use is followed by a pastoral phase before regeneration of forest are more limited in their regional distribution. The episodes last from a few decades to several hundred years and apparently affect larger areas of land than the earlier, transient, ones.

To understand the hydrological impact it is as important to know how the land was managed within the cleared areas affecting the pollen rain of any particular site, but it is much more difficult to investigate this than to estimate changes in overall forest cover. For most of the palynologically recognized neolithic clearances we have little or no archaeological evidence of settlement location, and none of them have associated evidence for field systems. Neolithic people used ploughs of the simple ard type, but buried soils showing plough furrows are too limited in extent to be useful in studying the boundaries which might indicate permanent fields rather than frequent intake of newly cleared patches in progressive or rotational patterns. We do not know whether ploughing was in autumn for winter cereals or in late winter for spring sowing, leaving stubble and aftermath to protect and stabilize soil surfaces. The mode of integration of animal husbandry with arable agriculture, through stubble grazing, folding, dunging, and the growing of fodder crops affects not only the condition of soil surfaces but also the area of land required and the dynamics of rotation.

Permanent fields, which are known in Ireland at this period (Caulfield, 1978) but not so far in Britain, may be in use long enough for nutrient depletion and loss of organic matter to lead to failure of soil structure, increasing susceptibility to erosion, but field boundaries will trap moving soil, and systems of gates and trackways control runoff patterns.

Land use strategy also affects interception and infiltration parameters by determining details of the mosaic of clearance, regeneration complex, managed woodland and full forest, which will be quite different between fixed field agriculture and various possible systems involving more transient cultivation of cleared areas.

Although we may suspect that mesolithic people did operate some regular system in their exploitation of woodland for food and for fuel and building materials, it is during the neolithic period that we have the first real evidence, from timber used to make hurdling and preserved in the trackways in the Somerset Levels (Rackham, 1977; Morgan, 1982). Whether coppice woodland developed from primary forest via its utilization as a timber and fodder source, or under a woodland grazing regime, or was a management product of secondary woodland after clearance and agriculture, we do not know. The coppice cycle imposes characteristic cyclic patterns on the hydrological regime whose overall impact depends on the length of the cycle, the size of the coppice panels and the spacing of standards, giving ratios of areas at different stages of canopy development. The relationship between the scale of the drainage network and that of the coppice panels is also significant in assessing impact in individual catchments. Integration of grazing with coppicing and the establishment of wood pasture impose their variants on the hydrological regime.

Neolithic forest clearance lead to progressive and permanent deforestation in some areas, particularly in the chalklands, where evidence from land snail faunas in buried soils (Evans, 1972) indicates sequences of clearance, arable land use and pasture, with little or no regeneration of woodland. Thomas (1982) however, shows by land snail analysis a more varied pattern of chalkland forest clearance, and pollen evidence too (Godwin, 1962; Dimbleby and Evans, 1974; Waton, 1982) suggests rather more persistent woodland and that only certain areas became permanently open as early as the neolithic period.

The indications from lake stratigraphy that neolithic activity resulted in soil erosion (Pennington, 1970) are matched in the chalklands. At Brooke, Kent, indications of clearance and the initiation of colluvial fill in a dry valley is dated to the middle of the third millennium bc (Kerney *et al.*, 1964; Burleigh and Kerney, 1982). Such evidence adds support to a suggestion (Limbrey, 1975) that some of the buried soils beneath neolithic monuments are already to some extent truncated. There is evidence that some soils even on base-rich parent materials had by then suffered sufficient acidification for clay dispersal and eluviation to occur, but the extent to which this was a consequence of increased base depletion through cropping and grazing and the imposition of the leaching regime of an open landscape in place of that of forest remains uncertain. The development of argillic soils, whether natural or hastened by man would undoubtedly have led to enhanced susceptibility to erosion under agriculture and to reduced permeability.

Neolithic people in Britain did not exploit actual or potential areas of lowland heath to any great extent, though they may have used them for hunting and for rough grazing. Many upland areas show only transient and scattered evidence of neolithic activity, but use of already open areas for pasture, as well as further clearance, is seen in some places, and is implicated in soil degradation and initiation of peat formation (Moore, 1975).

There is evidence for a period of change in mode or intensity of exploitation of the landscape in the middle part of the third millennium bc. A number of clearance phases in pollen diagrams terminate with full forest regeneration, and there are fewer new clearings. Palynological, molluscan, and archaeological evidence all suggest a trend towards a more pastoral economy and there may have been some increase in the use of wild resources. Whittle (1978) attributes these changes to degradation of the more fragile soils under the impact of agriculture, but this would imply that all areas of soils which were both workable with the available techniques and not susceptible to degradation were already being exploited, which may seem unlikely, given that extended clearance episodes had only affected relatively restricted areas and much of Britain remained under forest still little affected by human activity.

The limitations of archaeological survey and the destruction of traces of earlier settlement by the long history of later cultivation on the soils of greatest inherent resilience makes it difficult to test this hypothesis. Support for it, however, comes from an area where detailed study of artifact distribution has been considered in the light of palynological evidence. Hawke-Smith (1979), studying the area of the Dove–Derwent interfluve in the southern Pennines, sees demand for arable land pushing cereal cultivation up onto the limestone plateau, between 200 and 400 m, in the late third millennium bc, the gritstone uplands being used for grazing. The Trent valley had a more scattered population utilizing the river terraces, but the intervening gritstone and shale foothills were not apparently exploited. We cannot judge what part social factors may have played in territorial constraints, but we do know that there was sufficient intercommunity stress in the neolithic period for sites to be defended and battles fought. As detailed surveys and analyses of artifact distributions are carried out in more areas we may well find that a very uneven distribution of population encompassed sufficient concentrations to put heavy demand on the limited areas of easily worked but degradable soils in particular localities.

By the later part of the third millenium bc forest clearance was again accelerating and there are indications of diversification of agriculture, with evidence of more upland land use than formerly. Stone clearance cairns begin to be formed at this time, suggesting persistence of exploitation, whether pastoral or arable, leading to sufficient erosion for increased quantities of stone to be exposed in surface horizons of soils, and the necessity for clearing it.

The Bronze Age

The introduction of metallurgy to Britain about 2000 bc may not be of any immediate agricultural significance, but the effects on communication and on relative prosperity and intercommunity relationships of the exploitation of highly valued resources of limited geographical distribution must be reflected to some extent in land use strategies. A new factor was also introduced into the demands made upon forests for fuel.

The agricultural expansion which was underway by the end of the neolithic period gained momentum in the succeeding period, and permanent land boundaries and field systems were established very widely in upland and lowland areas. Palynological evidence shows the land, particularly in the south, being cleared of forest to an increasing extent, and the cleared areas remaining open, either because of continued occupation or, in the uplands, because of soil degradation and the spread of blanket peat. Land organization, imposed in planned units over very large areas, changed the character of the landscape and must have had a very great hydrological impact. We see the pattern clearly only in those areas where subsequent land use has been less intense and traces of boundaries have persisted under a more open pastoral system.

The most detailed evidence for early land organization comes from Dartmoor and from the Wessex chalklands. On Dartmoor, survey of the 'reaves' shows that all land below an altitude which varies between 1,300 and 1,500 feet was incorporated into the system (Fleming, 1978). Reaves are stone walls, sometimes with associated ditches. They divide large areas in what is now the lower part of the moorland and the upper part of the presently enclosed land into rectangular fields, and above the fields enclose larger areas, each associated with particular patches of fields. The fields show some evidence of having been cultivated, there being traces of plough marks and some lyncheting, but this evidence is slight, and it is thought that the cereal growing, for which there is evidence in the pollen record and in crop remains preserved in settlements, was largely carried on in lower and more sheltered areas where traces of the fields have been obliterated by later land use. The use of the reave field systems for intensive stock management is supported by the patterns of gates and trackways which give access to the larger enclosures above and then to the upper moorland, which was never subdivided and would have been a summer grazing resource.

The Dartmoor reave systems were established by the middle of the second millenium bc, and structural modifications to them indicate a long and complex history. Dartmoor may be a special case because of its tin resources, but there is widespread evidence of a similar kind and intensity of land organization in other upland areas in the same period (Bowen and Fowler, 1978).

In the chalklands of southern Britain very large areas of fields, again in planned systems, also date from this period (Bowen, 1978). These fields are delimited by well-developed lynchets, indicating considerable soil movement under cultivation, and development of colluvial soil accumulations in dry valleys, with included artifacts and snail shells giving evidence of date and of land use, provides further evidence of cultivation and its consequent erosion (Bell, 1982). Arable farming was probably, on the evidence of plant and animal remains from excavated sites, integrated with stock keeping exploitation the remaining unenclosed land, and the long use of the fields for crop growing suggests that folding and dunging were practised.

From the beginning of the first millenium bc, the late Bronze Age sees widespread reorganization of the agricultural landscape. Some of the field systems continued in use, and indeed in lowland parts of western Britain some of the stone-walled fields have continued in use until today, but many of the upland fields ceased to be maintained and their area was absorbed into the rough grazing of the moorlands. In the chalklands and elsewhere in lowland Britain, new land boundaries in the form of linear ditches cut across the earlier fields and divided up unenclosed areas, forming a patttern of larger units. It has been suggested that these ditches, which become associated with hill forts, whose construction begins in the late Bronze Age, define the land units of a cattle ranching economy.

The Bronze Age field systems which are known in detail are those in landscapes whose exploitation after these fields went out of use has been predominantly pastoral. Traces of them also show up in aerial photographs in areas with a long history of arable land use on the light soils responsive to the formation of crop marks. We have very much less idea of the extent of farming in those very large areas of Britain where neither subsequent history nor the character of the geological substrate have promoted either preservation or detection of field system. Artifact distribution, again in the Trent basin (Vine, 1982), suggests expansion in the middle Bronze Age onto a range of light and potentially acid soils in the midland plateau lands. In and around Birmingham, large numbers of 'burnt mounds' of this period (Barfield and Hodder, 1981, and see below) suggest a considerable population, and can be correlated with evidence of soil erosion and sedimentation in stream valleys, in a region where soils which in recent times have been maintained as productive farmland by heavy fertilization had earlier, possibly during the Bronze Age, podzolized.

The Iron Age

By the middle of the first millennium bc we see great intensification of land use and probably a steeply rising population. Defensive sites become fully integrated with what appears to be intensive arable agriculture, the existing field systems are extended and spacing of settlements suggests expansion and infilling such that large areas of lowland Britain must have been fully exploited. Aerial photography of the lowland river terraces, whose soils are so responsive to crop mark formation, shows them to be densely occupied by farmsteads and their fields. Where sufficient field survey has been done it is found that there is also intensive use of other lowland soils, including clays (Cunliffe, 1976; Bradley, 1978). The farmstead and field boundaries are ditched, even on the terrace soils where the drainage function is not required, a feature which may help to correct the bias which aerial photography imposes on the evidence, since widespread use of field edge ditches may indicate a leading role of agricultural innovation on poorly draining soils at this time.

In the upland margins, farmsteads are more widely spaced, perhaps selecting the more productive soils for arable use and requiring access to larger areas of grazing to maintain an economy adapted to the lower extractive potential. There is some reoccupation of areas abandoned in the late Bronze Age upland retrenchment, though not to the altitude previously reached by enclosures, and the spacing of farmsteads and defensive sites suggests a fully utilized landscape.

Pollen evidence indicates that it was during the Iron Age that the acceleration of forest clearance and its transition from cyclic to permanent clearance spread from the southern areas where it is seen in the Bronze Age into most of Northern England (Turner, 1979, 1981). The combination of archaeological and palynological evidence suggests that remaining woodland must have been under considerable pressure and it can only have continued to supply fuel for domestic and industrial use under careful management. Domestic needs may have been supplied to a large extent from hedgerows, and charcoal from hearths in excavated sites is dominated by rosaceous species, giving the best indication we have that there were indeed hedges in the landscape. The demand for charcoal for metallurgical use must have been high and uniformly distributed, since many sites show evidence of at least forging. We have become unfamiliar with woodland managed as an essential resource for every community only recently; it is probable that the management regimes documented historically and discussed in detail by Rackham (1980), and glimpsed archaeologically already in the neolithic period had encompassed all remaining woodland by the later Iron Age.

The iron-shod plough, with its greater weight and cutting power, facilitated cultivation of the heavier soils, and the use of such soils and of poorly drained land is confirmed by the species of weeds whose seeds are found in crop remains. Study of weed seeds also shows that winter cereals were now in use, and the introduction of spelt, which may have been the winter wheat, and of rye and oats, suggests diversification onto poorer, already degraded, soils and a potential for tolerating climatic adversity (Jones, 1981). Legumes are now an important crop, indicating rotational management of arable land.

The Romano-British period

The Roman invasion of Britain may not have had such a great and immediate impact on the landscape as has sometimes been supposed. Many of the agricultural innovations once attributed to the Romans came earlier, and with the density of settlement and high population now accepted for the Iron Age the demands in food for the army and in tax may have been low in relation to the existing level of production and the sophistication of organization of the established market economy. Local stress, from increased demand and from military disruption, certainly occurred, but in much of the country life on the Iron Age farms went on as before, and prosperity and trade were sufficient to distribute imported pottery and other artifacts through the population quite rapidly.

Intensification of agriculture is certainly seen in some areas, with new settlements and field systems being established and existing ones spreading, and forest clearance continued in those northern areas still quite heavily wooded in the Iron Age. Villa estates were imposed upon the existing farmland, sometimes superseding traditional farms on the same site or nearby, sometimes apparently setting up management over land farmed from several earlier sites, but the distribution of villas is very uneven, related in some way to the development of some, but not all, urban centres (Miles, 1982). In some areas, for example the Cotswolds, a very high level of exploitation by villa estates were established, while in the neighbouring Avon valley the native farmsteads continue their activity undisturbed by the new land organization. It may be that the greater potential productivity of the heavier soils offered a better basis for the larger scale farming system of the villas and that now the river terraces, with their twin disadvantages of leaching and drought, were the less prosperous zone (Esmonde Cleary, 1982).

Some of the increased production went to feed not people but pests. Buckland (1978) discusses the impact of the grain weevil and its spread with the Roman habit of above ground storage. The zone of pit storage in the soft terrains of southern England had previously formed a *cordon sanitaire*, limiting its spread. Rats, too, now appear in Britain to take their share of the crop (Rackham, 1979).

Though staple crop introductions were largely an Iron Age phenomenon, it is in Romano-British times that livestock improvements occur, with considerable increase in size of cattle over their remarkably small predecessors, and some increase in sheep size, with evidence of more efficient meat production (Maltby, 1981). Increased ploughing efficiency is brought in with the true plough, probably in late Roman times, giving power to sever root mats, turn a furrow, and disrupt soil profiles to a much greater extent than before.

During Romano-British times modifications to waterways were carried out for a large number of purposes. Water mills must have been common; sites of several are known, though few have been studied in detail, an exception being at Haltwhistle Burn Head, Northumberland (Simpson, 1976). Water transport involved stabilization of river banks, the building of quays, and dredging and straightening water courses. At Worcester, the creation of a harbour by excavating marshland at the confluence of a minor brook and the Severn is postulated (Carver, 1980). Roman London provides evidence of massive waterfront construction, and many inland riverside towns must have had similar, if smaller, features: on the Severn, a quay was excavated at Gloucester (Hurst, 1974). Multi-pier stone bridges were built over the larger rivers,

London probably having the largest but the best known archaeologically being over the Tyne at Newcastle. The Severn must have been bridged at Gloucester, Worcester and Wroxeter, and possibly other points. The Worcester bridge may have survived in part as the five stone piers of the mediaeval bridge, which were demolished 'with the utmost difficulty' in 1781 (Carvet, 1980). Many smaller rivers and streams would also have had bridges; their known and possible sites are discussed by Dymond (1961). Fords were also common and were sometimes paved. Thus, water courses were controlled, banks stabilized, and channels restricted and pinned. Where river crossings were approached across land that was wet or liable to flood, causeways were built, again limiting channel migration and modifying flood regimes. In the Vale of Wigmore, on the border of Shropshire and Herefordshire, the extravagant meanders of the river Teme were confined to a smaller area than they had formerly exploited by the causewayed Roman road across the basin, and deposition of alluvium there since Roman times has been limited to the same area.

Britain does not have surviving aqueducts on the scale of other parts of the Roman empire, but there is evidence for canalized and piped water supplies being brought considerable distances to towns. Lincoln, for instance, seems to have got its water from a spring about twenty five km away across a valley, using clay pipes sealed in concrete. Gloucester probably brought water from the Cotswolds. The quantity of water in use in towns, derived from wells and piped supplies, and the sophistication of drainage and sewerage systems suggests considerable impact on local hydrology.

Water was also used for a variety of industrial purposes, including mining. At Dolaucothi in mid-Wales, water was brought seven miles in channels cut in and built on the hillsides, so as to create sufficient head for ore winning and washing processes in the gold mines (Jones *et al.*, 1960).

The Anglo-Saxon period

There is much uncertainty, and considerable controversy, over the condition of the landscape and the history of settlement in the post-Roman period. Much evidence suggests a sudden drop in population, and there is certainly widespread, though not universal, discontinuity in rural settlement. Most of the villas and the native settlements are apparently not occupied after the 4th century ad, and though continuity can be demonstrated in towns, it is a modified kind of urban life, and there is evidence in many towns for subsistence farming within the urban zone, deep deposits of dark soil accumulating over public spaces and the remains of buildings (Macphail, 1981). Collapse of the distribution network and urban markets may have removed the stimulus from a countryside which had depended on production of surplus since Iron Age times. Continuity of land holdings is, however, demonstrable in many areas, as boundaries established in the bronze Age persist into Saxon times and on into the present.

Many Saxon farms and villages are established on new sites, and for the first time we can glimpse in their names an actual description of the countryside. It is difficult to escape the conclusion that, in spite of evidence for continuity and some evidence for continued prosperity, in areas where Saxon settlement was late much of the land farmed efficiently and productively until late Roman times had tumbled down to waste by the time it occurred and population began to increase again. Pollen analysis supports this view, with widespread evidence for woodland regeneration. Early place names show evidence of selection of dry islands in wet valley floors (Gelling, 1978), and it may be that failure of Roman river management and drainage schemes contributed to exacerbation of such conditions. Topographic placenames of hydrological interest indicate the value of riverine hay meadows (-ham) and of certain particularly wide areas of floodplain bounded by meander loops ('was'), though we do not know whether it was the morphology of the 'was' or a feature of its groundwater regime or its use which is being denoted (Gelling, pers.comm.). It is interesting that 'was' features are still conspicuous, indicating minimal progress in meander migration at these places since Saxon times (e.g. Broadwas and Buildwas on the Severn, Wasperton on the Warwickshire Avon).

Saxon towns were usually established on rivers, either continuing the use of Roman sites, usually bridging points or fords, or in new situations, common among which are meander loops and confluences where marshy ground completes a watery circuit, possibly for security or at least control of access. Thus, the 'town marsh' is a common feature, some persisting into recent times but often built over after a long history of stabilization by casual or deliberate dumping. River frontages too are encroached upon by urban development and waterfront construction, confining river channels and exacerbating risk of flooding.

Very large numbers of mills were constructed on rivers and streams, many parishes having several flour mills, and with the growth of the wool industry fulling mills add to their number in sheep-rearing areas. The Domesday survey gives numbers but not siting of mills and few parishes have been sufficiently well surveyed for all the mills and their associated modifications of water courses to be found. Diversion of streams and building of dams to make fish ponds is another hydrological modification which begins in the Roman period and is taken up in Saxon times at monastic sites. Again, field work has only defined the full extent of such systems in a few cases. Fish farming was practised in certain towns in the Norman period, Lichfield and Stafford both having evidence of creation of ponds by modification of the marshes within or adjacent to the towns (Carver, pers. comm.).

The Domesday survey of the late 11th century gives a lot of information about the condition of the countryside under the full development of the Saxon farming system, and synthesis of documentary and field evidence can be used to show the development of the system during the Saxon period and to attempt quantification of land under different uses. Much work of this kind is in progress. In the case of woodland, however, quantification from the Domesday survey is only possible to a limited extent, depending on the way in which the information was recorded. Rackham (1980) arrives at figures which range from 40 per cent of the area of the county in Worcestershire, where the Wyre forest covered a much larger area than it does now, to some 3 to 4 per cent for a number of counties in the east and the southwest and nil for part of Lincolnshire. Those counties for which woodland was recorded in terms of pannage for swine, rather than by some measure of dimension or area, yield much less usable data, kinds of woodland which provide no pannage being invisible and the parameters of the pigs per acre equation being subject to immense variation. In some counties the information appears at one further remove from reality, in 'swine-rent' terms, and the ratio of herd size to rent must have been very variable. Other areas, particularly in the north, yield very little evidence. In the Severn basin, Worcestershire, Warwickshire and east Gloucestershire have information of some usefulness, west Gloucestershire and Herefordshire have very little, Shropshire is patchy, and there is of course no record for Wales. Of England as a whole, Rackham estimates that some 15 per cent was woodland in 1086 (See Note on page 212).

ARCHAEOLOGICAL SITES AND ALLUVIAL STRATIGRAPHY

Artifacts of Late Devensian age may occur in the surface soils on river terraces, but have rarely been found stratified in alluvial deposits, other than in buried channels in the lower reaches of rivers, e.g. at Sproughton, Suffolk (Wymer *et al.*, 1975). In the Holocene, however, we begin to find mesolithic sites in deposits below modern floodplains, reflecting the advantageous characteristics of the riverine zone in a forested landscape and illustrating the nature of the valley floor.

Broxbourne

The Lea and Colne valleys, tributary to the lower Thames, have yielded numerous mesolithic sites overlying gravels and sands and beneath, or in the lower part of, peat deposits of the Boreal period. The sites exposed by gravel works at Rickhof's Pit, Broxbourne, on the Lea (Warren *et al.*, 1934) are the best published example; recent work in the same area by R. Jacobi, with further dating and stratigraphic evidence is unpublished. Comparable sites in the Colne valley are reported by Lacaille (1963).

At Broxbourne, large numbers of flint implements and waste from flint working were found on a bank of interstratified gravels and calcareous sandy silts, part of a complex of gravel, sand and mud banks overlying Late Glacial gravels of the Ponders End stage. A layer of coarse sand and fine gravel overlying the bank is interpreted as a deflation and weathering surface, and it was within this that the flints lay. The bank was truncated to leave a steep slope, with evidence of reworking of the gravels, suggesting encroachment by a river channel; the channel was then abandoned and peat accumulation began. Peat growth began in Late Boreal times in the channel and had overwhelmed the area of occupation still within that period, giving way to deposition of alluvial clay in the early part of the Atlantic period.

The flint assemblage is of Maglemosian type, a culture found all around the North Sea basin. There are numerous isolated finds of artifacts of this type in the London area, commonly recorded as below peat, in/on gravels below peat, or below alluvium, indicating occupation of the Lower Thames and its tributaries passing well below present sea level, in continuity with occupation of the basin itself, from which artifacts have been obtained in dredging.

Thatcham

The Kennet valley, tributary to the middle Thames, also has sub-peat mesolithic sites, of which those at Thatcham have the most detailed dating and stratigraphic evidence (Churchill, 1962). The sites, having hearths, bones, and flint artifacts, occur along the edge of a low terrace of the Kennet at its confluence with the Moor Brook. On the terrace the archaeological material is found below a humic silty clay deposit, which is overlain by a grey silt layer and then peat. The peat surface suffered desiccation, cracking and erosion and was then buried by a layer of redeposited algal marl. On the floodplain, the algal marl is *in situ* overlying undated peats and gravels, and its upper part, continuous with the marl deposit on the terrace, is reworked. Reedswamp peat overlies the marl on the floodplain. Dating of charcoal from the hearths on the terrace and of wood from the marl indicates occupation covering several hundred years in the early 9th and late 8th millennium bc. Pollen analysis of the marl shows its accumulation to begin in the pre-Boreal period and continue into at least the late Boreal. The peat overlying the terrace formed during the Atlantic period (Dimbleby, 1959). At an unknown time, following drying out and intense humification of the peat, flooding reworked the marl and deposited the layer over the terrace.

It has been suggested that the ponding up of water in the Kennet valley, resulting in clay and silt deposition and peat accumulation above the level of the present reedswamp, could be attributed to beaver dams; beaver bones were found on the archaeological sites. Whether or not the beaver is implicated here, its presence in Britain should not be ignored in discussion of river deposits; it is thought to have survived until the 12th century AD in Wales, probably becoming extinct rather earlier in England (Coy, 1982).

Runnymede Bridge

At Egham, Surrey, a sequence of road constructions have revealed Bronze Age occupation in the locality over a considerable area, and in 1978 the construction of a motorway bridge over the Thames at Runnymede revealed a settlement site and river frontage interstratified with floodplain and channel deposits. Archaeological excavation was carried out down to about one metre below present river level behind a coffer dam built for the bridge works (Needham and Longley, 1980).

Though the settlement site and river frontage features are of late Bronze Age date, established soon after 800 bc, the excavation exposed alluvial stratigraphy covering a longer period and including neolithic remains. During salvage recording neolithic artifacts had appeared to be associated with a deliberately laid brushwood layer, and an excavation nearby in 1980 revealed an occupation at a higher level on a gravel surface beneath fine alluvium. During the coffer dam excavation, water level could not be kept low enough for the neolithic stratum to be pursued and only redeposited artifacts, which are of the middle neolithic, in the early part of the third millennium bc, and considerable quantities of bone, were found.

The lowest deposit exposed was a bed of calcareous tufa, the neolithic material lying above this in interbedded sands, fine gravels, and organic silts, deposited in layers and lenses of varying thickness forming a shallow bank and channel structure. Above the neolithic material the deposits fined upwards, becoming predominantly organic silt with minor lenses of sand and fine gravel which then die out. About half a metre above the tufa, the organic content diminished and the clay content increased, and a further metre of uniform calcareous silty clay was deposited. The deposit became well aerated, signs of gleying diminishing upwards, and no trace of sedimentary structure remaining. As deposition slowed and eventually ceased, a humic soil developed, and the Bronze Age settlement was established on the floodplain.

Before occupation of the settlement, the river encroached upon this part of the floodplain, cutting a steep face through the entire sequence exposed, and then depositing a great variety of silts, silty clays, sands and gravels, intercalated with some slumping from the bank. It was during the filling of this channel that occupation of the Bronze Age site began, and burnt branches and masses of charcoal were incorporated into the deposit. When the channel margin was already silted up to within less than a metre of the level of the floodplain surface, a timber structure was built by driving two rows of closely spaced sharpened piles into the channel margin deposits. The tops of the piles and any superstructure they carried have decayed, but the lower parts, where the deposits have remained continuously waterlogged, are well preserved. It is possible that a landing place was built to give access from the firm ground of the floodplain across the increasingly silted up channel margin to deeper water beyond, but the circumstances of the excavation did not allow more than an area some ten metres wide to be examined and in that area no evidence of deeper water or of any dredging activity was seen. The present river bank lies between fifteen and thirty metres out from the timber structure, traces of which, in a series of shallow arcs, have been recorded over a distance of some fifty metres.

Silting up of the channel margin continued while the piles were in place, and eventually there are signs of flooding across the adjacent settlement area. After the site was abandoned a further half metre of floodloam similar to that underlying the settlement accumulated to establish the present floodplain, covering the site, the river frontage and the silted up channel margin, but the chronology of this later deposit is unknown.

Sedimentological work on the deposits by the present author and P. Fisher is in progress. Results so far of pollen analysis by J. R. A. Greig show that during the period of channel silting the area was predominantly pasture, with little evidence of woodland. Studies of coleoptera and mollusca from the deposits are also in progress.

The occurrence of tufa at Runnymede, and of algal marl at Thatcham are examples of a widespread phenomenon of calcareous deposition in rivers and around springs in the calcareous terrains of lowland Britain in mid-Holocene times (Evans, 1972; Evans *et al.*, 1978). Gibbard and Hall (1982) construct a cross section of the valley of the Colne Brook, which enters the Thames almost opposite the Runnymede Bridge site and which may have contributed to evolution of channel morphology and deposits there, and report no tufa. Their information comes from bore hole records, and it is worth noting that the bore hole records for the bridge site itself include no identifiable tufa, even though they span the excavated area where a thick, hard, continuous layer of it was exposed.

The neolithic site at Runnymede is one of very few well documented examples of an occupation site of the period in a floodplain situation. The occupation was on a gravel surface above active channel deposition of the period, and was later buried by the fine, uniform silty clay of the pre- late Bronze Age floodloam. The river was not apparently subject to over-bank flooding until after the middle neolithic period. Traces of neolithic occupation beneath a Bronze Age burial mound in the Trent floodplain indicate a similar situation there, with some indication of flooding between the neolithic occupation and the early Bronze Age mound which is itself later partially buried by floodloam (Reavy, 1968).

Upper Thames sites

In the Upper Thames valley the history of floodplain conditions and the chronology of the alluvial sequence can be assembled from a number of sites.

Since the alluvium is initially calcareous, decalcification of soils on the floodplain can be taken as indication of periods of reduced accretion, and on the lowest terrace pits and ditches associated with archaeological sites reach down to levels at which organic materials may be preserved when the water table is high but not when it is low. With rather less confidence we can make some assumptions about the dryness of the floodplain in summer or all the year round, from evidence of occupation and field systems.

In the early Holocene the river occupied a complex of channels in the surface of the gravels of Late Devensian age which underlie much of the floodplain. Some of the channels silted up, but, as in the Lower Thames and the Trent, there does not seem to have been overbank flooding to any great extent before the later neolithic or early Bronze Age. The Drayton cursus near Abingdon was sited on a surface which carried only a thin loam over gravel. The loam may be a soil rather than an alluvial deposit, since the limestone gravels yield a stone-free loam on weathering, but this remains to be determined. This site, together with early Bronze Age burial sites in Oxford (Palmer, 1980) show that the present floodplain was a dry enough area then to be selected for ritual activities, and though it is sometimes argued that such sites would be placed on the less productive land, it seems unlikely that for burial, at least, wet or seasonally flooded ground would be chosen. Organic material was not preserved in the ditches of the cursus or burial mounds, confirming at least seasonally low water levels then or since.

Such floodloam as was deposited, which nowhere exceeded about 30 cm, had been decalcified by the end of the Bronze Age. Pits and ditches of Bronze Age date cut into the gravels of the first terrace do not contain preserved organic material, in strong and consistent contrast to later features of the same depth, indicating that groundwater levels were lower before the Iron Age than they have ever been since. Such preservation in Iron Age sites on the first terrace shows that the water table had risen significantly by 200 bc, and by 150 bc flooding was bringing riverine mollusca into ditches on the floodplain. Between about 150 bc and the early part of the Roman period, about 0.5 m of fine alluvium was deposited over much of the floodplain. Deposition ceased or slowed down sufficiently for the deposit to be partially decalcified before a second period of deposition from late Saxon times to the end of the 13th century.

Much of the evidence for conditions in Iron Age and Romano-British times comes from the site of Farmoor, Oxfordshire (Lambrick and Robinson, 1979), and I am indebted to M. A. Robinson for providing the general outline above. At Farmoor, occupation in the early Iron Age was on the first terrace, which has its surface about 2 m above the present floodplain surface. In the middle Iron Age, the floodplain itself was occupied, with houses and enclosures, and the presence of riverine mollusca and seeds of aquatic plants in enclosure ditches indicates that the site was subject to flooding; it is therefore suggested that people lived there only in summer, for exploitation of the grazing. There is no evidence of flooding up onto the terrace at that time. For the Romano-British period there is no evidence of occupation on the floodplain, but there is a trackway onto it from a settlement and its field system on the first terrace. It is suggested, on the basis of riverine organisms found in the Roman wells, that flooding did reach the terrace then.

Beckford

In the Severn basin there have been numerous finds of prehistoric artifacts in floodplain deposits. Few have adequate stratigraphic records, but of the numerous Bronze Age finds many seem to occur on gravels below fine alluvium, a situation in accord with the radiocarbon dates available from equivalent stratigraphic positions (Shotton, 1978). No excavations of prehistoric sites in floodplain locations have been carried out, and in most areas there is little indication from survey so far that such sites exist, but aerial survey in the Upper Severn valley indicates that a number of farmsteads of Iron Age to Romano-British date are within the floodplain area (J. Collens, work in progress). At Beckford, on the Carrant Brook, tributary to the Avon just above its confluence with the Severn, excavation of a site on the lowest terrace shows flooding up onto the terrace in post-Roman times, and investigations on the floodplain provide evidence of alluvial chronology.

The terrace at Beckford grades downstream into Avon No. 2 terrace, the first terrace being entirely beneath the floodplain in this part of the valley. The gravels are capped by a wind-blown sand deposit of variable depth (Briggs, 1975), and the land surface, now planed by its long agricultural history, was undulating. The earliest archaeological feature is a Bronze Age ditch, of the kind recognized as forming the major land boundaries of the period. In the Iron Age a farmstead was established and the land continued to be farmed in the Romano-British period, when a villa may have been built nearby. There is a hillfort on Bredon Hill, at the foot of which the site lies, and aerial photography has shown that the pattern of Iron Age and Romano-British fields extends all along the terrace with farmsteads at intervals. The slope soils are not responsive to crop mark development, so the extent of agriculture off the terraces is not known.

Excavation at Beckford (J. Wills, report in progress) showed that after the farmstead and its enclosures went out of use agriculture continued and a layer of dark grey plough soil accumulated over the occupation levels, beginning to fill the declivities in the terrace surface. Overlying this soil, a deposit very similar to the alluvium of the adjacent floodplain occurs, filling remaining hollows. Since this deposit has its base above the level of the present floodplain surface and there is a distinct bluff at the terrace edge, trenches were cut in the floodplain to investigate its relationships. It was found that the deposit did form a continuous mantle, about 50 cm thick, from the floodplain onto the terrace, but that it is not everywhere present on the floodplain itself. The grey post-Roman plough soil passes over the terrace edge to form a buried lynchet, beyond which it merges with the upper part of a peat deposit. Radiocarbon dating shows that organic accumulation began in the early second millennium bc and continued, with some intervention of mineral inwash, until about 1000 ad, when it was covered by the alluvial layer which passes up onto the terrace. Ridge and furrow of the mediaeval field system of Beckford village was then imposed on the terrace surface. Study of the sediments at Beckford is in progress by the present author; pollen analysis is being done by S. Colledge.

Alcester

At Alcester, a town established in Roman times at the confluence of the Alne and the Arrow, Warwickshire, part of thc Roman town lies on the gentle slope of a terrace edge and has not been built over in later times. An aerial photograph shows the town plan becoming submerged beneath the floodplain alluvium. In the river bank nearby, a humic horizon indicates a period of slow or no accretion, and burials of the Roman period now exposed by river erosion allow us to date this stable surface to that time. Such a soil horizon has not been observed elsewhere in the notably uniform upper floodplain deposits of the lowland part of the Severn basin; it does suggest a correlation of the period of minimal floodplain accretion here with that of the Roman and earlier Saxon period in thc Upper Thames, and the Beckford evidence suggests that in this region too further accretion occurred in late Saxon to early mediaeval times.

Burnt mounds

Mounds of fire-cracked stones and charcoal occur very widely in Britain, usually sited with access to water and often on the banks of small streams. They have a fairly wide range of dates, but dates in the Bronze Age predominate. Those that have been excavated generally have pits or stone or wood tanks into which water could have been led or carried from the stream, and though some have little else of structural significance, others have post holes indicating some kind of wooden structure. Traditionally they have been interpreted as cooking places, where water was heated by dropping hot stones into it and food was steamed or boiled. This does work, but the hypothesis has been criticized by Barfield and Hodder (1981), who point out the rarity of bones at some of the sites and the lack of pottery or other artifacts to be associated with food preparation. They propose instead that they are sauna baths.

Barfield and Hodder have been searching stream banks in Birmingham and find these sites at intervals of 2 km or less on all the minor streams. Their excavation of one at Cob Lane, on Griffin's Brook, showed the mound to overlie about a metre of silty clay, which itself overlies organic deposits and gravels. The stream was encroaching on the mound during its use and there is evidence of timber revetment of the bank. After the site went out of use it was partly buried by a further deposit of alluvial clay. The site is dated to the early second millennium bc. Bones of domestic animals in the organic deposit below the fine alluvium under the mound show that there were farmers in the area before that, suggesting soil disturbance in the catchment of the stream. Study of organic deposits by S. Colledge is in progress.

DISCUSSION

From the point of view of someone working mainly in the archaeological field, correlation between widespread changes in alluvial regime and agricultural developments detected archaeologically and palynologically is tempting. The numerous dates obtained on organic materials from floodplain situations that can be gleaned from date lists in the journal *Radiocarbon* support the chronology suggested by the information from archaeological sites. There is clearly much variation within and between valleys, but dates from organic lenses high in the complex of coarse channel deposits which generally underlies the upper body of fine alluvium in our lowland valleys tend to cluster in the second millennium bc, those from lower levels in such deposits in the third millenium. Of such dates, only Shotton's (1978) were on samples selected specifically to date as near the top of the coarse deposits as possible, and these are in the first millenium bc.

Land use changes which may be significant are the change from small or localized and transient forest clearance to more widespread and permanent clearance, the setting up of field systems and changes in their pattern or their use, the adoption of winter sowing of cereals and legumes, increasing exploitation of heavy soils, the adoption of the true plough, and the establishment and then the abolition of the open field systems. As well as the developments in agricultural technology, changes in population and in the balance between arable land and pasture are important. Evidence for soil profile developments such as podzolization, clay eluviation and changes in permeability, and for soil erosion may be correlated with particular phases of human activity.

In considering the impact of climatic change we have to take into account its indirect influence through the farming community's reaction to crop response and soil changes as well as the direct impact, and it is here that the difficulty of quantification from the archaeological evidence is a serious obstacle. A possible approach is through detailed investigation of buried soils, colluvial accumulations, and alluvial deposits in small catchments having high quality archaeological data, and this is being attempted in the Upper Thames basin.

It is unfortunate that the Severn basin has not so far provided as much useful data in the archaeological field as the Thames. This is in part because of the greater amount of development and the high population of archaeologists as one approaches London, but also in part because of the different characteristics of the valleys. The Thames floodloam is thinner and sites within or under it are more likely to be detected, but even so the Severn floodplain does not appear to have been occupied to the same extent as that of the Thames, and the organic preservation sequence giving information on groundwater levels in the lowest terrace has not been found in the Severn basin. The impact in the Severn basin of the increasing differentiation in land use between the highland and lowland zones, as soil differentiation progressed and as the market economy allowed increasing specialization into the stock rearing and the cereal growing areas favoured by climatic factors, has no counterpart in the Thames, which may be an easier system to deal with.

ACKNOWLEDGEMENTS

I am grateful to A. S. Esmonde Cleary for discussion of the Romano-British period, and to S. Needham, M. A. Robinson, J. Wills, and L. H. Barfield for additional information about their excavations.

NOTE

The figure of 40 per cent woodland for Worcestershire suggested on pge 201 is probably too high; it leaves insufficient room for the farmland of known settlements, and J. Bond (personal communication) suggests, from the dimensions of woods whose Saxon boundaries can be identified, that a non-standard unit of measurement was in use in the county.

REFERENCES

Barfield, L. H., and Hodder, M. 1981. Birminghamn's Bronze Age. *Current Archaeology,* No. 78, 198–200.

Bell, M. 1982. The effects of land-use and climate on valley sedimentation. In Harding, A. (Ed.), *Climatic Change in Later Prehistory*, Edinburgh, University Press, 127–142.

Bowen, H. C. 1978. 'Celtic' fields and 'ranch' boundaries in Wessex. In Limbrey, S., and Evans, J. G. (Eds.), *The Effect of Man on the Landscape: the Lowland Zone*. London, Council for British Archaeology, Research Report No. 21, 115–123.

Bowen, H. C., and Fowler, P. J. (Eds.). 1978. *Early Land Allotment in the British Isles: a Survey of Recent Work*, Oxford, British Archaeological Reports, British Series No. 48.

Bradley, R. 1978. *The Prehistoric Settlement of Britain*, London, Routledge and Kegan Paul.

Briggs, D. J. 1975. Origin, depositional environment and age of the Cheltenham Sand and Gravel and related deposits. *Proceedings of The Geologists Association,* **86**, 333–348.

Buckland, P. C. 1978. Cereal production, storage and population: a caveat. In Limbrey, S., and Evans, J. G. (Eds.), *Effect of Man on the Landscape: the Lowland Zone*, London, Council for British Archaeology, Research Report No. 21, 43–45.

Burleigh, R., and Kerney, M. P. 1982. Some chronological implications of a fossil molluscan asemblage from a neolithic site at Brook, Kent, England. *Journal of Archaeological Science,* **9**, 29–38.

Carver, M. O. H. 1980. Mediaeval Worcester: an archaeological framework. *Transactions of the Worcestershire Archaeological Society*, 3rd Series, **7**, 1–335.

Caulfield, S. 1978. Neolithic fields: the Irish evidence. In Bowen, H. C., and Fowler, P. J. (Eds.), *Early Land Allotment in the British Isles: a Survey of Recent Work*, Oxford, British Archaeological Reports, British Series No. 48, 137–143.

Churchill, D. M. 1962. The stratigraphy of the Mesolithic Sites III and V at Thatcham, Berkshire, England. *Proceedings of the Prehistoric Society,* **28**, 361–370.

Coy, J. 1982. Woodland mammals in Wessex — the archaeological evidence. In Bell, M., and Limbrey, S. (Eds.), *Archaeological Aspects of Woodland Ecology,* Oxford, British Archaeological Reports, International Series No. 146, 287–296.

Cunliffe, B. W. 1976. The origins of urbanism in Britain. In Cunliffe, B. W., and Rowley, R. T. (Eds.), *Oppida: the Beginnings of Urbanism in Barbarian Europe*, Oxford, British Archaeological Reports, International Series No. 11, 135–161.

Dimbleby, G. W. 1959. Thatchem—pollen analysis. *Berkshire Archaeological Journal*, **57**, 25–33.

Dimbleby, G. W., and Evans, J. G. 1974. Pollen and land-snail analysis of calcareous soils. *Journal of Archaeological Science,* **1**, 117–133.

Dymond, D. P. 1961. Roman bridges on Dere Street, County Durham. *The Archaeological Journal,* **118**, 136–164.

Esmonde Cleary, A. S. 1982. Romans and river gravels. *West Midlands Archaeology,* **25**.

Evans, J. G. 1972. *Land Snails in Archaeology*, London, Seminar Press.

Evans, J. G., French, C., and Leighton, D. 1978. Habitat change in two Late-glacial and Post-glacial sites in southern Britain. In Limbrey, S., and Evans, J. G. (Eds.), *The Effect of Man on the Landscape: the Lowland Zone*. London, Council for British Archaeology, Research Report No. 21, 63–75.

Fleming, A. 1978. The prehistoric landscape of Dartmoor. Part 1. South Dartmoor. *Proceedings of the Prehistoric Society,* **44**, 97–123.

Gelling, M. 1978. The effect of man on the landscape: the place-name evidence in Berkshire. In Limbrey, S., and Evans, J. G. (Eds.), *The Effect of Man on the Landscape: the Lowland Zone*, London, Council for British Archaeology, Research Report No. 21, 123–125.

Gibbard, P. L., and Hall, A. R. 1982. Late Devensian river deposits in the lower Colne valley, West London, England. *Proceedings of the Geologists Association,* **93**, 291–300.

Godwin, H. 1962. Vegetational history of the Kentish chalk downs as seen at Wingham and Frogholt. *Veröfflichungen des Geobotanischen Instituts, Zurich,* **37**. 83–99.

Hawke-Smith, C. F. 1979. *Man—Land Relationships in Prehistoric Britain: the Dove-Derwent Interfluve; a study in Human Ecology*, Oxford, British Archaeo logical Reports, British Series No. 64.

Hurst, H. 1974. Excavations at Gloucester, 1971–1973: second interim report. *Antiquaries Journal,* **54**, 8–52.

Jones, M. 1981. The development of crop husbandry. In Jones, M., and Dimbleby, G. W. (Eds.), *The Environment of Man: The Iron Age to the Anglo-Saxon Period*, Oxford, British Archaeological Reports, British Series No. 87, 95–127.

Jones, G. D. B., Blakey, I. J., and Macpherson, E. C. F. 1960. Dolaucothi: the Roman aqueduct. *Bulletin of the Board of Celtic Studies,* **19**, 71–84.

Keefe, P. A. M., Wymer, J. J., and Dimbleby, G. W. 1965. A mesolithic site on Iping Common, Sussex, England. *Proceedings of the Prehistoric Society,* **31**, 85–92.

Kerney, M. P., Brown, E. H., and Chandler, T. J. 1964. The late-glacial and post-glacial history of the chalk escarpment near Brook, Kent. *Philosophical Transactions of the Royal Society of London*, Series B, **248**, 135–204.

Lacaille, A. D. 1963. Mesolithic industries beside Colne waters in Iver and Denham, Buckinghamshire. *Records of Buckinghamshire,* **17**, 143–181.

Lambrick, G., and Robinson, M. 1979. *Iron Age and Roman Riverside Settlements at Farmoor, Oxfordshire*, London, Council for British Archaeology, Research Report No. 32.

Limbrey, S. 1975. *Soil Science and Archaeology*, London, Academic Press.

Limbrey, S. 1978. Changes in quality and distribution of soils. In Limbrey, S., and Evans, J. G. (Eds.), *The Effect of Man on the Landscape: the Lowland Zone*, London, Council for British Archaeology, Research Report No. 21, 21–27.

Macphail, R. 1981. Soil and botanical studies of the 'dark earth'. In Jones, M., and Dimbleby, G. W. (Eds.), *The Environment of Man: the Iron-Age to the Anglo-Saxon period*, Oxford, British Archaeological Reports, British Series No. 87, 309–331

Maltby, M. 1981. Iron Age, Romano-British and Anglo-Saxon animal husbandry: a review of the faunal evidence. In Jones, M., and Dimbleby, G. W. (Eds.), *The Environment of Man: the Iron Age to the Anglo-Saxon period.* Oxford, British Archaeological Reports, British Series No. 87, 155–203.

Miles, D. (Ed.) 1982. *The Romano-British Countryside: studies in Rural Settlement and Economy*, Oxford, British Archaeological Reports, British Series No. 103.

Moore, P. D. 1975. Origin of blanket mires. *Nature (London),* **256**, 267–269.

Morgan, R. A. 1982. Current tree-ring research in the Somerset levels. In Bell, M., and Limbrey, S. (Eds.), *Archaeological Aspects of Woodland Ecology*, Oxford, British Archaeological Reports, International Series No. 146, 261–278.

Needham, S., and Longley, D. 1980. Runnymede Bridge, Egham: a late Bronze Age riverside settlement. In Barrett, J., and Bradley, R. (Eds.), *Settlement and Society in the British Later Bronze Age*, Oxford, British Archaeological Reports, British Series No. 83, 397–436.

Palmer, N. J. 1980. A beaker burial and mediaeval tenements in The Hamel, Oxford. *Oxoniensia,* **45**, 124–225.

Pennington, W. 1970. Vegetation history in the north west of England: a regional synthesis. In Walker, D., and West, R. G. (Eds.), *Studies in the Vegetational History of the British Isles*, Cambridge, University Press, 41–79.

Rackham, D. J. 1979. *Rattus rattus*; the introduction of the black rat into Britain. *Antiquity,* **53**, 112–120.

Rackham, O. 1977. Neolithic woodland management in the Somerset Levels: Garvin's, Walton Heath and Rowland's tracks. *Somerset Levels Papers*, **3**, 65–72.

Rackham, O. 1980. *Ancient Woodland*, London, Arnold.

Reavy, D. 1968. Beaker burials in south Derbyshire. *Derbyshire Archaeological Journal,* **88**, 68–81.

Rowley-Conway, P. 1982. Forest grazing and clearance in temperate Europe with special reference to Denmark: an archaeological view. In Bell, M., and Limbrey, S. (Eds.), *Archaeological Aspects of Woodland Ecology*, Oxford, British Archaeological Reports, International Series No. 146, 199–216.

Shotton, F. W. 1978. Archaeological inferences from the study of alluvium in the lower Severn–Avon valleys. In Limbrey, S., and Evans, J. G. (Ed.), *The Effect of Man on the Landscape: the Lowland Zone*, London, Council for British Archaeology, Research Report No. 21, 27–32.

Simmons, I. G. 1964. Pollen diagrams from Dartmoor. *New Phytologist*, **68**, 165–180.

Simmons, I. G., Dimbleby, G. W., and Grigson, C. 1981. The Mesolithic. In Simmons, I. G., and Tooley, M. J. (Eds.), *The Environment in British Prehistory*, London, Duckworth, 82–124.

Sims, R. E. 1978. Man and vegetation in Norfolk. In Limbrey, S., and Evans, J. G. (Eds.), *The Effect of Man on the Landscape: the Lowland Zone*. London, Council for British Archaeology, Research Report No. 21, 57–62.

Simpson, G. (Ed.) 1976. *Watermills and Military Works on Hadrian's Wall: Excavations in Northumberland, 1907–1913 by F. Gerald Simpson,* Kendal, Titus Wilson and Son.

Smith, A. G. 1981. The Neolithic. In Simmons, I. G., and Tooley, M. J. (Eds.), *The Environment in British Prehistory*, London, Duckworth, 125–209.

Thomas, K. D. 1982. Neolithic enclosures and woodland habitats on the South Downs in Sussex, England. In Bell, M., and Limbrey, S. (Eds.), *Archaeological Aspects of Woodland Ecology*, Oxford, British Archaeological Reports, International Series No. 146, 147–170.

Turner, J. 1979. The environment of northeast England during Roman times as shown by pollen analysis. *Journal of Archaeological Science,* **6**, 285–290.

Turner, J. 1981. The Iron Age. In Simmons, I. G., and Tooley, M. J. (Eds.), *The Environment in British Prehistory,* London, Duckworth, 250–281.

Vine, P. M. 1982. *The Neolithic and Bronze Age Cultures of the Middle and Upper Trent Basin*, Oxford, British Archaeological Reports, British Series No. 105.

Warren, S. H., Clark, J. G. D., Godwin, H., Godwin, M. E., and Macfadyen, W. A. 1934. An early mesolithic site at Broxbourne sealed under boreal peat. *Journal of the Royal Anthropological Institute,* **64**, 101–128.

Waton, P. V. 1982. Man's impact on the chalklands: some new pollen evidence. In Bell, M., and Limbrey, S. (Eds.), *Archaeological Aspects of Woodland Ecology,* Oxford, British Archaeological Reports, International Series No. 146, 75–92.

Whittle, A. W. R. 1978. Resources and population in the British Neolithic. *Antiquity,* **52**, 34–42.

Wymer, J. J., Jacobi, R. M., and Rose, J. 1975. Late Devensian and early Flandrian barbed points from Sproughton, Suffolk. *Proceedings of the Prehistoric Society,* **41**, 235–241.

Background to Palaeohydrology
Edited by K. J. Gregory

9

The reflection of hydrologic changes in the fluvial environment of the temperate zone during the last 15,000 years

Leszek Starkel

Institute of Geography, Polish Academy of Sciences, Cracow

This paper presents some preliminary results of the fluvial subproject of the IGCP Project No. 158 'Palaeohydrology of the temperate zone in the last 15,000 years' (*cf*. Starkel, In Press, Starkel and Thornes, 1981). The aim of this subproject is to reconstruct changes of the fluvial environment in the temperate zone with special attention paid to the changes in the hydrological regime and climate in the late glacial and Holocene. The results will be compared with the data from lakes and rivers (analysed by subproject B), and from glaciers and sea level changes. The response of rivers during the same climatic trend of the last 15,000 years was not the same in two former pleniglacial zones, which were the extraglacial or periglacial (with permafrost), and the glacial ones (covered by extensive ice sheets).

Reconstruction of the fluvial environment is based on the detailed examination of sediments and morphology in many selected river basins. One also has to take into account the extensive theoretical knowledge and all the data contained in numerous contributions which are not connected directly with the subproject. Therefore, the theoretical models presented by Soergel (1921), Schumm (1965, 1977), Knox (1975), and others are discussed in this paper in order to develop a working hypothesis which should be useful for future work and could later be accepted or rejected.

Various methods are available for the reconstruction of changes of the fluvial environment. Geomorphological methods help to distinguish channels, floodplains, and terraces. The division into braiding, meandering, and transitional rivers is based on the detailed characteristics of channel parameters: width (w), depth (d), length of meander (λ), river gradient (s) and sinuosity (P). Formulae and equations involving those parameters (Leopold *et al.*, 1964; Dury, 1977a; Schumm, 197) are suitable for the reconstruction of the palaeohydrologic regime. The analysis of transverse sections of the valley floor enables one to explain changes caused by lateral erosion, aggradation, and avulsion of channels. The examination of the longitudinal profile of the floodplain allows one to evaluate the influence of eustatic and tectonic factors in fluvial history.

Sedimentological methods are used to distinguish various facies of fluvial deposits (Shancer, 1951; Lawrushin, 1966; Allen, 1977; Church, 1978). The facial setting is different in braided and meandering systems with the dominance of channel facies in braided rivers. For detailed stratigraphic reconstruction the undisturbed deposits of palaeochannels are of high value. The granulometric characteristics of facial units, the presence of fills of different age, and the recognition of the rate of deposition, can assist in the reconstruction of many parameters of the past environment.

Other complementary methods involved in the reconstruction of the fluvial environment include palaeopedological methods (Brunnacker, 1978; Kowalkowski and Starkel, 1977), clustering of the fossil oaks dated by dendrochronological methods (Becker and Schirmer, 1977), percentage of the rebedded Tertiary pollen in diagrams indicating variation in the intensity of erosion (Mamakowa and Starkel, 1977), deposition phases of the calcareous tufa (Jäger, 1982; Lozek, 1975), and archaeological and historical methods.

Contributions cited in this paper refer to the river basins studied under the IGCP 158 project (e.g. Rhine, Danube, Morawa, Tisza, Vistula, Warta, Oulanka, and Saskatchewan Rivers) as well as to those distributed all around Europe, Siberia and North America (Starkel, 1982].

GENERAL CONCEPTS OF CHANGE IN THE TEMPERATE ZONE

In the former glacial and periglacial zones, the main parameters causing changes of the fluvial system are influenced both by climate, especially temperature and precipitation, and by vegetation.

The parameters related to the river regime are mean annual discharge, type of flood (snowmelt, ice jam, rainy), and, on a longer time scale, the frequency of extreme floods. Equally important are sediment load, particularly mean annual sediment load, maximum sediment load during floods, and the relation of the bedload to the suspended load. Both river discharge (Q_w) and sediment load (Q_s) are not only reflected in the shape and pattern of the river channel but also in the river gradient (S) which simultaneously reflects the other factors such as tectonic and glacitectonic movements, glaciation, and sea level changes. The length of the river channel and river valley (L) vary in time due to eustatic sea variation or shifting of the ice front.

The main trends of change

Concepts of valley floor formation during the late glacial and Holocene are derived from field studies or from theoretical models. Among the earliest, Soergel's concept (1921) should be mentioned. It provides an explanation of the erosion in interglacial periods and aggradation in the cold stages by changes in the river discharge and supply of material from slopes exemplified by the foreland of the Alps and surrounding areas. The existence of two different fills, the periglacial and Holocene ones, is a background for the theory of Trevisan (1949) and Jahn (1956) on two phases of aggradation separated by two phases of erosion during the early glacial and late glacial.

In his substantial papers on Quaternary palaeohydrology, Schumm (1965, 1977) discusses the main trends of change during the glacial–interglacial cycle in various morphogenetic zones. Special attention is paid to the reflection of changes in Q_w (mainly channel-forming discharge) and Q_s in the parameters of the channel. Schumm's assumptions are based on the following relations:

$$Q_w = \frac{wd\lambda}{sP} \tag{1}$$

$$Q_s = -\frac{w\lambda s}{dP} \tag{2}$$

Taking into account basins partially occupied by ice sheets or valley glaciers, as well as basins located in the periglacial zone during cold phases, Schumm states after Zeuner (1959) that the main phases of erosion were in late glacial and early interglacial times, following the deposition in the pleniglacial period and succeeded by the stable phase during the full interglacial period or during the Holocene. Therefore, the main trend of change in the fluvial system during the last 15 000 years can be expressed by:

$$Q_w^- \, Q_s^- = w^- \, d^{-}_{+} \, \lambda^- \, s^- \, P^+ \tag{3}$$

It explains the tendency to incision and meandering. However, during the moister phases of the Holocene, and especially during the periods of increased human impact, one has to take into consideration the possibility of higher river discharges. In that case, the opposite tendency is observed:

$$Q_w^- \, Q_s^- = w^+ \, d^- \, \lambda^+ \, s^+ \, P^- \tag{4}$$

If one does not consider the rivers fed earlier by melting ice sheets in the former periglacial zone (i.e. the present temperate zone), a general climatically-controlled trend may be observed. The rise of temperature, precipitation, and the expansion of forest vegetation occur in parallel with the decrease in the percentage of meltwater, frequency of floods, and retreat of permafrost. According to Dury (1977b), Lamb (1977), and Starkel (1982), this picture is much more complicated in relation to the general trend of aridity during late glacial and Eoholocene and to overlapping with climatic rhythms of a shorter duration. Metachronous deforestation and the introduction of agriculture make the above pattern even more complicated. Based on the data collected in numerous river basins of the temperate zone we find that both erosional and aggradational effects are reported for the periods with a tendency to braiding as well as to meandering. The general trend from braiding in the pleniglacial towards meandering, from dominance of channel facies to overbank ones, can be exhibited by rivers whose channels either go down or rise up. In this context the phases of higher erosional and depositional river activity are simultaneous (Klimek and Starkel, 1974) and are related to periods of higher frequency of channel-forming floods.

If a model of changes of the channel–floodplain system during the last 15 000 years is to be constructed, that time period has to be considered as the period of continual variations. The fluvial system tends to reach a threshold value at a certain time interval. At each phase a leading factor exists which pushes the whole system towards downcutting or aggradation and the downcutting or upbuilding leads to the equilibrium between Q_w and Q_s. This leading factor can be a change in river discharge as well as in sediment load. Both of them reflect environmental changes in the river basin (Figure 9.1).

The discussion below does not apply to the uppermost parts of river channels and to the lowermost reaches entering seas. Then, four cases of change are presented.

The leading role of the decrease in Q_s before Q_w ($Q_w^- < Q_s^-$) is observed during the phase of rapid invasion of forests in the Eoholocene against the background of decreasing frequency of floods and this is expressed in the tendency to meandering and incision (*Em*). The leading role of decrease in water discharge ($Q_w^- > Q_s^-$) in a more continental climate with a slow decrease of periglacial slope processes is expressed in aggradation with a slow change from braiding to meandering (*Am*). The increase in flood frequency and high bedload transport occurring several times during the Holocene can be the result of a rise of river discharge ($Q_w^+ > Q_s^+$) under natural conditions, causing the deepening and straightening of channel, or overloading of rivers during periods of higher flood frequency ($Q_w{}^+ < Q_s{}^+$) in agricultural areas with a dense pattern of cart-tracks.

The discussion above elucidates the characteristic changes of channels related to the four formulae distinguished. All of the formulae (5, pp.218) reflect the whole range of changes of the equilibrium profile from the pleniglacial braided rivers to the Holocene meandering ones which were later disturbed by human impact. The change of channel equilibrium depends on variations in channel-forming discharge and on the relation between bedload and suspended load (Figure 9.1):

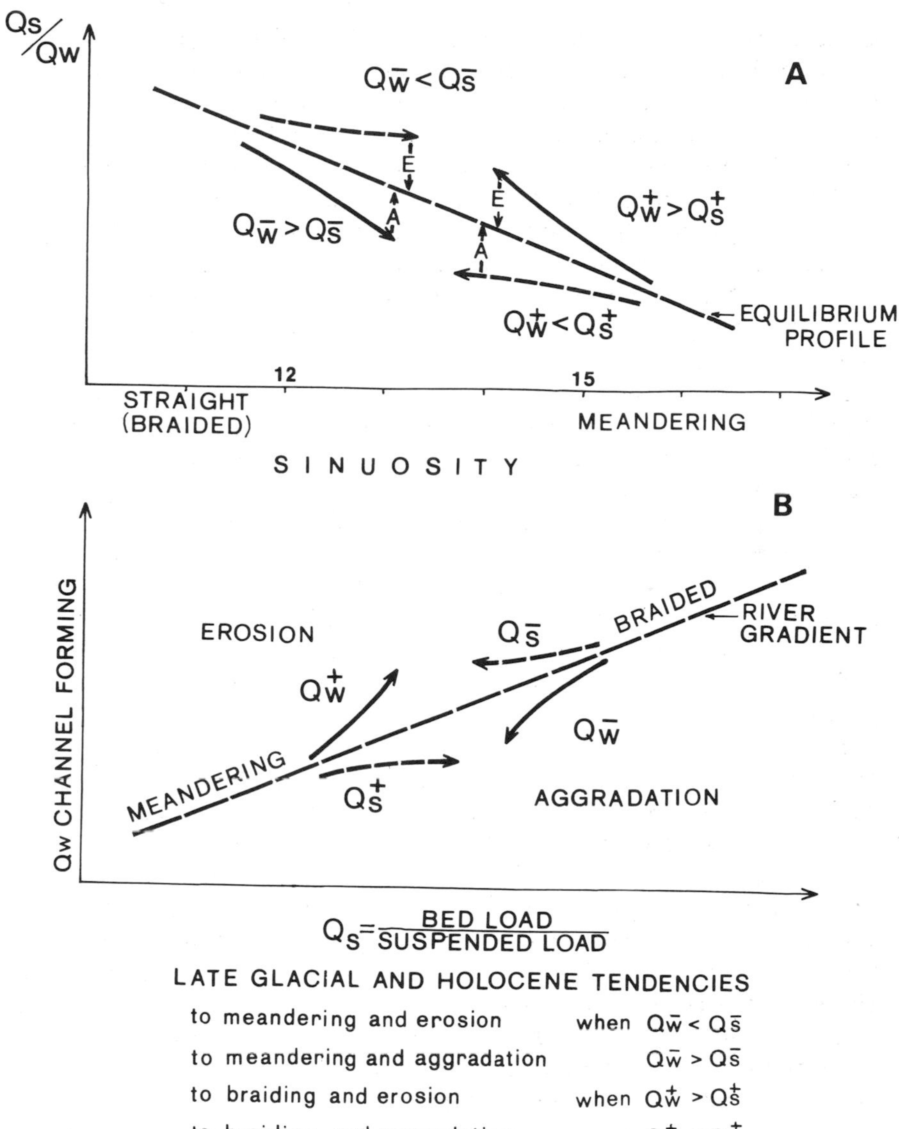

FIGURE 9.1 MODEL OF CHANGES OF THE FLUVIAL SYSTEM DURING THE COLD STAGE–INTERGLACIAL CYCLE

These diagrams illustrate the main tendencies of change and indicate four leading factors (decrease in Q_w or in Q_s, rise in Q_w or in Q_s) causing a trend towards erosion or aggradation

These formulae are as follows:

$$Q_w^+ < Q_s^- = w^- \, d^+ \, \lambda^- \, s^- \, P^+ \, (Em)$$
$$Q_w^- > Q_s^- = w^- \, d^- \, \lambda^- \, s^- \, P^+ \, (Am) \qquad (5)$$
$$Q_w^+ > Q_s^+ = w^+ \, d^+ \, \lambda^+ \, s^+ \, P^- \, (Eb)$$
$$Q_w^+ < Q_s^+ = w^+ \, d^- \, \lambda^+ \, s^+ \, P^- \, (Ab)$$

The letters in brackets indicate the tendency to erosion (E), aggradation (A) with a simultaneous increase of river sinuosity (m—meandering), or decrease of river sinuosity (b—braiding). The various stages of river development are exemplified and discussed below.

The key position of the late glacial phase

The first 5000–6000 years of the period under consideration will be called generally 'late glacial', although the stratigraphically defined 'Late glacial' covers only the time interval between 13 000–10 000 a B.P. (Mangerud *et al*, 1974). The increasing continentality superimposed on general warming supported the incision and change from braided to meandering channels and this is reported from many valleys in Poland (Falkowski, 1975; Kozarski and Rotnicki, 1977), Belgium (de Smedt, 1973; Paulissen, 1973), Germany (Brunnacker, 1978), Great Britain (Rose *et al.*, 1980), and other countries.

Underfit steams with wide valley floors and narrow sinuous Holocene channels are typical for the whole of the former periglacial zone, especially for lowland valleys, not only for the valleys originally fed by glacial meltwaters. In the earlier glaciated areas this contrast is greater and the rate of incision as well. That aspect is discussed later.

In the whole of the former periglacial zone a supplementary role was played by a very low position of sea level which rose at that time from *ca.*100 m to 40 m below the present sea level. During the late glacial period *sensu lato* all pleniglacial alluvial bodies have been dissected up to 10 m and more also in the upper reaches as in the Carpathian valleys (Starkel, 1977) as depicted in Figure 9.2. The lower erosional benches were covered by the deposits dated between 12 000 and 10 000 a B.P. Instead of the braided pattern of the higher pleniglacial terraces, large scars of meanders exist on the lower benches. The radius of those meanders is 3–4 times greater than that of the Holocene meanders. This meandering trend continued until the Eoholocene when the incised meanders of a smaller size started to form (Szumański, 1982). This indicates the lowering of the channel-forming discharge for some Polish rivers by up to 5–10 times and agrees very well with the calculations made by Dury (1977) who assumed that Q_w and Q_s decreased 3–6 times during the time interval between 12 000 and 10 000 a B.P.

During that period there was a short break, namely the cooling of the Younger Dryas (10 900–10 250 a B.P.), reflected in the higher activity of rivers. This activity was manifested in the coarsening of material in channel

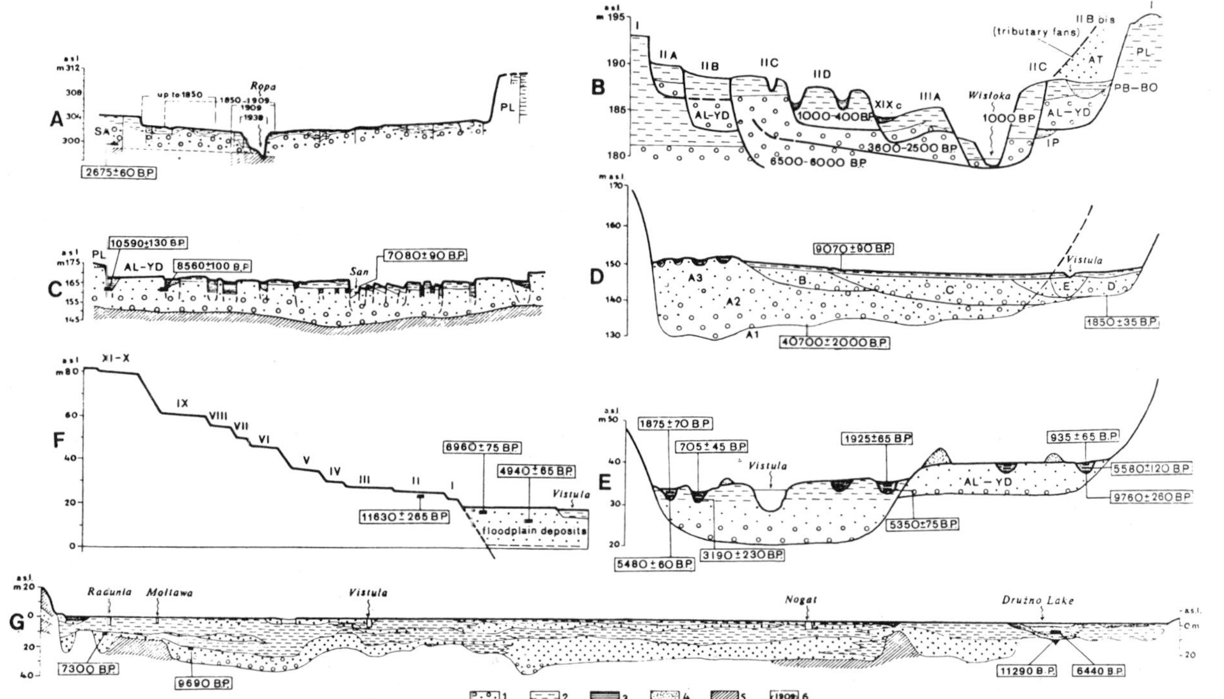

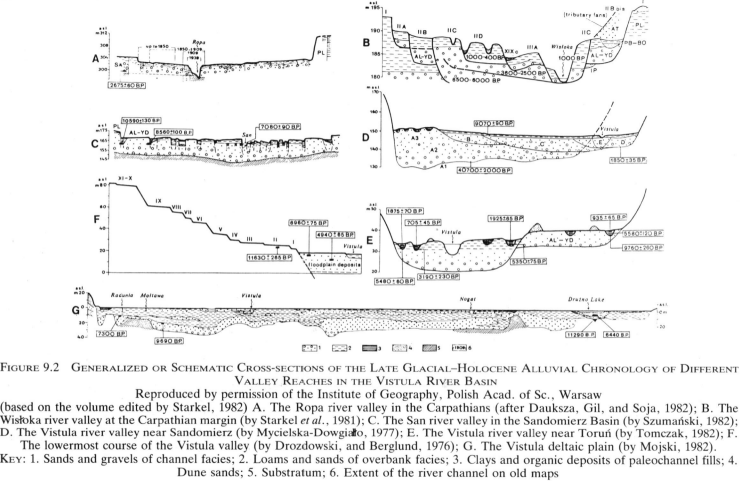

FIGURE 9.2 GENERALIZED OR SCHEMATIC CROSS-SECTIONS OF THE LATE GLACIAL–HOLOCENE ALLUVIAL CHRONOLOGY OF DIFFERENT VALLEY REACHES IN THE VISTULA RIVER BASIN

Reproduced by permission of the Institute of Geography, Polish Acad. of Sc., Warsaw (based on the volume edited by Starkel, 1982) A. The Ropa river valley in the Carpathians (after Dauksza, Gil, and Soja, 1982); B. The Wisłoka river valley at the Carpathian margin (by Starkel *et al.*, 1981); C. The San river valley in the Sandomierz Basin (by Szumański, 1982); D. The Vistula river valley near Sandomierz (by Mycielska-Dowgiałło, 1977); E. The Vistula river valley near Toruń (by Tomczak, 1982); F. The lowermost course of the Vistula valley (by Drozdowski, and Berglund, 1976); G. The Vistula deltaic plain (by Mojski, 1982).

KEY: 1. Sands and gravels of channel facies; 2. Loams and sands of overbank facies; 3. Clays and organic deposits of paleochannel fills; 4. Dune sands; 5. Substratum; 6. Extent of the river channel on old maps

bars and levees in the Dijle valley (Munault and Paulissen, 1973), the Gipping valley (Rose *et al.*, 1980), Wisłoka valley (Starkel *et al.*, 1981) and elsewhere. It was probably the tendency to braiding caused by more frequent ice drives (*cf.* Smith, 1981). A distinct change of the grain size and turn from channel to floodplain swampy deposits was found at the very beginning of the Holocene and dated in Southern Poland between 10 300 and 9900 a B.P. (Figure 9.3).

The late glacial trend was not uniform throughout the extraglacial zone. According to Kozarski and Rotnicki (1977) the shift from braided to meandering channels in the Polish Lowland was not simultaneous and followed by the continuous decrease of the channel size occurring at the same time as the incision of shifting channels. The late glacial rise of temperature induced the growth of the active layer and retreat of permafrost in the extreme continental climate of Siberia. Kartashov (1972) reports the period of aggradation due to the great activity of solifluction and thermokarst in the East and Central Siberian river valleys. In the Aldan valley, Mochanov (1976) shows only a formation of parallel fills without a change of the base level during the whole Late glacial and Holocene.

Cyclic variations in the fluvial regime during the Holocene

Cyclic variations of river discharge and sediment load during the Holocene are superimposed on the general trend of the decrease of Q_w and Q_s in the forest zone. The cycles *ca.* 2 000 years long are reflected well in the alluvial fills side by side. The bases of those fills are composed of gravels and sands of channel facies with coarse pebbles in the bottom horizon. Fine overbank facies and fills of meandering palaeochannels occur on top of the channel facies. Studying the sequence of changes from the seventeenth until the nineteenth century in the Carpathian valleys (Starkel *et al.*, 1981), the author came to the conclusion that each new fill started to form during a phase with a higher flood frequency. When the threshold value has been reached it causes the straightening of the existing channels or avulsion combined with incision and deposition of new fills (Figure 9.4). These phases of intensive river activity are separated by longer periods of relative stability and accompanied by the formation of meandering channels due to lateral erosion and accretion and most of the preserved palaeomeanders are related to the stable phases.

Parallel Holocene fills are found in numerous European river valleys, namely the Maas (Paulissen, 1973), Rhine (Brunnacker, 1978), Main (Becker and Schirmer, 1977), Danube (Fink, 1977), Weser (Lüttig, 1960), Wisłoka (Starkel *et al.*, 1981), Kama (Goreckij, 1964); in Siberian river valleys (Kind, 1967); and North American ones: Bow river valley (Jackson *et al*, 1981) or Pomme de Terre river valley (Brakenridge, in press). These fills are shown in Figure 9.2.

The simultaneity of some phases with a higher flood frequency is present in

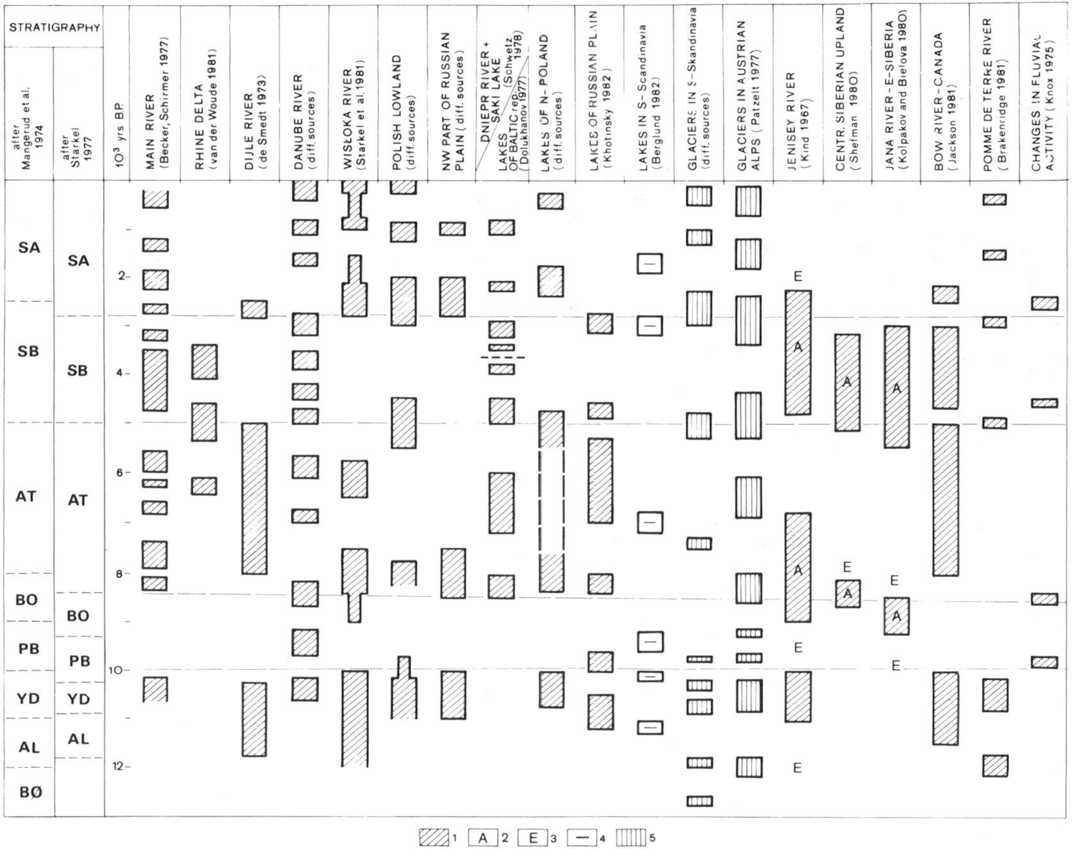

STRATIGRAPHY

after Mangerud et al. 1974: SA, SB, AT, BO, PB, YD, AL, BØ

after Starkel 1977: SA, SB, AT, BO, PB, YD, AL

10^3 yrs B.P.: 2, 4, 6, 8, 10, 12

MAIN RIVER (Becker, Schirmer 1977)

RHINE DELTA (van der Woude 1981)

DIJLE RIVER (de Smedt 1973)

DANUBE RIVER (diff. sources)

WISŁOKA RIVER (Starkel et al. 1981)

POLISH LOWLAND (diff. sources)

NW PART OF RUSSIAN PLAIN (diff. sources)

DNIEPR RIVER + SAKI LAKE (Schwetz 1978) / LAKES OF BALTIC rep. (Dolukhanov 1977)

LAKES OF N-POLAND (diff. sources)

LAKES OF RUSSIAN PLAIN (Khotinsky 1982)

LAKES IN S-Scandinavia (Berglund 1982)

GLACIERS IN S-Skandinavia (diff. sources)

GLACIERS IN AUSTRIAN ALPS (Patzelt 1977)

JENISEY RIVER (Kind 1967)

CENTR. SIBERIAN UPLAND (Shefman 1980)

JANA RIVER-E-SIBERIA (Kolpakov and Bielova 1980)

BOW RIVER-CANADA (Jackson 1981)

POMME DE TERRE RIVER (Brakenridge 1981)

CHANGES IN FLUVIAL ACTIVITY (Knox 1975)

1 A 2 E 3 — 4 5

FIGURE 9.3 CHANGES OF HYDROLOGICAL REGIME IN THE TEMPERATE ZONE IN THE LATE GLACIAL AND HOLOCENE Compilation based on information on river activity compared with changes of lake levels and fluctuations of mountain glaciers. KEY: 1. Increased fluvial activity or high lake water level: 2. Aggradation phase (after author): 3. Downcutting phase (after author): 4. Low lake water level: 5. Phase of glacial transgression

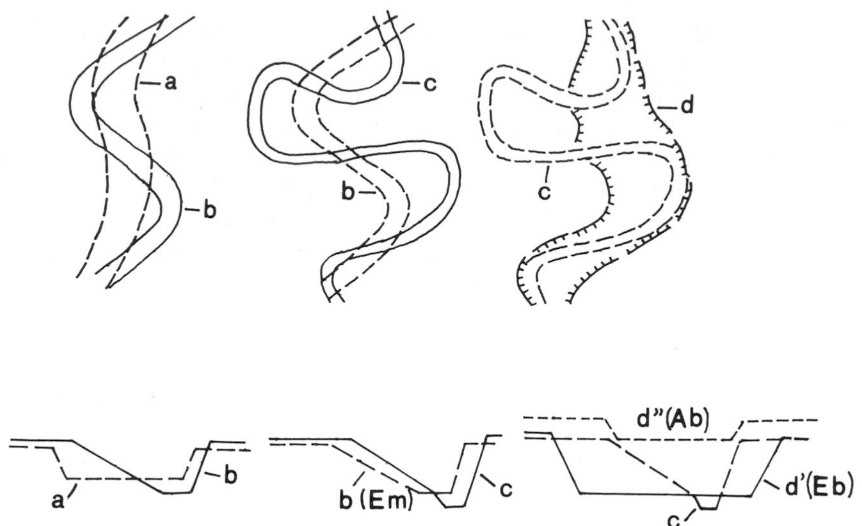

FIGURE 9.4 CHANGES OF CHANNEL PATTERN AND CROSS-SECTION DURING CYCLIC VARIATIONS OF HYDROLOGICAL REGIME
KEY: a. Straight channel of braided river; b. Sinuous channel; c. Deeper meandering channel; d. Straight braided channel formed during the phase with higher flood frequency transformed by incision (*Eb*) or by aggradation (*Ab*)

Figure 9.3. Among them a very distinct rise in the river activity is visible all around Europe between 8500 and 7500 a B.P. This latter phase is reflected also in the avulsion of many river channels, for example in Hungarian Plain (Borsy and Felegyhazi, 1982). The subsequent unstable phase took place at 6 500–5 900 a B.P. and at the Atlantic/Subboreal transition (5 100–4 500 a B.P.). The fills younger than Atlantic ones are recorded also from Eastern and Central Siberia (Kolpakov and Bielowa, 1980; Schafman, 1980) and from Bow River (Jackson *et al.*, 1981). Neoholocene synchronous aggradational phases are reported from many localities and dated from the beginning of the Subatlantic and Roman times (2800–2200 a B.P.) and from the Mediaeval period after a relatively stable condition in late Roman and early Mediaeval times (Havliĉek, 1980; Starkel *et al.*, 1981). However, starting from *ca.* 3500 a B.P. many fills are not synchronous, probably due to overlapping with landnam phases and the growing input of suspended load from arable land.

These effects of increased fluvial activity may be detected as clusterings of buried oaks (Becker, 1975). Similar phases of increased flood frequency (8500–7500, 6800–5800, 5300–5000 a B.P. and many younger ones of the Bronze Age) have been established by means of dendrochronological methods. Analysis of the distribution of radiocarbon dates in alluvial deposits (Knox, 1975) supports the above statements (Figure 9.3). Very good indicators of the moister periods are the phases of gullying and deposition of small aluvial fans found in Southern Poland (8400–7500 a B.P.) which are described by Starkel *et al.* (1981) and in Scandinavia where the main phase took place at 8400–7700 (Sonstegaard and Mangerud, 1977). The existence of humid phases at the beginning and at the end of the Atlantic period results in the formation of new landslides in the Pennines (Franks and Johnson, 1964) and in the Flysh Carpathians (Gil *et al.*, 1974). Similar phases based on historical data have been described from the sixteenth until the eighteenth century in Northern Europe (Grove, 1972) and coincide with the Little Ice Age.

The role of extreme events and rapid changes

In the analysis of phases of higher flood frequency, the role of individual events cannot be overlooked. In reality, each phase of increased activity is composed of many extreme floods. The clustering of these events causes the fluvial system to reach its external threshold (Schumm, 1977) and a new alluvial fill starts to form. It is possible to count beds or layers related to individual floods in the alluvial fans, as in the Wisłoka valley for the period between 8 400 and 7 500 a B.P. (Starkel *et al.*, 1981). A similar effect can be achieved by one individual event which has a recurrence interval of hundreds or even thousands of years. Such an event causes a rapid change in the existing system and a slow return of the river to the former stable phase. This is exemplified by the Cimarron River which changed its channel width rapidly from 15 to 360 m in 1914 and since that time has been changing back slowly (Schumm, 1977). This is a kind of self regulation of the fluvial system and many avulsions known from mountain forelands are related to such extreme floods (Smith, 1981).

In a semiarid climate, these very rare events are a basic feature of the acting fluvial system (Leopold *et al.*, 1964). Slack water deposits in the rocky canyons help to count the number of extreme floods (Baker, in press). Costa (1978), studying the deposits of the 1972 year flood in the Big Thompson canyon in the Frontal Range, has found only two similar events in the Holocene, dated 10 210 and 7 420 a B.P.

On a global scale, the most extreme event was the famous Missoula flood connected with transfluence of a glacial lake, when probably several times during the decline of the pleniglacial the discharge of river reached 9 million m^3 s^{-1} (Baker, 1978). Some rivers exhibit relatively rapid but permanent changes if the discharge diminishes as the supply of glacial meltwaters ceases. The existence of icing and ice jams during the lateglacial meant that different types of extreme events were relatively frequent at that time and played a dominant role in the transformation of the fluvial system.

Change in the longitudinal profile of the river valley

A very simple model indicates that erosion upstream should be reflected in aggradation downstream. The concept of evolution of the longitudinal profile of the river valley is much more complicated than is outlined generally in Woldstedt's paper (1952). Eustatic and tectonic factors are of great importance as well as supply of glacial meltwaters (Starkel *et al.*, 1982). In mountain areas the upper courses of gravel bed rivers show a permanent tendency to braiding (Werritty and Ferguson, 1980) but even there incision takes place after the pleniglacial (Baumgart-Kotarba, 1980). In small basins the role of the extreme single event can be so great that it makes a study of the continuity of climatic trends impossible. On the contrary, the meandering channels typical of Holocene have a tendency to disappear in the lower reaches due to very low gradients. The tendency to downcutting or aggradation is related to the variations of the sea level (Hageman, 1969; Drozdowski and Berglund, 1976).

Therefore, the best conditions for the study of the effects of climatic changes on the fluvial environment are in the middle course of the river valley. In his theoretical paper Kadar (1955) presented a model of continuous alternation of erosional and aggradational phases in the middle course. In fact, well expressed sequences of fills and palaeomeanders of different age could be found only in these reaches together with well-developed meandering channels and floodplains.

The alluvial fans in the mountain or upland forelands represent a particular case. They are characterized by sequences of incised fan fills (Troll, 1957; Brunnacker, 1960) and by the shifitng or avulsion of channels well known from subsiding margins of the Hungarian Plain (Borsy and Felegyhazi, 1982; Vaŝkovska *et al.*, 1979), the Po Plain (Cremaschi, 1981) or from the Subcarpathian Foreland (Starkel, 1977).

The data collected in selected river basins are not as yet sufficient to say that phases of downcutting and aggradation were simultaneous at most parts of the longitudinal profile. However, it is probable that long-term phases of increased fluvial activity were simultaneous in the whole river basin, although the effects in the channel and alluvial plain formation were different.

General concept of the sequence of change

It is too premature to propose a general concept of fluvial system change which could be applicable to the whole temperate zone. The data cited above, and the relatively high level of detail obtained from study of the Central European river valleys, permits one to summarize preliminary results excluding mountain headwaters and lowermost reaches (Figure 9.3). Following the sequence of changes of Q_w and Q_s and trying to find the leading factor causing the tendency to erosion or aggradation we can find the sequence of events:

$$Ab \rightarrow (Am) \rightarrow Em \rightarrow Ab \rightarrow (Am \rightarrow Em \rightarrow Eb)^n \sim (Ab \rightarrow Am)^n \rightarrow Eb \qquad (6)$$

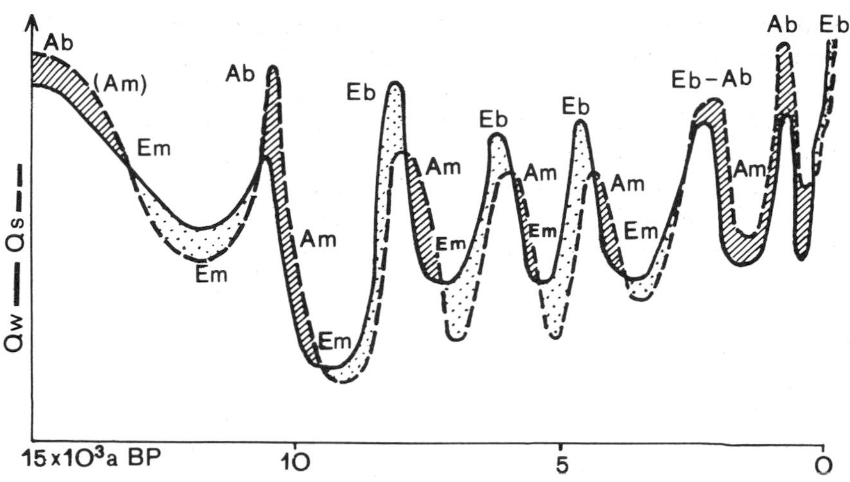

Four types of changes of Qw and Qs

I (Em)	$Q\bar{w} < Q\bar{s}$	$= w^-, d^+, \lambda^-, s^- P^+$	E – erosion
II (Am)	$Q\bar{w} > Q\bar{s}$	$= w^-, d^-, \lambda^- s^- P^+$	A – aggradation
III (Eb)	$Q\overset{\pm}{w} > Q\overset{\pm}{s}$	$= w^+, d^+, \lambda^+ s^+ P^-$	m – tendency to meandering
IV (Ab)	$Q\overset{\pm}{w} < Q\overset{\pm}{s}$	$= w^+, d^-, \lambda^+ s^+ P^-$	b – tendency to braiding (λ^-)

Typical sequence in Central Europe:

Ab → (Am) → Em → Ab → (Eb → Am → Em)^x → Ab → Eb

FIGURE 9.5 TYPES OF SEQUENCE OF CHANGE IN THE HYDROLOGICAL REGIME AND SEDIMENT YIELD TYPICAL OF CENTRAL EUROPEAN VALLEYS DURING THE LAST 15000 YEARS

The general idea behind this sequence (see Figure 9.5) is a shift between the braided and the meandering channel system. The pleniglacial period (*ca.* 15 000 a B.P.) with a braiding pattern is characterized by a surplus of sediment yield and that caused the aggradation phase (*Ab*). The subsequent increase of continentality and warming of climate supported erosion and a trend to meandering. In the Alleröd phase a horizontal shift (*Am*) or incision by meandering rivers (*Em*) can be noticed. The next episode of the increased braiding tendency and deposition of coarser material is related to the younger Dryas (*Ab*), followed by Eoholocene decrease both in discharge and sediment load (*Am*, later *Em*). At the Boreal/Atlantic transition the flood frequency increases rapidly under dense forest cover. Those types of moist phase alternate with the more stable ones. This causes the following sequence of change:

$$Eb \rightarrow Am \rightarrow Em \qquad (7)$$

A repetition of those phases is typical for the period of the existence of the dense forest communities before disturbance by man. Since Neolithic or Bronze Age times or more often in the Roman or Mediaeval periods (Falkowski, 1975; Jäger, 1962; Knox, 1977; Becker and Frenzel, 1977; Butzer, 1980; Havliĉek, 1980), both kinds of phases with higher or lower flood frequency are characterized by overloading of rivers mainly by suspended load. Therefore, we observed a dominance of aggradation in cyclic changes ($Ab \rightarrow Am \rightarrow Ab$) so that the floodplain builds up. This trend expressed in the increased size of meanders has been disturbed in the last 2–3 centuries. Introduction of potatoes, very frequent floods, and artificial cuts of meanders induced a tendency to braiding and simultaneous downcutting (*Eb*). This phase has been well described from the San river valley (Szumański, 1982).

OTHER FACTORS MODIFYING THE CONCEPT PRESENTED

Some other factors like glaciation, sea level changes and tectonic movements can disturb this general picture of change or can accelerate the main trend of evolution related to the zonal climatic variations.

Glaciation in the river basin

Glaciation can act in various forms changing not only Q_w and Q_s but also the slope and length of the channel and valley floor or even, after deglaciation, create a new river pattern. The most typical situation exists when the river is temporarily fed by glacial meltwaters crossing the watershed (transfluence) or flowing from the glaciated headwaters. The first case can be exemplified by the Mississippi which had a rapidly diminishing discharge ca. 11 500 a B.P. (Kennett, 1974) or the Volga in which the supply of meltwaters stopped *ca.* 14 500 a B.P. causing the great regression of the Caspian Sea (Kvasov, 1976). In the Volga catchment, the Q_w of some tributary rivers had to decrease up to 100–200 times (Sladkopievcev, 1976). In such cases the contrast between past and present channel and floodplain is extremely distinct (Dury, 1977a). The retreat of glaciers in the headwaters caused the incision downstream, especially in the piedmont zones with the extensive fans. The system of cut and fill fans from the Alpine foreland has been described by Troll (1957). Detailed examination of change of the sandur terraces of 13 000 a B.P. in N. Esk river valley in Scotland is described by Maizels (in press) who found changes of sinuosity from 1.022 to 1.068, of braiding index from 5.50 to 2.31, and decreasing gradient of the four terrace steps. The present-day N. Esk river has a gradient of only 0.0037 and this contrasts with the 0.035 typical for the late glacial. Present maximum discharge Q_{max} of 262 $m^3\ s^{-1}$ also contrasts with that of 2 400 $m^3\ s^{-1}$ calculated for the time when the river was fed by melting ice.

A similar study of four fills on Baffin Island (Church, 1978) has shown that during the deposition of the oldest body (*ca.* 5 700–4 300 a B.P.) the mean water discharge was ten times higher, maximum discharge 50 per cent higher and sediment load as much as 18 times higher than those at the present time.

Ice wastage followed by rapid incision due to the lowering of the base level was very typical in the river valleys inclined to the north. The series of steps of erosional terraces are common in the basin of the lower Vistula (Galon, 1934; Wiśniewski, 1976), Niemen (Voznyachuk and Valczyk, 1977), and other valley systems. The downcutting which was reflected in a staircase of *ca.* 10 terraces reached 30–50 m in *ca.* 3 000–4 000 years. In the case of the Prosna and Warta river valleys the incision caused by braiding started to change to meandering rivers *ca.* 13 000–12 000 a B.P. (Kozarski and Rotnicki, 1977). The change to meandering channels is observed in the Niemen river valley at the end of Younger Dryas.

The similar sequence of glacial dammed lakes has been described for Saskatchewan and other rivers (St. Onge, 1972). The base level was lowered more than 200 m and this tendency is continuing; the main alluvial body of the 19.5 m high terrace was dated at 6 000 a B.P., the 10 m terrace at 2 800 a B.P. In central Finland, free of ice in the Eoholocene, the rate of erosion in the Oulanka valley reached 25 m during 3–4 millenia and later a relatively stable meandering system has been formed (Koutaniemi, 1979].

The greatest changes occurred in the relatively flat, ice-free areas where a new river pattern started to form. These 'young rivers', using the terminology of Falkowski (1975), form new longitudinal profiles due to erosion or aggradation in different reaches. In such river valleys the climatic variations are not reflected simply in the shape of channels and floodplains. There are many examples from Northern Poland or Finland where the joining of dead ice depressions and lakes in a new river system is continuing (Koutaniemi and Rachocki, 1981). Moreover, the captures of rivers and shifting of watersheds are typical in those areas. The Hinkua river in Southern Finland which had a specific runoff up to 30 per cent lower in Atlantic time due to decrease in precipitation, is a good example (Aario and Castren, 1969).

The Flandrian transgression

Rise of sea level is usually reflected in the lowermost courses of rivers, namely in the shifting of the perimarine facies (Hageman, 1969). The growth of deltas usually coincides with the very high sea level. The formation of lobes of the great Mississippi delta started *ca.*7 500 a B.P. (Kolb and van Lopik, 1966), and the formation of the small Vistula delta followed the transgression of 6 600 a B.P. (Mojski, 1982). Van der Woude (1981) suggests increasing fluvial activity *ca.* 6 400–6 100, 5 300–4 600 and 4 100–3 400 a B.P. and associates these phases with regressional phases of the North Sea. The coincidence of the two first phases with the increased fluvial activity in the whole of central Europe is remarkable. How far does the influence of sea level extend upstream? During the low sea level the regressive erosion can go very far and the low level of the Baltic ice lake or Yoldia Sea is visible in the incision below the present channel up to 400 km upstream (Wiśniewski, 1976). In the smaller basins of the Mecklemburgia with transfluent lakes the phases of Baltic regression are reflected in peat horizons (Richter, 1968).

Glaciostatic and tectonic movements

Formerly glaciated areas exhibit glacial rebound and in the highlands of Greenland, Scotland, or Scandinavia this is reflected in the rising of the river channel gradient in opposition to the climatic trend. As a result we find erosional terraces and deeper incision. This is especially pronounced in reaches of overlap with long term tectonic uplift and can be expressed, for example, in warping of the Holocene terraces (Miescheriakov and Fiedorova, 1961). The opposite tendency, the rising of the lower river course, is observed around the Hudson Bay or Bothnian Gulf. Seas are regressing and some rivers can be captured in the opposite direction (e.g. Hinkua river).

In the young orogenic zones, the longitudinal profile of the valley bottom shows much more contrast. In the mountains the climatic variations are reflected in different rates of incision and Holocene river channels are cut in the bedrock (Starkel, 1977). On the other hand alluvial fans in foreland show a tendency to aggradation, depending on the rate of subsidence.

Specific character of limestone regions

In areas underlain by limestone and other rocks rich in $CaCO_3$, the moist phases with higher rainfall are reflected in the increase of leaching. In the closed basins the late glacial phase of the retreating permafrost and surficial denudation is marked by lacustrine chalk and calcareous gyttia (Berglund and Malmer, 1971; Gerlach *et al.*, 1972). At the Boreal/Atlantic transition the warming of climate and increase of humidity induced deposition of calcareous tufa in river valleys and lasted with some breaks until the end of the Subboreal (Lożek, 1975; Jäger, 1982). The increased precipitation of carbonates is reflected in the synchronous growth of stalagmites in the cave of Zelze (Franke and Geyh, 1971). The calcareous deposition in the loess uplands occurred at the same period and formation of chernozem-like soils is dated at *ca.* 8 500 a B.P. (Jersak, 1977). The breaks in the deposition are associated with drier phases when rendzina soils have been formed as dated by archaeological methods.

COMPARISON WITH OTHER ENVIRONMENTS AND CONCLUSIONS

The main difficulty in fluvial lithostratigraphy is the lack of continuous sequences, as some phases are represented by an erosional break or a hiatus. The most active periods are represented by cut surfaces or by coarse lag subfacies. Some selected data on lakes or glaciers which reflect trends similar to those reconstructed in the fluvial environment are presented in Figure 9.3.

Some lakes are incorporated in fluvial systems. In most Polish lakes the rise *ca.* 8 000 a B.P. is spectacular (Niewiarowski, 1978). However, it is not exactly known to what extent that rise is related to degradation of the dead ice (*cf.* Koutaniemi and Rachocki, 1981). In the Baltic republics the phases of high lake levels of ca. 7 200–6 000 a B.P. and 5 000–4 500 a B.P. are well expressed (Dolukhanov, 1977). A rise about *ca.* 400–200 a B.P. is clearly visible in many lakes. The presence of intercalations of sand in numerous lake cores of Central and Eastern Europe at the break Boreal/Atlantic is of great value for correlations (Jakovleva *et al.*, 1981. Pawlikowski *et al.*, 1982). These lake level fluctuations coincide well with low lake levels in Scandinavia (Berglund, in press). The Saki lake on the northern coast of the Black Sea is a curious example and the varved lake sediments when correlated with the discharge of the Dniepr river show an increase of floods *ca.* 2 150–1 950 a B.C. 400–200 a B.C. 900–1 160 A.D. and in the nineteenth century (Schwetz, 1978).

Analyses of peat/bog profiles reflect similar variations. The Southern Scandinavian mires show the increase of humidity *ca* 5 000, 4 500, 3 800, 2 600, 2 200, 1 600, 1 200 and 500 a B.P. (Berglund, in press). Among the older phases the break Boreal/Atlantic and inter-Atlantic moist phases (*ca.* 6 500 a B.P.) are well expressed (Lamb, 1977; Ralska-Jasiewiczowa, 1980). It should be mentioned that the climatic reconstruction of the Atlantic in Eastern Europe (Klimanov, 1978) shows a tendency to a slight change towards oceanic climate in the east and warming to the north. This indicates that the synchroneity of changes does not mean a similar amplitude of variations. During some periods, such as the Atlantic, the gradient of precipitation decrease to the East was lower and cyclonic circulation could more easily reach the eastern margin of Europe. Different parts of Siberia show various trends (Khotinsky, 1977).

The other type of environment which coincides remarkably with fluvial history are the mountain glaciers. In the Holocene, glaciers in the Alps (Patzelt, 1977), Scandinavia, Alaska, and Rocky Mountains (Denton and Karlen, 1973) show similar fluctuations with well-expressed humid phases with transgressions of *ca.* 8 000, 7 000–6 000, 5 000, 3 000–2 000, 1 200–1 000, and 400–100 a B.P.

The presence of similar fluctuations in other environments tests our information collected in the river valleys of the temperate zone. The second order variations showing cyclic tendencies to repetition are superimposed on the main trend of change from the cold relatively dry pleniglacial to the humid and warm temperate climate. The alternation of phases with higher and lower flood frequency is the means of destruction of the existing equilibrium after reaching a threshold. This alternation is reflected in alluvial fills deposited side by side and vertical lowering of channels expressed by a staircase of terraces. The human impact in the fluvial environment of the temperate zone has changed not only the second order cyclicity but also the main trend causing the increase in aggradation and later downcutting due to the direct intervention in the shape of river channels.

Studies of the impact of climatic fluctuations on the fluvial system are continuing. In a few years the IGCP Project will give new conclusions, because the picture will be better documented. However, at the half-way stage of our collective work we should remember that there is a great difference between maritime and extremely continental climate with existing permafrost in the so-called temperate zone. The most distinct contrast between the pleniglacial, late glacial and Holocene fluvial systems existed in the areas, which were under the direct influence of ice sheets and their meltwaters.

ACKNOWLEDGEMENTS

The author expresses cordial thanks to Mrs. Teresa Mrozek for help with translation into English and to Mrs. Maria Klimek for the drawings. He also extends his thanks, for fruitful discussions, to colleagues attending the presentation of this paper at the meeting of the Quaternary Commission in the Cracow Branch of the Polish Academy of Sciences.

REFERENCES

Aario, R., and Castren, V. 1969. The northern discharge channel of ancient Päijänne and the palaeohydrology of the Atlantic period. *Bulletin Geological Society, Finland,* **41**, 3–20.

Allen, J. R. L. 1977. Changeable rivers: some aspects of their mechanics and sedimentation. In Gregory. K. J. (Ed.), *River Channel Changes*, Wiley, Chichester, 15–45.

Baker, V. R. in press. Paleoflood hydrological analysis from slack-water deposits. *Proceedings of the Holocene Symposium, Quaternary Studies in Poland,* **4**, Poznań.

Baker, V. R., and Nummedal, D. (Eds.) 1978. *The Channelled Scabland Guide to the Geomorphology of the Columbia Basin*, Washington, Comparative Planetary Geology Field Conf. June 5–8, 1978.

Baumgart-Kotarba, M. 1980. Braided channel changes at chosen reaches of the Białka river. *Studia Geomorphologica Carpatho-Balcanica*, **14**, 113–134.

Becker, B. 1975. Dendrochronological observations of the postglacial river aggradation in the Southern part of central Europe. *Biuletyn Geologiczny, UW,* **19**, Warszawa, 127–136.

Becker, B., and Frenzel, B. 1977. Paläoökologische Befunde zur Geschichte postglazialer Flussauen im südlichen Mitteleuropa. In *Dendrochronologie und postglaziale Klimaschwankungen in Europa*, Erdwis. Forschung 13, Wiesbaden, 43–61.

Becker, B., and Schirmer W. 1977. Palaeoecologic study of the Holocene valley development of the River Main, Southern Germany. *Boreas,* **6**, 303–321.

Berglund, B. in press. Palaeoclimatic changes in Scandinavia and on Greenland—a tentative correlation based on lake- and bog-stratigraphical studies. *Quaternary Studies in Poland,* **4**, Poznań.

Berglund, B. E., and Malmer, N. 1971. Soil conditions and late-glacial stratigraphy. *Geol. Fören. Förhandlingar,* **93(3)**, 575–586.

Borsy, Z., and Felegyhazi, E. 1981. The development of network of rivers at the north-eastern part of the Great Hungarian Plain from the end of Pleistocene up till now. *Symposium 'Palaeohydrology of the Temperate Zone'*. Abstracts of papers, Poznań, 16–17.

Brakenridge, G. R. in press. Late Quaternary floodplain sedimentation along the Pomme de Terre River, Southern Missouri. *Quaternary Research.*

Brunnacker, K. 1960. Zur Kenntniss des Spät- und Postglazials in Bayern. *Geologia Bavarica,* **43**, 74–150.

Brunnacker, K. 1978. Der Niederrhein im Holozän. *Fortschritte der Geologie Rheinland und Westfalen,* **28**, 399–440.

Butzer, K. W. 1980. Holocene alluvial sequences: problems of dating and correlation. In Cullingford, R. A. *et al.* (Eds.), *Timescales in Geomorphology*, Wiley, 131–142.

Chebotarieva, N. S., Malgina, E. A., Devirts, A. L., and Dobkina, E. I. 1965. On the age of river terraces in the northwestern part of the Russian plain (in Russian). In *Upper Pleistocene and Holocene Palaeogeography and Chronology in the Light of Radiocarbon Dating*, Moscow, 51–66.

Church, M. 1978. Palaeohydrological reconstructions from a Holocene valley fill. *Fluvial Sedimentology,* **5**, 743–772.

Costa, J. E. 1978. Holocene stratigraphy in flood frequency analysis. *Water Resources Research,* **14**, 626–632.

Cremaschi, M. 1981. Il quadro geostratigrafico dei depositi archeologici del fiume Parano. *Il Neolitico e Èta del Rame, Ricerca a Spilamberto e S.Cesario 1977–1980,* Cassa di Risparmio di Vignola, 29–41.

Cremaschi, M., and Marchesini, A. 1978. L'evoluzione di un tratto di Pianura Padana (prov. Reggio e Parma) in rapporto agli insediamenti ed alla struttura geological tra il XV sec. A.C. ed. il sec. XI D.C., *Archeologia Medievale,* **5**, 542–556.

Denton, G. H., and Karlen, W. 1973. Holocene climatic variations—their pattern and possible cause. *Quaternary Research,* **3(2)**, 155–205.

Dolukhanov, P. M. 1977. The Holocene history of the Baltic sea and ecology of prehistoric settlement. *Baltica,* **6**, 227–244.

Drozdowski, E., and Berglund, B. E. 1976. Development and chronology of the lower Vistula River valley, North Poland. *Boreas,* **5**, 95–107.

Dury, G. H. 1977a. Peak flows, low flows, aspects of geomorphic dominance. In Gregory, K. J. (Ed.), *River Channel Changes*, Wiley, Chichester, 61–74.

Dury, G. H. 1977b. Underfit stream: retrospect, perspect and prospect. In Gregory, K. J. (Ed.) *River Channel Changes*. Wiley, 281–293.

Falkowski, E. 1975. Variability of channel processes of lowland rivers in Poland and changes of the valley floors during the Holocene. *Biuletyn Geologiczny UW.*, **19**, Warszawa, 45–78.

Fink, J. 1977. Jüngste Schotterakkumulationen im österreichischen Donauabschnitt. In *Dendrochronologie und postglaziale klimaschwankungen in Europa*, Erdwiss, Forschung 13, Wiesbaden, 190–211.

Franke, H. W. and Geyh, M. A. 1971. ^{14}C-Datierungen von Kalksinter aus slovenischen Höhlen. *Der Aufschluss,* **22(7/8)**, 234–237.

Franks, J. W., and Johnson, R. H. 1964. Pollen analytic dating of a Derbyshire landslip. The Cown Edge Landslides, Charlesworth. *New Phytologist,* **63**, 209–216.

Galon, R. 1934. Dolina dolnej Wisly jej kształt i rozwój na tle dolnego Powiśla. *Badania Geograficzne,* **12/13**, Poznań.

Gerlach, T., Koszarski, L., Koperowa, W., and Koster, E. A. 1972. Sediments lacustres postglaciaires dans la depression de Jaslo-Sanok. *Studia Geomorphologica Carpatho-Balcanica,* **6**, 37–62.

Gil, E., Gilot, E., Kotarba, A., Starkel, L., and Szczepanek, K. 1974. An early Holocene landslide in the Niski Beskid and its significance for paleogeographical reconstructions. *Studia Geomorphologica Carpatho-Balcanica,* **8**, 69–83.

Gorecki, G. J. 1964. *Alluvial deposits of the great anthropogenic rivers of the Russian Plain* (in Russian), 249 pp, Moscow.

Grove, J. M. 1972. The incidence of landslides, avalanches, and floods in western Norway during the Little Ice Age. *Arctic and Alpine Research,* **4**, 131–138.

Hageman, B. P. 1969. Development of the western part of the Netherlands during the Holocene. *Geologie en Mijnbouw,* **48**, 373–388.

Havliĉek, P. 1980. Vyvoj terasoveho systemu reky Moravy v hradistskem prikopu. *Antropozoikum,* **13**, 93–125.

Jackson, L. E., MacDonald, G. M., and Wilson, M. 1981. Paraglacial and nonglacial fluvial origins for fluvial terraces and terrace sediments in the Bow River Valley, Alberta, Canada, *Conference: Modern and Ancient Fluvial Systems*, Abstracts, University of Keele, UK., p. 59.

Jäger, K. D. 1962. Über Alter und Ursachen der Auelehmablagerung Türingischer Flüsse. *Praehistorische Zeitschrift,* **40(1/2)**, 1–59.

Jäger, K. D. 1982. Stratigraphische Belege für Klimawandlungen im mitteleueropäischen Holozän. *Zeitschrift für geologische Wissenschaften,* **10**, 6, Berlin, 799–809.

Jahn, A. 1956. The action of rivers during the Glacial Epoch and the stratigraphic significance of fossil erosion surfaces in Quaternary deposits. *Przeglad Geograficzny,* **28**, Suppl. 101–104.

Jakovleva, L. V., Chomulova, V. L., Siergiejeva, L. V., and Drabkowa, V. G. 1981. Postglacial lacustrine deposition in Latvia and their change under the human impact (in Russian). *Paleolimnologiczeskij podchod k izuczeniu antropogennogo wozdiejstwija na oziera*, iGeograf.Obszcz. SSSR, Akad. Nauk SSSR, Leningrad 1981, Martinson, G. G., and Dawydova, N. N. (Eds.), 36–51.

Jersak, J. 1977. The late Pleistocene and Holocene deposits in side valleys of the Kunów region. *Folia Quaternaria,* **49**, 15–22.

Kadar, L. 1955. Das Problem der Flussmäander. *Abhandlungen aus d. Geogr. Inst. in Debrecen,* **21**.

Kartashov, J. P. 1972. Main regularities of the geological activity of rivers in the mountain regions (in Russian). *Transactions of Academy of Sc.*, Moscow, 184 pp.

Kennett, J. P. 1974. Latest Pleistocene melting of the Laurentide ice sheet recorded in deep sea cores from the Gulf of Mexico. *Geological Society of America, Abstracts, Annual Meeting*, 1015.

Khotinsky, N. 1977. *The Holocene of the Northern Eurasia* (in Russian), Moskva.

Kind, N. W. 1967. Erosion and aggradation in the Yenisey river valley. *Biuletin Komissi po Izutchenju Chetviertichnogo Perioda,* **34**, Moskva, 40–49.

Klimanov, W. A. 1978. Palaeoclimatic condition of the Russian Plain and the climatic optimum of the Holocene (in Russian). *Doklady Akad. Nauk SSSR*, **242**, 4, Moskva, 902–904.

Klimek, K., and Starkel, L., 1974. History and actual tendency of floodplain development at the border of the Polish Carpathians. *Nachrichten Akad. Göttingen* in *Report of Commission on present-day Processes*, IGU, 185—196.

Knox, J. C. 1975. Concept of the graded stream. In Melhorn, W. N., and Flemal, R. C. (Eds.), *Theories of Landform Development*, Binghamton, New York, 169–198.

Knox, J. C. 1977. Human impact on Wisconsin stream channels. *Annals Association American Geographers,* **76**, 323–342.

Kolb, C. R., and Van Lopik, R. R. 1966. Depositional environments of the Mississippi River deltaic plain, Southeastern Louisiana. In Shirley, M. L., and Ragsdale, J. A. (Eds.), *Deltas and their Geologic Framework*, Houston Geological Society, Texas, 17–61.

Kolpakov, W. W., and Bielowa, A. P. 1980. Radiocarbon datings in the glaciated part of the Verchoyansk Mts. and its surroundings (in Russian). *Geochronologia chetviertichnogo perioda*, Nauka, Moskva, 230–235.

Koutaniemi, L. 1979. Outline of the development of relief in the Oulanka river valley, North-eastern Finland. *Acta Universitatis Ouluensis*, A, **82**, Geol. 3. 29–38.

Koutaniemi, L., and Rachocki, A. 1981. Palaeohydrology and landscape development in the middle course of the Radunia basin. North Poland. *Fennia,* **159**, 2, 335–342.

Kowalkowski, A., and Starkel, L. 1977. Different age of soil cover at the Holocene terraces in Carpathian valleys. *Folia Quaternaria,* **49**, 63–74.

Kozarski, S., and Rotnicki, K. 1977. Valley floors and changes of river channel pattern in the North Polish Plain during the Late-Würm and Holocene. *Quaestiones Geographicae,* **4**, Poznań 51–93.

Kozarski, S., and Tobolski, K. (Eds.) 1981. *Guide-book of Excursions, Symposium 'Paleohydrology of the Temperate Zone'*, Poznań, Adam Mickiewicz Univ. 120 pp.

Kvasov, D. D. 1976. Paleohydrology of Eastern Europe during Valdai Period (in Russian). In *Problemy paleogidrologii*, Moskva, 260–266.

Lamb, H. H. 1977. *Climate: Present, Past and Future*, Methuen, London, vol. II, 835 pp.

Lawrushin, J. A. 1966. The characteristics of differentiated formations of lowland river sediments in different climatic zones (in Russian). *Biulletin Kom. po Izutch. Chetwiertichnogo Perioda,* AN SSSR, 162–175.

Leopold, L. B., Wolman, M. G. and Miller, J. P. 1964. *Fluvial Processes in Geomorphology*, San Francisco, Freeman, 522 pp.

Ložek, V. 1975. Zur Problematik der Landschaftsgeschichtlichen Entwicklung in verschiedenen Höhenstufen der Westkarpaten während des Holozäns. *Biuletyn Geologiczny,* UW, **19**, Warszawa, 72–92.

Lüttig, ◀. ◀. 1960. Towards a classification of the haugh loams (Auelehm) of the Weser river. Eiszeitalter und Gegenwart 11, 39–50.

Maizels, J. K. (in press). Channel changes, paleohydrology and deglaciation: evidence from some late glacial sandur deposits, northeast Scotland. *Proceedings of the Holocene Symposium, Quaternary Studies in Poland,* **4**, Poznań.

Mamakowa, K., and Starkel, L. 1977. Stratigraphy of the Eo- and Mesoholocene alluvia in Podgrodzie upon Wisloka river. *Studia Geomorphologica Carpatho-Balcanica,* **11**, 101–110.

Mangerund, J., Andersen, Sv. T., Berglund, B. E., and Donner, J. J. 1974. Quaternary stratigraphy of Norden, a proposal for terminology and classification. *Boreas,* **3**, Oslo, 109–126.

Miescheriakov, J. A., and Fiedorova, R. W. 1961. Age and genesis of the terraces of Western Dvina river (in Russian). *Materialy Wsiesoj, Soviesch, po izutch, chetvertich, period,* **2**, Moskva, 32–41.
Mochanov, Y. A. 1976. The palaeolithic of Siberia (in Russian). *Beringia in Cenozoik*, Reports of Symposium, Vladivostok, 540–565.
Mojski, J. E. 1982. Geologic section across the Holocene sediments forming the northern and eastern parts the Vistula Deltaic Plain. *Prace Geograficzne IGiPZ PAN,* special issue 1, Warsaw, 149–169.
Munaut, A. V., and Paulissen, E. 1973. Evolution et palèo-ècologie de la vallèe de la petite Nèthe au cours du post-Würm (Belgique), *Annaels de la Sociète Gèologique de Belgique,* **96**, 301–348.
Mycielska-Dowgiallo, E. 1977. Channel pattern changes during the last Glaciation and Holocene, in the Northern part of the Sandomierz Basin and the middle part of the Vistula Valley, Poland. In Gregory, K. J. (Ed.), *River Channel Changes*, Wiley, Chichester, 75–87.
Niewiarowski, W. 1978. Fluctuations of water-level in the Gopła Lake and their reasons. *Polskie Archiwum Hydrobiologii,* **25**. 1/2, 301–306.
Patzelt, G. 1977. Der zeitliche Ablauf und das Ausmass postglazialer Klimaschwankungen in den Alpen. In *Dendrochronologie und postglaziale Klimaschwankungen in Europa, Erdwiss. Forschung 13*, Wiesbaden. 249–259.
Paulissen, E. 1973. De morfologie en de kwartairstratigrafie van de Maasvallei in Belgisch Limburg. *Verhandel. Kon. Vlaamse Acad. Kl. Wetenschappen,* **35**, 127, Brussel, 266 pp.
Pawlikowski, M., Ralska-Jasiewiczowa, M., Schönborn, W., Stupnicka, E., and Szeroczyńska, K. 1982. Woryty near Gietrzwałd, Olsztyn Lake District, NE Poland—vegetational history and lake development during the last 12 000 years. *Acta Palaeobotanica,* **22**, 85–116.
Ralska-Jasiewiczowa, M. 1980. *Late-glacial and Holocene Vegetation of the Bieszczady Mts.* (Polish Eastern Carpathians). Inst. Botaniki PAN. Kraków, 202 pp.
Richter, G. 1968. Fernwirkungen der Litorinen Ostseetransgression auf tiefliegende Becken und Flusstäler. *Eiszeitalter und Gegenwart,* **19**, 48–72.
Rose, J., Tutner, C., Coope, G. R., and Bryan, M. D. 1980. Channel changes in a lowland river catchment over the last 13 000 years. In Cullingford, R. A. *et al.* (Eds.), *Timescales in Geomorphology*, Wiley, Chichester, 160–175.
Schafman, I. L. 1980. Geochronology and palaeogeography of the Late Quaternary of the extraglacial zone. north-eastern part of the Sibirian Plateau (in Russian). *Geochronologia chetviertichnogo perioda*, Nauka, Moskva, 223–230.
Schumm, S. A. 1965. Quaternary palaeohydrology. In Wright, H. E., and Frey, D. G. (Eds.), *The Quaternary of the United States*, Princeton Univ. Press, 783–794.
Schumm, S. 1977. *The Fluvial System,* Wiley–Interscience Publ., New York. 338 pp.
Schwetz, G. I. 1978. *Multi-centennial Changes of the Runoff of the Dniepr River* (in Russian), Gidrometeoizdat, Leningrad, 82 p.
Shancer, E. V. 1951. Alluvia of the lowlands of the temperate zone and their role for knowledge of the regularities of alluvial beds formation (in Russian). *Trudy Geol. Akademii Nauk SSSR,* **135**, Moscow.
Sladkopievcev, S. A. 1976. On run-off changes in rivers in the Pleistocene (in Russian). In *Problemy paleogidrologii*, Akad. Nauk SSSR, Moskva, 241–245.
Smedt, P. 1973. Palaeogeografie en kwartairgeologie van het confluentiegebied Dijle–Demer. *Acta Geographica Lovaniensia,* **11**, 141 pp.
Smith, D. G. 1980. River ice processes—thresholds and geomorphic effects in northern and mountain rivers. In Coates, D. R. and Vitek, J. D. (Eds.), *Thresholds in Geomorphology*, George Allen and Unwin, London, 323–344.

Smith, D. G. 1981. Anastomosed fluvial styles, a comparison in intramontane and plains geomorphic settings. *Conference: Modern and Ancient Fluvial Systems*, Abstracts, University of Keele, U.K. 111.

Soergel, W. 1921. *Die Ursachen der diluvialen Aufschotterung und Erosion*, Berlin, 1–74.

Sønstegaard, E., and Mangerund, J. 1977. Stratigraphy and dating of Holocene gully sediments in Os, western Norway. *Norsk Geologisk Tidsskrift,* **57**, 313–346.

Starkel, L. 1977. Last glacial and Holocene fluvial chronology in the Carpathian valleys. *Studia Geomorphologica Carpatho-Balcanica,* **11**, 33–51.

Starkel, L. in press. Progress of research in the IGCP–Project No. 158. subproject A: fluvial environment. *Quaternary Studies in Poland,* **4**, Poznań.

Starkel, L. (Ed.), Alexandrowicz, S. W., Klimek, K., Kowalkowski, A., Mamakowa, K., Niedzialkowska, E., Pazdur, M., Starkel, L. 1981. The evolution of the Wisloka valley near Debica during the lateglacial and Holocene. *Folia Quaternaria,* **53**, Kraków, 91 pp.

Starkel, L. (Ed.) 1982. Evolution of the Vistula river valley during the last 15 000 years, part I. *Prace Geogr. IGiPZ PAN*, special issue 1, Warsaw.

Starkcl, L., and Thornes, J. B. (Ed.) 1981. Palaeohydrology of river basins. Guide to Subproject A of IGCP Project No. 158, *BGRG, Technical Bulletin,* **28**, 107 pp.

St. Onge, D. A. 1972. Sequence of glacial lakes in north–central Alberta. *Geological Survey of Canada, Bulletin,* **213**, 16 pp.

Szumański, A. 1982. The evolution of the Lower San river valley during the lateglacial and Holocene. *Prace Geograficzne IGiPZ Pan*, spec. issue, **1**, Warsaw, 57–78.

Trevisan, L. 1949. Genese des terrasses fluviales en relation aves lec cycles climatiques. *C. C. Congr. Internat. Geogr. Lisbon*, vol. 2.

Troll, C. 1957. Tiefenerosion, Seitenerosion und Akkumulation der Flüsse im fluvioglazialen und periglazialen Bereich. *Geomorphologische Studien, Machatschek-Festschrift*, Gotha, 213–226.

Vaŝkovska, E., Vaŝkovsky, I., and Schmidt, Z. 1979. Formation, structure and composition of Holocene sediments of the Zitny Ostrov island, Danube Lowland, Czechoslovakia. *Acta Universitatis Ouluensis*, Series A, 82, *Geologica,* **3**, 155–163.

Voznyachuk, L. N., and Valczyk, M. A. 1977. *Morphology, geology and development of the Niemen Valley in Neopleistocene and Holocene* (in Russian), Inst. of Geochemistry and Geophysics Minsk, 210 pp.

Werritty, A., and Ferguson, R. I. 1980. Pattern changes in a Scottish braided river over 1, 30, and 200 years. In Cullingford, R. A. *et al.* (Eds.), *Timescales in Geomorphology*, Wiley, Chichester, 53–68.

Wiśniewski, E. 1976. Rozwój geomorfologiczny doliny Wisły pomiędzy Kotliną Płocką, a Kotliną Toruńską, Prace Geograficzne IGiPZ PAN. 119, Warszawa.

Woldstedt, P. 1952. Probleme der Terassenbildung, *Eiszeitalter und Gegenwart,* **2**. Öhringen.

van der Woude, J. D. 1981. *Holocene Paleoenvironmental Evolution of a Perimarine Fluviatile Area*, Vrije Universitet Amsterdam, 112 pp.

Zeuner, F. E. 1959. *Pleistocene Period*, Hutchinson, London.

Background to Palaeohydrology
Edited by K. J. Gregory

10

Palaeohydrological studies in lakes and mires — a palaeoecological research strategy

BJÖRN E BERGLUND

Department of Quaternary Geology, Lund, Sweden

Hydrological changes in the past in temperate and subarctic regions are often documented in the deposits of lakes and mires. These habitats may be regarded as natural gauges for water-level and humidity changes. In lakes the sedimentation limit and the shore vegetation belts can be traced by sedimentological studies. Shore displacement can then be correlated with ground-water fluctuations. In mires, especially in the ombrotrophic types, peat humification changes, and bog transgressions can be identified and correlated with small changes in humidity. Changes of lake water-level and of periods of bog formation seem to correspond to large-scale hydrological waves throughout time. Humification changes of bogs can reflect very small fluctuations in humidity which occur on these waves. An advantage with lake and mire habitats is that the time resolution of the deposits is high and this makes it possible to establish an accurate chronology for palaeoecological and palaeohydrological events. The situation is less favourable in fluvial environments but such environments contribute other important information on the hydrological regime in the past.

It is therefore possible to identify hydrological changes in time and space but the causality of the changes is very complex. When the sites are carefully chosen lake fluctuation as well as the precipitation factor is involved, and it is often difficult to separate the effects of these two variables from each other. In addition, it is also possible that deforestation caused by man can lead to a disturbance of the hydrological balance and therefore to raised water-level and increased bog formation. Local factors, such as dead-ice melting, increase of lakes, and incision of nearby rivers, have to be considered as 'noise' in the general hydrological trend. Therefore several sites within a selected area have to be studied palaeohydrologically before the regional pattern can be identified.

Palaeohydrological research therefore has to be combined with palaeo climatic and palaeoecological studies. Regional hydrological changes are the response to climate (temperature, precipitation) and to human impact. This is the background for the geological correlation project IGCP 158B — Palaeohydrological changes in the temperate zone in the last 15 000 years, Subproject B. This subproject, based on studies in lake and mire environments, is briefly described in this paper. Since there are still many geographical gaps in the coverage obtained it is not yet possible to present a general review of results. Instead the potential of this research strategy is illustrated by two regional examples, one from southern Sweden and one from eastern USA and this complements the review in Chapter 9 which relates mainly to the fluvial subproject (158A)

AIMS AND METHODS OF PROJECT IGCP 158B

Subproject B of IGCP 158 is complementary to subproject A which deals with changes of the fluvial water regime. The B subproject concentrates upon environmental changes in lakes and mires and in their catchment areas.

The aims can be summarized as:

1. To provide palaeoecological and stratigraphical research with continental reference sites related to an absolute chronology covering the last 15 000 years.
2. To apply a variety of uniform palaeoecological methods to obtain fully accurate information on biotic and environmental conditions on the continents during the last 15 000 years.
3. To describe biotic changes in both time and space. Biotic changes refer to local ecosystems of lakes and mires as well as to regional, terrestrial ecosystems of selected reference areas.
4. To describe hydrological and limnological changes in both time and space. These changes include quantitative aspects of the hydrological cycle, such as water level changes, as well as qualitative aspects of lake ecosystems, such as nutrient status and productivity.
5. To describe climatic changes in both time and space and to assess their impact on biological and hydrological systems.
6. To describe human activity in both time and space and to assess its impact on biological and hydrological systems.
7. To correlate hydrological changes in lake catchment areas with conditions in fluvial environments in order to describe the total hydrological regime and its relation to climatic and human factors.

Geographically, the whole project is intended to cover the temperate and the subarctic regions of the North American and the Eurosiberian continents. This is extremely ambitious, it is more realistic to expect coverage of the main parts of Europe, and Canada–USA and of some parts of East Asia. The relation palaeohydrology–palaeoclimatology–palaeoecology (see Chapter 1) has meant that we have adopted the following research strategy for selecting field areas. Firstly, reference areas for palaeoecology sites, including palaeohydrology and palaeoclimatology, are so-called primary reference areas, and form a sparsely spaced network of areas which include one large or several small river catchments and ideally contain several lakes and mires. A great variety of methods should be applied to achieve rich information on biotic as well as on physical environmental changes. Water-level changes of lakes and changes of bog humification are emphasized in these studies. When possible, the human impact is also studied in cooperation with archaeologists and other researchers in these primary reference areas. Often a team of coworkers has to collaborate in order to cover the methodological demands for a primary reference area. Therefore the number of these areas will necessarily be rather small. Secondly, reference sites for chronology and palaeovegetation, are the so-called secondary reference sites, which form a densely spaced network of well-dated, continuous sediment profiles with good pollen diagrams. There ought to be one reference site in each type region (that is an area which is uniform as regards geology, morphology, climatology, and biology). These sites are the basis for computerized correlation and for the mapping of biotic zones and their shifts. By using transfer functions such data may give information on climate and human impact in time and space. This is a complement to the multidisciplinary studies of the primary reference areas. The two sets of data will interact to provide the basis for explaining environmental changes in the past.

From a *methodological* point of view we have emphasized the importance of selecting uniform field sites and of applying uniform methods. Otherwise regional and continental correlations will be difficult and may yield incorrect results. Therefore methodological recommendations have been published in a series of project guidebooks (Berglund, 1979–1982) which will be the basis for a revised handbook on Quaternary palaeoecology. A brief survey of the project is presented by Berglund (1979).

The relation between primary reference areas and secondary reference sites is demonstrated by the project map for Scandinavia where selected areas and sites are plotted in relation to the main climatological/biotic regions (Figure 10.1). A similar survey map is under compilation for Europe north of the Mediterranean region.

The project was developed as a general palaeoecological project for Scandinavia (Berglund and Digerfeldt, 1976), but it is easy to extend its idea and methods to all glaciated areas which have a humid climate because there is always access to lakes and mires with deposits which are useful for

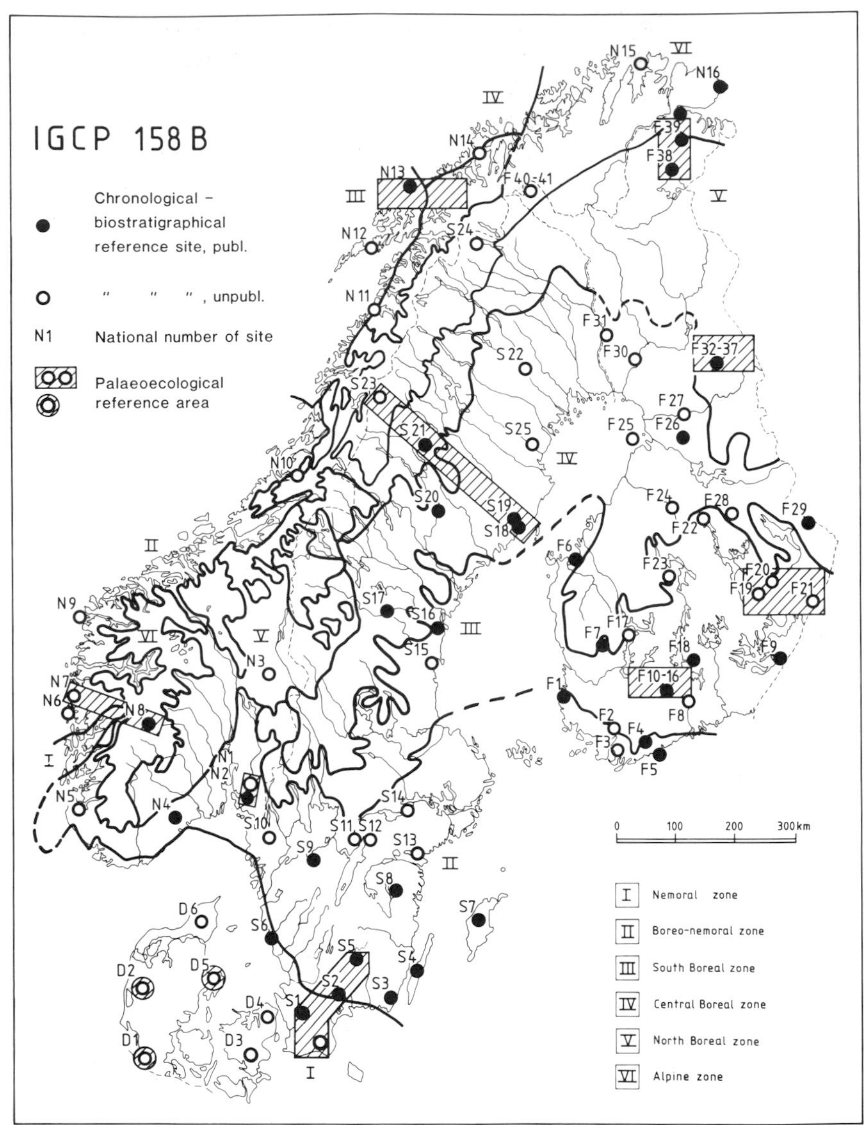

FIGURE 10.1 RESEARCH PLAN FOR IGCP 158B IN SCANDINAVIA
Selected reference areas/sites, here plotted in relation to the biotic zones. In some areas there are still reference sites missing, e.g. in central Sweden and central Norway

palaeoecological research. Problems arise in lowland regions like southern England and central Europe where lakes and mires are scarce or have been exploited in some way. This means that the project requirements have to be adapted to the physical conditions presented by different areas. There will therefore be a varying density of sets of field sites, and this has to be borne in mind when the conclusions are drawn from correlation diagrams and from maps.

Research results from a reference area/site should be used and interpreted on several different levels. Firstly, locally, based on one site/area which is hopefully representative of one type region; secondly, regionally, based on correlations between several sites/areas each representative of one uniform climatological–biotic zone; and thirdly, continentally, based on correlations between all sites/areas available on a continent like Europe or North America.

SOUTH SWEDEN — A REGIONAL PERSPECTIVE

In the primary reference area of South Sweden, in the provinces of Skåne and Småland, palaeoecological studies have been performed for many years, but systematic research on *lake level fluctuations* was initiated by Gunnar Digerfeldt in the 1960's. His research strategy was to trace low lake levels stratigraphically in sections through shallow bays, by means of sedimentological analyses, plant macrofossil and pollen analyses together with radiocarbon dating. To date, five basins have been studied within the area indicated by Figure 10.1 and a stratigraphical example is illustrated in Figure 10.2 derived from Digerfeldt's study of Lake Växjösjön (Digerfeldt, 1975). The four sand layers present can obviously be correlated with lowering of the sedimentation limit. The sedimentological changes are confirmed by measuring the minerogenic content in each sediment column. The chronological correlation of the sand layers is based on pollen diagrams constructed from three profiles. This dates the stages of lowering in the following way: number 1 around 9 500–9 300 B.P., number 2 around 7 000 B.P., number 3 around 3 000 B.P., and number 4 around 2 000 B.P.

Another stratigraphical example is shown in Figure 10.3, and is derived from the nearby Lake Trummen (Digerfeldt, 1972). Seven sediment columns near the shore have been correlated on the basis of macrofossil and pollen analysis. The macrofossil diagrams indicate the displacements of the shore vegetation and these changes are correlated and dated by means of pollen diagrams. The conclusion is that two major stages of lowering are registered in Lake Trummen and they correspond chronologically to number 1 and number 3 in Lake Växjösjön.

A regional synthesis of the lake-level changes recorded in South Sweden has been compiled by Digerfeldt (Berglund *et al.*, 1982, includes a complete reference list) and a scheme is reproduced in Figure 10.4. Eight water-level lowerings seem to have culminated about 11 200–11 400, 10 000–10 200, 9 300–9 500 (major lowering), 8 000–8 200, 8 600–8 800, 6 900–7 100, 2 900–3 100 (major lowering), 1 600–1 800 B.P. respectively. A minimum

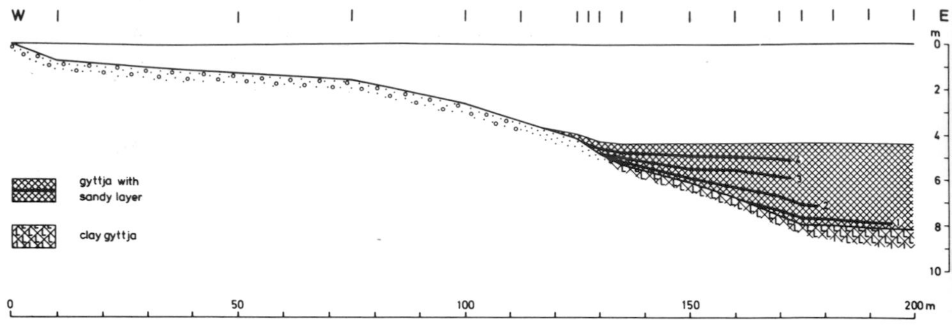

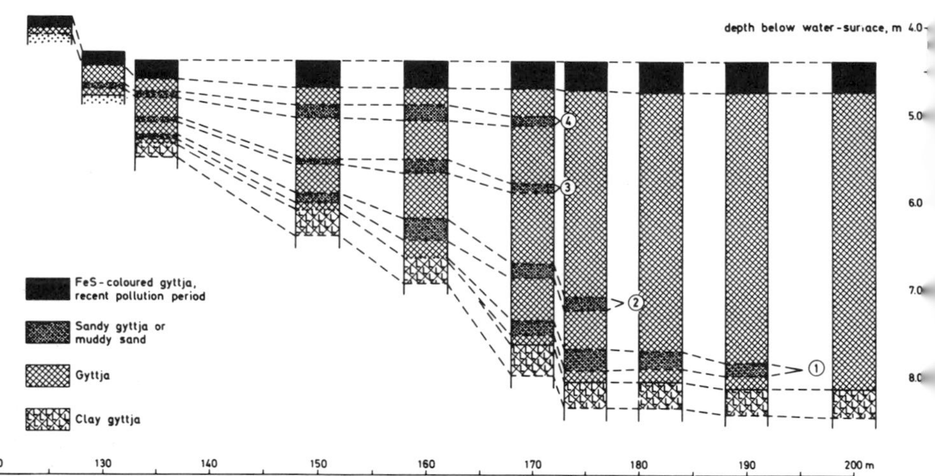

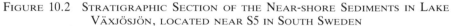

FIGURE 10.2 STRATIGRAPHIC SECTION OF THE NEAR-SHORE SEDIMENTS IN LAKE VÄXJÖSJÖN, LOCATED NEAR S5 IN SOUTH SWEDEN
The sediments cover a full Holocene sequence. The sandy layers 1–4 are explained as resulting from lowering of the water level and the sedimentation limit. From Digerfeldt (1975)

lake level was reached in most lakes during the lowering around 9 500 B.P. In general the mean water level seems to have been higher in the Holocene after 5 000 B.P.

Bog humification changes have been studied in several earlier bog-stratigraphical works in South Sweden. The most carefully-studied raised bog of the area is that of Ageröd's Mosse where Tage Nilsson identified and dated seven so-called recurrence surfaces (Nilsson, 1964). These together with four others recorded in other bogs have been plotted in the survey scheme of Figure 10.4. Rapid bog growth in the late Holocene seems to have coincided with the period of higher water level in lakes.

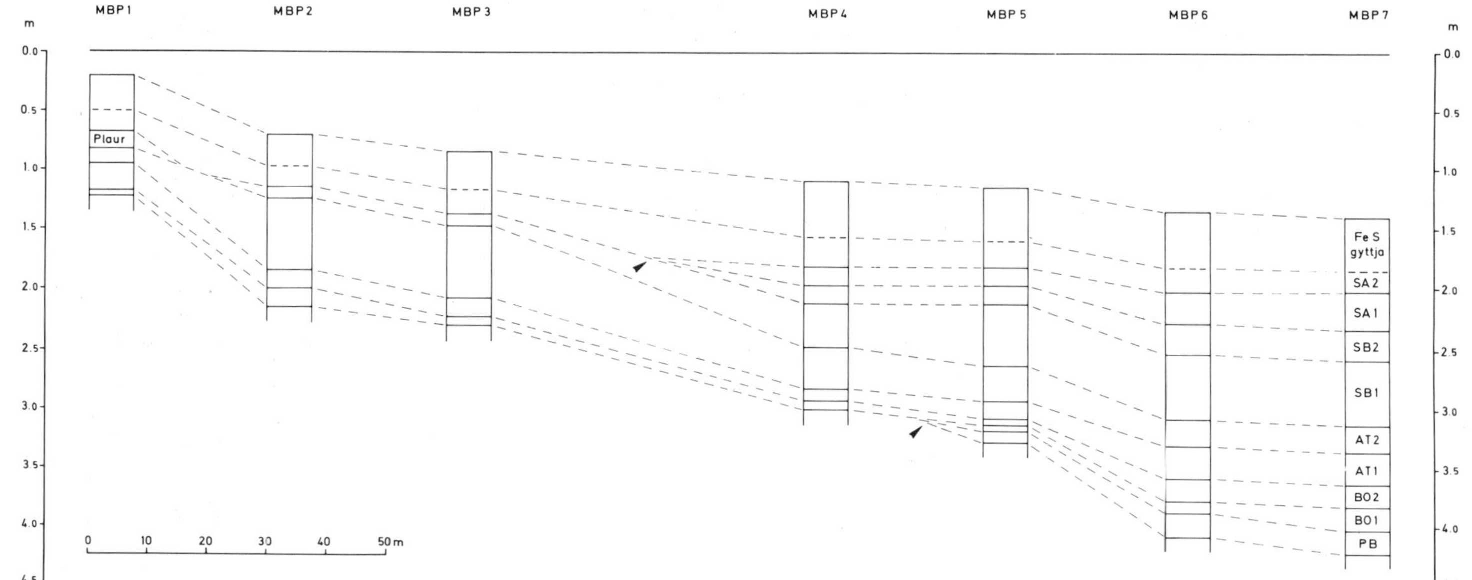

FIGURE 10.3 STRATIGRAPHIC SECTION OF THE NEAR-SHORE SEDIMENTS IN LAKE TRUMMEN, SITE S5 IN SOUTH SWEDEN
The sediments cover a full Holocene sequence with the pollen zones PB–SA2, but pollen diagrams from each sediment core indicate two hiatus horizons near the shore (arrows in the figure). Macrofossil analyses confirm that these correspond to low lake-levels, correlated with lowerings 1 and 3 at Lake Växjösjön (Fig. 10.2). From Digerfeldt (1972)

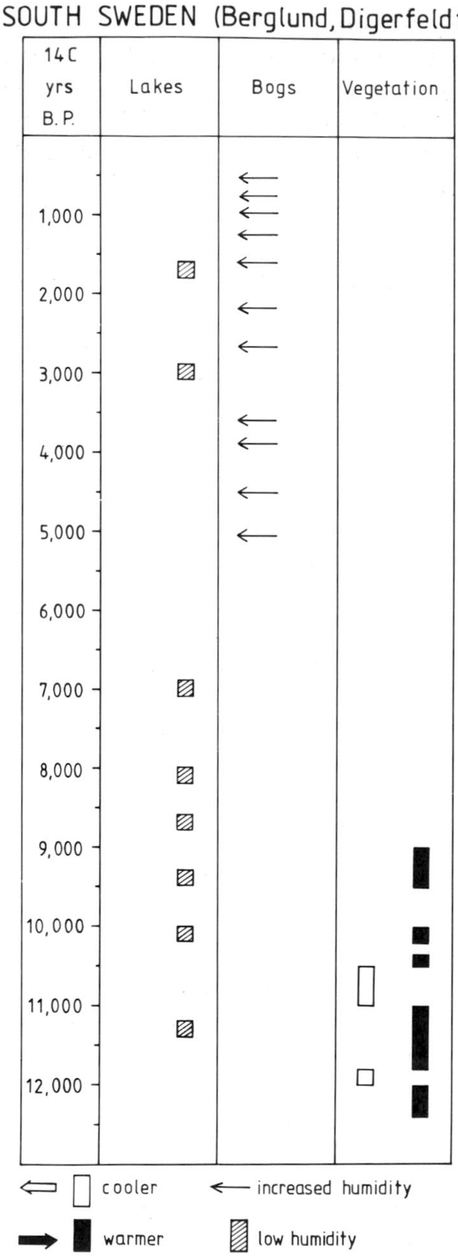

SOUTH SWEDEN (Berglund, Digerfeldt)

14C yrs B.P.	Lakes	Bogs	Vegetation

1,000
2,000
3,000
4,000
5,000
6,000
7,000
8,000
9,000
10,000
11,000
12,000

cooler — increased humidity

warmer — low humidity

FIGURE 10.4 SURVEY DIAGRAM SHOWING PALAEOCLIMATIC AND PALAEOHYDROLOGIC CHANGES IN SOUTH SWEDEN

Based on water-level changes in lakes, humification changes in bogs and vegetation changes. The chronology is based on conventional radiocarbon years. From Berglund *et al.* 1982

More detailed bog-stratigraphical studies have been performed in Denmark by Bent Aaby and a survey with references has been presented by Aaby (see Berglund *et al.*, 1982). Aaby found that humidity changes occurred during the late Holocene with a periodicity of about 250–270 yrs (Aaby, 1976).

Terrestrial vegetation changes may be a response to climatic shifts. Such climatically-conditioned changes are best traced when the arctic-subarctic environment is dealt with. The reason for this is that the tundra/forest ecotone is distinctively reflected in pollen diagrams. In South Sweden this is the case only for Late Weichselian time. In the scheme of Figure 10.4 the cool/warm shifts are indicated for the period 12 500 to 9 000 B.P. There seems to bc a rather good correlation between the warm phases, indicated by the pollen diagrams, and the dry phases indicated by water-level studies. Of special interest is the correlation of the major water-level lowering about 9 500–9 300 B.P. and the slight expansion of dryland vegetation about 9 500 to 9 000 B.P.

EUROPE — A CONTINENTAL PERSPECTIVE

The palaeoclimatic evidence presented above is based on several local studies within one rather uniform region. The correlation scheme of Figure 10.4 is a preliminary attempt to correlate the most obvious climatic and hydrological changes. A future task will be to correlate these changes on the continental scale. Most of the research within the framework of our project is in progress, but a few references to recent publications may illustrate the potential of future continental correlations concerning Holocene palaeoecology in general.

Pollen maps for each millenium covering most of Europe are now available in the extensive work by Huntley and Birks (1982). These are of great palaeobotanical as well as palaeoclimatological interest. They also reveal the gaps in our knowledge about palaeovegetation which hopefully will be closed by our project. Regional pollen maps will also assist in understanding palaeoenvironmental changes, such as the revegetation after deglaciation and the human impact in the Late Holocene, for example in Poland (Ralska-Jasiewiczowa, 1983).

For East Europe and great parts of Asia the vegetation zones have been mapped for different Holocene time periods (Khotinsky, 1977, 1983, INQUA lecture, 1982). Such maps together with palaeoecological transect studies through the biotic/climatic zones in European USSR (e.g. Serebryannaya, 1982) may help in understanding the vegetational response to climatic changes. But there is a need to base such correlations on better dated and more homogeneous sediment cores.

For Central Europe and Scandinavia a preliminary survey of the significance of human impact is in the process of elaboration on the basis of a transect of pollen diagrams (Berglund and Ralska-Jasiewiczowa in prep; INQUA lecture

1982). Phases of increased activity with clearing, grazing, and agriculture are correlated in time and space and can be compared with archaeological settlement periods. For Scandinavia a series of preliminary palaeoclimatic syntheses for different regions has been presented by the national project leaders (Berglund *et al.*, 1982). These are similar to the regional synthesis presented above for South Sweden and based on evidence from lakes, bogs, and palaeovegetation. For Great Britain, Peter Moore and collaborators have studied blanket mire formation in the Holocene and they have related it to three ecological factors namely climate, soil maturation, and human impact (Moore, 1975; Moore *et al.*, 1983; Wiltshire and Moore, this volume Chapter 19). In many areas deforestation caused by man around and after 5 000 B.P. seems to have played a determining role. Moore (1983) has presented the causal background to blanket peat formation in a scheme showing the interaction of different factors and this scheme also seems to be valid in other areas when mire formation is discussed. The question arises how and when human impact caused mire formation in other areas of Europe. Detailed studies of humification changes in raised bogs have also been performed in England by Barber and others (e.g. Barber, 1981).

In Switzerland the environmental background to the lake-shore settlements, pile-dwelling sites from the Neolithic and the Bronze Age, has been reinterpreted and can now be correlated with low water-levels, which probably occurred due to a climatic cause (Joos, 1982, with further references). A series of wet/dry phases have been postulated for the time period 7 000 to 3 000 B.P. Intense palaeolimnological research has now been initiated within the framework of IGCP 158B by G. Lang, B. Ammann, and others. In Poland the research strategy of IGCP 158B was introduced at an early stage by the national leader Magdalena Ralska-Jasiewiczowa. The first national report publication also came from Poland and had contributions from eight reference sites and these related to Holocene vegetation, lake development and human impact (Ralska-Jasiewiczowa, 1982). In one lake of Pommerania the water-level changes were studied with the same methods as used in South Sweden by Hjelmroos-Ericsson (1981). A series of low-water periods was established and correlated with other evidence in Poland and with Digerfeldt's results from Sweden.

EASTERN NORTH AMERICA — LOCAL STUDIES IN A CONTINENTAL PERSPECTIVE

Shifts of the vegetation belts in Eastern North America during the Holocene are well documented in several palaeoecological surveys and have been explained as being the response to climatic changes (e.g. Davis, 1976; Wright, 1971, 1976; Bernabo and Webb, 1977; Delcourt and Delcourt, 1981). This is illustrated by the palaeovegetation map sequence published by Delcourt and

Delcourt (1981), which is however here restricted to only three Holocene time intervals, 10 000, 5 000 and 200 yrs B.P. (Figure 10.5 a,b,c). The most obvious displacement of vegetation belts is the eastward expansion of the prairie followed by shrinkage of the forest belts during the Early Holocene. The largest eastward shift occurred just after 10 000 B.P. but the prairie/forest border reached its maximum extent about 7 000 B.P. (Wright, 1968, 1976; Bernabo and Webb, 1977). This is also illustrated here in two ways; firstly, by a north–south Minnesota transect of pollen diagrams (Figure 10.6, from Wright and Watts, 1969) and secondly, by prairie border isochrones (Figure 10.7, from Bernabo and Webb, 1977).

This extensive displacement of vegetation belts has as its cause optimal temperature and dryness of the Hypsithermal, dated to 8 000 to 6 000 B.P. in southeastern USA. The environmental response in local watersheds and lake basins has rarely been studied and discussed in detail. Two good sites have been chosen for this demonstration, namely Kirchner Marsh in southeast Minnesota (Watts and Winter, 1966) and Cahaba Pond in central Alabama (Delcourt, Delcourt, and Spiker, 1983). The situation of the sites is indicated on the map sequence in Figure 10.5. Kirchner Marsh is situated in the oak forest region but close to the prairie border, while Cahaba Pond is located in the oak–pine forest region of Alabama. In both cases pollen and macrofossils of sediment cores have been analysed and the results interpreted ecologically.

The sequence of the upland vegetation around Kirchner Marsh as reflected in the pollen diagram as shown in the transect of Figure 10.6. This shows that a herb-rich pollen zone with oak indicates prairie conditions during the time interval 7 100 to 5 100 B.P. In the macrofossil diagram this zone matches with peaks of shallow-water aquatics and even with annuals growing on the exposed lake mud. The evidence of a regional drought period is supported by biostratigraphic similarities at other sites in Minnesota. Further important palaeolimnological and palaeohydrological results can be expected from a continuing project in this region (Limnological Research Center, Minnesota). The pollen data from Kirchner Marsh has also been calibrated in terms of temperature and precipitation by using multivariate statistical analysis of modern pollen and climatic data (Webb and Bryson, 1972; Webb, 1980). The palaeoclimate diagram which is reproduced in Figure 10.8 shows extremely low precipitation values for the period 8 000 to 5 000 B.P. in good correlation with the traditional interpretation.

The Cahaba Pond is a very small lake (0.2 ha) in a small watershed, but this means that the local environmental changes are reflected in the sediment column of the pond throughout Holocene. On the surrounding slope a zonation of forest ecosystems from hydric to xeric has been postulated. Hydrological changes in the watershed will result in displacements of these forest zones (Figure 10.9). Of great interest is the climatic change around 10 000 B.P. resulting in a low water-level of the pond until 8 400 B.P. The

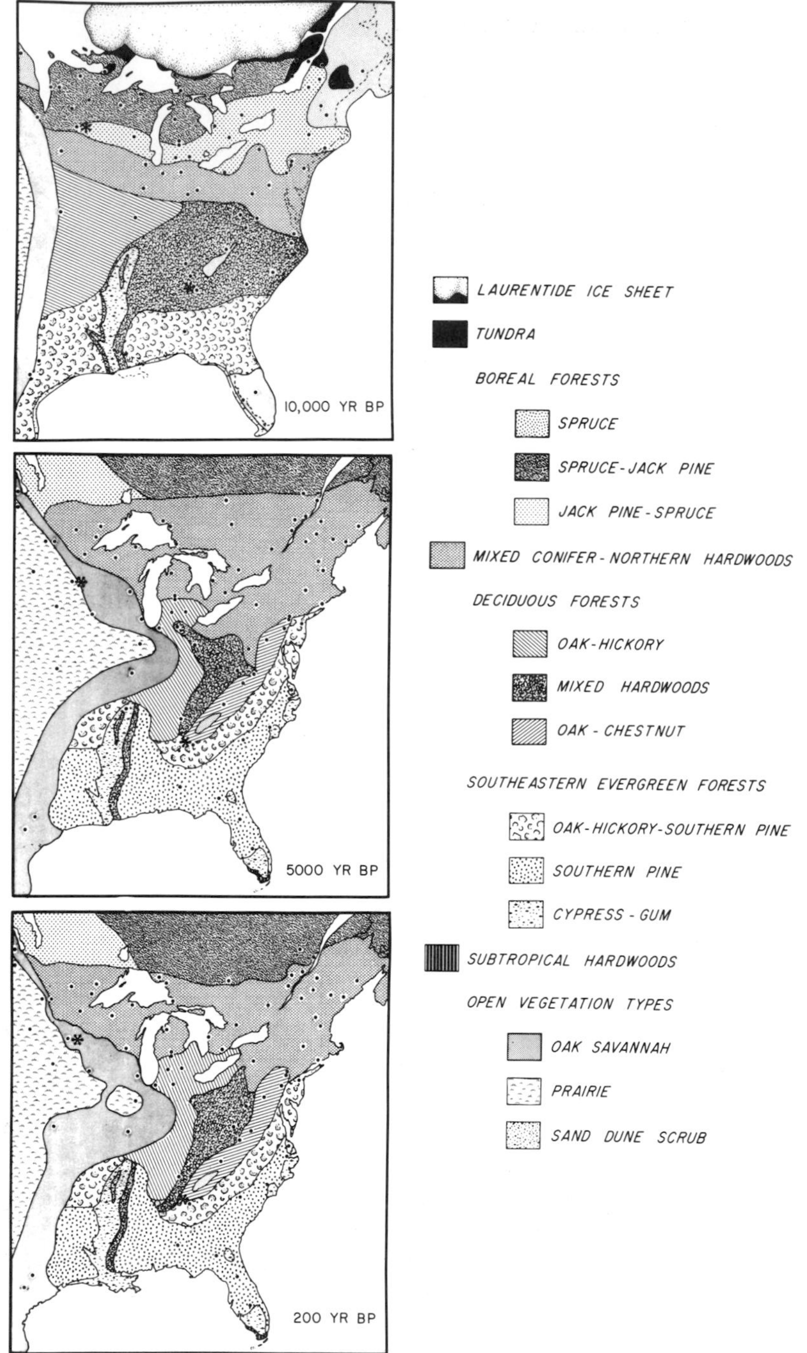

FIGURES 10.5 a, b, c. PALAEOVEGETATION MAPS FOR EASTERN NORTH AMERICA REPRESENTING EARLY, MIDDLE AND LATE HOLOCENE

The map sequence illustrates the climatically conditioned displacements of vegetation belts. Sites discussed in text indicated. From Delcourt & Delcourt (1981). (Reproduced with permission)

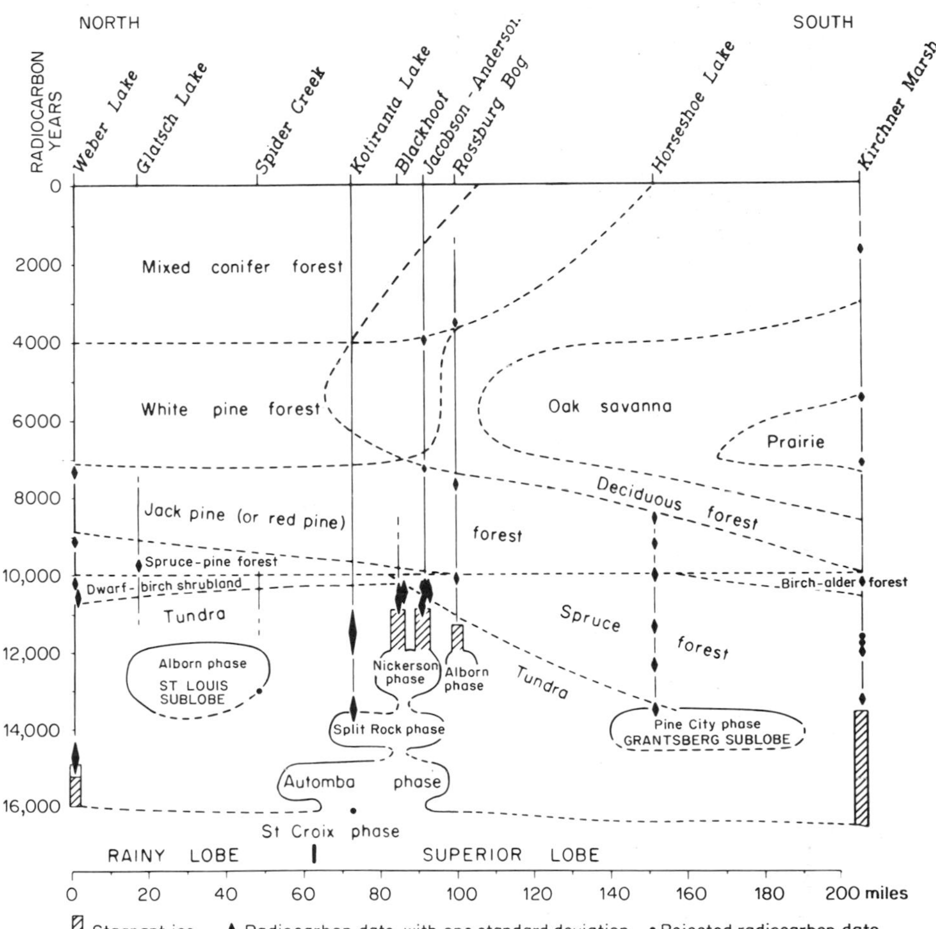

FIGURE 10.6 TIME-SPACE DIAGRAM FROM NORTH TO SOUTH IN EASTERN MINNESOTA
Based on radiocarbon-dated pollen profiles from the sites indicated. The diagram illustrates the development of different vegetation types. From Wright & Watts (1969)

background is supposed to be regional aridity causing an expansion of the xeric upland forests at the expense of the mesic ones. Around 8 400 B.P. there is evidence of a distinct rise of the water-level causing an expansion of the hydric forests. As a whole the hydrological change at Cahaba Pond is an early response to the Holocene climatic improvement which culminated later in Minnesota.

CONCLUSIONS

The IGCP project 158B has sometimes been criticized for being too ambitious in its attempt to achieve many aims of palaeoecology. But the keys for

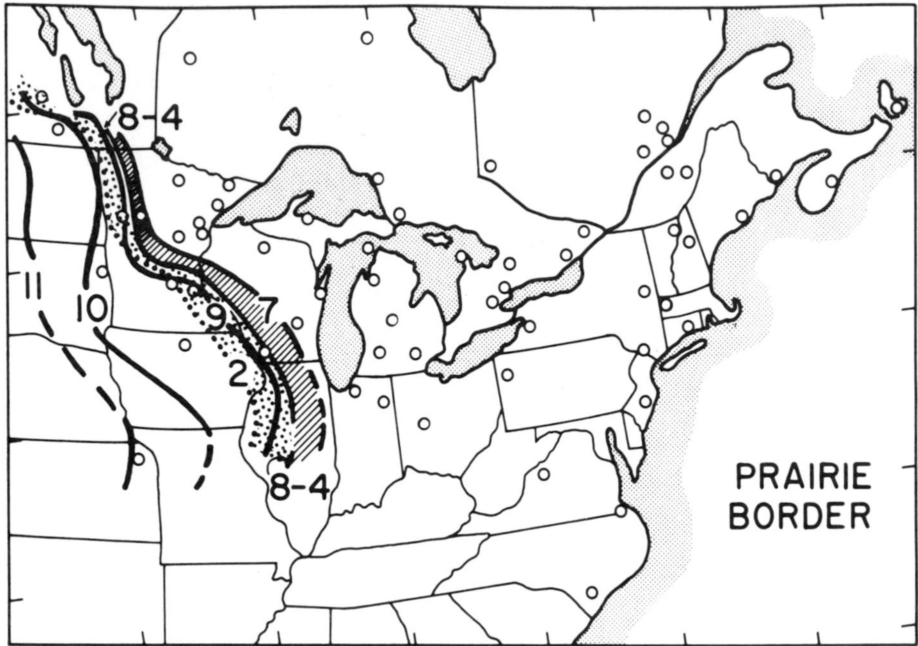

FIGURE 10.7 ISOCHRONES FOR EACH MILLENIUM B.P. ILLUSTRATING THE MOVEMENT OF THE PRAIRIE/FOREST ECOTONE IN NORTHEASTERN NORTH AMERICA Shaded areas indicate the prairie retreat after reaching its Holocene maximum extent at 7000 B.P. From Bernabo & Webb (1977). (Reproduced by permission of Academic Press, Inc.)

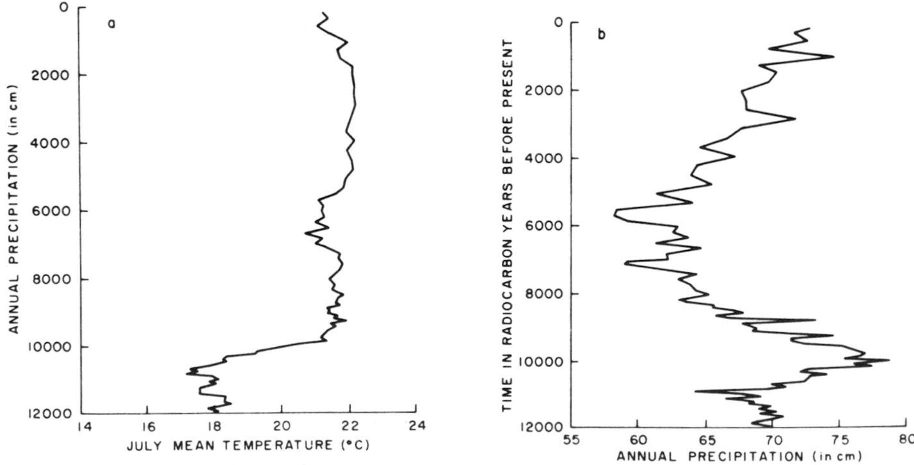

FIGURE 10.8 ESTIMATES OF TEMPERATURE AND PRECIPITATION DERIVED FROM CALIBRATED POLLEN DATA OF THE COMPLETE SEDIMENT SEQUENCE IN KIRCHNER MARSH IN MINNESOTA. From Webb (1980). (Reproduced by permission of Academic Press, Inc.)

D : 8400 TO 0 YEARS BP

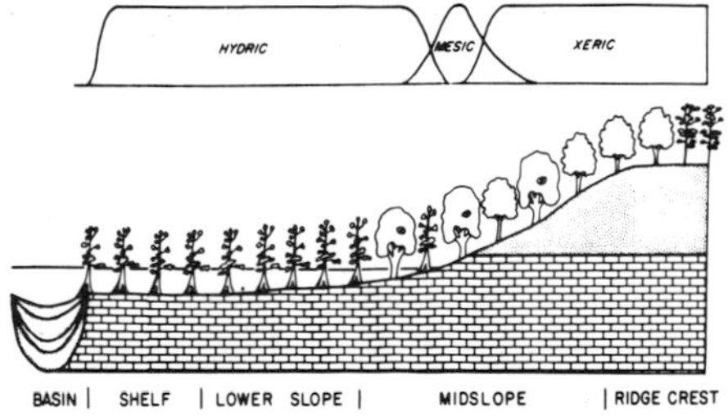

C : 10 000 TO 8 400 YEARS BP

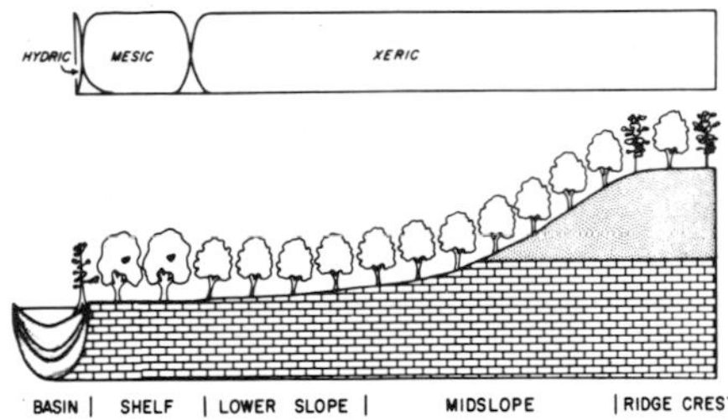

B : 10 200 TO 10 000 YEARS BP

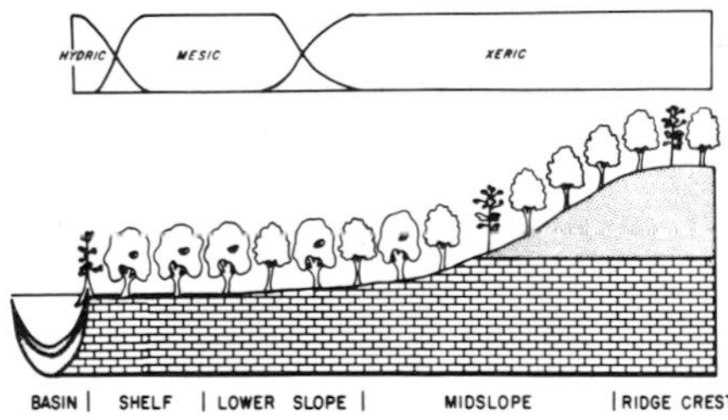

A : 12 000 TO 10 200 YEARS BP

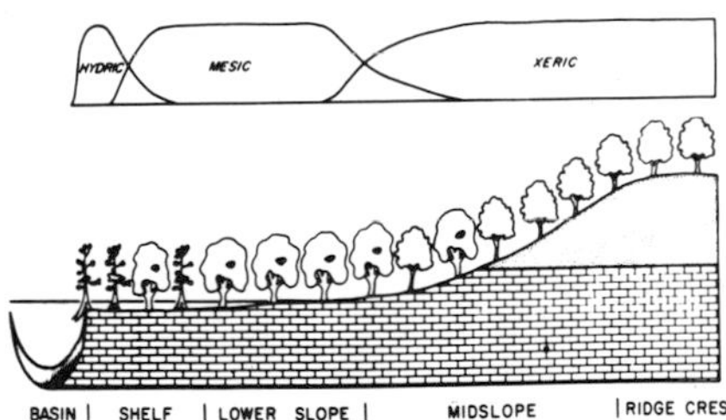

FIGURE 10.9 SCHEMATIC PROFILES FOR THE CAHABA POND WATERSHED Illustrating Holocene changes in the pond water-level and the distribution of hydric, mesic and xeric forest communities. From Delcourt *et al*. (1983). (Copyright © 1983, the Ecological Society of America)

understanding the past environment lie in geological deposits and it is a waste of resources to restrict sampling and analyses to those methods which will answer only single questions, for example those related to hydrological changes. These single questions could be enlightened by concentrating on lake-level studies and bog humification changes only within the primary reference areas. But there is always an interrelation between palaeoclimate, palaeohydrology, and palaeobiota. Therefore there is a need for palaeoenvironmental mapping of countries and continents, an inventory research with a palaeoecological goal. The International Geological Correlation Programme gives the opportunity for palaeoecologists to cooperate by applying a uniform strategy for sampling, analyzing, correlating and interpretating results. The need for this strategy and the potential of the project has hopefully been demonstrated by the present paper.

ACKNOWLEDGEMENTS

I wish to thank all colleagues and publishers who have given permission to reproduce figures for this paper. I owe special thanks to Drs. Gunnar Digerfeldt, and Paul and Hazel Delcourt for stimulating discussions concerning palaeoenvironmental changes in their field areas which are used here as case studies.

REFERENCES

Aaby, B. 1976. Cyclic climatic variations in climate over the past 5 500 yrs reflected in raised bogs. *Nature*, **263**, 281–284.

Barber, K. E. 1981. *Peat Stratigraphy and Climatic Change. A Palaeoecological Test of the Theory of Cyclic Peat Bog Regeneration*, A. A. Balkema, Rotterdam, 219 pp.

Berglund, B. E. (Ed.) 1979–82. *Palaeohydrological Changes in the Temperate Zone in the last 15 000 years. Subproject B. Lake and Mire Environments*, I–III, Department of Quaternary Geology, Lund, 140+340+163 pp.

Berglund, B. E. 1979. Presentation of the IGCP project 158B Palaeohydrological changes in the temperate zone in the last 15 000 years—lake and mire environments. *Acta Universitatis Ouluensis A 82*, **1979**, Geol 3, 39–48.

Berglund, B. E., and Digerfeldt, G. 1976. Environmental changes during the Holocene — a geological correlation project on a Nordic basis. *Newsl. Stratigr.*, **5**, 80–85.

Berglund, B. E., Aaby, B., Digerfeldt, G., Fredskild, B., Huttunen, P., Hyvärinen, H., Kaland, P. E., Moe, D. and Vasari, Y. 1982. Palaeoclimatic changes in Scandinavia and on Greenland — a tentative correlation based on lake and bog stratigraphical studies. *Quaternary Studies in Poland*, **4**.

Bernabo, J. C., and Webb, T. 1977. Changing patterns in the Holocene pollen record of northeastern North America. A mapped summary. *Quaternary Research*, **8**, 64–96.

Davis, M. B. 1976. Pleistocene biogeography of temperate deciduous forests. *Geoscience and Man*, **13**, 13–26.

delcourt, P. A. and Delcourt, H. R. 1981. Vegetation maps for eastern North America: 40 000 yr B.P. to the present. In Romans, R. C. (Ed.), *Geobotany II*, Plenum Publ. Corp., 123–165.

Delcourt, H. R., Delcourt, P. A. and Spiker, E. C. 1983. A 12 000-year record of forest history from Cahaba Pond, St. Clair County, Alabama. *Ecology*, **1983**.

Digerfeldt, G. 1972. The post-glacial development of Lake Trummen. Regional vegetation history, water-level changes and palaeolimnology. *Folia Limnol. Scand.*, **16**, 1–96.

Digerfeldt, G. 1975. Post-glacial water-level changes in Lake Växjösjön, central southern Sweden. *Geol. Fören. Stockh. Förh.*, **97**, 167–173.

Hjelmroos-Ericsson, M. 1981. Holocene development of Lake Wielkie Gacno area, northwestern Poland. *Univ. of Lund, Dept. Quat. Geol. Thesis*, **10**, 101 pp.

Huntley, B. and Birks, H. J. B. 1982. *An Atlas of Past and Present Pollen Maps for Europe: 0–13 000 years ago*, Cambridge Univ. Press. 667 pp.

Joos, M. 1982. Swiss Midland-lakes and climatic changes. In Harding, A. (Ed.). *Climatic Change in Later Prehistory*, Edinburgh Univ. Press, 44–51.

Khotinsky, N. A. 1977. *Holocene of the Northern Eurasia*, Publ. House Nauka, 200 pp.

Khotinsky, N. A. 1983. Development of Holocene vegetation. In Velichko, A. A. (Ed.) *Late Quaternary Environments of the Soviet Union*, in press.

Moore, P. D. 1975. Origin of blanket mires. *Nature*, **256**, 267–269.

Moore, P. D., Merryfield, D. L., and Price, M. D. R. 1983. The vegetation and development and blanket mires. In Moore, P. D. (Ed.), *European Mires*, Acad. Press, London, in press.

Nilsson, T. 1964. Entwicklungsgeschichtliche Studien im Ageröds Mosse, Schonen. *Lunds Univ. Årsskr. N. F.*, 2, 59, 8, 1–34.

Ralska-Jasiewiczowa, M. (Ed.) 1982. Polish national contributions to IGCP 158B. *Acta Palaeobotanica*, **22**, 1, 161 pp.

Ralska-Jasiewiczowa, M. 1983. Isopollen maps for Poland: 0–11 000 years B.P. *New Phytologist*, in press.

Serebryannaja, T. A. 1982. On the forest–steppe zone dynamics at the central part of the Russian Plain during Holocene. In *Evolution of the Environment at the USSR Territory during Late Pleistocene and Holocene*, Publ. House Nauka, 179–186.

Watts, W. A., and Winter, T. C. 1966. Plant macrofossils from Kirchner Marsh, Minnesota — a palaeoecological study. *Geological Society of America Bulletin*, **77**, 1339–1360.

Webb, T. 1980. The reconstruction of climatic sequences from botanical data. *Journal of Interdisciplinary Hist.*, **10**, 4, 749–772.

Webb, T., and Bryson, R. A. 1972. Late- and postglacial climatic change in the Northern Midwest, USA; quantitative estimates derived from fossil pollen spectra by multivariate statistical analysis. *Quaternary Research*, **2**, 70–115.

Wiltshire, P. E. J., and Moore, P. D. 1983. Palaeovegetation and palaeohydrology in upland Britain. In Chapter 19 this volume.

Wright, H. E., Jr. 1968. History of the prairie peninsula. In *The Quaternary of Illinois*, Univ. of Ill. Coll. of Agric. Spec. Publ. 14, 78–88.

Wright, H. E., Jr. 1971. Late Quaternary vegetational history of North America. In Turekian, K. K. (Ed.). *The Late Cenozoic Glacial Ages*, Yale Univ. Press, 425–464.

Wright, H. E., Jr. 1976. The dynamic nature of Holocene vegetation. A problem in palaeoclimatology, biogeography and stratigraphic nomenclature. *Quaternary Research*, **6**, 581–596.

Wright, H. E., and Watts, W. A. 1969. *Glacial and Vegetational History of North Eastern Minnesota*. Minnesota Geological Survey Special Publication 11, 59 pp.

PERSPECTIVES

Background to Palaeohydrology
Edited by K. J. Gregory

11

Drainage networks and palaeohydrology

V. Gardiner

University of Leicester

The drainage network of a fluvial system may be envisaged as a morphological response to processes operating within the system. Relationships between various hydrological parameters and aspects of drainage networks have been shown to exist within contemporary systems and it can therefore be suggested that it should be possible to use form attributes of prior networks as at least an indication of previous hydrological processes. River networks are produced by complex interactions between many characteristics of their hydrological environment, including climate, vegetation, rock type, soil cover, and land use, since these all determine or modify both the amount of water reaching river channels and its temporal distribution. The network may therefore be regarded as an effective 'integrator' of such attributes, and information on a palaeonetwork should tell us something about the palaeoenvironment in overall hydrological terms. This contrasts with other lines of evidence concerning palaeohydrology, which give more precise, but more specific, information. For example, size of sediment suggests hydraulic properties of the channel, irrespective of its environment, and palaeobiological evidence may represent the response of organisms to predominantly a single environmental stimulus, such as temperature in the case of coleoptera. In view of these arguments it is perhaps surprising that evidence concerning the form and extent of former river networks has been comparatively rarely used as evidence of palaeohydrological change.

The practical and conceptual difficulties which have made this so are discussed further below; the chapter also examines some of the attempts which have been made in the United Kingdom to exploit the potential of this approach to palaeohydrological investigations, despite the difficulties identified. Information concerning drainage networks may be divided into that relevant to the density of the network (length per unit area) and that concerning its structure, or the manner in which individual lengths of stream forming the network are arranged with respect to one another, and the composition of these links in terms of type of channel. Studies which have examined network density in relation to hydrological inputs are examined first, and studies which have concentrated on aspects of network composition, structure or typology are then considered. As a precursor to the discussion of studies of palaeonetwork density a brief review of relationships which have been identified between network density and hydrological inputs and outputs in contemporary systems is also included.

NETWORK DENSITY

Contemporary relations

Drainage density may be viewed as an index which links hydrological inputs and outputs, as the extent of the network responds to precipitation input yet it also governs runoff output. Input must be regarded in this context as effective input, produced by interaction between climatic elements such as precipitation, radiation, and wind speed. The density of drainage networks is undoubtedly at least partly dependent upon climate. Gregory (1976) has provided a comprehensive review of the ways in which climate and drainage networks are related, and Gregory and Gardiner (1975) examined relationships between climate and drainage density at three spatial scales. Despite the range of sources, conventions, and scales of examination that have been employed, general and tentative conclusions may be drawn from some of the data so far available (Figure 11.1). Areas with low annual rainfall have high values of drainage density. These semiarid areas have episodic high-intensity rains and sparse vegetation, and hence rapid runoff. In humid mid-latitudes drainage densities are lower and there is a tendency for values to be positively related to mean annual rainfall (Gregory and Gardiner, 1975). The range of values reported from these areas is relatively restricted. Humid tropical zones are characterized by somewhat higher values than the temperate zones; they also have a higher range of values reported. Data from periglacial areas are few but it is thought that drainage densities are in general fairly low, although with a wide range represented. For example, Gregory's (1976) data from Alaska suggest mean drainage densities of between 1 and 6 km km^{-2}, the lower figure being for the mapped stream network, the higher for valley networks. It is probable that the value for the functioning network lies between these extremes.

It should of course be noted that the impact of climate on drainage density is not only felt directly but also via its control of vegetation and soil development, which in turn affect runoff generation (Gardiner, 1982a). Chorley (1958) has related drainage density to a vegetation index. Climate may also be assessed by

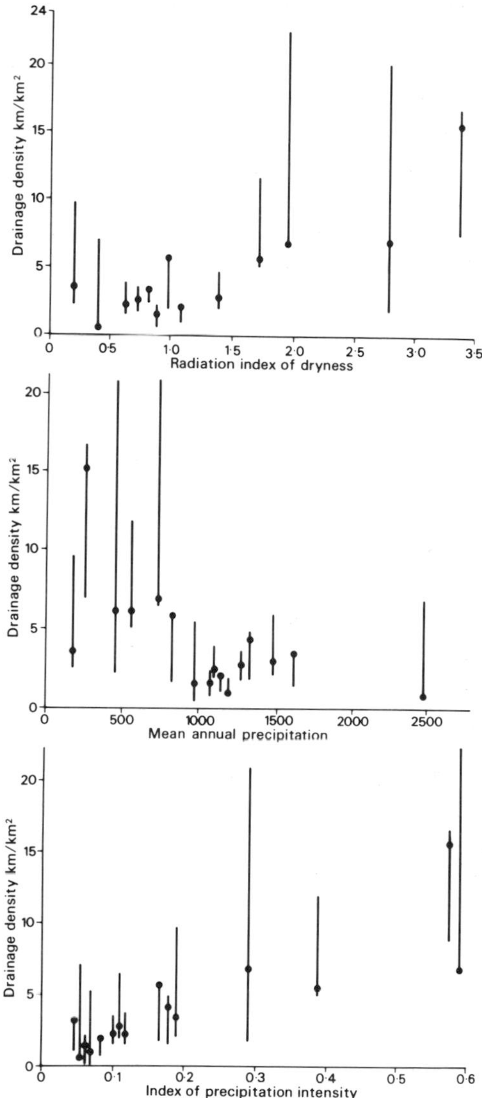

FIGURE 11.1 SOME CLIMATE-DRAINAGE DENSITY RELATIONSHIPS
(After Gregory, 1976; Gardiner, 1982a; Gregory and Gardiner, 1975. Reproduced by permission of John Wiley & Sons Ltd)

a variety of indices. For example, Melton (1957) and Madduma Bandara (1974) related drainage density to the Thornthwaite potential evaporation index and Gregory (1976) employed a radiation index of dryness (Budyko, 1958) and precipitation intensity (Figure 11.1).

Relations have also been sought between drainage density and runoff. High drainage densities result from low infiltration capacities, short, steep slopes, thin soils and low amounts of vegetation cover, and result in rapid concentrations of runoff and a high flood potential (Patton and Baker, 1976). The opposite environmental conditions result in low drainage density and therefore low flood potential. Empirical relations have therefore been sought between peak discharge (Q_p) and drainage density (D), as tabulated by Gardiner and Gregory (1982). Carlston (1963, 1965) used the Jacob water table model to provide an explanation as to why the relationship

$$Q_p \propto D^2$$

resulted from his empirical study. The weaknesses of this approach were indentified by Gregory and Walling (1968). They pointed out that drainage density varies temporally during storm events, whereas Carlston had assumed that a single value was appropriate. Much observational data now exists to support their view of drainage density as a temporally dynamic quantity (e.g. Day, 1978; Gurnell, 1978; Dunne, 1976; Gregory and Gardiner 1979).

Other aspects of discharge may also be related to drainage density. For example Hickok *et al.* (1959) and Murphey *et al.* (1977) suggested both lag time and flood hydrograph peakedness to be controlled by drainage density. Trainer (1969) examined the relationship between base flow and drainage density. Carlston (1963) also reported that low flows (Q_b) were related to drainage density, as:

$$Q_b \propto D^{-2}$$

This he again justified theoretically by reference to the Jacob water table model, although the same weakness as noted above must again apply. Gregory and Walling (1968) also pointed out that the existence of two such distinct relationships is itself unlikely and some general relationship should apply to all discharges, including those of intermediate magnitude. Gardiner and Gregory (1982) noted that Gurnell (1978) showed that basins with low drainage densities at low flows tend to have high rates of network expansion, whereas those with higher drainage densities at low flows tend to have low rates of expansion. They then reconciled Gurnell's conclusions and Carlston's peak flow relationship using a model in which the base flow/drainage density relationship is of a similar positive form to that for peak flows, as suggested by Gregory and Walling (1968). Rodda (1969) has reported a positive relationship of this form between flood discharge and drainage density. Palaeohydrological applications of such relationships should take account of the dynamic nature of drainage density wherever possible, as a particular network can only be meaningfully envisaged as the product of a particular flow magnitude.

Palaeohydrological application

The potential of applying network data and relationships to palaeohydrological interpretation has not been fully realized for a number of reasons, which concern the nature of the drainage network and its relation to flow magnitude, and the nature of the evidence available. As noted above attempts which have been made to relate discharge to drainage density have not as yet satisfactorily incorporated the variability of drainage density into the relationships proposed, and their application to palaeohydrological retrodiction must therefore be at least questioned.

Networks are also characterized by an uneven spatial distribution of the intensity of geomorphic activity, and hence landform development. For simplicity it is usually assumed that a fossil valley network is a single entity formed at a single date. However it has been shown that for contemporary systems the rate of geomorphological development may be very uneven, with active network development by headward extension and gullying occurring at the same time as other parts of the network are experiencing relative stability. Change to an external stimulus is propagated through the system according to the pattern and arrangement of links in the network, and changes can have both downstream and upstream effects. For example, increased rates of erosion by a river will liberate sediment, causing changes downstream, and will also lead to increased rates of erosion by its own tributary streams, since their effective local baselevel will have been lowered by the erosional activity. This will in turn liberate pulses of sediment into the main river, and pulses of activity in different parts of the basin may interact with one another in an intricate fashion, as conditioned by the spatial arrangement of the network, in a manner which Schumm (1973, 1977) has termed Complex Response. Such complexity must inevitably apply to palaeohydrological situations, but the sedimentological evidence for its recognition and interpretation is rarely available.

The use of networks for palaeodischarge estimation is also attended by problems concerning the magnitudes of flow responsible for the networks observed. Considerable debate has occurred concerning the discharge magnitudes and associated recurrence intervals of flows responsible for channel form (e.g. Williams, 1978). Functioning networks, too, are presumably produced by and are in equilibrium with a particular range of flows, yet few attempts have been made to investigate this relationship. Such investigations are hindered by the presence in the landscape of features in disequilibrium with the present hydrological regime. The persistence of features produced by high magnitude events has been investigated, by, for example, Anderson and Calver (1977), who studied in the mid-1970's the landscape features produced by the 1952 Exmoor floods, and concluded that landscape form depends upon both the frequency distribution of geomorphic events and the order of most recent events.

Features may also have been produced by catastrophic events or by processes which functioned in a way very differently from the present because of extreme changes in environmental conditions. For example, features may have resulted from catastrophic discharges occasioned by unusual events as a series of papers collected by Baker (1981) for the Channelled Scablands of America suggests. Seemingly fluvial features may owe their origins to non-fluvial processes. Stephens (1966) described anomalous elements in the valley network of Exmoor, North Devon, such as the Valley of the Rocks, as being glacial submarginal channels. This claim which has been later disputed by Dalzell and Durrance (1980) who, from the evidence of a geophysical subsurface investigation, thought a fluvial origin followed by very active cliff retreat to be a more likely explanation. Many fossil valley networks in Britain are thought to have formed under conditions of continuous permafrost. For example, Richards and Anderson (1978) used soil mechanical and infiltration properties of sediments from the Cromer–Holt ridge in North Norfolk to suggest that the dense dry valley networks dissecting the ridge could only have formed under frozen ground conditions. However relationships between discharge and network characteristics in contemporary systems have rarely been proposed in the context of permafrost, as relatively few studies of high latitude fluvial processes and forms have occurred until recently. Such fluvial processes and their related landforms may differ very markedly from those of more temperate areas, as described by Gardiner (1982a). Thus attempts to apply contemporary relationships derived from temperate areas to palaeohydrological situations in which permafrost may have occurred are likely to be of questionable value.

Finally, use of networks for palaeohydrological interpretation is hindered by a deficiency of evidence. Former networks may be obscured by an infill of later deposits. Substantial buried drainage systems have been reported in many areas, including those described by Beaumont (1970) in the Wear basin, County Durham, and by Hollis and Reed (1981) in the upper Severn valley; on a smaller scale Waters (1966) found head-filled gullies in the Haldon Hills, Devon. Sedimentological evidence of network expansion in the form of correlative deposits is often absent, since sediment is transported completely out of the fluvial system in the estuarine and marine areas; this is discussed further below (pp. 270). Finally, even where networks may be observed, their analysis for large areas requires data to be derived from cartographic sources. Such data are subject to many limitations, conditioned by the scale of the map, the conventions and techniques of the mapping agency, and the measurement techniques employed. These problems are discussed further in the general context of morphometric analysis by Gardiner (1981, 1982b).

Network contraction

Network contraction has received much attention in geomorphological research in Britain, and a large literature exists concerning 'dry' valleys. However this literature has until recently represented an uneven appreciation of the problem for two reasons. First, attention has been focused on the dry valley networks of calcareous rocks, particularly the Cretaceous Chalk of southeast England (e.g. Reid, 1887; Bull, 1940; Kerney *et al.*, 1964; Sparks and Lewis, 1957; Brown, 1969; Small, 1965; Morgan, 1971) and Yorkshire (e.g. Lewin, 1969), and Carboniferous Limestone of the Pennines (e.g. Warwick, 1964). Secondly, relatively few attempts have been made to extend investigations beyond modes or times of origin of the features to interpretation of palaeohydrological processes in detail.

Only in comparatively recent times has there been expressed a realization that fossil valley networks occur on most rock types within the British Isles, and not only on the calcareous rocks where they are most obvious. For example, Gregory (1966) felt it necessary to point out that the use of contour crenulation defined drainage networks in Britain would necessarily include a substantial fossil element. Dry valley networks have now been examined on a variety of rock types including Mercia Mudstone (J.A.A. Jones, 1968; P.F. Jones, 1979), Sherwood Sandstone (Gardiner, 1983), Permian breccias (Clayden and Manley, 1962), and sands and gravels of Tertiary age (Brown, 1964) and Quaternary age (Richards and Anderson, 1978). An indication of the widespread occurrence of an extensive valley network in addition to the currently active drainage networks was given by Gregory (1971), who for a sample of basins covering 28 per cent of Devon and Cornwall, and therefore embracing a variety of rock types, derived relationships:

$$L_s = 2.62\,A^{0.89}$$

and

$$L_v = 4.61\,A^{0.84}$$

where L_s and L_v represent the total length of streams and valleys in the basins, of area A (Imperial units). Thus an overall average of 2.0 ml ml^{-2} (1.25 km km^{-2}) is indicated to exist i.e. ($L_v - L_s$ for $A = 1$). A more detailed examination within the same area is shown by Figure 11.2, from Gardiner (1971). The three major rock types of the area centred on the Dartmoor granite massif are granite, Carboniferous (Culm) shales and sandstones, and Devonian sediments (mainly sandstones and shales). The stream and valley densities for a random sample of 40 1 km grid squares from each of these major lithological groups are shown, the valley densities (Dd_v) being the additional network indicated by contour crenulations to exist beyond the limits of the blue-line functioning network as shown on Ordnance Survey Provisional Edition maps. On all of the rock types a substantial dry network is found, of about 1.3–1.6 km km^{-2}. This is in contrast with the markedly different values of currently-functioning network density (Dd_s) indicated by the blue lines and for which similar values are reported by Brunsden (1969). This suggests that whereas the contemporary network is conditioned by runoff

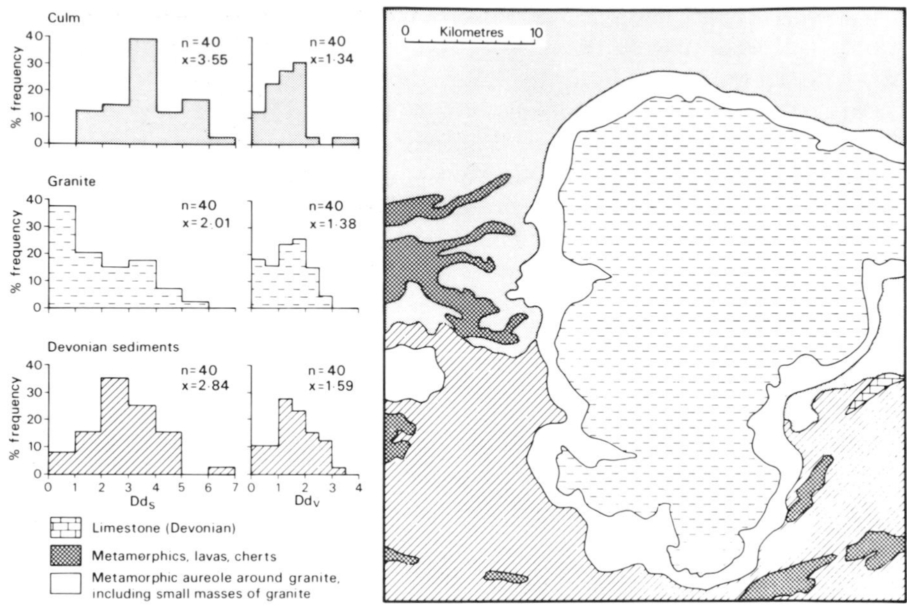

FIGURE 11.2 GEOLOGY AND NETWORK DENSITY IN THE DARTMOOR REGION (after Gardiner, 1971). Densities of the stream network (Dd_s) and valley network (Dd_v) are given in km km^{-2}. For further explanation see text. (Reproduced by permission of the Council of the Devonshire Association)

processes which are controlled by lithological variations, through their impact upon soil type vegetation and land use, the fossil networks are likely to have been produced when lithological differences were made largely irrelevant by permafrost which rendered the surface impermeable and hence much more nearly hydrologically homogenous.

A similar demonstration of the widespread occurrence of dry valleys on various lithologies within part of southwest England was made by Gregory (1971). In the Otter basin (Figures 11.3 and 11.5) of Mercia Mudstone, Cretaceous Upper Greensand and Pebble Beds, and sandstones within the Sherwood Sandstone Formation, the extent of a dry valley network was assessed from morphometric evidence. In this case the frequency distribution of dry valley density was suggested to be bimodal, and field mapping of detailed morphological characteristics (Figure 11.5) suggested that the networks may be divided into two age groups, as discussed further below (p. 269).

The existence of asymmetric slope development in valley systems has also been used to support the hypothesis of permafrost being responsible for

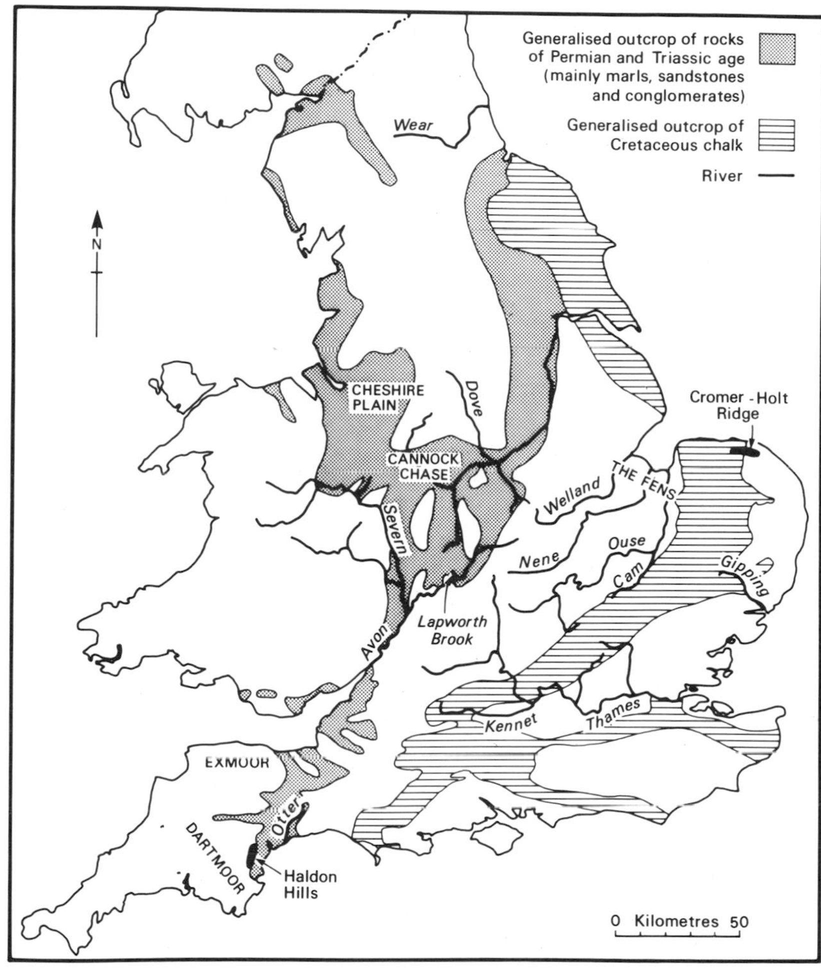

FIGURE 11.3 LOCATION OF MAIN STUDIES REFERRED TO IN THE TEXT

creating conditions suitable for valley development, since valleys in areas underlain by permafrost are often asymmetric in cross profile (French, 1976). Gardiner (1983) suggested that the dry valleys of Cannock Chase (Figure 11.3; Figure 11.5), on the Sherwood Sandstone Formation, were formed under permafrost conditions, and used as evidence the well-marked valley asymmetry in which the south-facing slopes are steeper than those facing north. Unfortunately such arguments are attended not only by problems of definition of asymmetry (Kennedy, 1976) but also by debate concerning its nature or even its very existence in particular areas. For example, Clayden (1971) reports that in the Exeter district valleys are frequently asymmetrical, with slopes having a

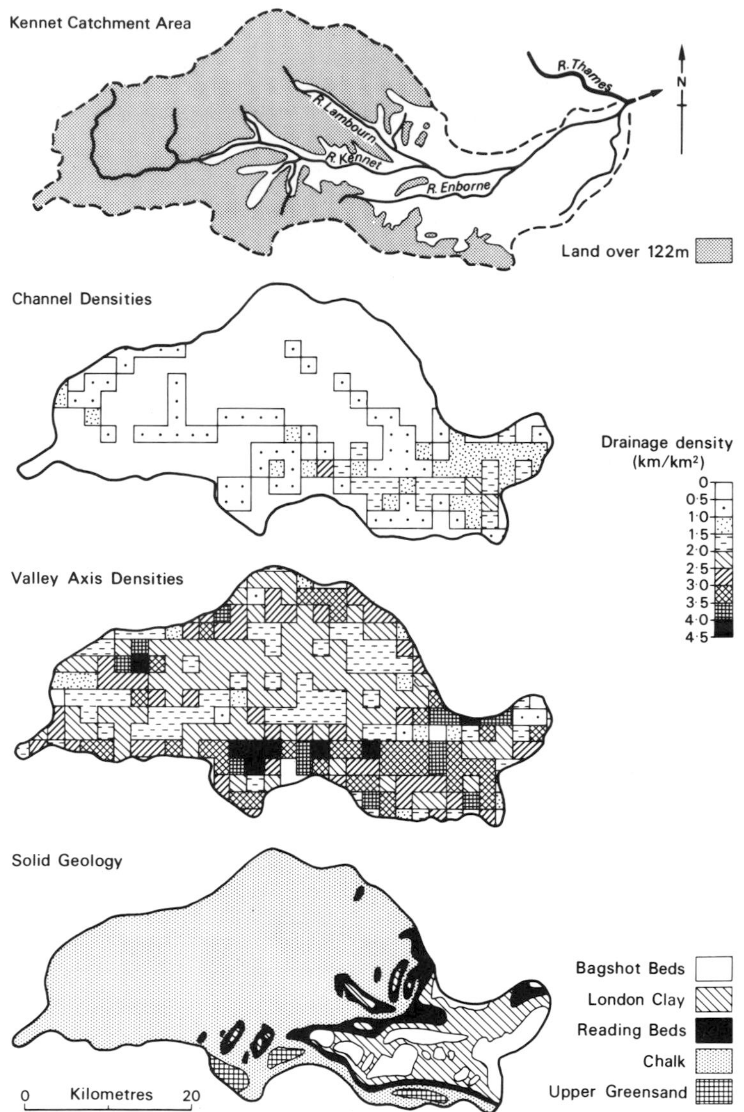

FIGURE 11.4 NETWORK DENSITY, AND RELIEF OF THE KENNET BASIN
(From Cheetham, 1980. With permission from *The Shaping of Southern England*. Copyright 1981, The Institute of British Geographers)

westerly-facing component being considerably steeper than those opposite. However Cullingford (1982), in a review of Quaternary landscape development in Devon as a whole, mentions that asymmetric valleys exist, with their steeper slopes facing to the northeast. This difficulty is further compounded by

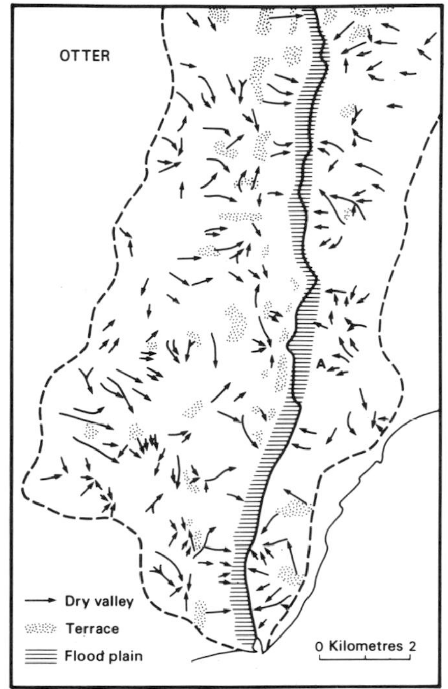

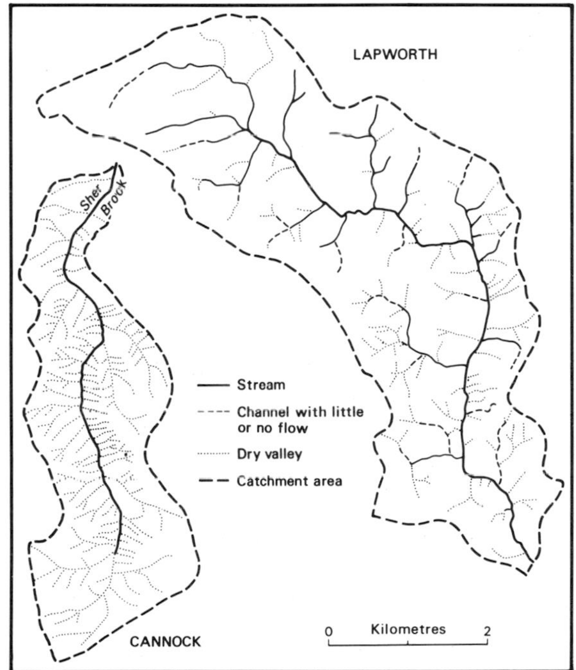

FIGURE 11.5 EXAMPLES OF FIELD MAPPING OF DRY VALLEY NETWORKS. From the Otter Valley (Gregory, 1971), Cannock Chase (Gardiner, 1983), and the Lapworth Brook (Jones, 1968). (Reproduced by permission of The University of Exeter; Geo Books; and *The Geographical Journal*)

alternative explanations of asymmetry in particular locations, such as explanations invoking geological structure.

Although the existence of permafrost has often been invoked as an explanation for the existence of a contracted drainage network it must be borne in mind that in particular cases other explanations may also or alternatively apply. The dry valleys of Cannock Chase have been at least partly ascribed to a former wetter climate or as the product of low frequency runoff (Ojany, 1966), and to a lowered water table in the area due to pumping for water supply as well as to the existence of permafrost as described above. In addition Gardiner and Dackombe (1980) have described small-scale fluvial activity as still occurring in the 'dry' valleys, under extreme meteorological conditions. Gregory (1971) provided a general model in which changes in drainage density may be conditioned by a whole range of explanatory factors, operating at a range of timescales. These factors may be divided into two major groups, depending on whether the fluvial system is viewed as functioning in essentially the same way as at present, or alternatively, environmental change may be so extreme that the mode of hydrological response is radically altered; the existence of permafrost provides an obvious example in this second category. Less dramatic mechanisms for hydrological and hence network change include reduced precipitation totals, a reduced effective precipitation because of increased evapotranspiration losses, changes in vegetation, and transmission of water through the hydrological cycle, and changes in the timing of runoff, if not the absolute amount. In this final explanation it is obviously necessary to determine what recurrence interval of discharge is likely to be responsible for network development, as discussed above (pp. 261).

The investigation by Cheetham (1976, 1980) represents what is one of the more fruitful attempts to date to link palaeonetwork density to palaeodischarge in the United Kingdom, and it will therefore be described in some detail. Within the study area of the Kennet catchment (Figure 11.3; Figure 11.4), a tributary of the Thames, dry valleys occur on all lithologies present (Figure 11.4), although most prominently on the Cretaceous Chalk. Cheetham examined the spatial distribution of both the valley density and currently functioning channel density, as derived from Ordnance Survey 1:25 000 maps. It can be seen (Figure 11.4) that there is a considerable extent of now dry network, and by detailed comparison of the two maps a difference in their spatial organization becomes apparent.

The functioning network has an overall density of 0.45 km km^{-2}, with very low values on the permeable lithologies being balanced by values ranging up to 1.5 km km^{-2} on the impermeable Tertiary sediments, such as the London Clay in the valley of the Enborne. The overall valley density is 2.34 km km^{-2}, and it has much less spatial variation than does the stream density. The influence of lithology is much more subdued. It is therefore suggested, in the same way as for Dartmoor (pp. 263), that the palaeonetwork was developed when lithological differences were made irrelevant by permafrost, which produced a widespread substratum of hydrological uniformity. In addition, many of the dry valleys grade to a river terrace rather than to the present floodplain, and it was therefore argued that the valleys must be inherited features from a former hydrological regime.

In an effort to determine former discharges in this palaeosystem Cheetham made the assumption that a redetermination of Carlston's (1965) relationship between discharge and drainage density (D) is valid for permafrost conditions. This relationship is:

$$Q = 0.063\, D^{1.59}$$

where Q is the mean annual flood per km^2 in $m^3\ s^{-1}$ and D is in km km^{-2}. For the fossil network considered this suggests a palaeodischarge of between 153 and 418 $m^3\ s^{-1}$ ($p = 0.95$) for the former mean annual flood, which is 4–10 times greater than that experienced today in the basin. The lower figure is more than twice the highest flow recorded during a 13 year gauging record. However the upper value is still less than that suggested by Dury's (1965) examination of channel and valley forms in England, and lends some support to those studies (e.g. Tinkler, 1971) which have attempted to revise Dury's estimate in a downwards direction.

A similar approach to that of Cheetham has also been adopted by Jones (1982) for parts of the upper Severn basin, in an as yet unpublished study. For these areas the inclusion of the complete valley network in a Carlston-type relationship does not even extend the estimated discharge beyond the range of annual maxima already recorded, and greatly changed environmental conditions are preferred as an explanation of now decreased runoff.

Network expansion

It must be remembered when discussing palaeonetworks that now dry networks such as those described above have been produced not only by network contraction and valley abandonment but also by an earlier phase or phases of network expansion responsible for the excavation of the valley system. Evidence for repeated phases of erosion associated with alternating morphogenetic regimes, possibly correlated with stadials and interstadials, may come from morphological evidence. Thus Gregory (1971) used form to assist in distinguishing two generations of dry valleys in the Otter valley, Devon (Figure 11.5). The larger, steeper-sided features are graded to the lowermost river terrace, and therefore predate the smaller convexoconcave-sided dells which often dissect this terrace. More frequently sedimentological evidence is used to distinguish phases of erosive activity. For example, Thompson and

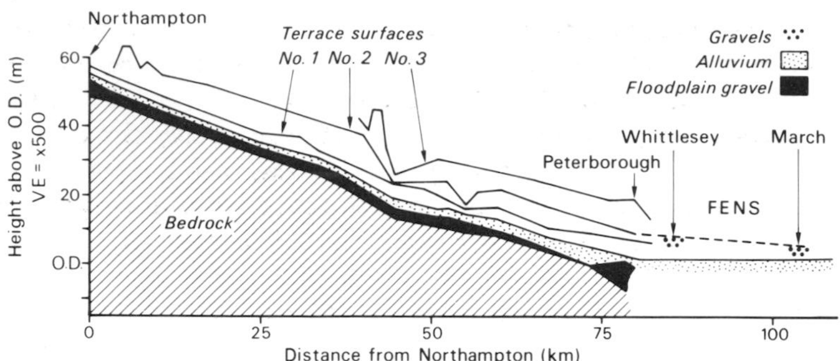

FIGURE 11.6 SCHEMATIC ILLUSTRATION OF POSTULATED RELATIONSHIPS BETWEEN TERRACE DEPOSITS OF THE RIVER NENE AND FEN DEPOSITS (after Castleden, 1980b. Reproduced by permission of the East Midlands Geological Society)

Worsley (1967) showed how in the northeast Cheshire Plain periods of ventifact formation were interspersed by short fluvial episodes, and Rose *et al.* (1980) used sedimentological evidence to identify periods of relative stability and instability in the Gipping basin, Suffolk.

Castleden (1977, 1980a) has developed a general theory of fluvial valley development in which planes of lateral corrasion are produced by rivers at valley floor level under periglacial conditions, and vertical incision, producing terrace sequences (Castleden, 1980b), is confined to the short warmer phases. He argued that the very seasonal discharges and catastrophic floods resulting from snowmelt in periglacial phases gave rise to a valley floor with much coarse sediment produced by frost activity, and a braided channel pattern, probably underlain by permafrost and hence resistant to downward erosion. By contrast during these phases the valley slopes thawed and retreated actively, producing a valley-floor fluvioperiglacial pediplain.

Sedimentological evidence for alternating phases of network expansion and contraction is usually absent, as the eroded material is transported from the fluvial system into the sea and lost to examination or recognition. However, in the case of the Fens of Eastern England (Figure 11.3), the post-glacial regression has preserved deposits which have been interpreted as correlatives of fluvial terrace deposits (Castleden, 1980b; Horton *et al.*, 1974). A series of deposits variously known as the Fen, Fen-edge, or March Gravels (Figure 11.6) may once have been continuous with the terraces of the rivers bordering the area. Although wholly marine origins have been proposed (Baden-Powell, 1934), they are now generally thought to represent debouchment aprons formed as streams such as the Welland, Nene, Great Ouse, and Cam left their narrow valleys and entered the lower-lying Fenland (Horton, 1981). At the margin of the Fens they are undoubtedly fluviatile, but they become thinner and more marine in character when traced eastwards. Some of the gravels have been correlated with the first terrace gravels of the River Nene (Horton, 1981) whilst others, including the March Gravels *sensu stricto*, are thought to be of Second Terrace age (Castleden, 1980b). The fluvial history of the area during the Devensian is interpreted in general terms as one of progressive downcutting, interrupted by at least three stages of deposition. It is presumed that each phase would have associated network expansion, providing the sediment forming the deposit.

In interpreting such sequences, however, some caution must be exercised; in the same way as in the case of network contraction many mechanisms may be responsible, not all of which are hydrological in nature. For example, pulses of tectonic activity or stripping off of resistant rocks to reveal less resistant ones may produce sedimentary sequences which represent pulses of network expansion and increased sediment transport, but which do not reflect changed hydrological conditions, at least as narrowly defined.

NETWORK STRUCTURE

The second major aspect of networks which may be used in palaeohydrological interpretation is the structure of the network, irrespective of its density. Structure encompasses both the topological arrangement of the individual links that make up the network as well as their composition or type of channel. Theoretical research directed towards network topology (e.g. Smart, 1978; Jarvis, 1977) has resulted in a disappointingly small impact on palaeohydrological interpretation in Britain. Jones (1968) examined the Horton (1945) relationship between link numbers and stream orders for the Lapworth Brook in Warwickshire (Figures 11.3; 11.5), for both the functioning and abandoned valley networks. The complete network gives an almost linear plot, which is interpreted as being indicative of formation as a fully integrated fluvial system. However if the now dry components of the network are excluded the resulting relationship appears somewhat deficient in first order elements, being convex upwards rather than concave upwards as is more normal for fluvial systems (Schumm, 1956). This Jones interpreted as indicating a network still in disequilibrium.

Link length, as well as link numbers, is also a significant aspect of network structure. Again, a considerable literature has been devoted to the determination and explanation of link length distributions, but with little application of the results to interpretation of real world palaeohydrological problems. Abrahams (1976) attempted to develop a model which explained changes in both interior and exterior link lengths as stream abstraction from the network occurred. This incorporated an increase in the mean length of both interior and exterior links and a consequent decrease in link frequency and drainage density as erosional reduction of relief and hence stream abstraction occurred. Such non-hydrological change in both structural and density aspects of the network must be carefully discounted or allowed for in palaeohydrological interpretations based on network information.

Spatial arrangement of the elements forming the network may also be important in certain cases, as already discussed above in the case of Complex Response (pp. 261). A further illustration of this is the case of the Otter valley in Devon (Figure 11.5). The course of the main channel of the Otter today bears some relationship to the distribution of dry valley density within the basin. The underlying rocks dip to the east and throughout much of its length the channel is at or near the eastern side of the small floodplain, where a rock cliff is developed in places on the eastern bank. However at several locations the channel is displaced to the west, as at A in Figure 11.5, where a small rock cliff is cut on the western bank, suggesting that this westerly position has long been occupied by the main channel. The locations where the channel is displaced to the west are in areas where the density of dry valleys on the valley's eastern flank is exceptionally high, and it has therefore been suggested that the volume of material issuing from these coombes was sufficient to build up alluvial fans which forced the river to adopt a course in the western part of the floodplain. This is supported by subsurface investigations carried out in the mouth of the coombe system at location A, which revealed a coarse rubbly gravel derived from Cretaceous rocks beneath floodplain alluvium, the boundary between these rising towards the mouth of the coombe.

The final aspect of network structure which must be considered is its composition, as expressed by the types of channel which form the network. Gregory (1979) pointed out that network metamorphosis due to changing hydrological conditions may occur not only in terms of drainage density or network topology but also by changes in the type of channel making up the network. Ovenden and Gregory (1980) showed how, for the case of the River Dove in the South Pennines, comparison of maps from 1870 and 1970 suggested that vegetation-filled features of high resistance to flow (flushes) had been transformed into definite channels following intensification of land use. Since 1970 further changes in channel type had also occurred, consequent upon Man's drainage activities. This specific interpretation of the cartographic evidence has been challenged by Burt and Gardiner (1982), but the general principle is undoubtedly valid. Some attempts have been made to develop quantitative descriptions of channel type based upon resistance coefficients such as Manning's n (Ovenden and Gregory, 1980) or the Darcy–Weisbach f (Bevan *et al.*, 1979, Gardiner and Gregory, 1982), but there has as yet been little explicit application of the concept to palaeohydrological interpretation.

CONCLUSIONS

One obvious but nevertheless important conclusion which may be reached from this necessarily partial and selective review is that although the application of network information to palaeohydrological interpretation and retrodiction holds considerable potential, this potential has yet to be fully realized. Its exploitation would appear to depend upon initiatives being taken to turn the applicable into the applied (Gregory, 1982). Studies of contemporary systems are now beginning to yield results which are capable of being used in retrodiction and some fresh initiatives are needed in order to apply these results.

Future progress would seem to depend upon efforts being directed in at least four principal directions. First, a greater understanding of contemporary systems, including for example an awareness of the dynamic nature of drainage density and of Complex Response within fluvial systems, will lead to much greater confidence in relationships identified between fossil networks and former hydrological characteristics. Second, the full understanding of the complexity of development of drainage networks will be aided by the use of evidence derived from correlative deposits found beneath floodplains and in estuarine and marine areas. Third, studies in Britain to date have produced an uneven spatial distribution of information (Figure 11.3), with most studies concentrating on the southwest peninsula, which was outside the generally accepted limits of Quaternary glaciation, or on calcareous or Permo-Triassic rocks for which network contraction has been most marked. Further studies must seek to examine areas on other rock types and in more northern areas, in order that more comprehensive and more generally applicable models may be developed. Finally, relatively little attention has yet been paid to metamorphosis of networks in terms of their structure and composition, even in contemporary environments, and its palaeohydrological significance has yet to be investigated. Such avenues of research in Britain should also derive from and contribute to efforts being made by both British workers and others, to tackle similar problems elsewhere. Studies of network adjustment to changing hydrological conditions in areas such as Malta (Park, 1977) and southwest USA (Cooke and Reeves, 1976) must inevitably result in greater understanding of the processes responsible for British landscapes, and it can be argued that the fullest understanding of British drainage networks will, perhaps paradoxically, only be gained by examining palaeohydrological analogues elsewhere.

ACKNOWLEDGEMENTS

I gratefully acknowledge the collaboration of R.V. Dackombe during fieldwork in both Cannock Chase and the Otter valley. The latter work was supported by a research grant from the British Geomorphological Research Group.

REFERENCES

Abrahams, A. D. 1976. Evolutionary changes in link lengths: further evidence for stream abstraction. *Transactions, Institute of British Geographers. New Series*, **1**, 225–230.

Anderson, M. G., and Calver, A. 1977. On the persistence of landscape features formed by a large flood. *Transactions, Institute of British Geographers, New Series*, **2**, 243–254.

Baden-Powell, D. F. W. 1934. On the marine gravels at March, Cambridgeshire. *Geological Magazine*, **71**, 193–219.

Baker, V. R. (Ed.) 1981. *Catastrophic Flooding: the Origin of the Channelled Scabland*, Academic Press.

Beaumont, P. 1970. Geomorphology. In Dewdney, J. C. (Ed.), *Durham County and City with Teesside*, British Association, Durham, 26–45.

Beven, K., Gilman, K., and Newson, M. D. 1979. Flow and flow routing in upland channel networks. *Hydrological Sciences Bulletin*, **24**, 303–325.

Brown, E. H. 1964. Some aspects of the geomorphology of south-east England. In Clayton, K. M. (Ed.), *Guide to London Excursions*, 113–118.

Brown, E. H. 1969. Jointing, aspect and the orientation of scarp-face dry valleys, near Ivinghoe, Buckinghamshire. *Transactions, Institute of British Geographers*, **48**, 61–73.

Brunsden, D. 1969. *Dartmoor*, The Geographical Association.

Budyko, M. I. 1958. *The Heat Balance of the Earth's Surface*, U.S. Dept of Commerce.

Bull, A. J. 1940. Cold conditions and landforms in the South Downs. *Proceedings, Geologists' Association*, **51**, 63–71.

Burt, T. P., and Gardiner, A. T. (1982). The permanence of stream networks in Britain: some further comments. *Earth Surface Processes*, **7**, 327–332.

Carlston, C. W. 1963. *Drainage Density and Streamflow*, United States Geological Survey Professional Paper 422–C.

Carlston, C. W. 1965. The effect of climate on drainage density and streamflow. *International Association for Scientific Hydrology, Bulletin*, **11**, 62–69.

Castleden, R. 1977. Periglacial pediments in central and southern England. *Catena*, **4**, 111–121.

Castleden, R. 1980a. Fluvioglacial pedimentation: a general theory of fluvial valley development in cool temperate lands illustrated from western and central Europe. *Catena*, **7**, 135–152.

Castleden, R. 1980b. The second and third terraces of the River Nene. *Mercian Geologist*, **8**, 29–46.

Cheetham, G. H. 1976. Palaeohydrological investigations of river terrace gravels. In Davidson, D. A., and Shackley, M. (Eds.), *Geoarchaeology: Earth Science and the Past*, 335–343. Duckworth, London.

Cheetham, G. H. 1980. Late Quaternary palaeohydrology: the Kennet valley case-study. In Jones, D. K. C. (Ed.), *The Shaping of Southern England*, I.B.G. Spec. Pub. 11, Academic Press, 203–223.

Chorley, R. J. 1958. Climate and morphometry. *Journal of Geology*, **65**, 628–638.

Clayden, B. 1971. *The Exeter District*, Memoir of the Soil Survey of Great Britain.

Clayden, B., and Manley, D. J. R. 1962. Devonshire. *Soil Survey of Great Britain, Report*, **15**, 22–24.

Cooke, R. U., and Reeves, R. W. 1976. *Arroyos and Environmental Change in the American South-west*, Clarendon Press, Oxford.

Cullingford, R. A. 1982. The Quaternary. In Durrance, E. M., and Laming, D. J. C. (Eds.), *The Geology of Devon*, 249–290.

Dalzell, D., and Durrance, E. M. 1980. The evolution of the Valley of the Rocks, north Devon. *Transactions, Institute of British Geographers, New Series*, **5**, 66–79.

Day, D. G. 1978. Drainage density changes during rainfall. *Earth Surface Processes*, **3**, 319–326.

Dunne, T. 1976. Field studies of hillslope flow processes. In Kirkby, M. J. (Ed.), *Hillslope Hydrology*, Wiley, Chichester, 227–294.

Dury, G. H. 1965. *Theoretical Implications of Underfit Streams*, United States Geological Survey Professional Paper 452–C, 43 pp.

French, H. M. 1976. *The Periglacial Environment*, Longman, London.

Gardiner, V. 1971. A drainage density map of Dartmoor. *Reports and Transactions of the Devonshire Association for the Advancement of Science*, **103**, 167–180.

Gardiner, V. 1981. Drainage basin morphometry. In Goudie, A. (Ed.), *Geomorphological Techniques*, George Allen and Unwin, London, 47–55.

Gardiner, V. 1982a. The impact of climate on fluvial systems. In Sharma, H. S. (Ed.), *Perspectives in Geomorphology*, vol. II., Concept Publishing Company, New Delhi, 19–40.

Gardiner, V. 1982b. Drainage basin morphometry — quantitative analysis of drainage basin form. In Sharma, H. S. (Ed.), *Perspectives in Geomorphology*, vol. II, Concept Publishing Company, New Delhi, 107–142.

Gardiner, V. 1983. The relevance of geomorphometry to studies of Quaternary morphogenesis. In Briggs, D. J., and Waters, R. S. (Eds.), *Studies in Quaternary Geomorphology*, Geo-Abstracts, 1–17.

Gardiner, V., and Dackombe, R. V. 1980. Gullying on Cannock Chase. In Doornkamp, J. C., Gregory, K. J., and Burn, A. S. (Eds.), *Atlas of Drought in Britain 1975–6*, Institute of British Geographers, 67–70.

Gardiner, V., and Gregory, K. J. 1982. Drainage density in rainfall–runoff modelling. In Singh, V. (Ed.), *Rainfall–Runoff Relationship*, Water Resources Publications, Colorado, 449–476.

Gregory, K. J. 1966. Dry valleys and the composition of the drainage net. *Journal of Hydrology*, **4**, 327–340.

Gregory, K. J. 1971. Drainage density changes in south-west England. In Ravenhill, W. L. D., and Gregory, K. J. (Eds.), *Exeter Essays in Geography*, University of Exeter, Exeter, 33–54.

Gregory, K. J. 1976. Drainage networks and climate. In Derbyshire, E. (Ed.), *Geomorphology and Climate*, Wiley, London, 289–318.

Gregory, K. J. 1979. Changes of drainage network composition. *Acta Universitatis Ouluensis, Ser. A.*, **82**, 19–28.

Gregory, K. J. 1982. Fluvial geomorphology: less uncertainty and more practical application? *Progress in Physical Geography*, **6**, 427–438.

Gregory, K. J., and Gardiner, V. 1975. Drainage density and climate. *Zeitschrift für Geomorphologie*, **19**, 287–298.

Gregory, K. J., and Gardiner, V. 1979. Comment on 'Drainage density and streamflow: a closer look' by S.L. Dingman. *Water Resources Research*, **15**, 1662–1664.

Gregory, K. J., and Walling, D. E. 1968. The variation of drainage density within a catchment. *International Association for Scientific Hydrology, Bulletin*, **13**, 61–68.

Gurnell, A. M. 1978. The dynamics of a drainage network. *Nordic Hydrology*, **9**, 293–306.

Hickok, R. B., Keppel, R. Y., and Rafferty, B. R. 1959. Hydrograph synthesis for small arid land watersheds. *Agricultural Engineering*, **40**, 608–611, 615.

Hollis, J. M., and Reed, A. H. 1981. The Pleistocene deposits of the southern Worfe catchment. *Proceedings, Geologists Association*, **72**, 59–74.

Horton, A. 1981. The Peterborough area — glacial deposits. In Douglas, T. D. (Ed.), *Field Guide to the East Midlands Region*, Quaternary Research Association, 27–35.

Horton, A., Lake, R. D., Bisson, G., and Coppock, B. C. 1974. *The Geology of Peterborough*, Report 73/12, Institute of Geological Sciences.

Horton, R. E. 1945. Erosional development of streams and their drainage basins: hydrophysical approach to quantitative morphology. *Bulletin Geological Society of America*, **56**, 275–370.

Jarvis, R. S. 1977. Drainage network analysis. *Progress in Physical Geography*, **1**, 271–295.

Jones, J. A. A. 1968. Morphology of the Lapworth Valley, Warwickshire. *Geographical Journal*, **134**, 216–226.

Jones, M. D. 1982. *Palaeogeography and Palaeohydrology of the River Severn, Shropshire, during the Late Devensian Glacial Stage and the Early Holocene*, unpublished M.Phil. thesis, University of Reading.

Jones, P. F. 1979. The origin and significance of dry valleys in south-east Derbyshire. *The Mercian Geologist*, **7**, 1–18.

Kennedy, B. A. 1976. Valley-side slopes and climate. In Derbyshire, E. (Ed.), *Geomorphology and Climate*, John Wiley and Sons, London, 171–202.

Kerney, M. P., Brown, E. H., and Chandler, T. J. 1964. The late-glacial and post-glacial history of the Chalk escarpment near Brook, Kent. *Philosophical Transactions of the Royal Society, Series B*, **248**, 135–204.

Lewin, J. 1969. *The Yorkshire Wolds*, University of Hull Press.

Madduma Bandara, C. M. 1974. Drainage density and effective precipitation. *Journal of Hydrology*, **21**, 187–190.

Melton, M. A. 1957. *An Analysis of the Relations Amongst Elements of Climate, Surface Properties and Geomorphology*, Office of Naval Research, Geography Branch, Project NR 389–042, Tech. Rpt. 11, Columbia University.

Morgan, R. P. C. 1971. A morphometric study of some valley systems on the English Chalklands. *Transactions, Institute of British Geographers*, **54**, 33–44.

Murphey, J. B., Wallace, D. E., and Lane, L. J. 1977. Geomorphic parameters predict hydrograph characteristics in the south-west. *Water Resources Bulletin*, **13**, 25–38.

Ojany, F. F. 1966. The denudation chronology of the Cannock Chase area, Staffordshire. *East Midland Geographer*, **4**, 77–87.

Ovenden, J. C., and Gregory, K. J. 1980. The permanence of stream networks in Britain. *Earth Surface Processes*, **5**, 47–60.

Patton, P. C., and Baker, V. R. 1976. Morphometry and floods in small drainage basins subject to diverse hydro-geomorphic controls. *Water Resources Research*, **12**, 941–952.

Park, C. C. 1977. Dry valley network density in Malta. *Revue de Geomorphologie Dynamique*, **26**, 49–58.

Reid, C. 1887. On the origin of dry Chalk valleys and of Coombe rock. *Quarterly Journal of the Geological Society of London*, 364–373.

Richards, K. S., and Anderson, M. G. 1978. Slope stability and valley formation in glacial outwash deposits, North Norfolk. *Earth Surface Processes*, **3**, 301–318.

Rodda, J. C. 1969. The significance of characteristics of basin rainfall and morphology in a study of floods in the United Kingdom. *UNESCO Symposium on Floods and their Compilation, IASH-UNESCO-WMO*, **2**, 839–845.

Rose, J. R., Turner, C., Coope, G. R., and Bryan, M. D. 1980. Channel changes in a lowland river catchment over the last 13 000 years. In Cullingford, R. A., Davidson, D. A., and Lewin, J. (Eds.), *Timescales in Geomorphology*, Wiley, Chichester, 159–175.

Schumm, S. A. 1956. The evolution of drainage systems and slopes in badlands at Perth Amboy, New Jersey. *Bulletin of the Geological Society of America*, **67**, 597–646.

Schumm, S. A. 1973. Geomorphic thresholds and complex response of drainage systems. In Morisawa, M. (Ed.), *Fluvial Geomorphology*, Pubs. in Geography, State University of New York, Binghampton, 299–310.

Schumm, S. A. 1977. *The Fluvial System*, Wiley, Chichester, 356pp.

Small, R. J. 1965. The role of spring-sapping in the formation of Chalk escarpment valleys. *Southampton Research Series in Geography*, **1**, 3–30.

Smart, J. S. 1978. The analysis of drainage network composition. *Earth Surface Processes*, **3**, 129–170.

Sparks, B. W., and Lewis, W. V. 1957. Escarpment dry valleys near Pegsdon, Hertfordshire. *Proceedings of the Geologists Association*, **68**, 26–38.

Stephens, N. 1966. Some Pleistocene deposits in north Devon. *Biuletyn Peryglacjalny*, **15**, 103–114.

Thompson, P., and Worsley, P. 1967. Periods of ventifact formation in the Permo-Triassic and Quaternary of the North East Cheshire Plain. *The Mercian Geologist*, **2**, 279–298.

Tinkler, K. J. 1971. Active valley meanders in south central Texas and their wider implications. *Bulletin, Geological Society of America*, **81**, 1783–1799.

Trainer, F. W. 1969. Drainage density as an indicator of baseflow in part of the Potomac River. *United States Geological Survey Professional Paper 650–C*, C177–83.

Warwick, G. T. 1964. Dry valleys in the southern Pennines. *Erdkunde*, **15**, 116–123.

Waters, R. S. 1966. The Exeter symposium. *Biuletyn Peryglacjalny*, **15**, 123–149.

Williams, G. P. 1978. Bank-full discharge of rivers. *Water Resources Research*, **14**, 1141–1154.

Background to Palaeohydrology
Edited by K. J. Gregory

12

The arroyo problem — palaeohydrology and palaeohydraulics in the short term

William L. Graf

Department of Geography and Center for Southwest Studies
Arizona State University

The problem of the origin and dynamics of arroyos has generated more interest for geomorphic investigators than any other research question in arid environments. The last decade has seen significant progress in the development of cumulative generalizations that may lead to theory formulation, and the time may now be opportune to assess progress. The general objectives of this paper are to briefly review the voluminous literature on the arroyo problem, to focus on the recent advances that hold promise for theory building, and to suggest possible profitable avenues for further work toward an integrated dynamic theory for arroyos.

Research into the origin and dynamics of arroyos has been significant in the science of geomorphology because the subject of arroyos embodies many of the important considerations of the wider field. The geomorphic response to general environmental changes, the nature of the connections between geomorphic and hydroclimatic systems, the distributions of the processes of erosion, transportation, and deposition, and the chronology of adjustment are foci of investigative interest shared by the specific research into arroyos and the general endeavours of geomorphological work.

The analysis of arroyo dynamics involves explanation of the hydrologic processes in channels coupled with study of the connection between channel adjustments and hydroclimatic changes, so that the arroyo problem has many points of similarity to the studies of palaeohydrology and palaeohydraulics. The arroyo problem entails events of shorter duration and in the more recent past than the events of concern in the 'palaeo' problems, but the similarities are stronger than the differences. Palaeohydrologic and palaeohydraulic studies have been remarkably successful in generating a variety of basic law-like statements, and as evidenced by other papers in this volume the development of integrated theories in these fields seems close at hand. That workers addressing the arroyo problem can and should make similar theoretical efforts is a primary thesis for this paper.

ARROYOS

Because the first step in successful theory building is the establishment of a set of primary or primitive terms, the remainder of this paper employs the following definitions. An arroyo is a trench with a roughly rectangular cross-section excavated in valley-bottom alluvium with a major stream channel on the floor of the trench. The word has been part of the Spanish literature since at least 775 in the sense of a river or stream bed. It first appeared in an English context in writings about western North America in 1862, and Clarence King, a US government surveyor and geologist, used it in his descriptions of the western American landscape in 1872 (Murray *et al.*, 1933, vol. 1, p. 465). Dodge (1902) was the first to employ the term in the scientific literature in the context of a research problem, and Bryan (1925) and Antevs (1952) subsequently proposed definitions essentially along the outline of the one used here.

A gully is a V- or U-shaped trench in unconsolidated materials with a minor channel on the bottom. Gullies are not generally colocated with major stream channels. The English language has associated the term gully with natural excavations by water on hillsides since 1657 (Murray *et al.*, 1933, vol. 4, p. 505), but its modern usage relates to the fact that gullies are larger than rills and the characterization that they 'cannot be crossed by wheeled vehicles or eliminated by ploughing' (Gregory and Walling, 1973, p. 369–370). In their earliest uses in English both terms referred to features that were barriers to travel. In the interests of theory building, it is preferable to eliminate the scale dependent factor (Church and Mark, 1980), so that arroyos (the subject of the following discussion) are here differentiated from gullies (which are not considered) by the fact that the former occur in valley alluvium and that they are colocated with major through-flowing stream channels. In general this distinction follows the suggested scheme of Schumm (1980b), except that I have substituted 'through-flowing' for Schumm's 'pre-existing' stream channels in recognition of the problem of determining after incision whether or not a channel existed in a given location before the arroyo. Use of 'through-flowing' also eliminates trenches excavated into colluvium, and follows the original sense of the words.

A commonly recognized series of events occurs in the history of an arroyo. The assumed starting condition or initial phase is a stream channel on the upper surface of an alluvial deposit usually in an arid or semiarid environment. A change in the hydroclimatic system controlling inputs of water and sediment to the stream or an internal adjustment in the fluvial system initiates erosion and channel incision in an erosional phase, creating a trench that is rectangular in cross section. Bed and/or bank erosion continues until a non-erodible boundary halts the process or until an equilibrium condition prevails among water discharge, sediment discharge, and channel configuration. Another change in hydroclimatic controls or internal adjustments in the fluvial system triggers a depositional phase wherein the trench is refilled with sediment until the channel again rests on the upper surface of alluvial fill in a final phase that has conditions similar to the initial phase. Any successful arroyo theory must account for the mechanisms by which the fluvial system switches from one phase to the other and must be dynamic in the sense of including all the phases in its explanation of the phenomena.

GEOMORPHOLOGY AND THE ARROYO PROBLEM

In reviewing the progress of geomorphology and the conduct of arroyo research, I have adopted Kuhn's perspective of the structure of scientific activities (Kuhn, 1970). Although critics have expressed some reservations about this perspective (Shapere, 1964; Masterman, 1970), there has been wide acceptance of Kuhn's ideas (Harvey, 1969), and his precepts appear to fit the development of geomorphology through the last century (reviewed in part by Chorley, Dunn, and Beckinsale, 1964; Chorley, Beckinsale, and Dunn, 1973). The history of geomorphology has demonstrated a pre-paradigm period when exploration and data gathering were dominant and when there was intense competition among world views (e.g., Davis, 1909; Penck, 1953). Eventually one paradigm grew to dominance in a period of normal science ruled by Davisian evolutionary geomorphology in the early 1900's, but this approach left many unanswered questions leading to a scientific crisis in the field.

A scientific revolution in geomorphology about 1950 emphasized time-independent processes and resolved the crisis by overthrowing the evolutionary paradigm. A period of normal science again prevailed from the early 1950's to the 1970's, but geomorphology may again be entering a crisis stage where the emphasis on time-independent processes has exhausted the available puzzles susceptible to solution by the equilibrium approach. Many pressing issues appear to demand new spatial- and temporal-dependent solutions that are capable of incorporating the successful aspects of the evolutionary and equilibrium concepts, but that are also capable of accounting for switches from one operating mode to another and for spatial variation in process operation.

Attempts to solve the arroyo problem as outlined in the remainder of this paper reflected the direct influence of the events in the general science of geomorphology. The next (or second) section provides a brief review of the original paradigm in arroyo studies which emphasized the problem of initial causes of erosion, including an explanation of why that paradigm had collapsed by the 1970's. The third section provides a more detailed review of the paradigm of process in the arroyo problem, a paradigm that has been immensely successful in some aspects but that is now entering a crisis because of its inability to accommodate events that occur in specific spatial and temporal frameworks requiring significant attention to disequilibrium. The fourth section explores the possible dimensions of a new paradigm which attempts to solve the arroyo problem by combining the successful parts of evolutionary and equilibrium approaches with newer ideas in a unified spatial–temporal theory flexible enough to account for all phases of the development of arroyos, for conditions of equilibrium and disequilibrium, and for process reversals. The emerging paradigm has remarkable similarities to palaeohydrology and palaeohydraulics. A concluding section provides speculations on the future course of research in the arroyo problem.

THE PARADIGM OF ORIGIN

When Dodge (1902) first analysed arroyos in the scientific literature the major research problem entailed explaining why arroyos developed. Observers viewed arroyos as anomalous features of the landscape, and as undesirable, destructive changes in stream systems that brought economic ruin to previously productive agricultural valleys, an extension of older values regarding the 'perfect' agrarian landscape (Davies, 1969). The loss of productive croplands and valuable grazing areas, high rates of sedimentation in areas downstream from arroyos, the creation of new channels that were difficult to dam for irrigation, and declining water tables resulting from channel entrenchment provided economic incentives for over half a century of research aimed at the question of arroyo origin. The resulting body of literature is immense (and immensely confusing), but it produced a limited number of explanations for the initiation of arroyo cutting. Each of the general causes is outlined briefly below, along with a review of the advantages and disadvantages of each explanation. Given the magnitude of the literature which represents the paradigm of origin, citations are only made for representative papers, and the geographic area of interest is limited to the interior American southwest, the area of primary interest in arroyo literature.

Grazing

The introduction of large numbers of cattle to the American southwest in the late 1800's has been the most frequently cited reason for the regional development of arroyos (citations summarized in Table 12.1). Many early researchers argued that overgrazing stripped vegetation from hillsides and valley floors and that trampling weakened channel banks so that increased runoff from slopes excavated unconsolidated alluvial materials. Researchers recognized the role of floods in accomplishing rapid incision of channels, but

TABLE 12.1 ARROYO STUDIES EMPHASIZING GRAZING AS AN INITIATING CAUSE

Reference	Study area
Dodge, 1902	Southeastern Utah
Hough, 1906a	Southern Colorado Plateau
Hough, 1906b	Southern Colorado Plateau
Dodge, 1910	Southeastern Utah
Thornber, 1910	Southern Arizona
Rich, 1911	Western New Mexico
Duce, 1918	Southern Colorado
Leopold, 1921	New Mexico and Arizona
Ross, 1923	Western Arizona
Miser, 1924	Southeastern Utah
Swift, 1926	Eastern Arizona
Winn, 1926	Eastern Arizona
Bailey *et al.*, 1934	Northern Utah
Bailey, 1935	Southern Utah
Brady, 1936	Northern Arizona
Colton, 1937	Northeastern Arizona
Cooperider and Henricks, 1937	New Mexico
Gregory, 1938	Southeastern Utah
Knechtel, 1938	Southwestern Arizona
Cottam and Stewart, 1940	Southwestern Utah
Calkins, 1941	New Mexico
Thornthwaite *et al.*, 1941	Northeastern Arizona
Thornthwaite *et al.*, 1942	Northeastern Arizona
Bailey *et al.*, 1947	Utah
Peterson, 1950	Southwestern USA
Marston, 1952	Northern Utah
Antevs, 1952	New Mexico
Melton, 1965	Southern Arizona
Cooke and Reeves, 1976	Southern Arizona and California
Womack and Schumm, 1977	Northwestern Colorado
Cooley, 1979a	Southern Colorado Plateau
Masannat, 1980	Southwestern Arizona
Dobyns, 1981	New Mexico and Arizona
Alford, 1982	Southwestern New Mexico

most assumed that under natural conditions without domestic livestock the floods would not have caused the catastrophic erosion. A notable variation on this theme was by Melton (1965) who postulated from his analysis of the sedimentary evidence that grazing on hillslopes caused movement of surface materials from hillslopes onto the valley alluvium. The new material choked valley floor vegetation and altered the valley cross-sections by restricting the available channel area and thus producing deeper, more erosive flows.

The popularity of grazing as a major causal factor rested in large part on the temporal coincidence of overgrazing and arroyo cutting: arroyos seemed to appear only after cattle were introduced. But as Quaternary researchers investigated the palaeohydraulic evidence, they found that arroyos had cut and filled frequently during the thousands of years before the introduction of livestock (Bryan, 1940, 1941; Gregory, 1950). As research into historical stocking levels progressed, it became apparent that in New Mexico and Arizona significant numbers of livestock had been introduced by Mexican herders four decades before the Anglo-American influx, yet the earlier periods of overgrazing did not result in arroyo development (Cooperider and Hendricks, 1937; Denevan, 1967). By the 1920's and 1930's the grazing hypothesis alone clearly could not explain arroyo cutting, and although later workers frequently identified grazing as a contributing factor, climatic controls received increasing attention.

General climatic conditions

Hypotheses to explain arroyo cutting by reference to general climatic conditions fall into three general categories: cutting initiated by a shift from drier conditions to wetter conditions; by a shift from wetter conditions to drier conditions; or by a change in frequency, intensity, or seasonality of precipitation events. The first climatic hypothesis was that in times of increasing moisture more runoff would be available for erosion, resulting in arroyo development (climate-related works are summarized in Table 12.2). At the time of its proposal, this hypothesis fit observations in the Colorado Plateau in the 1860's and 1870's when researchers assumed that conditions reflected increasing moisture (Dutton, 1882). The explanation was attractively simple, but it relied on the assumption that more precipitation resulted in more runoff. More precipitation might, for example, result in more vegetation which would inhibit flood peaks and retard runoff, so that the basic assumption might be invalid. The lack of a widely applicable dynamic model prevented accurate testing of the increasing moisture hypothesis.

Bryan (1922) accepted this first hypothesis, but later changed his opinion to reverse the basic assumption. His work, related primarily to archaeological studies, led him to postulate that when climate shifted from wetter to drier conditions vegetation cover declined, permitting increasing runoff and resulting in channel erosion (e.g., Bryan, 1954). By the early 1900's instrumental records documented a shift from generally more moist conditions to drier conditions in many areas of the American southwest, disproving earlier assumptions of increasing moisture. Also, unlike the grazing hypothesis, Bryan's climatic hypothesis was attractive because it easily extended to Quaternary timescales. Richardson (1945) offered one resolution of the competition between hypotheses predicated on increasing or decreasing moisture by suggesting that shift in either direction would cause arroyo development. By the mid-1940's the development toward cumulative generalizations concerning arroyo causes was obviously at a standstill. Later Haynes (1968) could state that climate was a controlling factor in arroyo initiation, but that it was impossible to determine exactly how the connection operated.

As instrumented records became longer and more complete in their areal coverage, significant temporal and spatial variability in moisture became apparent in the arid and semiarid southwest. A number of workers discarded the clumsy general climatic hypotheses in favour of statements centred on changes in the frequency, intensity, or seasonality of precipitation events (citations reviewed in Table 12.2). It might be argued, for example, that during periods of arroyo cutting, rare intense storms occurred in summer when little protective vegetation was present. Periods of arroyo filling might be characterized by frequent gentle rains that fostered dense vegetation and sedimentation on valley floors. Unfortunately the statistical analysis of precipitation records has been a major issue in the testing of these hypotheses (Knox, 1978), and they remain largely unproven. As with all climate-based hypotheses there are significant conflicting data (Tuan, 1966), and for periods greater than a century techniques such as palynology, dendrochronology, and sedimentology lack the precision required for adequate testing (Heede, 1976).

Researchers frustrated by the limitations of the grazing and climatic hypotheses proposed a widely accepted resolution of the problem of arroyo initiation by suggesting that overgrazing acted as a trigger mechanism in disrupting a system that was potentially unstable because of climatic changes (Huntington, 1914). This 'trigger mechanism' hypothesis had the advantage of incorporating much of the available evidence, and could be extended to prehistoric arroyos by assuming the action of some other 'trigger mechanism' such as fire (citations reviewed in Table 12.2). Unfortunately the 'trigger mechanism' hypothesis was untestable because the erosion processes of arroyo development erased indications of the causes of initiation. Also the compromise hypotheses failed in those cases where there was no coincidence of overgrazing, climatic change, and arroyo development.

Catastrophic events

Many popular accounts of arroyo development such as those in newspapers noted that large floods appeared to accomplish the work of initial incision (Cooke and Reeves, 1976). Many researchers therefore emphasized the occasional large flood (return interval greater than 100 years) with deep flow and high shear stress as a solution to the arroyo problem (citations reviewed in Table 12.3). The large flood was an event that might or might not be coincidental with overgrazing or climatic change. The flood hypothesis also subsumed in part those who held to a climatic hypothesis because the flood in question could result from climatic change or from a purely random meteorological event. The problem with the flood hypothesis was that floods of

TABLE 12.2 ARROYO STUDIES EMPHASIZING CLIMATIC CAUSES

General thesis	Reference	Study area
Change to wetter conditions	Dutton, 1882	Southern Utah and Northern Arizona
	Johnson, 1900	Southwestern USA
	Huntington, 1914	Southeastern Arizona
	Bryan, 1922	Southern Arizona
	Gregory and Moore, 1931	Southern Utah
	Richardson, 1945	Southwestern USA
	Quinn, 1957	Southwestern USA
	Martin, 1963	Southern Arizona
Change to drier conditions	Bryan, 1925a	New Mexico and Arizona
	Bryan, 1925b	Southern Arizona
	Bryan, 1928a	New Mexico
	Bryan, 1928b	Northwestern New Mexico
	Bryan, 1929	Arizona and New Mexico
	Hack, 1939	Northern Arizona
	Bryan, 1940	New Mexico
	Bryan, 1941	Northwestern New Mexico
	Hack, 1942	Northern Arizona
	Richardson, 1945	Southwestern USA
	Leopold and Snyder, 1951	Northwestern New Mexico
	Judson, 1952	Southwestern USA
	Euler *et al*, 1979	Southern Colorado Plateau
	Leopold and Bull, 1979	Southeastern Utah

Change in frequency, intensity, or seasonality	Visher, 1913	Southern Arizona
	Leopold, 1951	Northern New Mexico and Arizona
	Miller and Wendorf, 1958	New Mexico
	Martin *et al.*, 1961	Arizona and New Mexico
	Cooley, 1962	Southern Colorado Plateau
	Schoenwetter and Eddy, 1964	New Mexico and Arizona
	Martin, 1963	Arizona and New Mexico
	Cooke, 1974	Southern Arizona
	Cooke and Reeves, 1976	Southern Arizona and California
Climate change with the grazing 'trigger mechanism'	Huntington, 1914	Southeast Arizona
	Gregory, 1917	Northeast Arizona
	Miser, 1924	Southeastern Utah
	Richardson, 1945	Southwestern USA
	Bryan, 1928b	Northwestern New Mexico
	Gregory and Moore, 1931	Southern Utah
	Bryan, 1940	Arizona and New Mexico
	Bryan, 1941	Northern Mexico
	Leopold, 1951a	New Mexico and Arizona
	Judson, 1952	Southwestern USA
	Bryan, 1954	Northwestern New Mexico
	Cooke and Reeves, 1976	Southern Arizona and California

TABLE 12.3 ARROYO STUDIES EMPHASIZING CATASTROPHIC FLOODS AS CAUSAL MECHANISMS

Reference	Study area
Dellanbaugh, 1912	Southwestern USA
Gregory, 1917	Northeastern Arizona
Reagan, 1922	Southern Colorado Plateau
Reagan, 1924a	Southern Colorado Plateau
Reagan, 1924b	Southern Colorado Plateau
Gregory and Moore, 1931	Southern Utah
Brady, 1936	Northern Arizona
Senter, 1937	Northwestern New Mexico
Wooley and Alter, 1938	Utah
Thornthwaite *et al.*, 1941	Northeastern Arizona
Thornthwaite *et al.*, 1942	Northeastern Arizona
Wooley, 1946	Utah
Gregory, 1950	Southwestern Utah
Hunt *et al.*, 1953	Southcentral Utah
Hastings, 1959	Southeastern Arizona
Hastings and Turner, 1965	Southern Arizona
Tuan, 1966	New Mexico
Mosley, 1972	Northern Colorado
Cooke and Reeves, 1976	Southern Arizona and California
Love, 1979	Northwestern New Mexico

similar magnitudes on similar streams did not always produce channel entrenchment (e.g., Swift, 1926). Flood events played a large role in arroyo initiation but they provided only unsatisfactory partial explanations.

Human activities other than grazing slowly came to light as important variables influencing the geomorphic processes associated with floods in the initiation of arroyos. Roads, bridges, canals, ditches, and railroad embankments that restricted floodways concentrated flood waters to depths capable of channel entrenchment and enlargement. Verification may be difficult in some cases because identification of the responsible structure may be impossible after the system adjusts, especially if the arroyo enlarges to consume the structure that initiated the phenomena. An arroyo initiated by overgrazing fifty years ago may have the same appearance now as one initiated by a poorly-designed bridge. Because erosion may have removed most of the original alluvial surface, the modern geomorphologist may have insufficient data to address the question of origin. Evidence for human or natural causes may be later obscured by channel migration as well as vertical cutting (La Marche, 1965).

Crisis in the paradigm of origin

Kuhn (1970) contended that when a paradigm encounters large numbers of cases or puzzles where its explanation is unworkable, normal science activities shift to a crisis stage wherein the old paradigm collapses and a replacement appears. He contended that scientists may even abandon the basic questions of the old paradigm as either irrelevant, unimportant, or unsolvable. Geomorphologic approaches to the arroyo problem closely fit this postulated series of events. Several forces brought about the demise of the paradigm of origin. First, by the 1960's the search for original causes was obviously not producing widely applicable cumulative generalizations. Second, the advent of open systems analysis in geomorphology encouraged process-oriented analyses instead of investigations into problems of initiation and evolutionary changes. Palaeohydraulics and palaeohydrology flourished under this newly-accepted approach, and research into the arroyo problem underwent drastic changes in perspective. Third, the increasing importance of land management by public and private decision makers produced a need for understanding the dynamics of arroyos and their channels (see Toy, 1982, for general commentary). Faced with the unhappy fact of the existence of arroyos, managers and engineers were not so much concerned with why the arroyos were initiated as with how they behaved once established (e.g., Cooley, 1979a, 1979b).

Finally, in 1976 two publications appeared that effectively sounded the death knell for the paradigm of origin. Cooke and Reeves (1976) addressed the basic research audience while Heede (1976) addressed the applied specialists, but their message was the same: arroyos are the end products of a wide variety of original causal mechanisms, and given the final form it is often impossible to determine the exact initial cause. In short, the question of the origin of arroyos was declared to be essentially unanswerable in general terms. Burkham (1981) later reached similar conclusions.

The shift to a new process-based paradigm was not sudden nor did it present new ideas, but it did represent a wider acceptance and a more detailed examination of processes than the paradigm of origin. For example, Gregory's (1917, 1938, 1950) work consistently described arroyo development in terms of the spatial rearrangement of sediments in a fashion that would be comfortable for specialists in geomorphology several decades later. By the 1960's the majority of published works addressing the arroyo problem were concerned with how the form of the arroyo system interacted with fluvial processes. The paradigm of process had replaced the paradigm of origin and the arroyo problem became a question of applying palaeohydraulics and palaeohydrology in the short term.

THE PARADIGM OF PROCESS

The paradigm of process derived many of its basic concepts from applied physics, but the concept of systematic investigation of fluvial forms and processes using basic precepts of physics was not new to geomorphology in the 1960's. Playfair (1802) had called on the principle of equilibrium to specify that the sizes of valleys were generally adjusted to the sizes of the rivers that flowed through them. Dana (1850) found in his investigations of the landforms of the South Pacific islands that simplistic concepts from physics were useful in explaining fluvial processes. Gilbert (1877) applied the principles of physical mechanics to the slope and channel systems of the Henry Mountains, and later extended his reasoning to more detailed investigations of channel processes in laboratory settings (Gilbert, 1914). Gilbert was unable to compete with the evolutionary paradigms of Davis, however, and it was not until the collapse of Davisian geomorphology in the mid-twentieth century that the applied physics approach reappeared. In two seminal papers directly applicable to the arroyo problem Strahler (1952, 1956) made a strong case for the establishment of a new paradigm in geomorphology based on the application of physics and mathematics. One of the physical principles that attracted increased research effort was an expected tendency toward equilibrium states.

The equilibrium approach

The concept of equilibrium was so attractive that it became the guiding principle of the early years of the paradigm of process. Its major expression appeared in the concept of the graded stream (Mackin, 1948; Dury, 1966; Knox, 1975), hydraulic geometry (Leopold and Maddock, 1953), dynamic equilibrium (Hack, 1960), and a sense of balance between the resistance of channel sediments and resulting forms (Schumm, 1961). The introduction and implicit (occasionally explicit) use of open systems concepts provided a useful framework for the application of equilibrium concepts, and further discredited approaches based on closed system evolutionary ideas (Chorley, 1962; Howard, 1965; Chorley and Kennedy, 1971). The open systems framework also fits well with the demise of the paradigm of origin in research on the arroyo problem: '...in terms of analysing the causes of phenomena which exhibit a marked steady-state tendency, considerations regarding previous history became not only hypothetical, but largely irrelevant' (Chorley, 1962, p. B6). The equilibrium version of the process paradigm was suitably enshrined in textbooks from the most elementary to the most sophisticated levels (Morisawa, 1968; Leopold, Wolman, and Miller, 1964).

The 1960's also brought extensive development of hydraulics and palaeohydrology. The research of Schumm and Lichty (1963), Schumm (1968a, 1968b), and Birkeland (1968) showed that a quantitative analysis of forms, deposits, and processes could be a useful approach to understanding the operation of streams over long time periods. Their work was later successfully extended to arroyo environments (e.g., Burkham, 1971; Baker, 1974; Andres, 1980). As investigations considered shorter timescales and more restricted spatial scales, detailed analysis of arroyos began to yield data on rates of change which could accurately characterize process operation (Malde and Scott, 1977; Gerson, 1977; Ploy, 1977; Pickup, 1975).

Alternatives to equilibrium approaches

The equilibrium version of the process paradigm was not entirely successful in attacking the arroyo problem for three reasons: the nature of the processes involved, the persistence of evidence of cyclic behaviour of the systems, and questions about the utility of equilibrium concepts in general. First, it was difficult to determine how equilibrium would be established between the discontinuous processes of discharge and the continuous forms of the arroyos, and it was clear that arroyos were part of a tremendously complex channel system. Vegetation in and near the channel, for example, strongly influences fluvial processes (Hadley, 1961; Burkham, 1972; Graf, 1981). Early in the development of the science of geomorphology, Gilbert (1886, p. 286–287) remarked on the complexity: 'Phenomena are arranged in chains of necessary sequence. In each chain each link is the necessary consequent of that which precedes, and the necessary antecedent of that which follows... If we examine any link...we find it has more than one antecedent and more than one consequent.'

Schumm (1973) resurrected these views, labelled them 'Complex Responses', and combined them with the concept of geomorphic thresholds to explain gully and arroyo processes. Based on earlier work with discontinuous gullies and arroyos (Schumm and Hadley, 1957; Leopold, Wolman, and Miller, 1964, p. 448–453; Mosley, 1972) he suggested that the phenomena resulted from rapid erosion after channel systems surpassed some failure threshold. In subsequent work he and his associates investigated thresholds and complex responses in arroyo-like fluvial systems (Patton and Schumm, 1975; Womack and Schumm, 1977), and from these efforts the concepts diffused to other geomorphic research (Coates and Vitek, 1980). Summary statements followed (Schumm, 1977, 1980a, 1980b). Work on the arroyo problem had changed from being influenced by trends in geomorphology to originating the trends and disseminating them to the larger field.

The persistence of cyclic evidence in arroyo systems has also limited applications of equilibrium concepts which appeared in some cases to be simply old wine in new intellectual bottles. Many workers had cast the problem of origin in a framework of cyclic erosion whereby cutting and filling of channels were limited reflections of Davis' more comprehensive geographic cycle (Gregory, 1917, 1938, 1950; Miser, 1924; Gregory and Moore, 1931; Bailey, 1935). Some later workers used the cyclic concept with the trappings of equilibrium states (Schumm, 1976, 1977; Schumm and Hadley, 1957; Womack and Schumm, 1977), so that the cyclic approach has remained viable through the emergence of the 'new' process paradigm (Heede, 1974; Patton and Schumm, 1981).

Finally, the equilibrium approach to the process paradigm in arroyo research enjoyed only limited success because the processes and forms might not tend toward equilibrium in the usual sense of the word (Bull, 1975a, 1975b). Once initiated, arroyo development seemed to proceed until an entirely new system of processes and forms took over from the prior conditions (Wertz, 1970). Bull (1980a, 1980b) showed that Schumm's view of thresholds was definable with established notions of shear stress in flowing streams.

Analyses of equilibrium and disequilibrium showed a need for improved interpretive models to relate system variables to time. Knox (1972) proposed the use of step functions to describe the radical shifts in a geomorphic system from one mode to another. Using evidence from an arroyo system Graf (1977) later proposed the exponential rate law as a model of temporal change. Evidence from probable shear stresses in montane arroyos that had developed over a century confirmed that once the adjustment began with the transgression of a threshold, change took place rapidly at first, but at decreasing rates thereafter (Graf, 1979b).

These new studies of the variation of force through time paralleled research into variation across space (e.g., Faulkner, 1974; Van Arsdale, 1982). Graf (1979a) showed that spatial variation in arroyo cutting was linked directly to spatial variation of shear stress and resistance by vegetation. The spatial variation had demonstrable regularity that could be described by distance decay functions, but variation in geologic materials affected the functions (Graf, 1982b). Further work indicated that temporal and spatial variation in arroyo processes could be described and generalized in an integrated system that included limited concepts of equilibrium but that primarily emphasized the changes associated with disequilibrium (Graf, 1982a). In the most recent approaches, equilibrium plays a minor role in explanation while the major role is played by the mapping and comparison of forces and resistances in the sense of applied physics. Palaeohydraulics and palaeohydrology has also come to rely heavily on analyses of fluvial forces as revealed in sedimentary evidence (Novak, 1973; Baker and Ritter, 1975; Bridge, 1978; Church, 1978).

Integrated theory

Graf (1977) and Leopold (1978) have called for the development of an integrated dynamic theory for arroyo development that encompasses temporal and spatial variation. The theory should be broad enough to provide explanation for all the depositional as well as erosional phases of arroyo development, and must be flexible enough to account not only for reversal of processes in the same place (noted by Leopold, 1976, 1978), but for opposite results in seemingly similar situations (Dunne and Leopold, 1977, p. 705). A successful unified theory must accommodate these opposing situations and could include the concepts of thresholds, complex responses, equilibrium, rate laws, distance decay functions, reversibility, and the comparisons of force and resistance. Based on experiences in the western United States Graf (1979c) proposed that catastrophe theory provided a useful framework for an integrated theory and presented a successful pilot test of the method for arroyos. Thornes (1980) arrived at similar conclusions from his experiences with arroyos in Spain. The major advantage of catastrophe theory is that it accounts for gradual as well as abrupt changes and avoids the pitfalls of overly restrictive assumptions sometimes associated with uniformitarianism (reviewed by Shea, 1982).

At present (1982) efforts toward the development of an integrated theory for arroyos continue along several lines that mix basic and applied research. W. B. Bull (funded in part by the US Geological Survey) is applying the models of temporal and spatial change in channel erosion to the problems of tectonic geomorphology where channel erosion is initiated by vertical movement of faults across channels. S. A. Schumm (funded by the US Department of Agriculture) is investigating channel incision in agricultural areas with the hope of fruitful generalizations. W. L. Graf (funded by the National Science Foundation) is approaching the arroyo problem using catastrophe theory applied to arroyos on canyon floors that produce sedimentation problems for downstream reservoirs. Other researchers are investigating arroyos or arroyo-like features in a variety of worldwide environments (e.g., Eyles, 1977).

CONCLUSIONS

A widely accepted, well integrated theory for arroyos has not yet developed because of three major problems: limited testing, lack of sediment data, and lack of long-term historical records. Limited testing is particularly a problem with the concepts of rate laws and to a lesser degree with thresholds. The limited analytical studies that introduced the concepts need to be replicated in a variety of situations. Data on sediment characteristics need to be coupled with analyses of shear stress to complete the development of the force versus resistance approach to processes. Finally, early studies of arroyo development could investigate only the erosional phase of the process, but in the 1980's nearly a century of observations is available for some arroyo systems. It may be useful to return to some classic arroyo localities and to update earlier studies (e.g., as in Emmett, 1974; Leopold, 1976).

In future efforts, researchers investigating the arroyo problem can profit from the example of their colleagues in palaeohydraulics and palaeohydrology who have modified engineering approaches for application to natural science problems. The natural environment is rarely susceptible to analysis by conventional engineering techniques because of highly variable conditions and numerous questionable assumptions, but some basic engineering and physical principles can be usefully applied to the arroyo problem (e.g., Chang, 1982).

Geomorphology and engineering are complementary (Shen, 1979, p. 20/10), but do not represent the same perspective. For example, engineering approaches characterize processes as products of broadly defined statistical series of events. Within any 100-year period, there will be one 100-year flood, two 50-year events, and so on. On the other hand, geomorphologic approaches must account for the effects of the actual order of the events as well as the frequency. A reach of a stream that has experienced a 100-year event followed by two 50-year events and then four 25-year events will exhibit much different end conditions than a similar reach that experiences first the smaller events and then the largest. In addition to the temporal context the geomorphologist also must include the spatial context in explanation. Where the engineer usually considers cross-sections or limited reaches of a stream, the geomorphologist must consider the operation of spatially variable processes on a network-wide basis.

The requirement for spatial and temporal contexts in explanation raises a significant question: are there basic underlying spatial and temporal laws governing arroyo processes? It is unlikely that such laws exist in nature in a strictly defined classical sense because they would have to be universally applicable (Kitts, 1977). In theory construction the geomorphologist will therefore have to be content with law-like statements that are incomplete, a common compromise in science (Harvey, 1969). Recent research has produced several suggested basic spatial and temporal law-like statements. The assembly of these statements into testable theoretical structures for wide application, similar to theories in palaeohydraulics and palaeohydrology, represents the next step to explanation for the long-standing arroyo problem.

ACKNOWLEDGEMENTS

Background research for this paper was made possible by grants from the National Science Foundation (EAR–7727698 and EAR–8119932), the National Geographic Society (63876 and 101757), the US Department of Agriculture (58–9AHZ–0–542), and the US Army Corps of Engineers (DACW09–79–C0059). This paper is based in part in useful discussions with V. R. Baker, W. B. Bull, C. B. Hunt, J. C. Knox, and S. A. Schumm.

REFERENCES

Alford, J. J. 1982. San Vicente arroyo. *Annals of the Association of American Geographers*, **72**, 398–403.

Andres, W. 1980. On the palaeoclimatic significance of erosion and deposition in arid regions. *Zeitschrift für Geomorphologie (Supplementband)*, **36**, 112–122.

Antevs, E. 1952. Arroyo cutting and filling. *Journal of Geology*, **60**, 375–385.

Bailey, R. W. 1935. Epicycles of erosion in the valleys of the Colorado Plateau Province. *Journal of Geology*, **43**, 337–355.

Bailey, R. W., Craddock, C. W., and Croft, A. R. 1947. Watershed management for summer flood control in Utah. *U.S. Department of Agriculture, Miscellaneous Publication 639*, 24 pp.
Bailey, W., Forsling, C. L., and Becraft, R. J. 1934. Floods and accelerated erosion in northern Utah. *U.S. Department of Agriculture Miscellaneous Publication 196*, 36 pp.
Baker, V. R. 1974. Palaeohydraulic interpretation of Quaternary alluvium near Golden, Colorado. *Quaternary Research*, **4**, 94–112.
Baker, V. R., and Ritter, D. F. 1975. Competence of rivers to transport coarse bedload material. *Bulletin of the Geological Society of America*, **86**, 975–978.
Birkeland, P. W. 1968. Mean velocities and boulder transport during Tahoe-age floods of the Truckee River, California–Nevada. *Bulletin of the Geological Society of America*, **79**, 137–141.
Brady, L. F. 1936. The arroyo of the Rio de Flag, a study of an erosion cycle. *Museum of Northern Arizona, Museum Notes*, **6**, 33–37.
Bridge, J. S. 1978. Palaeohydraulic interpretation using mathematical models of contemporary flow and sedimentation in meandering channels. In Miall, A. D. (Ed.), *Fluvial Sedimentology*, Canadian Society of Petroleum Geologists, Calgary, 723–742.
Bryan, K. 1922. Erosion and sedimentation in the Papago County, Arizona. *U.S. Geological Survey Bulletin 730–B*, 19–90.
Bryan, K. 1925a. Date of channel trenching (arroyo cutting) in the arid southwest. *Science*, **62**, 338–344.
Bryan, K. 1925b. The Papago county, Arizona. *U.S. Geological Survey Water-Supply Paper 499*, 436 pp.
Bryan, K. 1928a. Historic evidence on changes in the channel of the Rio Puerco, a tributary of the Rio Grande in New Mexico. *Journal of Geology*, **361**, 265–282.
Bryan, K. 1928b. Change in plant associations by change in ground water level. *Ecology*, **9**, 474–478.
Bryan, K. 1929. Flood-water farming. *Geographical Review*, **19**, 444–456.
Bryan, K. 1940. Erosion in the valleys of the southwest. *New Mexico Quarterly Review*, **10**, 227–232.
Bryan, K. 1941. Pre-Columbian agriculture in the southwest as conditioned by periods of alluviation. *Annals of the Association of American Geographers*, **31**, 219–242.
Bryan, K. 1954. The geology of Chaco Canyon, New Mexico in relation to the life and remains of the prehistoric people of Pueblo Bonito. *Smithsonian Miscellaneous Collections 122, Number* 7, 65 pp.
Bull, W. B. 1975a. Landforms that do not tend toward a steady state. In Melhorn, W. H., and Flemal, R. C. (Eds.), *Theories of Landform Development*, State University of New York, Binghamton, 111–128.
Bull, W. B. 1975b. Allometric change of landforms. *Bulletin of the Geological Society of America*, **89**, 1489–1498.
Bull, W. B. 1980a. Geomorphic thresholds as defined by ratios. In Coates, D. R., and Vitek, J. D. (Eds.), *Thresholds in Geomorphology*, George Allen and Unwin, London, 259–263.
Bull, W. B. 1980b. The threshold of critical power in streams. *Bulletin of the Geological Society of America*, **93**, 453–464.
Burkham, D. E. 1972. Channel changes of the Gila River in Safford Valley, Arizona, 1846–1970. *U.S. Geological Survey Professional Paper 655–G*, 24 pp.
Burkham, D. E. 1981. Uncertainties resulting from changes in river form. *Journal of the Hydraulics Division, American Society of Civil Engineers*, **107**, HY5, 593–610.
Calkins, H. G. 1941. Man and gullies. *New Mexico Quarterly Review*, **11**, 69–78.

Chang, H. H. 1982. Mathematical model for erodible channels. *Journal of the Hydraulics Division, Proceedings of the American Society of Civil Engineers*, **108**, 678–689.

Chorley, R. J. 1962. Geomorphology and general systems theory. *U.S. Geological Survey Professional Paper 500–B*, 10 pp.

Chorley, R. J., Beckinsale, R. P., and Dunn, A. J. 1973. *The History of the Study of Landforms or the Development of Geomorphology: Volume II, Life and Work of William Morris Davis*, Methuen and Wiley, London, 874 pp.

Chorley, R. J., Dunn, A. J., and Beckinsale, R. P. 1964. *The History of the Study of Landforms or the Development of Geomorphology: Volume I, Geomorphology Before Davis*, Methuen and Wiley, London, 874 pp.

Chorley, R. J., and Kennedy, B. A. 1971. *Physical Geography, A Systems Approach*, Prentice-Hall International, London, 370 pp.

Church, M. 1978. Palaeohydrological reconstructions from a Holocene valley fill. In Miall, A. D. (Ed.), *Fluvial Sedimentology*, Canadian Society of Petroleum Geologists, Calgary, 743–772.

Church, M., and Mark, D. D. 1980. On size and scale in geomorphology. *Progress in Physical Geography*, **4**, 342–390.

Coates, D. R., and Vitek, J. D. 1980. *Thresholds in Geomorphology*, George Allen and Unwin, London, 498 pp.

Colton, H. S. 1937. Some notes on the original condition of the Little Colorado River: a side light on the problems of erosion. *Museum of Northern Arizona, Museum Notes*, **10**, 17–20.

Cooke, R. U. 1974. The rainfall context of arroyo initiation in southern Arizona. *Zeitschrift für Geomorphologie (Supplementband)*, **21**, 63–75.

Cooke, R. U., and Reeves, R. W. 1976. *Arroyos and Environmental Change in the American South-West*, Clarendon Press, Oxford, 213 pp.

Cooley, M. E. 1962. Late Pleistocene and recent erosion and alluviation in parts of the Colorado River system, Arizona and Utah. *U.S. Geological Survey Professional Paper 450–B*, 48–50.

Cooley, M. E. 1979a. Depths of channels in the area of the San Juan Basin regional uranium study, New Mexico, Colorado, Arizona, and Utah. *U.S. Geological Survey Open-File Report 79–1526*, 38 pp.

Cooley, M. E. 1979b. Effects of uranium development on erosion and sedimentation in southern San Juan Basin, New Mexico. *U.S. Geological Survey Open-File Report 79–1496*, 21 pp.

Cooperider, C. K., and Hendricks, B. A. 1937. Soil erosion and stream flow on range and forest lands of the upper Rio Grande watershed in relation to land resources and human welfare. *U.S. Department of Agriculture Technical Bulletin 567*, 88 pp.

Cottam, W. P., and Stewart, G. 1940. Plant succession as a result of grazing and meadow desiccation since settlement in 1862. *Journal of Forestry*, **38**, 613–626.

Dana, J. D. 1850. On denudation in the Pacific. *American Journal of Science*, **9**, 48–62.

Davies, G. L. 1969. *The Earth in Decay; a History of British Geomorphology*, American Elsevier Publishing Company, New York, 390 pp.

Davis, W. M. 1909. In Johnson, D. W. (Ed.), *Geographical Essays*, Facsimile Reprint, 1954, Dover Publications, New York, 777 pp.

Dellenbaugh, F. S. 1912. Crosscutting and retrograding of streambeds. *Science*, **35**, 656–658.

Denevan, W. M. 1967. Livestock numbers in nineteenth-century New Mexico, and the problem of gullying in the southwest. *Annals of the Association of American Geographers*, **57**, 691–703.

Dobyns, H. F. 1981. *From Fire to Flood: Historic Human Destruction of Sonoran Desert Riverine Oases*, Ballano Press, Socorro, 222 pp.
Dodge, R. E. 1902. Arroyo formation. *Science*, **15**, 746.
Dodge, R. E. 1910. The formation of arroyos in adobe filled valleys in the southwestern United States (abstract). *British Association for the Advancement of Science Report*, **79**, 531–532.
Duce, J. T. 1918. The effect of cattle on the erosion of canyon bottoms. *Science*, **67**, 450–452.
Dunne, T., and Leopold, L. B. 1977. *Water in Environmental Planning*, W. H. Freeman, San Francisco, 818 p.
Dury, G. H. 1966. The concept of grade. In Dury, G. H. (Ed.), *Essays in Geomorphology*. American Elsevier Publishing, New York, 211–234.
Dutton, C. E. 1882. Tertiary history of the Grand Canyon District. *U.S. Geological Survey Mongraph 2*, 264 p.
Emmett, W. W. 1974. Channel aggradation in western United States as indicated by observation at Vigil Network sites. *Zeitschrift für Geomorphology (Supplementband)*, **21**, 52–62.
Euler, R. C., Gumerman, Karlstrom, J., Karlotran, T. N. V., Dean, J. S., and Hevly, R. N. 1979. The Colorado Plateaus: cultural dynamics and palaeoenvironment. *Science*, **205**, 1089–1101.
Eyles, R. J. 1977. Changes in drainage networks since 1820, Southern Tableland, N.S.W. *Australian Geographer*, **13**, 377–386.
Faulkner, H. 1974. An allometric growth model for competitive gullies. *Zeitschrift für Geomorphologie (Supplementband)*, **21**, 76–87.
Gerson, R. 1977. Sediment transport for desert watersheds in erodible materials. *Earth Surface Processes*, **2**, 343–361.
Gilbert, G. K. 1877. *Report on the Geology of the Henry Mountains*, U.S. Geographical and Geological Survey of the Rocky Mountain Region, Washington, D. C., 160 pp.
Gilbert, G. K. 1886. The inculcation of scientific method by example. *American Journal of Science*, **31**, 284–299.
Gilbert, G. K. 1914. The transportation of debris by running water. *U.S. Geological Survey Professional Paper 86*, 263 pp.
Graf, W. L. 1977. The rate law in fluvial geomorphology. *American Journal of Science*, **277**, 178–191.
Graf, W. L. 1979a. Mining and channel response. *Annals of the Association of American Geographers*, **69**, 262–275.
Graf, W. L. 1979b. The development of montane arroyos and gullies. *Earth Surface Processes*, **4**, 1–14.
Graf, W. L. 1979c. Catastrophe theory as a model for change in fluvial systems. In Rhodes, D. D., and Williams, G. P. (Eds.), *Adjustments of the Fluvial System*, Kendall/Hunt Publishers, Dubuque, 13–32.
Graf, W. L. 1981. Channel instability in a braided, sand bed river. *Water Resources Research*, **17**, 1087–1094.
Graf, W. L. 1982a. Spatial variation of fluvial processes in semi-arid lands. In Thorn, C. E. (Ed.), *Space and Time in Geomorphology*, George, Allen and Unwin, London, 193–217.
Graf, W. L. 1982b. Distance decay and arroyo development in the Henry Mountains region, Utah. *American Journal of Science*, **282**, 1541–1554.
Gregory, H. E. 1917. Geology of the Navajo country. *U.S. Geological Survey Professional Paper 93*, 161 pp.
Gregory, H. E. 1938. The San Juan country. *U.S. Geological Survey Professional Paper 188*, 123 pp.

Gregory, H. E. 1950. Geology and geography of the Zion Park region, Utah and Arizona. *U.S. Geological Survey Professional Paper 220*, 200 pp.

Gregory, H. E., and Moore, R. C. 1931. The Kaiparowits region: a geographic and geologic reconnaissance of parts of Utah and Arizona. *U.S. Geological Survey Professional Paper 164*, 161 pp.

Gregory, K. J., and Walling, D. E. 1973. *Drainage Basin Form and Process: a Geomorphological Approach*, Wiley, New York, 456 pp.

Hack, J. T. 1939. Late Quaternary history of several valleys of northern Arizona, a preliminary announcement. *Museum of Northern Arizona, Museum Notes*, **11**, 63–73.

Hack, J. T., 1942. The changing physical environment of the Hopi Indians of Arizona. *Papers of the Peabody Museum of America Archeology and Ethnology, Harvard University 35*, 85 pp.

Hack, J. T. 1960. Interpretation of erosional topography in humid temperate regions. *American Journal of Science*, **258**, 80–97.

Hadley, R. F. 1961. Influence of riparian vegetation on channel shape, northeastern Arizona. *U.S. Geological Survey Professional Paper 424–C*, 30–31.

Haible, W. W. 1980. Holocene profile changes along a California coastal stream. *Earth Surface Processes*, **5**, 249–264.

Harvey, D. 1969. *Explanation in Geography*, St. Martin's Press, New York, 521 pp.

Hastings, J. R. 1959. Vegetation change and arroyo cutting in southeastern Arizona. *Journal of the Arizona Academy of Sciences*, **1**, 60–67.

Hastings, J. R., and Turner, R. M. 1965. *The Changing Mile*, University of Arizona Press, Tucson, 317 pp.

Haynes, C. V., Jr. 1968. Geochronology of late-Quaternary alluvium. In Morrison, R. B., and Wright, H. E. Jr. (Eds.), *Means of Correlation of Quaternary Successions*, University of Utah Press, Salt Lake City, 591–631.

Heede, B. H. 1974. Stages of development of gullies in western United States of America. *Zeitschrift für Geomorphologie (Supplementband)*, **18**, 260–271.

Heede, B. H. 1976. Gully development and control: the status of our knowledge. *U.S. Forest Service Research Paper RM–169*, 42 pp.

Hough, W. 1906a. Pueblo environments. *American Association for the Advancement of Science Proceedings*, **55**, 447–454.

Hough, W. 1906b. Pueblo environment. *Science*, **23**, 865–869.

Howard, A. D. 1965. Geomorphological systems — equilibrium and dynamics. *American Journal of Science*, **263**, 303–312.

Hunt, C. B., Averitt, P., and Miller, R. L. 1953. Geology and geography of the Henry Mountains region, Utah. *U.S. Geological Survey Professional Paper 228*, 234 pp.

Huntington, E. 1914. The climatic factor as illustrated in arid America. *Carnegie Institution of Washington Publication 192*, 341 pp.

Johnson, W. D. 1900. The high plains and their utilization. *U.S. Geological Survey, 21st Annual Report*, 609–741.

Judson, S. 1952. Arroyos. *Scientific American*, **187**, 71–76.

Kitts, D. B. 1977. *The Structure of Geology*, Southern Methodist University, Dallas, 180 pp.

Knechtel, M. M. 1938. Geology and ground-water resources of the valley of Gila River and San Simon Creek, Graham County, Arizona. *U.S. Geological Survey Water-Supply Paper 796–F*, 181–222.

Knox, J. C. 1972. Valley alluviation in southwestern Wisconsin. *Annals of the Association of American Geographers*, **62**, 401–410.

Knox, J. C. 1975. Concept of the graded stream. In Melhorn, W. N., and Flemal, R. C. (Eds.), *Theories of Landform Development*, State University of New York, Binghamton, 169–198.

Knox, J. C. 1978. Arroyos and Environmental Change in the American south-west (book review). *Annals of the Association of American Geographers*, **68**, 137–139.
Kuhn, T. S. 1970. *The Structure of Scientific Revolutions* (second edition), The University of Chicago Press, Chicago, 210 pp.
La Marche, V. C. 1966. An 800-year history of stream erosion as indicated by botanical evidence. *U.S. Geological Survey Professional Paper 550–D*, 83–86.
Leopold, A. 1921. A plea for recognition of artificial works in forest erosion and control policy. *Journal of Forestry*, **19**, 267–273.
Leopold, L. B. 1951a. Rainfall frequency: an aspect of climatic variation. *Transactions of the American Geophysical Union*, **32**, 347–357.
Leopold, L. B. 1951b. Vegetation of southwestern watersheds in the nineteenth century. *Geographical Review*, **61**, 295–316.
Leopold, L. B. 1976. Reversal of erosion cycle and climatic change. *Quaternary Research*, **6**, 557–562.
Leopold, L. B. 1978. El Asunto del Arroyo. In Embleton, C., Brunsden, D., and Jones, D. K. C. (Eds.), *Geomorphology: Present Problems and Future Prospects*, Oxford University Press, Oxford, 25–39.
Leopold, L. B., and Bull, W. B. 1979. Base level, aggradation, and grade. *Proceedings of the American Philosophical Society*, **123**, 168–202.
Leopold, L. B., and Maddock, T. R. 1953. *The Hydraulic Geometry of Stream Channels and some Physiographic Implications*. U.S. Geological Survey Professional Paper 252, 55 pp.
Leopold, L. B., and Miller, J. P. 1953. Ephemeral streams — hydraulic factors and their relation to the drainage net. *U.S. Geological Survey Professional Paper 282–A*, 37 pp.
Leopold, L. B., and Snyder, C. I. 1951. Alluvial fills near Gallup, New Mexico. *U.S. Geological Survey Water-Supply Paper 1110–A*, 1–19.
Leopold, L. B., Wolman, M. G., and Miller, J. P. 1964. *Fluvial Processes in Geomorphology*, W. H. Freeman, San Francisco, 522 pp.
Love, D. W. 1979. Quaternary fluvial geomorphic adjustments in Chaco Canyon, New Mexico. In Rhodes, D. D., and Williams, G. P. (Eds.), *Adjustments of the Fluvial System*, Kendall/Hunt Publishers, Dubuque, 277–308.
Mackin, J. H. 1948. Concept of the graded river. *Bulletin of the Geological Society of America*, **59**, 463–512.
Malde, N. E., and Scott, A. G. 1977. Observations of contemporary arroyo cutting near Santa Fe, New Mexico, U.S.A. *Earth Surface Processes*, **2**, 39–54.
Marston, R. B. 1952. Ground cover requirements for summer storm runoff control on aspen sites in northern Utah. *Journal of Forestry*, **50**, 303–307.
Martin, P. S. 1963. *The Last 10,000 Years: A Fossil Pollen Record of the American Southwest*, University of Arizona Press, Tucson, 87 pp.
Martin, P. S., Schoenwetter, J., and Arms, B. C. 1961. *Palynology and Prehistory, the Last 10,000 Years*, Geochronology Laboratories, University of Arizona.
Masannat, Y. M. 1980. Development of piping erosion conditions in the Benson area, Arizona, U.S.A. *Quarterly Journal of Engineering Geology*, **13**, 53–61.
Masterman, M. 1970. The nature of a paradigm. In Lakatos, I., and Musgrave, A. (Eds.), *Criticism and the Growth of Knowledge*, M.I.T. Press, Cambridge, 431–454.
Melton, M. A. 1965. The geomorphic and palaeoclimatic significance of alluvial deposits in southern Arizona. *Journal of Geology*, **73**, 1–38.
Miller, J. P., and Wendorf, F. 1958. Alluvial chronology of the Tesuque Valley, New Mexico. *Journal of Geology*, **66**, 177–194.
Miser, H. D. 1924. The San Juan Canyon, southern Utah. *U.S. Geological Survey Water-Supply Paper 538*, 80 pp.

Morisawa, M. 1968. *Streams: Their Dynamics and Morphology*, McGraw-Hill Book Company, New York, 175 pp.

Mosley, M. P. 1972. Evolution of a discontinuous gully system. *Annals of the Association of American Geographers*, **62**, 655–663.

Murray, J. A. H., Bradley, H., Craige, W. A., and Onions, C. T. 1933. *The Oxford English Dictionary, Being a Corrected Re-Issue with an Introduction, Supplement, and Bibliography of a New English Dictionary on Historical Principles*, Clarendon Press, Oxford, 13 volumes.

Novak, I. D. 1973. Predicting coarse sediment transport: the Hjulstrom curve revisited. In Morisawa, M. (Ed.), *Fluvial Geomorphology*, State University of New York, Binghamton, 13–25.

Patton, P. C., and Schumm, S. A. 1975. Gully erosion, northwestern Colorado: a threshold phenomenon. *Geology*, **3**, 88–89.

Patton, P. C., and Schumm, S. A. 1981. Ephemeral-stream processes: implications for studies of Quaternary valley fills. *Quaternary Research*, **15**, 24–43.

Penck, W. 1953. *Morphological Analysis of Landforms*, translated by H. Czech and K. C. Boswell, Macmillan and Company, New York, 309 pp.

Peterson, H. V. 1950. The problem of gullying in western valleys. In Trask, P. P. (Ed.), *Applied Sedimentation*, Wiley, New York, 407–434.

Pickup, G. 1975. Downstream variations in morphology, flow conditions and sediment transport in an eroding channel. *Zeitschrift für Geomorphologie (Supplementband)*, **19**, 443–459.

Playfair, J. 1802. *Illustrations of the Huttonian Theory of the Earth*, Facsimile Reprint, 1956, Dover Publications, New York, 328 pp.

Ploy, V. de. 1977. Mechanical properties of hillslopes and their relation to gullying in central semi-arid Tunisia, *Zeitschrift für Geomorphologie (Supplementband)*, **21**, 177–190.

Quinn, J. H. 1957. Paired river terraces and Pleistocene glaciation. *Journal of Geology*, **65**, 149–166.

Reagan, A. B. 1922. Archeological notes on Pine River Valley, Colorado and the Kayenta–Tuba region, Arizona. *Transactions of the Kansas Academy of Science*, **30**, 262–267.

Reagan, A. B. 1924a. Stream aggradation through irrigation. *Pan American Geologist*, **42**, 335–344.

Reagan, A. B. 1924b. Recent changes in the plateau region. *Science*, **60**, 283–285.

Rich, J. L. 1911. Recent stream trenching in the semi-arid portion of southwestern New Mexico, a result of removal of vegetation cover. *American Journal of Science*, **32**, 237–245.

Richardson, H. L. 1945. Discussion: the significance of terraces due to climatic oscillation. *Geological Magazine*, **82**, 16–18.

Ross, C. P. 1923. The Lower Gila Region, Arizona. *U.S. Geological Survey Water-Supply Paper 498*, 237 pp.

Schoenwetter, J. 1962. The pollen analysis of eighteen archaeological sites in Arizona and New Mexico. In *Chapters in the Pre-history of Eastern Arizona*. Fieldiana. *Anthropology*, **53**, 168–209.

Schoenwetter, J., and Eddy, F. W. 1964. Alluvial and palynological reconstruction of environments, Navajo Reservation District. *Museum of New Mexico Papers in Anthropology*, **13**, 155 pp.

Schumm, S. A. 1961. Effect of sediment characteristics on erosion and deposition in ephemeral stream channels. *U.S. Geological Survey Professional Paper 352–C*, 31–70.

Schumm, S. A. 1968a. Speculations concerning palaeohydrologic controls of terrestrial sedimentation. *Bulletin of the Geological Society of America*, **79**, 1573–1588.

Schumm, S. A. 1968b. River adjustment to altered hydrologic regimen — Murrumbidgee River and Palaeochannels, Australia. *U.S. Geological Survey Professional Paper 598*, 65 pp.

Schumm, S. A. 1973. Geomorphic thresholds and complex response of drainage systems. In Morisawa, M. (Ed.), *Fluvial Geomorphology*, State University of New York, Binghamton, 299–310.

Schumm, S. A. 1976. Episodic erosion, a modification of the geomorphic cycle. In Melhorn, W. N., and Flemal, R. C. (Eds.), *Theories of Landform Development*, State University of New York, Binghamton, 69–86.

Schumm, S. A. 1977. *The Fluvial System*, Wiley–Interscience, New York, 338 pp.

Schumm, S. A. 1979. Geomorphic thresholds: the concept and its applications. *Transactions of the Institute of British Geographers N.S.4*, 485–515.

Schumm, S. A. 1980a. Some applications of the concept of geomorphic thresholds. In Coates, D. R., and Vitek, J. D. (Eds.), *Thresholds in Geomorphology*, George Allan and Unwin, London, 473–485.

Schumm, S. A. 1980b. Incised channels: type, morphology and evolution. *Phase 1 Report, Soil Conservation Service Project SCS–23–MS–80*, 7 sections.

Schumm, S. A., and Hadley, R. F. 1957. Arroyos and the semiarid cycle of erosion. *American Journal of Science*, **255**, 161–174.

Schumm, S. A., and Lichty, R. W. 1963. Channel widening and floodplain construction along Cimarron River in southwestern Kansas. *U.S. Geological Survey Professional Paper 352–D*, 71–88.

Senter, D. 1937. Tree rings, valley floor deposition, and erosion in Chaco Canyon, New Mexico. *American Antiquity*, **3**, 68–75.

Shapere, D. 1964. The structure of scientific revolutions. *Philosophical Review*, **73**, 383–394.

Shea, J. H. 1982. Twelve fallacies of uniformitarianism. *Geology*, **10**, 455–460.

Shen, H. W. 1979. *Modelling of Rivers*, Wiley–Interscience, New York, 20 sections.

Strahler, A. N. 1952. Dynamic basis of geomorphology. *Bulletin of the Geological Society of America*, **63**, 923–928.

Strahler, A. N. 1956. The nature of induced erosion and aggradation. In Thomas, W. L. Jr. (Ed.), *Man's Role in Changing the Face of the Earth*, University of Chicago Press, Chicago, volume 2, 621–638.

Swift, T. T. 1926. Rate of channel trenching in the southwest. *Science*, **63**, 70–71.

Thornber, J. J. 1910. The grazing ranges of Arizona. *Arizona Agricultural Experiment Station Bulletin*, **65**, 245–360.

Thornes, J. B. 1980. Structural instability and ephemeral channel behaviour. *Zeitschrift für Geomorphologie (Supplementband)*, **36**, 233–244.

Thornthwaite, C. W., Sharpe, C. F. S., and Dosch, E. F. 1941. Climate of the southwest in relation to accelerated erosion. *Soil Conservation*, **6**, 298–304.

Thornthwaite, C. W., Sharpe, C. F. S., and Dosch, E. F. 1942. Climate and accelerated erosion in the arid and semi-arid southwest, with special reference to the Polacca Wash drainage basin, Arizona. *U.S. Department of Agriculture Technical Bulletin 808*, 134 pp.

Toy, T. J. 1982. Accelerated erosion: process, problems, and prognosis. *Geology*, **10**, 524–529.

Tuan, Y. F. 1966. New Mexican gullies: a critical review and some recent observations. *Annals of the Association of American Geographers*, **56**, 573–597.

Van Arsdale, R. 1982. Influence of calcrete on the geometry of arroyos near Buckeye, Arizona. *Bulletin of the Geological Society of America*, **93**, 20–26.

Visher, S. S. 1913. The history of the bajadas of the Tucson bolson. *Science*, **37**, 459.

Wertz, J. B. 1970. The start of an ephemeral stream. *Zeitschrift für Geomorphologie (Supplementband)*, **14**, 96–102.
Winn, F. 1926. The West Fork of the Gila River. *Science*, **64**, 16–17.
Womack, W. R., and Schumm, S. A. 1977. Terraces of Douglas Creek, northwestern Colordo: an example of episodic erosion. *Geology*, **5**, 72–76.
Wooley, R. R. 1946. Cloudburst in Floods in Utah 1850–1938. *U.S. Geological Survey Water-Supply Paper 994*.
Wooley, R. R., and Alter, J. C. 1938. Precipitation and vegetation. *Transactions American Geophysical Union*, **19**, 604–607.

Background to Palaeohydrology
Edited by K. J. Gregory

13

Changes of channel patterns and floodplains

JOHN LEWIN

University College of Wales, Aberystwyth

A decade or so ago two models for floodplain sedimentation were dominant (Allen, 1965; Leopold and Wolman, 1957). The *meandering model* emphasized the role of lateral accretion of coarser sediments in point bars overlain by finer overbank sediments, with the coarser element generally believed to be the more significant. Fining-upward sequences were developed in a quantitative model by J.R.L. Allen (1970), and this model has subsequently been applied with some success to contemporary point bar sedimentation by J.S. Bridge (1975). The *braiding model* emphasized the accretion of bed sediment in mid-channel bars in shifting channels, with only a very minor role for finer overbank sedimentation.

More recently, empirical studies have shown that these models need to be both diversified and supplemented if the full range of real-world alluvial environments is to be properly understood (Miall, 1978; Collinson and Lewin, 1983). Interpretation of alluvial sections and the surface patterning of present and former floodplain environments must now depend on the appreciation that there are different kinds of single and multithread channels, and that their deposits are varied. It may even be true that the association between sedimentation pattern and channel planform is itself misleading (Jackson, 1978) because sedimentation processes are not uniquely associated with specific channel planform types.

Floodplains are formed by fluvial processes in several contrasted ways. *Accretion* of moving bed material may take place, usually in the form of channel bars of assorted types (so-called 'lateral accretion' deposits), together with *sedimentation, sensu stricto*, from suspension either within or beyond river channels ('vertical accretion' or 'overbank' deposits). The opportunity for sedimentation is further provided by the formation, abandonment and then infilling of *channels*, whilst the localized build-up of sediment barriers or ridges (bars or levées, for example) may also create topographic depressions behind them by *sediment occlusion*. These may temporarily be floored by backwater, lake, fen or swamp.

Conventionally these process-products have been associated with certain morphological forms and then channel types: levées, point pars and cutoffs with meandering channel systems, for example. But studies over the last decade have emphasized both that vertical build-up of suspended sediment *within* channels is important (e.g. Nanson, 1980), and that the relative proportions of accretion and sedimentation deposits in sedimentary bodies is environmentally sensitive and not a matter over which overall generalization is advisable (see Jackson, 1978). Some rivers (both single and multichannel types) with virtually non-migrating channels have floodplains dominated by overbank deposits and floodbasin organic fills; other rapidly-shifting meandering channels produce floodplains dominated by the accretion of laterally-stacked bed sediments with only pockets of suspension deposits and organic fills, these being where cutoffs or sediment occlusions have created localized floodplain depressions. Thus the improving field appreciation of a fuller range of floodplain processes and types suggests that oversimplified planform types should be abandoned. It is now known, for example, that the deposits of braided and meandering streams can be very difficult to distinguish in the geological record. Bluck (1971, 1979) has described fining-upward sequences in braided-stream deposits and coarsening upward ones in meandering-stream deposits. He suggested that the simplest criterion for meandering deposition lies in the presence of inner accretionary bank sediments which are one of several types of large-scale cross stratification, developed in finer sediment on top of point bar deposits.

This kind of improved field knowledge of floodplain sedimentation processes has some important implications for palaeohydrological reconstructions. One apparently successful means of estimating former channel discharges has been to associate the hydraulic dimensions or meander geometry of abandoned channels with formative discharges using contemporary analogues (for a review see Dury, 1977a; Ethridge and Schumm, 1978). Another method is the use of largest clast size present to reconstruct former stream velocities, bed shear stress, or discharge (Bradley and Mears, 1980). Many authors have stressed the extreme caution with which apparently precise estimates should be treated: Church (1978), for example, cautiously concluded that only order–of–magnitude discharge reconstruction could reasonably be derived from his approach based on the Shields entrainment function.

As pointed out in the IGCP Project 158 Guide (Starkel and Thornes, 1981, pp. 56, 59), these reconstructions are complicated by the fact that erosion often destroys evidence so that the sedimentary record is highly incomplete, whilst effects of changes in runoff regime may be confused with changes in sediment supply. We can also add that some of the basic data on which reconstructions have been based may be unreliable. For example, lateral accretion surfaces (epsilon cross stratification) are not always present in meandering stream deposits, yet they may be developed at more than one level in a single stream; coarse member thickness is not necessarily equivalent to a fixed proportion of former channel depth, and variations in channel width, depth, and sedimentary facies within single bends may make results from single cross-sections unrepresentative.

Thus the improving field knowledge of contemporary floodplain development suggests some caution before accepting some palaeohydraulic computations. But there is also the potential for some expansion in the scope of palaeohydrological reconstruction. It should be possible to say how rapidly some sedimentary bodies have been laid down or reworked in specific environments, and to specify more clearly what facies variability may mean in environmental terms. This may give an indication of the nature of former channel changes, downvalley meander migration for example, and of the mix of sediments in transit. It seems possible that river regime, dominated for instance by regular seasonal snow melt or by irregular precipitation events, might be interpreted from small-scale sedimentary sequences. Large-scale sedimentation contrasts resulting from such environmental changes have been identified from Holocene deposits in Wisconsin (Knox, McDowell, and Johnson, 1981). Finally, it has been shown that slack-water deposits may record flood magnitude and frequency (Kochel and Baker, 1982). The more that is known about contemporary channel and floodplain dynamics, the more will be available both to caution and to extend palaeohydrological reconstruction.

As a contribution to extending the scope of palaeohydrological work, this chapter next presents some information on floodplain development and channel changes in Wales and the Borderland. This is derived both from field survey and study, and from the analysis of historic maps and air photographs covering a period of about the last 200 years. Firstly the nature of channel changes and sedimentation, and then the rates at which such changes occurred, will be considered. As will be emphasized, the study of such contemporary environments may not provide all the appropriate analogues for earlier fluvial sedimentation, but it does show that the nature of sedimentation and channel development is systematically more complex than may yet have been appreciated, and this may be helpful to the interpretation of the preserved remains of earlier rivers elsewhere which might otherwise seem baffling.

SEDIMENTATION AND CHANGING CHANNEL PATTERNS

Figure 13.1 and Table 13.1 give details concerning recent areas of sedimentation on three Welsh rivers. The Twymyn has been characterized by downvalley meander loop translation with some switching of sedimentation from convex to

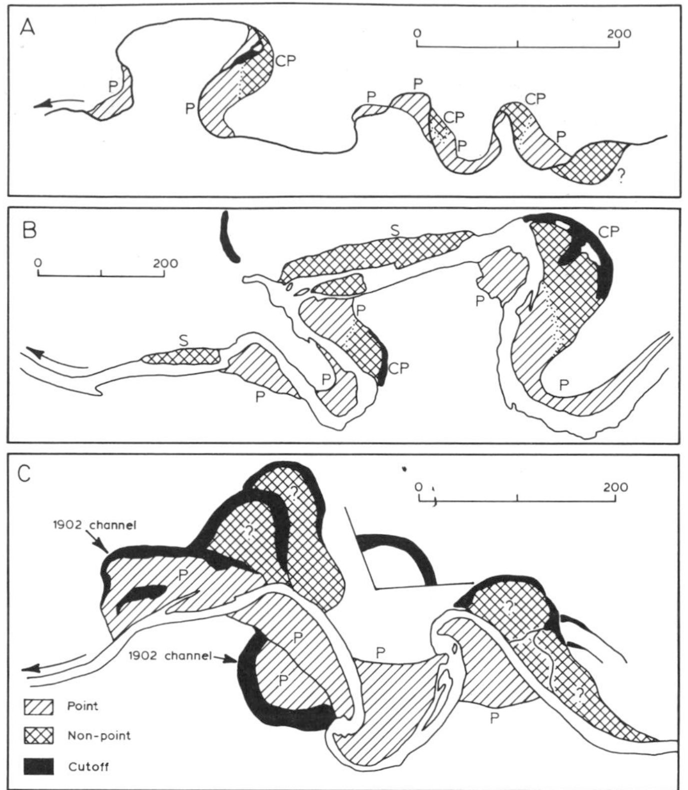

FIGURE 13.1 RECENT FLOODPLAIN SEDIMENTATION ON THREE WELSH RIVERS: A, THE TWYMYN (SN 885998); B, THE DEE (SH 980368); C, THE TEME (SO 384730). Shaded areas are those sedimented within approximately the last century

concave bends. On the Dee, a much larger river, floodplain development has taken place on both the inside and the outside of bends and in mid-channel, whilst on the Teme channel abandonment and infilling has been much more marked. These rivers, and others in Wales, are presently reworking the surfaces of thick Quaternary alluvial fills, so that much of the coarse component of available sediments is inherited from prior Pleistocene conditions (*cf.* Church and Ryder, 1972).

Active point bar development is now relatively well known on Welsh rivers (Blacknell, 1981; Lewin, 1976, 1978). Bars are commonly platform-like and may have inner or medial chutes. Sediments are generally coarse sandy gravels with structures (other than imbrication) that are rather poorly visible. Coarser bar head and finer bar tail sediments may be distinguishable, with the presence of discontinuous fine facies related to location with respect to bar morphology, seasonal vegetation growth, and flow deceleration sites. Grain size frequency

Table 13.1 Data for Channel Reaches Shown in Figure 13.1

A. SITES

		Floodplain slope ($m\ km^{-1}$)	Mean bankfull channel width (m)	Bankfull *W/D* ratio	Annual% channel shift	Mean annual flood (m^3s^{-1})
Twymyn	SN 885998	7.8	7.4	8.5	0.4	25
Dee	SH 980368	3.2	32.4	15.9	1.0	205
Teme	SO 384730	3.5	19.7	23.7	2.4	53

B. PERCENTAGE SEDIMENTATION

	Point	Counterpoint	Side	Mid-channel	Cutoff
Twymyn	65	19	16(?)	—	—
Dee	48	23	18	2	9
Teme	52	33(?)	—	1	14

C RELATIVE SIZE OF CONTIGUOUS SEDIMENTARY BODIES

	Point	Counterpoint	Side	Mid-channel	Cutoff
Twymyn	1.0	0.67	1.7	—	—
Dee	1.0	1.2	0.9	0.1	0.5
Teme	1.0	0.8	—	0.1	0.2

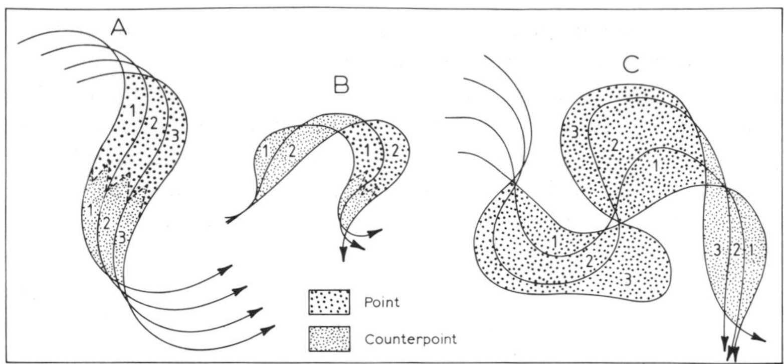

FIGURE 13.2 POINT AND COUNTERPOINT SEDIMENTS IN CONTRASTED EVOLVING PLANFORMS

distribution curves commonly exhibit truncation at the coarse end, suggesting competence limitation (Blacknell, 1981). But the variability of sediment sizes within the channel suggests that preserved sediments may reflect *local* competence within the channel, and it would be very difficult to reconstruct mean stream velocity or discharge using such sediment information.

Point bars have perhaps attracted the most attention from sedimentologists, but these are not the only form of accretion or sedimentation body present. Where channel loops are migrating downstream, rotating, or switching from convex to concave as a result of upstream unequal limb development (Figure 13.2), it is quite possible for channels to recede *away from* cut banks such that sedimentation occurs next to the outside of channel bends. Such sedimentation is here termed *counterpoint* sedimentation. It has previously been associated with eddy accretion deposits (Carey, 1969), concave-bank benches (Woodyer, 1975; Nanson and Page, 1983), and barbed channel planforms and irregular floodplain depressions (Lewin, 1978; Thorne and Lewin, 1979), but the variability of processes and forms in such areas appear to merit a new term balancing the commoner and much better understood, though again very variable, developments in point environments. Counterpoint sediments may consist of *accretionary* units in concave or ogive-shaped receding ridges, but also may be due to *sedimentation* in decaying vortices where flow separation occurs downstream of meander bends (Leeder and Bridges, 1975). In the horizontal, there may be a digitate, lingoid, or straight junction between coarser and finer fills, whilst vertical alternation between coarse and fine elements can occur where junctions migrate. These are important sites for the preservation of organic fills, derived both from floating debris and *in situ* vegetation growth on the finer sediment. Such fine sedimentation sites have been previously noted in studies concentrating on the coarser point sediments (e.g. Jackson, 1976, Bozeman bend; Wolman and Leopold, 1957). Here it should be noted that at such sites, the finer element of floodplain sedimentation is thicker than in point bars, and that it is possible to confuse such phenomena with true cutoffs in section exposures.

Cutoffs themselves are reasonably common on Welsh rivers. A survey (Lewis and Lewin, 1982) has revealed 145 cutoffs along 964 km of floodplain, 55 per cent being of simple chute type. In some areas, cutoff infills may be important components of near-surface floodplain sediments (14 per cent of the recent sedimentation on the Teme floodplain illustrated in Figure 13.1C, for example), but in others not at all. Chute cutoffs can only be expected where lateral migration is active and where loop development and floodplain erodibility during brief overbank flows is such as to allow the creation of new short-circuiting channels. In Wales this seems to be aided by the presence of irregular linear depressions in both point and counterpoint sedimentation surfaces, whilst cutoffs seem to be concentrated on the middle sections of Welsh rivers where alluvial bodies are laterally extensive and channel mobility is greater than both upstream and downstream. Cutoff type cannot be shown to be regionally differentiated. For example, on a short reach of the Severn at Caersws (Thorne and Lewin, 1979), both neck and chute cutoffs and the progressive abandonment of unfilled channel voids have all occurred within 2 km in the last 150 years.

These brief illustrations of the contrasting depositional patterns to be found within a small region emphasize the potential variability of information available for palaeohydrological reconstruction. Competence-controlled sedimentation is potentially present at all three sites, but it is worth recalling that for a comparable timespan in the region as a whole, only 25 per cent of a random sample of 100 channel reaches showed measurable channel change (Lewin, Hughes, and Blacknell, 1977). Without lateral migration or aggradation, sedimentation assignable to such a time span is not preserved and recognizable. Similarly, cutoffs are only found in quantity in certain areas: periods of some hundreds of years may produce few cutoffs so that the potential for reconstructing former discharges from palaeochannels is lacking. Where such cutoffs and channels are well preserved and not masked by later overbank sedimentation (Baker and Penteado-Orellana, 1977; Koutaniemi, 1979; Starkel, 1982), meander expansion and neck cutoffs appear to be more dominant than in Wales. Certain types of channel activity, which may be regionally and environmentally restricted in occurrence, seem to be more favourable for palaeohydrological work than others.

Sedimentation is also occurring in local environments that are not yet adequately documented. Counterpoint deposits are significant amongst sediments in all three examples illustrated in Figure 13.1. Particularly where meander loop translation is prevalent, preserved sediments may be dominated by those accreting or sedimenting on the upstream trailing edge of meander arcs. Figure 13.3 shows part of the Afon Dyfi floodplain where successive loops have moved down-valley abandoning dead channel arcs and a series of coarser

FIGURE 13.3 THE FLOODPLAIN OF THE AFON DYFI AT LLANWRIN (SH 785030). River flow is from right to left, and the reach shown is approximately 1 km in length

intrachannel lobes, contemporary equivalents of which may be seen on the present channel. Superficially the apparently branching channel system remaining on the floodplain resembles a braided channel system (*cf.* Cheetham, 1980, Figure 10), but these were *not* channels that were all simultaneously occupied. Again, age relationships of the coarser deposits can be confusing. Figure 13.1C shows a series of cutoffs, culminating in the 1902 channel, that are actually part of an episodically rotating sequence as a loop has moved jerkily downvalley — not a set of expanding loops each of which has grown from new. 'Point' sediments within the last loop can in fact be as old as ones within the first.

It is finally worth enquiring into the nature of floodplain sedimentation where channel migration has not been in evidence. Part of the Upper Severn is a useful case in point (Table 13.2). High average rates of lateral mobility exist above Welshpool, but between Welshpool (catchment area 800 km^2) and the Vyrnwy junction the often tortuous channel shows few signs of recent historic movement. It is also interesting to note that in this part of the Severn, the magnitude of discharges of recurrence intervals of between 1.01 and 20.0 years actually *drops* downstream (Hey, 1975), whilst the capacity of the low gradient channel here can be *less* than upstream (Table 13.2). Hey ascribed this discharge decrease to flood attenuation and overbank storage on the floodplain, and it seems that the floodplain here is naturally adjusted to take a higher proportion of high magnitude flows than upstream where the steeper channel is cut into coarser less cohesive sediment. Such conclusions have previously been reached following an analysis of small streams in New South Wales (Nanson and Young, 1981).

TABLE 13.2 CHARACTERISTICS OF THE UPPER SEVERN RIVER CHANNEL
Based on a field survey by G. W. Lewis

Catchment area (km^2)	Bankfull channel capacity (m^2)	W/D ratio	Average lateral change rate ($m\ yr^{-1}$)	Slope ($m\ m^{-1}$)	Unit stream power at bankfull ($\rho g Q S/W$)	BESMAF† ($m^3 s^{-1}$)
37	42	8	n.d.	0.003	80	32
122	22	25	0.19	0.0032	25	58
126	35	47	n.d.	0.0044	44	60
136	51	18	n.d.	0.0016	153	62
172	40	23	2.15	0.0081	220	61
181	62	56	1.28	0.0036	42	65
185	33	23	0.73	0.0038	52	72
324	53	31	0.47	0.0019	27	104
358	134	35	0.27	0.0074	39	116
415	112	15	0.15	0.0022	104	142
800	89	27	0.79	0.0009	12	275
876	140	10	—	0.0006	18	285
905	99	8	—	0.0006	17	270
983	160	8	0.08	0.0003	11	296

†The best estimate of mean annual flood (BESMAF) is derived using the method given in Institute of Hydrology 1975 *Flood Studies Report,* NERC, Wallingford.

Again, this has important implications for palaeohydrological interpretation. If the *channel* of a single stream is adjusted to transmit a variable proportion of total water conveyance in one event as one moves downstream, what do reconstructions of in-channel flows alone mean, without regard to the other controls and flows operating? It is at the least worth noting that channels in finer sediment (with lower stream power or lower power/resistance ratio) may be smaller than others. Such channels may be dominated by overbank sedimentation (in the more frequent overbank flows), with such deposits not removed by lateral migration which is itself inhibited by the more cohesive sediment. Here *vertical* age zoning of Holocene sediments in flood sequences may be preserved, in contrast to the *lateral* zoning according to age of mobile streams. Such rates of sedimentation may be very variable (Bridge and Leeder, 1979, Table 1), and reflect actual non-capacity suspended loads as well as the opportunity for overbank sedimentation.

The contemporary floodplain environments so far discussed have been ones of relatively low bed sediment flux, where material is transferred from cut bank to point bar, and total bed sediment movement through any particular channel cross-section may be little more than the flux of dispersed particles between largely local sites of erosion and deposition. Accretion represents a local response to the lateral migration of a scouring 'live' channel, and bed sediment supply rate is intrinsically determined by the location and rate of cut-bank development and bed scour. Bed sediment deposition typically occurs in locations that are stable with respect to evolving planforms, both on the inner margins of bends and where channels recede from their former cut banks. Erosion, transport and accretion may all be coincident in time at high flows, but spatially separated as competence varies along flow lines through bends. The same may also apply to many contemporary *braided* streams which can be incising (Williams and Rust, 1969) or rather stable in the vertical (Lewin and Weir, 1977). Here an important additional autogenic process for the localized redistribution of sediment lies in the creation of scour at channel confluences (see Ashmore, 1981; Mosley, 1982a).

By contrast, many Holocene and earlier fluvial sediments reveal thick multistorey sequences which suggest net aggradation. Bed material flux and build-up is regional, and more than just a function of cut-bank to adjacent bar movement. In an aggrading environment, bed sediment may be moving in mobile bars through river reaches, with laterally and downstream extending units only some of which can be closely identified with position in a prior sinuous channel planform.

Contemporary analogues for such activity are rarely available, but a case in point is provided by a reach of the River Ystwyth (Figure 13.4). This shows both the present low-sinuosity channel, and a digitate pattern of coarse sediment with fine sediment between on the adjacent floodplain. This pattern was produced in the nineteenth century in association with local metal mining

FIGURE 13.4 THE FLOODPLAIN OF THE AFON YSTWYTH NEAR TRAWSCOED (SN 660740). River flow is from right to left and the reach shown is approximately ½ km in length

activity when the floodplain aggraded by several metres. Since mining ceased, the channel has become incised again, such that this small example illustrates channel metamorphosis from meandering to braiding and back again, all within 200 years. The essential point to appreciate here is that sedimentation occurred in a series of overbank splays of coarse sediment; these splays have a highly crenulate margin, with the voids between and beyond filled subsequently by finer sediment. These fingering networks are again *not* contemporary channels but intersplay voids of various dates; they cannot be used as if they were simultaneously-operating channels to provide palaeohydrological estimates. It is likely that other apparent 'channels' in aggrading environments are similar in origin, whilst the hydraulic behaviour of braided streams in the field is in any case such as to make assumptions about the simultaneous operation of true branching channel systems hazardous (Mosley, 1982b).

A last point concerning aggrading sedimentary environments concerns the recently discovered importance of anastomosing stream systems which may have a distinctive sedimentary style (Smith, 1974; Smith and Smith, 1980). Briefly, anastomosing streams are multithread aggrading ones in which channels remain fixed in position with levées alongside and floodbasins beyond. Crevassing and avulsion (rather than lateral channel migration) leads to the abandonment of some channels and the development of new ones. Floodplains are underlain by ribbons of sub-channel sediment, often sandy, with thick interchannel organic deposits. Aggradation in the examples studied in Canada is caused by the downstream control of a rising sediment barrier, but rising Holocene sea levels could have similar effects, and this sedimentation style, too, may be much commoner in earlier deposits than is the case in contemporary environments.

RATES OF CHANNEL CHANGE

Evidence is now appearing that shows very variable rates of channel shift and sediment reworking in Britain. Highest lateral shift rates, and thus sediment accretion, are recorded in the upland west and north (Ferguson and Werritty, 1983; Hooke, 1980; Lewin, Hughes, and Blacknell, 1977), with lower rates in the southeast. Within upland Britain, rates of channel shift tend to increase downstream at first, then decrease towards river mouths, though rates may vary locally according to the growth and development of individual bends. Generalized rates may be illustrated by data from sites in Wales and the Borderland (Figure 13.5). Here change rates are spatial averages for river reaches of around 30 channel widths, and are derived from comparisons between plots from historic maps and air photographs over time periods ranging from 44–78 years. Thus the rates are not representative of maximum rates that can occur at single cross-sections, or during one year which may have one or more extreme flows. The pattern of averaged change rates rather crudely balances that of estimated gross bankfull stream power, as for unit stream power as previously shown in Table 13.2 for the River Severn. When plotted on a slope–discharge graph, rates of change are not, however, banded parallel to lines of equal stream power (Figure 13.6).

There may be several reasons for this. A set proportion of gross stream power may not be used for lateral bank erosion, and this may vary according to bend geometry. Thus Hickin and Nanson (1975) have argued that rates of migration decline once a critical value of meander radius is achieved. Begin (1981) has also shown that stream pattern itself may be related to stream power. A second point is that bank resistance must also be important, and variable. This is difficult to quantify as a single measure, though Schumm (1972) has used width–depth ratio as a surrogate for sediment character. For the sites mapped in Figure 13.5 this correlates well ($r = 0.71$, significant at 0.1 per cent) with change rate.

Pending firmer predictive models — probably involving measures of stream power from slope and discharge estimates, and bank resistance using sediment size or channel shape parameters (*cf.* Begin, 1981) — it can be concluded that contemporary channels in the study area described here are reworking

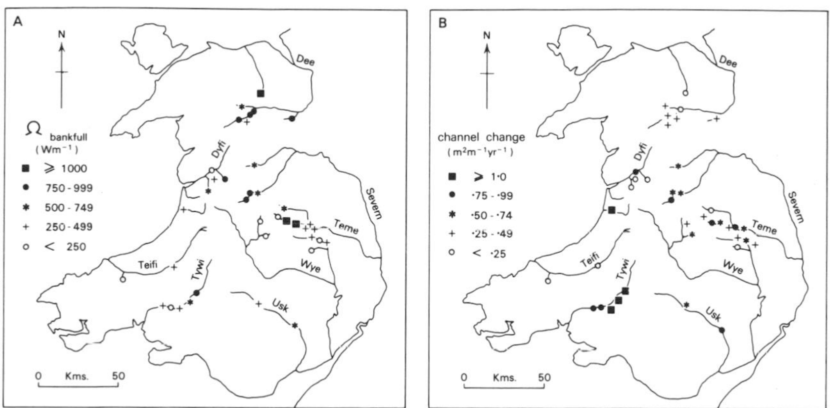

FIGURE 13.5 BANKFULL STREAMPOWER AND AVERAGED RATES OF LATERAL CHANNEL SHIFT FOR SELECTED REACHES OF WELSH RIVERS
Gross stream power ($\Omega = \rho g Q S$) is calculated from surveyed channel slope(S) and a Manning calculation of bankfull discharge (Q); change rates are derived from historic maps and air photographs, and are averages for several decades over lengths of around 30 channel widths

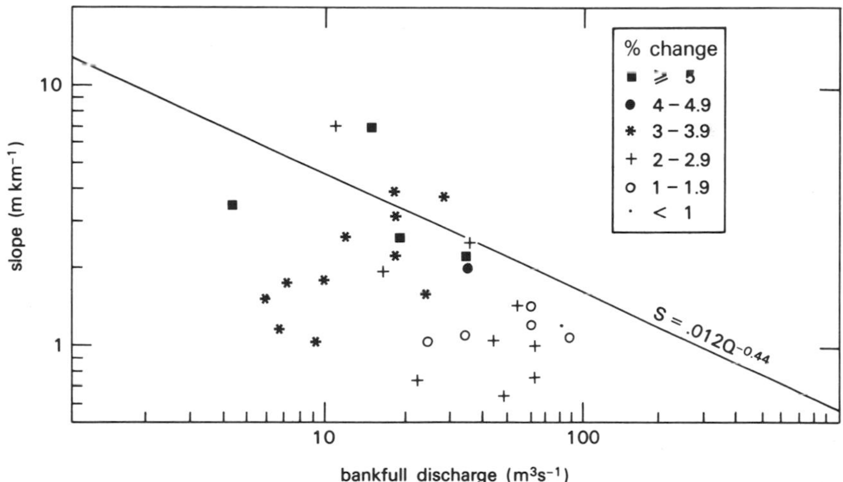

FIGURE 13.6 CHANNEL CHANGE RATES IN RELATION TO SLOPE AND DISCHARGE
Channel slope is based on field survey, and bankfull discharge calculated using the Manning formula. The relation $S = .012\ Q^{-0.44}$ has been used to discriminate braided (above the line) from meandering channels, though all those plotted are of single channel type. The discriminant line is approximately one of equal stream power

floodplain sediments at spatially and temporally averaged rates of up to 6.6 per cent of channel width, or 1.6 m, per year. At many sites, especially where channels are confined or of low power however, change rates are unmeasurable over many decades.

From the point of view of palaeohydrological reconstruction, this suggests two things. Firstly, in areas of high lateral mobility, it is extremely unlikely that early Holocene sediments will have been preserved in valley-bottom environments: these are much more likely to be found in either headwater areas, where small Holocene streams may only have removed a ribbon of infill deposits (and perhaps the fluvial facies of aggrading units, leaving colluvial facies along valley margins), or downstream areas (where vertical build-up has been significant), or where incision has left terrace fragments well clear of, and possibly protected by basal rock outcrops from, the activity of migrating streams. Thus where such sediments are preserved, they are intrinsically likely to be particular kinds of sediment, giving only a partial indication of what may once have been present along the river channels at some time past.

Secondly, the contemporary data does indicate how rapidly sedimentary bodies may accumulate even today. Thus a 200 m wide valley floor may be virtually reworked in a century, whilst terrace sediments can be thought of as being produced by lateral accretion at a rate measured in several metres a year. This does not include the possibility of catastrophic discharges. Clearly the assignment of sedimentary bodies of coarse materials to the centuries and millenia of Quaternary stages is but an indication of the partial preservation of bodies that must have been deposited and again lost during extended timespans.

DISCUSSION AND CONCLUSIONS

In presenting some information on the contemporary activity of Welsh streams, it has been necessary to point up the limitations and exploratory nature of present understanding. This is not an unfamiliar state of affairs to students of palaeohydrology, who must often confront field evidence and attempt reconstructions knowing full well that the really reliable tools with which they have to work are few indeed. Were evidence to be presented from other environments, the potential range of behaviour of alluvial rivers would be further increased. For example, in some environments single extreme floods are known to be capable of entirely reworking alluvial fills, and it is of course necessary to evaluate the role of such 'superfloods' in environments beyond those in which they were first described (*cf.* Baker, 1977; Dury, 1977b).

What the studies described above do show, however, is how much more complex the behaviour of alluvial rivers is than used to be appreciated, but equally how many more opportunities are coming into view for palaeohydrological work. Continuing study of the process of alluvial sedimentation suggests both new sites and features to examine, and ways of interpreting sedimentary bodies that did not previously exist. In particular, this chapter has drawn attention to sedimentation by meandering rivers other than in point bars; the interpretation of 'channels' and the similar or different sedimentation styles of single and multiple channel systems; and the downchannel variation in channel patterns and their sizes and activity rates which lead to expectable contrasts in the types of palaeohydrological information that have been preserved. Different parts of alluvial systems should be examined for different kinds of evidence. Thus counterpoint environments may be rich in organic materials and low power environments may preserve vertical age sequences. There is a rationale to the nature and availability of palaeohydrological evidence, and recent studies of contemporary channel and floodplain processes are now beginning to reveal this more clearly.

ACKNOWLEDGEMENTS

Some of the information here presented was obtained with the help of a Natural Environment Research Council Research grant; the survey work of D.A. Hughes, C. Blacknell and G.W. Lewis has also been invaluable.

REFERENCES

Allen, J. R. L. 1965. A review of the origin and characteristics of recent alluvial sediments. *Sedimentology*, **5**, 89–191.

Allen, J. R. L. 1970. A quantitative model of grain size and sedimentary structure in lateral deposits. *Geological Journal*, **7**, 129–146.

Ashmore, P. E. 1982. Laboratory modelling of gravel braided stream morphology. *Earth Surface Processes and Landforms*, **7**, 201–225.

Baker, V. R. 1977. Stream channel response to floods with examples from central Texas. *Geological Society of America Bulletin*, **88**, 1057–1071.

Baker, V. R., and Penteado-Orellana, M. H. 1977. Adjustment to Quaternary climatic change by the Colorado River in central Texas. *Journal of Geology*, **85**, 395–422.

Begin, Z. B. 1981. The relationship between flow-shear stress and stream power. *Journal of Hydrology*, **52**, 307–319.

Blacknell, C. 1982. Morphology and surface sedimentary features of point bars in Welsh gravel-bed rivers. *Geological Magazine*, **119**, 181–192.

Bluck, B. J. 1971. Sedimentation in the meandering River Endrick. *Scottish Journal of Geology*, **7**, 93–138.

Bluck, B. J. 1979. Structure of coarse grained braided stream alluvium. *Transactions of the Royal Society of Edinburgh*, **70**, 181–221.

Bradley, W. C., and Mears, A. I. 1980. Calculations of flows needed to transport coarse fraction of Boulder Creek alluvium at Boulder, Colorado: summary. *Geological Society of America Bulletin*, **91**, 135–138.

Bridge, J. S. 1975. Computer simulation of sedimentation in meandering streams. *Sedimentology*, **22**, 3–43.

Bridge, J. S. and Leeder, M. R. 1979. A simulation model of alluvial stratigraphy. *Sedimentology*, **26**, 617–644.

Carey, W. C. 1969. Formation of floodplain lands. *Journal of the Hydraulics Division, American Society of Civil Engineers*, **95**, HY3, 981–994.

Cheetham, G. H. 1980. Late Quaternary palaeohydrology: the Kennet Valley case-study. In Jones, D. K. C. (Ed.), *The shaping of Southern England*, London, Academic Press, London 274 pp., 203–223.

Church, M. 1978. Palaeohydrological reconstructions from a Holocene valley fill. In Miall, A. D. (Ed.), *Fluvial Sedimentology*, Canadian Society of Petroleum Geologists, Calgary, 743–772.

Church, M., and Ryder, J. M. 1972. Paraglacial sedimentation: a consideration of fluvial processes conditioned by glaciation. *Geological Society of America Bulletin*, **83**, 3059–3072.

Collinson, J., and Lewin, J. (Eds.) 1983. *Modern and Ancient Fluvial Systems*, International Association of Sedimentologists special publication no. 6, Blackwells, Oxford, 575 pp.

Dury, G. H. 1977a. Peak flows, low flows, and aspects of geomorphic dominance. In Gregory, K. J. (Ed.), *River Channel Changes*, Wiley, Chichester, 61–74.

Dury, G. H. 1977b. Underfit streams: retrospect, perspect and prospect. In Gegrory, K. J. (Ed.), *River Channel Changes*, Wiley, Chichester, 281–293.

Ethridge, F. G., and Schumm, S. A. 1978. Reconstructing palaeochannel morphologic and flow characteristics: methodology, limitations, and assessment. In Miall, A. D. (Ed.), *Fluvial Sedimentology*, Canadian Society of Petroleum Geologists, Calgary, 707–721.

Ferguson, R. I., and Werritty, A. 1983. Bar development and channel changes in the gravelly River Feshie, Scotland. In Collinson, J., and Lewin, J. (Eds.), *Modern and Ancient Fluvial Systems*, International Association of Sedimentologists special publication no. 6, Blackwells, Oxford, 181–193.

Hey, R. D. 1975. Design discharges for natural channels. In Hey, R. D., and Davies, T. D. (Eds.), *Science, Technology and environmental management*, Saxon House, Farnborough, 73–88.

Hickin, E. J., and Nanson, G. C. 1975. The character of channel migration on the Beatton River, Northeast British Columbia, Canada. *Geological Society of America Bulletin*, **86**, 487–494.

Hooke, J. M. 1980. Magnitude and distribution of rates of river bank erosion. *Earth Surface Processes*, **5**, 143–157.

Jackson, R. G. II. 1976. Depositional model of point bars in the Lower Wabash River. *Journal of Sedimentology Petrology*, **46**, 579–594.

Jackson, R. G. II. 1978. Preliminary evaluation of lithofacies models for meandering alluvial streams. In Miall, A. D. (Ed.), *Fluvial Sedimentology*, Canadian Society of Petroleum Geologists, Calgary, 543–576.

Knox, J. C., McDowell, P. F., and Johnson, W. C. 1981. Holocene fluvial stratigraphy and climatic change in the Driftless Area, Wisconsin. In Mahaney, W. C. (Ed.) *Quaternary Palaeoclimate*, Geobooks, Norwich, 107–127.

Kochel, R. C., and Baker, V. R. 1982. Paleoflood hydrology. *Science*, **215**, 353–361.

Koutaniemi, L. 1979. Late-glacial and post-glacial development of the valleys of the Oulanka River basin, North-eastern Finland. *Fennia*, **157**, 13–73.

Leeder, M. R., and Bridges, P. H. 1975. Flow separation in meander bends. *Nature*, **253**, 338–339.

Leopold, L. B., and Wolman, M. G. 1957. River channel patterns—braided, meandering and straight. *Professional Paper United States Geological Survey 282–B*.

Lewin, J. 1976. Initiation of bed forms and meanders in coarse-grained sediment. *Geological Society of America Bulletin*, **87**, 281–285.

Lewin, J. 1978. Meander development and floodplain sedimentation: a case-study from mid-Wales. *Geological Journal*, **13**, 25–36.

Lewin, J., Hughes, D., and Blacknell, C. 1977. Incidence of river erosion. *Area*, **9**, 177–180.

Lewin, J., and Weir, M. J. C. 1977. Morphology and recent history of the lower Spey. *Scottish Geographical Magazine*, **93**, 45–51.

Lewis, G. W., and Lewin, J. 1983. Alluvial cutoffs in Wales and the Borderland. In Collinson, J., and Lewin, J. (Eds.), *Modern and ancient fluvial systems*, International Association of Sedimentologists, special publication no. 6, Blackwells, Oxford, 145–154.

Miall, A. D. (Ed.). 1978. *Fluvial Sedimentology*, Canadian Society of Petroleum Geologists, Calgary, 859 pp.

Mosley, M. P. 1982a. Scour depths in branch channel confluences, Ohau River, Otago, New Zealand. *Transactions, Institute of Professional Engineers New Zealand*, **9**, 17–24.

Mosley, M. P. 1982b. Analysis of the effect of changing discharge on channel morphology and instream uses in a braided river, Ohau River, New Zealand. *Water Resources Research*, **18**, 800–812.

Nanson, G. C. 1980. Point bar and floodplain formation of the meandering Beatton River, northeastern British Columbia, Canada. *Sedimentology*, **27**, 3–29.

Nanson, G. C., and Page, K. 1983. Lateral accretion of fine-grained concave benches on meandering rivers. In Collinson, J., and Lewin, J. (Eds.), *Modern and ancient fluvial systems*, International Association of Sedimentologists special publication no. 6, Blackwells, Oxford, 133–145.

Nanson, G. C., and Young, R. W. 1981. Overbank deposition and floodplain formation on small coastal streams of New South Wales. *Paper presented to 2nd International Conference on Fluvial Sedimentology, Keele.*

Schumm, S. A. 1972. Fluvial palaeochannels. In Rigby, J. K., and Hamblin, W. K. (Eds.), *Recognition of Ancient Sedimentary Environments*, Society of Economic Palaeontologists and Mineralogists, special publication no. 16, 98–107.

Smith, D. G. 1974. Aggradation of the Alexandra–North Saskatchewan River, Banff Park, Alberta. In Morisawa, M. E. (Ed.), *Fluvial Geomorphology*, publications in Geomorphology, State University of New York, Binghamton, 201–209.

Smith, D. G., and Smith, N. D. 1980. Sedimentation in anastomosed river systems: examples from alluvial valleys near Banff, Alberta. *Journal of Sedimentary Petrology*, **50**, 157–164.

Starkel, L. (Ed.) 1982. Evolution of the Vistula River Valley during the last 15 000 years. *Polish Academy of Sciences, Institute of Geography and Spatial Organisation, Geographical Studies Special Issue No. 1*, 169 pp.

Starkel, L. and Thornes, J. B. (Eds.) 1981. *Palaeohydrology of River Basins*. British Geomorphological Research Group Technical Bulletin no. 28, 107 pp.

Thorne, C. R., and Lewin, J. 1979. Bank processes, bed material movement and planform development in a meandering river. In Rhodes, D. D., and Williams, G. P. (Eds.), *Adjustments to the Fluvial System*, Kendall-Hunt, Dubuque, Iowa, 117–138.

Williams, P. F., and Rust, B. R. 1969. The sedimentology of a braided river *Journal of Sedimentary Petrology*, **39**, 649–679.

Wolman, M. G., and Leopold, L. B. 1957. River floodplains: some observations on their formation. *Professional paper, United States Geological Survey 282–C.*

Woodyer, K. D. 1975. Concave-bank benches on Barwon River, N.S.W. *Australian Geographer*, **13**, 36–40.

Background to Palaeohydrology
Edited by K. J. Gregory

14
Modelling past discharges of meandering rivers

KAROL ROTNICKI

Quaternary Research Institute, Adam Mickiewicz University, Poznań, Poland

Stream flow is part of the permanent water circulation in the atmosphere–lithosphere system. On the surface of the Earth, it may be the most evident and the most important part of the continental phase of the hydrologic cycle of water circulation. No wonder that stream flow is the main research problem of continental hydrology, not only for theoretical–cognitive but for practical reasons as well. The rhythm and character of stream flow and water yield of a river are derivative of two groups of factors: firstly, climatic, which are variable in time; and secondly, physiographic, such as the area, morphology and relief energy of a drainage basin, the type of soil, geology, etc. (Wisler and Brater, 1959). In comparison with the great variability in time of climatic factors, physiographic ones can be taken as more or less constant over a long period. For this reason, a river is treated as a derivative of climatic factors.

The most important climatic factors directly determining the nature of river discharge are air temperature, precipitation, evaporation, and their annual sequence. These factors also influence runoff indirectly through the type of vegetation cover. The measurement of the amount of discharge is of primary importance (Biswas, 1970) and influences the formation of a particular channel pattern (Leopold and Wolman, 1957), the hydraulic geometry, and the accumulation of lithologic units. Mathematical and stochastic models of relationships between particular climatic elements and characteristic discharges of a river, and between geometric parameters of a river channel and these characteristic discharges, allow prediction of the order of magnitude of unknown discharges. This approach (see pp. 14) has featured in Quaternary palaeohydrology as proposed by Schumm (1965), and in estimates of the former discharges of channels on the Riverine Plain of Australia (Schumm,

1968). Since 1952, Dury had been developing an approach and when he formulated his theory of underfit streams in meandering valleys (Dury, 1964a, 1964b, 1965) he was the first to show the ways and means of estimating past river discharges from the dimensions of former river channels. The need for prediction or retrodiction (Dury, 1977), of former discharges is a primary purpose of palaeohydrology. Ability to calculate former river discharge makes it possible to determine the quantitative water balance of a drainage basin, the hydrologic regime of a river, and some elements of the climate, as well as to detect and specify quantitative changes of some hydrologic and climatic parameters in a given time-section of the Quaternary and, possibly for older geological periods too.

GENERAL METHODOLOGICAL PRINCIPLES

The key to the solution of palaeohydrologic problems, including the retrodiction of past river discharge, is in modern hydrology and in other disciplines such as climatology and sedimentology. The application of our knowledge of the present to the study of the past is made possible by the principle of geological uniformitarianism. However, it should be realized that there are thresholds at both ends of this procedure. The first consists of unknowns about the hydrology of a modern river and its relationships with a number of morphological elements, defined as hydraulic geometry, on the one hand, and with particular climatic factors and vegetation on the other. Recently, Gregory (1981) has drawn attention to this threshold.

A different threshold is the degree of preservation in geological space of measurable traces of former river discharge and their recognizability. This problem will be dealt with in a later part of this chapter.

Theoretically, there are two ways of determining past river discharge on the basis of its effects, firstly through analysis of deposits of past rivers, and secondly through analysis of preserved morphological effects caused by a given river discharge. Unfortunately, the connections between the deposit, its structure, and statistical parameters of grain-size distribution and the discharge are so complicated and obscure that they cannot be expressed in any quantitative models (see Chapter 5, pp. 103). Grain-size distribution and its parameters enable one to specify at most the local speed at which the deposition of a given sediment took place (Hjulström, 1935; Sundborg, 1967; Joppling, 1966).

The present state of knowledge gives us an understanding of the relations between characteristic discharges, such as mean annual discharge (q_m), bankfull discharge (q_b), and mean annual flood (q_{maf}), and morphological parameters including:

1. Channel cross-section parameters for bankfull discharge (q_b): channel width (w), maximum depth (d), hydraulic radius (r_h), channel slope (s), width/depth ratio (f), and cross-sectional area (a_c);

2. Channel pattern parameters: meander wavelength (l_m), meander belt width (w_m), and meander curvature radius (r_m).

It follows from the very nature of these relations that they are valid only for some types of rivers. The relations between discharge and channel cross-section parameters concern meandering and straight rivers, while the relations between discharge and channel pattern parameters can concern only meandering rivers. So far, this problem has not been dealt with successfully for braided rivers which must be excluded from the retrodiction of former river discharges. It can be argued that even if braided rivers had been measured in this way, there would still be the insurmountable problem in the form of the morphological and geological record of a past braided river; for it is not easily possible to reconstruct one coeval system of synchronically functioning channels.

It follows that we have the best basis for the retrodiction of meandering river discharge. The methodological justification of the construction of retrodictive models of past discharges follows from the random character of hydrologic phenomena and processes.

Relying on empirical examination in the study of river discharge, we are forced to introduce certain, sometimes quite significant, simplifications. We only examine the influence of basic factors ignoring secondary factors which are regarded as information noise, although both types of factor are random. For example, when studying the relation between meander wavelength and discharge, we realize that discharge is the main, though by no means the only, factor, modelling meander wavelength. It also depends on the amount of transported material, the type of this material, channel slope, etc. but these factors are treated as incidental. Sometimes information noise comes from factors whose existence we do not realize, or which we are unable to measure, such as turf roughness on the channel bank.

The probabilistic character of discharge results in the stochastic nature of relations between the discharge and its causes and effects. Realization of this fact makes it possible to understand the essence of past river discharge modelling and to evaluate the results. The stochastic character of past discharge modelling requires the fulfilment of certain preliminary conditions concerning the size and character of a tested sample. It must be statistically uniform (Kaczmarek, 1970), and its size big enough for the sample to be representative.

The general principles of modelling of past discharges are the same as those of statistical methods of forecasting or prediction. Forecasting is based upon an analysis of causal connections among natural phenomena under investigation. The basis of a forecast is a mathematically expressed relation between two kinds of variables: predictors and predictands. The former are variables on the basis of which the behaviour of forecast elements, or predictands, is determined (Kaczmarek, 1970). This mathematically-expressed relation may be in the form of a regression equation. Referring to past discharge modelling, or retrodiction, we could term the two groups of variables retrodictors and retrodictands, respectively.

STATE OF KNOWLEDGE

The beginnings of approaches to modelling past river discharges can be found in the borderland between geomorphology and hydrology. Since its very beginning, geomorphology has posed very broad and basic questions concerning the development of subaerial relief. In everyday practice it had to deal with the effects of river activity in the geological past, with the problems of climate conditioning great phases of river erosion and accumulation in the past, with the main tendencies of fluvial processes in various geological periods, and with different effects of river activity in different morphoclimatic zones. Geomorphology has long been using very simplified interpretations in the field of qualitative palaeohydrology, based partly on intuition and on the most basic observations. Thus, geomorphology had a strong need for a suitable methodological key with which to gain access to a new field of problems that would be created by quantitative formulation of issues connected with river activity in the past. However, as Thornes (1977, pp. 93) rightly observes: 'Geomorphology has been less strong in using analytical models to evaluate change, and most of this work has been in the hands of hydraulic engineers'.

The first foundations of past river discharge modelling were laid by the predecessors of a research trend that Allen (1977) called an empirical school of natural river system research. The trend is based on the search for mathematical relations between the geometrical parameters of a channel and the discharge. The first to formulate such a relationship was Jefferson (1902), whom Carlston (1965) called 'the grandfather of quantitative geomorphology'. These are the following dependences of geometrical features of a channel on the discharge (see Appendix pp. 351 for the explanation of the symbols):

$$w_m = 75q_m^{0.5} \quad \text{in ft and cfs} \tag{1}$$

$$l_m = 46q_m^{0.5} \quad \text{in ft and cfs} \tag{2}$$

$$w = 4.5q_m^{0.5} \quad \text{in ft and cfs} \tag{3}$$

Later investigations aimed to state more precisely the power function relationship between meander wavelength and discharge, which was given the more general form of:

$$l_m = cq^z \tag{4}$$

In 1941 and 1949 Inglis presented several specific versions of this form of dependence, of which the two best known are:

$$l_m = 29.6\, q_{max}^{0.5} \text{ in ft and cfs} \tag{5}$$

$$l_m = 36\, q_b^{0.5} \quad \text{in ft and cfs} \tag{6}$$

Leopold and Maddock (1953) confirmed Jefferson's (1902) earlier equations (2) and (3), proving that they have the form of a simple power function. When analysing the strength of relationships (equation 6), Leopold and Wolman (1957) stated that meander wavelength depends more directly on channel width than on discharge. They also found that channel width has a much better correlation with discharge than with meander wavelength. Finally, in 1965, which turned out to be the crucial year for the formulation of the issues of palaeohydrology (Schumm, 1965) and former discharge modelling (Dury, 1965), Carlston (1965) presented further versions of the dependence of the geometric parameters of a channel on discharge, at the same time providing the scale of error of these equations:

$$l_m = 106\, q_m^{0.46} \text{ with } SE = 11.8 \text{ per cent, in ft and cfs} \tag{7}$$

$$l_m = 8.2\, q_b^{0.62} \text{ with } SE = 25 \text{ per cent, in ft and cfs} \tag{8}$$

$$w_m = 65.8\, q_m^{0.47} \text{ with } SE = 23 \text{ per cent, in ft and cfs} \tag{9}$$

Carlston (1965) also quotes two unpublished formulae due to Thomas:

$$w = 7.0\, q_m^{0.46} \quad \text{in ft and cfs} \tag{10}$$

$$w = 2.95\, q_b^{0.47} \quad \text{in ft and cfs} \tag{11}$$

Dury had already recognized the existence of former great meanders and meander valleys in 1952, and in 1964 he formulated his 'stream underfitness' theory (Dury, 1964a). He found that a number of rivers of middle latitudes are commonly underfit, and the large inactive meanders at valley bottoms prove that channel-forming discharges used to be much greater than they are today. The statement that the numerical discharge-channel dimension relationship can be helpful in the reconstruction of former discharges was only a natural consequence of this discovery and a matter of time. In this way, Dury recognized former channel dimensions to be the predictors of past discharges. Thus, in 1965 he made use of the known dependence (4) to calculate former discharge, giving it the specific form of:

$$l_m = 30\, q_b^{0.5} \quad \text{in ft and cfs} \tag{12}$$

and after transformation:

$$Q_b = \left(\frac{L_m}{30}\right)^2 \quad \text{in ft and cfs} \tag{13}$$

This equation was worked out for 70 paired values of q_b and l_m of six rivers, but Carlston (1965) questions them as approximate data obtained on the basis of a regional graph and not from direct measurement. Dury (1965) also proposes computing former discharges by calculating the Q_b/q_b ratio by means of a nomogram based on Manning's equation. The use of this nomogram is possible for known values of A_c, a_c, R_h, r_h, S, and s for former and present channels.

In 1977 Dury presented several new equations for the relation between the main geometric parameters of a channel and the discharge, on the basis of which the final version of the equation for past discharge calculation was constructed. It has the following form:

$$Q_b = [(W/2.99)^{1.81} + (L/32.857)^{1.81} + 0.83\ A_c^{1.09}\ \Omega]/3 \text{ in m and m}^3\ \text{sec}^{-1} \tag{14}$$

Q_b is here a mean value of figures obtained from the following equations:

$$Q_b = (W/2.99)^{1.81} \text{ in m and m}^3\ \text{s}^{-1} \tag{14a}$$

$$Q_b = (L/32.857)^{1.81} \text{ in m and m}^3\ \text{s}^{-1} \tag{14b}$$

$$Q_b = 0.83\, A_c^{1.09} \text{ in m and m}^3\ \text{s}^{-1} \tag{14c}$$

each of which deals with a different parameter of the former channel (Dury, 1976, 1977).

A different approach to the problem under consideration was presented by Schumm (1968). He was aware that geometric parameters of a channel and its shape depend not only on the discharge but also on resistance to erosion of the deposit forming the channel perimeter (Schumm, 1960, 1968). Therefore he introduced into his equations value m, denoting the silt–clay percentage in the deposits composing the channel perimeter. The value of m is calculated as follows:

$$m = \frac{m_c + w + m_b + 2d}{w + 2d} \tag{15}$$

The values of m_c and m_b are obtained from 10–20 samples taken cross-sectionally from the subsurface layer from 2.5 to 10 cm in thickness. Next, Schumm (1960) examined the influence of the percentage of silt–clay on the shape of the channel expressed as the width/depth ratio = f. He obtained the following dependence:

$$f = 255\, m^{-1.08} \tag{16}$$

which shows that if m increases, then the width/depth ratio decreases, and conversely. Later, Schumm (1963) demonstrated that sinuosity is also dependent on the percentage of silt–clay in the following way:

$$\Omega = 0.94\, m^{0.25} \tag{17}$$

which means that the higher the percentage content of silt–clay in the channel perimeter, the greater the channel sinuosity. To be precise, one should mention that Dury (1965) was aware of the fact that the considerable scatter of the data about his regresion line given by equation (12) reflects also the influence of other variables, including deposit type. By the application of multiple regression analysis, Schumm obtained the following mathematical expressions of the relationships between characteristic discharges and the percentage of silt–clay on the one hand, and some geometric parameters of the channel on the other:

$$l_m = 438 \cdot (q_b^{0.43}/m^{0.47}) \text{ in ft and cfs} \tag{18}$$

where SE = 40.2 per cent average and 48 per cent maximum,

$$l_m = 1\,890 \cdot (q_m^{0.34}/m^{0.74}) \text{ in ft and cfs} \tag{19}$$

where SE = 37.7 per cent average and 44.5 per cent maximum,

$$w = 37 \cdot (q_m^{0.38}/m^{0.39}) \text{ in ft and cfs} \tag{20}$$

where SE = 32.8 per cent average and 38 per cent maximum,

$$l_m = 234 \cdot (q_{maf}^{0.48}/m^{0.74}) \text{ in ft and cfs} \tag{21}$$

where SE = 45 per cent average.

The above formulae, according to Schumm (1968), explain, respectively, 87, 88, 89 and 86 per cent of the variability of channel width or meander wavelength from the mean, with the correlation coefficients from +0.93 to +0.96. After transformation, these formulae were used for the estimation of bankfull discharge and mean annual discharge for an ancestral river and two prior streams of the Riverine Plain in Australia.

DISCUSSION OF AVAILABLE METHODS

Consideration of attempts to formulate mathematical relationships which could be used for determining former river discharges leads to a number of conclusions. Firstly, the approaches of Dury (1965, 1976, 1977) and Schumm (1968, 1969, 1971) are methodologically identical and they differ only in the number of samples used as the basis of modelling and in the areas from which the samples were taken. Secondly, the formulae established by Dury (equations 12, 13) and Schumm (equations 15–21) are empirical relationships which connect discharge and channel parameters especially meander wavelength and channel width although in the formulae proposed by Schumm (equations 18–21) the geometric parameters of the channel are dependent also upon the percentage silt–clay in the channel perimeter. Thirdly, the relationships usually use bankfull discharge (q_b) or mean annual discharge (q_m) but occasionally maximum discharge (q_{max}) and mean annual flood (q_{maf}) are used. The frequent use of q_b is justified because many authors believe it to be the channel-forming or effective (Leopold, Wolman, and Miller, 1964) discharge, and, according to Daniel (1971), the channel-forming discharge occurs between mean and bankfull discharge. The bankfull stage is the only characteristic stage that can be determined on the basis of a well-preserved fossil or subfossil cutoff meander channel on a floodplain, whereas water stages for the remaining characteristic discharges (q_m, q_{max}, q_{maf}) cannot be determined on the basis of fossil channels. Even where it is possible to give the approximate water stage connected with the former maximum flood (Baker, 1981; Kochel and Baker, 1982), this stage cannot be related to any particular fossil channel. Fourthly, the possible use of the formulae for past discharge retrodiction requires consideration of the size of sample on which each equation is based and the climatic zone from which the sample is taken.

The first formulae of Jefferson (1902) were based on only three paired variables q_m and l_m, and Carlston (1965) used 27–31 paired variables. Dury (1965) collected 70 paired variables for six rivers in 1965 and from 95 to 133 rivers in 1976. Schumm (1968) constructed his formulae on the basis of 36 paired data from 36 rivers. These sets of data are for rivers of very diverse sizes (q_b from 1 to 10^5 m^3 sec^{-1}). These are rivers of North America, Europe, Asia, and Australia, flowing in areas whose climates range from semiarid tropical to a continental forest climate with cool summers and to the humid climate of middle latitudes.

The high correlation coefficients, from +0.93 to +0.96, of Dury's equations (14a, 14b, 14c) and those of Schumm (equations 18–21) indicate a high significance level for the relation between the analysed variables, but it is only the value of the standard error which shows the probability and the range of precision with which variable Y, e.g. bankfull discharge, can be estimated on the basis of variable X, e.g. meander wavelength, or variables X and Z, e.g. meander wavelength and percentage of silt–clay. Unfortunately, only Carlston (1965) and Schumm (1968, 1969) give the value of this error for their equations. If standard error is to be a measure of usefulness of a given equation for former discharge modelling, then Dury's formulae cannot be evaluated in such terms because their error is unknown. It must be stated that apart from Carlston's equation (7) based on 31 paired observations and whose standard error is 11.8 per cent, his remaining equations have considerably higher errors (23–25 per cent), and the errors of Schumm's formulae are also very high, as they range from 32.8 per cent to 45 per cent. In principle, we should use the values of two standard errors, since it is only then that we get a 95 per cent probability level for the calculated limits bounding a modelled variable. On this basis Carlston's (1965) equations (7–9) would reach 45–50 per cent of error, and Schumm's equations (18–21) 65–90 per cent. A given equation, when used for predicting a value beyond the values for which it was constructed, may give an even greater error. This supposition is confirmed by Table 14.1, in which l_m, w, q_m and q_b were calculated by means of some familiar equations for the presently meandering channel of the Prosna river near Wieruszów (southeastern Wielkopolska, Poland), having the following parameters: q_m = 4.5 m^3

TABLE 14.1 COMPARISON OF REAL VALUES OF SOME GEOMETRIC PARAMETERS OF THE CHANNEL WITH THOSE CALCULATED BY MEANS OF VARIOUS EQUATIONS (See Appendix pp.351 for explanation of symbols)

Equation $n = 7.94\%$ (in analysed cross-section of the Prosna river channel)		Results l_m m	w m	q_m $m^3 sec^{-1}$	q_b $m^3 sec^{-1}$
Measured values		120.0	18.5	4.50	22.5
$l_m = 46\ q_m^{0.5}$	(Jefferson, 1902)	176.0		2.07	
$l_m = 36\ q_b^{0.5}$	(Inglis, 1949)	309.3			3.38
$l_m = 106\ q_m^{0.46}$	(Carlston, 1965)	332.5		0.48	
$l_m = 8.2\ q_b^{0.62}$	(Carlston, 1965)	157.0			14.42
$l_m = 30\ q_b^{0.5}$	(Dury, 1965)	257.7			4.87
$l_m = 32.857\ q_b^{0.55}$	(Dury, 1976)	182.1			10.40
$l_m = 438\ q_b^{0.43}/m^{0.47}$	(Schumm, 1968)	890.0			0.213
$l_m = 1\ 890\ q_m^{0.34}/m^{0.74}$	(Schumm, 1968)	696.0		0.03	
$w = 4.5\ q_m^{0.5}$	(Jefferson, 1902)		17.3	5.15	
$w = 2.99\ q_b^{0.55}$	(Dury, 1976)		16.6		31.18
$w = 37\ q_m^{0.38}/m^{0.39}$	(Schumm, 1968)		34.5	0.87	
$w = 7.0\ q_m^{0.46}$	(Thomas, after Carlston, 1965)		22.0;	3.07	
$w = 2.95\ q_b^{0.47}$	(Thomas, after Carlston, 1965)		22.2		17.23
$Q_b = [(w/2.99)^{1.81} + (L)32.857)^{1.81} + 0.83A_c\Omega]/3$	(Dury, 1976, 1977)				34.11

$\sec^{-1}$, q_b = 22.5 $m^3 \sec^{-1}$, w = 18.5 m, l_m = 130 m, s = 0.0004431, m = 7.95 per cent. It transpired that the results obtained differ considerably depending on the formula applied, and that they deviate substantially from the actual values.

The present state of former discharge modelling justifies the statement that the knowledge of the relation between bankfull and mean annual discharge on the one hand, and meander wavelength, channel width, and percentage of silt–clay in the channel perimeter on the other, is insufficient for the formulation of mathematical expressions that could be supraregional and supraclimatic in character. Starkel and Thornes (1981) consider Schumm's (1972) formulae to be inapplicable outside the range for which they were worked out and one might ask whether there are any universally applicable mathematical expressions of those relations?

We do not mean a general form of such a dependence as: $q_b = cl_m^z$, but a form in which coefficient c and exponent z have one and only one value, holding for all rivers of all climatic zones, regions and continents. It is possible that such a universal, empirical formula does not exist. Until we have answered this question, we cannot unreservedly use the equations discussed above as the universal formula for the modelling of former discharges of meandering rivers.

The modelling of former discharges of meandering rivers is difficult not only because of the hydrological and mathematical aspects of the theory available but also because considerable difficulties arise concerning the degree of preservation of measurable traces of past discharge in three-dimensional space, and the recognizability of those traces. As Dury (1976) states, errors in former banktop determination lead to complex errors in the estimation of other channel dimensions, and this makes it difficult or impossible to apply the Manning equation for discharge retrodiction (Dury, 1965). Also, the recognition of the bottom of a past fossil channel can be extremely difficult if the deposits into which the channel is cut do not differ from those filling the channel as can be seen, for example, in Schumm's work (1968). Another obstacle is the fact that, on the whole, palaeomeanders are preserved in such short segments that it is impossible to establish the meander wavelength precisely (Figure 14.1). For that reason, in most cases we cannot use a whole group of equations based on the relation between discharge and meander wavelength for discharge retrodiction. Moreover, if Schumm's equations (15–21) are to be applied, the method of collecting samples for determining the percentage of silt–clay in bank alluvium is not explicity stipulated.

Most probably because of these difficulties the retrodiction of former river discharges has been made sporadically (Padgett and Ehrlich, 1976; Klimek and Starkel, 1981; Baker, 1981; Kochel and Baker, 1982). Even Dury (1965, 1976, 1977) and Schumm (1968, 1969, 1971) used discharge retrodiction more as an illustration of the possiblities of a certain method and a certain approach rather than as a tool for systematic estimation of former discharges, their changes and

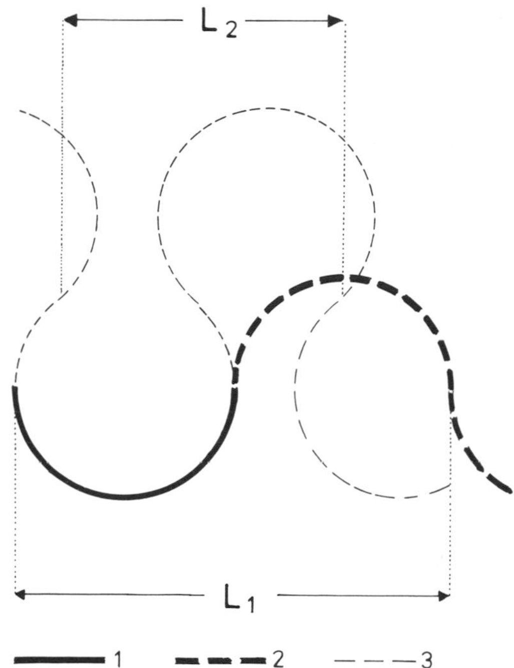

FIGURE 14.1 DIFFICULTIES OF DETERMINATION OF MEANDER WAVELENGTH ON THE BASIS OF A PRESERVED FRAGMENT OF A PALAEOMEANDER
1, the preserved fragment of the palaeomeander; 2, the first variant of interpretation (L_1); 3, the second variant of interpretation (L_2)

the changes in climatic conditions and hydrologic regime. It was only Schumm (1968) who showed examples of quantitative change of river discharge as an element of the change of the whole hydrologic cycle.

Since 1965 there has been an enormous growth of interest in the problems of channel-pattern changes, channel-size changes, and the evolution and the mechanism of valley transformation during the last 15 000 years. Apart from work by Schumm (1965) and Dury (1964a, 1964b, 1965), there appeared Falkowski's (1965, 1967, 1970, 1972) conception concerning channel-pattern changes under the influence of change of climatic conditions. This was elaborated by a number of authors (Mycielska-Dowgiałło, 1969, 1972, 1977; Szumański, 1972, 1981; Klimek and Starkel, 1974; Kozarski and Rotnicki, 1977; Rose *et al.*, 1980; Kozarski, 1981; Starkel, 1981). As can be seen, the development of the approach has been particularly fruitful in Poland. It is a pity that the problem of the evolution of valley bottoms and channel patterns is still often considered in terms of qualitative palaeohydrology.

MODELLING OF PAST DISCHARGE ON THE BASIS OF STEADY UNIFORM FLOW FORMULAE

Discussion of the existing methods of former discharge modelling shows that there are two basic groups of reasons why this modelling has not become, so far, a research tool for the solution of palaeohydrological problems. Firstly, because it is difficult to find a mathematical formula which permits the determination of past discharges with acceptable precision; and secondly, because the sphere of geological and geomorphological research has to supply such numerical data as can be applied to a suitable mathematical formula for determining former discharge. These two problems stimulated the present author to start investigations aiming to elaborate a method of modelling the past discharges of meandering rivers. The research was carried out in both spheres, theoretical and geological, simultaneously. In 1978 the studies were included within IGCP Project No. 158 concerning the palaeohydrology of the temperate zone during the last 15,000 years. The first results were presented at the XIth International INQUA Congress in Moscow in 1982 (Rotnicki, 1982). The present chapter outlines the main assumptions of the method, the way of testing it, and mathematical forms of the formulae obtained with examples of the first results. It also shows the range of geological research needed to obtain the data on the basis of which past discharges can be calcuated.

Mathematical–statistical basis of the method

If empirical formulae fail, then the main foundation and assumptions of modelling former discharges should be sought in the theoretical–physical principles of steady uniform flow, which is the main type of flow considered in open-channel hydraulics. The basis of this approach lies in the adaptation of Chézy's theoretical uniform flow formula for use in palaeohydrology. The formula concerns mean flow velocity and has the following form:

$$v = C\sqrt{r\,s} \qquad (22)$$

and inserted into the formula for discharge, it yields:

$$q = a_c C\sqrt{r\,s} \qquad (23)$$

It is surprising that so far nobody has attempted to test the usefulness of this formula for modelling past river discharges. Only Schumm (1968) applied Manning's formula, and Dury (1965) used it for the construction of a nomogram for determining the Q_b/q_b ratio, which, however, has not been used in palaeohydrological practice.

Several attempts have been made to determine Chézy's resistance factor C, for which five equations are best known. These are the empirical formulae of Manning, Ganguillet–Kutter, Bazin, Agroskin, and Pavlovskiĭ (Chow, 1959;

Lambor, 1971). These equations introduced into Chézy's formula have the following form of semi-empirical equations for metric units:

$$\text{Chézy's–Manning: } v = \frac{1}{n} r^{1/6} \sqrt{r.s} \tag{24}$$

Ganguillet–Kutter:

$$v = \frac{23 + \dfrac{1}{n} + \dfrac{0.00155}{s}}{1 + \left(12 + \dfrac{0.00155}{s}\right) \dfrac{n}{\sqrt{r}}} \sqrt{rs} \tag{25}$$

$$\text{Chézy–Bazin: } v = \frac{87}{1 + \dfrac{\gamma}{\sqrt{12}}} \sqrt{rs} \tag{26}$$

$$\text{Chézy–Agroskin: } v = \left(\frac{0.994}{n} + 17.72 \log r\right) \sqrt{rs} \tag{27}$$

$$\text{Chézy–Pavlovskiĭ: } v = \frac{1}{n} r^{25\sqrt{n} - 0.13 - 0.75\sqrt{r}(\sqrt{n} - 0.10)} \sqrt{rs} \tag{28}$$

The formulae can be of use for modelling both past mean flow velocities (V) in channel cross-section, and discharge (Q), because they contain variables which can be obtained through geological, geomorphological, and geodesic examinations of cutoff palaeomeander channels on flood plains.

In order to check the usefulness of these formulae for past discharge modelling and to choose the best one, they were tested on a set of hydrological observations collected from modern rivers. For that purpose, ten river gauging stations were chosen situated on five principal rivers of the Odra drainage basin in Poland (Figure 14.3). For each station, data were collected concerning all hydrometric measurements taken in the cross-section of the channel from at least several to more than ten times a year in order to estimate the discharge at various water stages. The basic set of data for each observation consisted of the following variables: channel width and depth, mean flow velocity, water stage, hydraulic radius, cross-sectional bottom profile, and discharge. For each station, from 80 to 200 data sets were collected covering the period 1951–1970. The total sample for all the stations amounted to 1352 sets of observations. The mean flow velocity occurring in each set, as well as the discharge calculated on the basis of measurements, will henceforth be called observed or real, velocity (v_r) and discharge (q_r). Next, each real observed velocity was assigned five velocities calculated by means of equations (24) to (28), and discharges were also calculated. These mean velocities (v_c) and discharges (q_c) were called calculated ones (Figure 14.2). In this way, for each river gauging station five

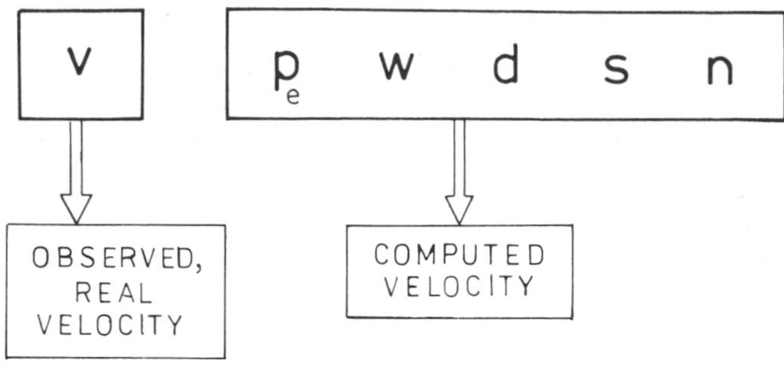

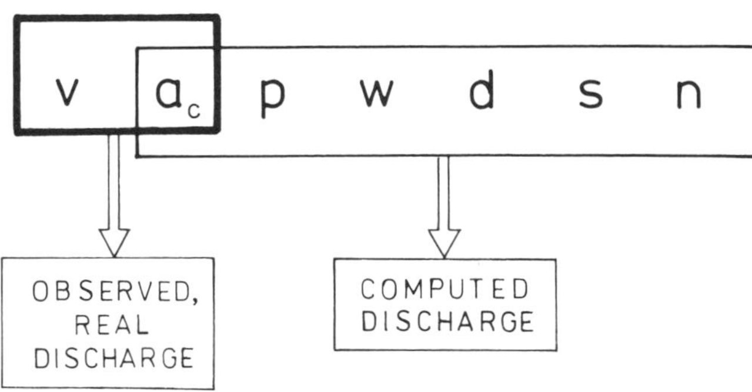

FIGURE 14.2 REAL AND CALCULATED FLOW VELOCITIES AND DISCHARGES (SEE APPENDIX FOR THE EXPLANATION OF SYMBOLS)

sets were obtained, each containing 80–200 pairs of real and computed mean flow velocities, and five sets containing the same numbers of pairs of real and computed discharges.

In order to calculate velocities and discharges using equations (24) to (28), it was necessary to estimate roughness coefficient n and to determine channel slope s. Coefficient n was estimated for the channel near each gauging station by means of Cowan's (1956) procedure. Though Richards (1973) is of the opinion that there is no objective method of assessing Manning's n, the results obtained were checked and for each observation n was calculated by means of equation (24). Next, for each station mean n and standard deviation were calculated (Table 14.2). It appeared that the differences between estimated and calculated n were insignificant. It must be noted, however, that the correct estimation of n requires some knowledge of hydraulics. Channel slope s was computed on the basis of a precise topographic map in the scale of 1:10,000 or 1:25,000. Both coefficients were then calculated in the way in which they would be later estimated for former channels on the basis of subfossil palaeomeanders.

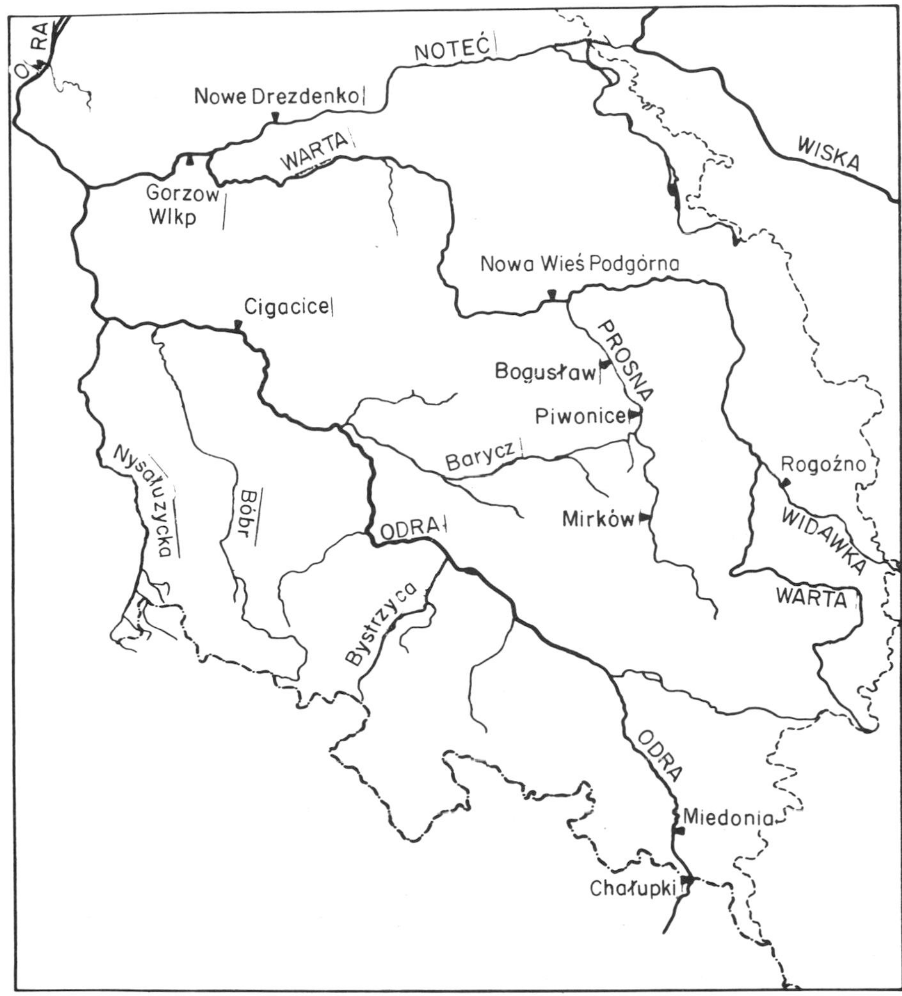

FIGURE 14.3 SELECTED GAUGING STATIONS OF THE ODRA DRAINAGE BASIN

In order to obtain the sets of pairs of computed and real mean velocities and discharges, correlations and regressions between the real and computed values were calculated. The aim of these studies was to answer the following questions: 1. What are the differences among the results obtained by means of five different formulae equations (24) to (28)? 2. What is the correlation between the real and calculated velocities and discharges, and which formula gives the best fit of these sets to one another? 3. What is the equation for this best fit and what are its errors? A formula that would fit v_r to v_c and q_r to q_c best, that is one that would show the smallest standard error, would constitute

TABLE 14.2 COMPARISON OF ROUGHNESS COEFFICIENT ESTIMATED USING COWAN'S (1956) EQUATION WITH THAT CALCULATED BY MEANS OF CHÉZY–MANNING EQUATION

Gauging station	n estimated	$\bar{n}$ computed	δn standard deviation of computed n
Mirków, Prosna river	0.045	0.048	0.0050
Piwonice, Prosna river	0.035	0.034	0.0070
Bogusław, Prosna river	0.035	0.033	0.0047
Nowa Wieś Podgórna, Warta river	0.025	0.026	0.0028
Gorzów Wlkp., Warta river	0.027	0.0289	0.004
Nowe Drezdenko, Noteć river	0.030	0.0306	0.0027
Miedonia, Odra river	0.024	0.021	0.005
Cigacice, Odra river	0.035	0.0345	0.006
Rogóźno, Widawka river	0.029	0.026	0.004

the basis for modelling former discharges and mean flow velocities. The obtained results are summarized as follows:

1. The correlation coefficients between calculated and real mean flow velocity for particular gauging stations range in value from average +0.56 to high +0.87. The differences arise for regional reasons rather than from the formula applied. At a given gauging station, the correlation is consistent for all the formulae, thus at Rogoźno on the Widawka river it amounts to 0.8358, 0.8367, 0.8365, 0.8335, and 0.8349 for equations (24) to (28), respectively.
2. There is a very strict correlation between the calculated and real discharge at all ten stations and for the five formulae (from +0.9210 to +0.9942). The equations of linear regression for the five formulae are very similar at all gauging stations (Figures 14.4 and 14.5). Only the line of Ganguillet–Kutter's formula diverges a little from the rest.
3. Particular formulae do not differ significantly as far as the standard error is concerned. Moreover, the differences are minimal, which is shown on the example of the Prosna river (Table 14.3).

At the next stage of the research, a regression analysis was made for the general population of 1352 pairs of v_r and v_c and 1352 pairs of q_r and q_c, where v_c and q_c were computed with the help of the Chézy–Manning formula. The regression equation for the relation between real and calculated mean flow velocity assumes the form:

$$V_r = 0.7908\, V_c + 0.1411 \qquad (29)$$

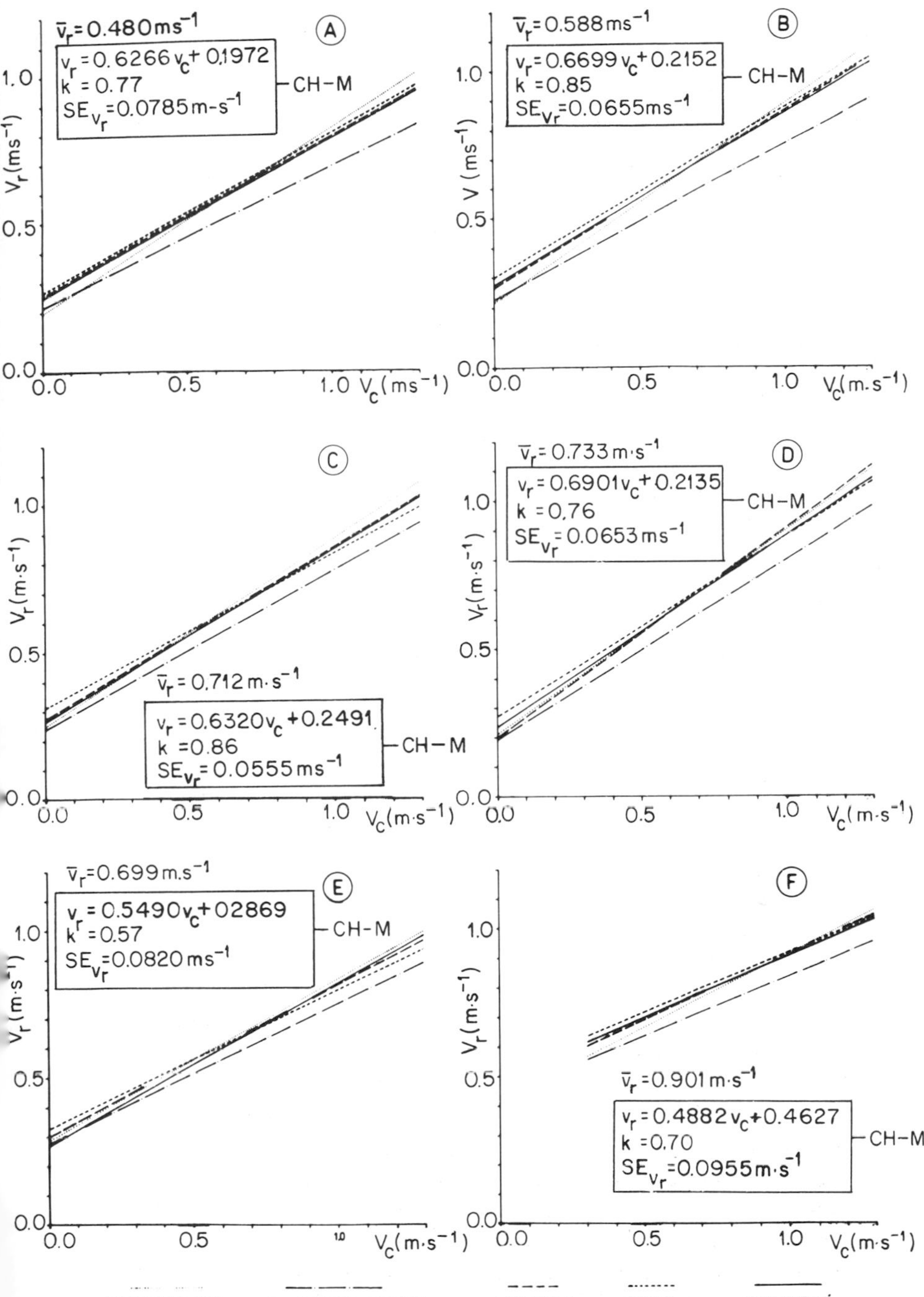

FIGURE 14.4 RELATION BETWEEN COMPUTED MEAN FLOW VELOCITY AND REAL FLOW VELOCITY

Gauging stations: A, Piwonice; B, Bogusław; C, Nowe Drezdenko; D, Nowa Wieś Podgórna; E, Gorzów; F, Miedonia

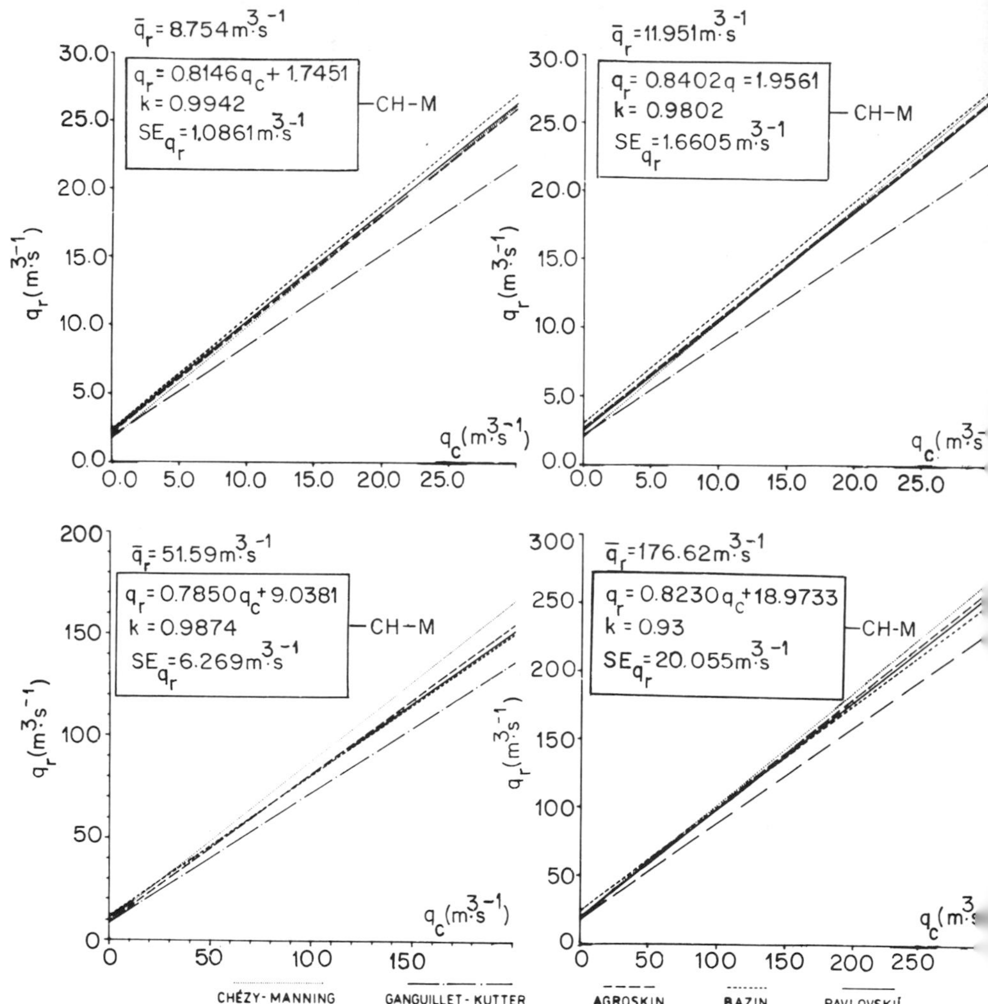

FIGURE 14.5 RELATION BETWEEN COMPUTED DISCHARGE AND OBSERVED DISCHARGE FOR SELECTED WATER GAUGING STATIONS IN THE ODRA DRAINAGE BASIN

Gauging stations: Widawka (q_r = 8.754 m^{3}sec^1); Bogusław (q_r = 11.951 m^{3}sec^{-1}); Miedonia (q_r = 51.590 m^{3}sec^{-1}); Gorzów (q_r = 176.62 m^{3}sec^{-1})

TABLE 14.3 STANDARD ERROR VALUE FOR MEAN VELOCITY AND DISCHARGE COMPUTED WITH THE HELP OF VARIOUS EQUATIONS FOR THE PROSNA RIVER, CENTRAL POLAND

Gauging station	Chézy–Manning		Ganguillet–Kutter		Agroskin		Bazin		Pavlovskiĭ	
	SE_v	SE_q	SE_v	SE_q	SE_v	SE_q	SE_v	SE_q	SE_v	SE_q
Mirków	0.073	1.03	0.073	1.06	0.073	1.06	0.074	1.06	0.072	1.06
Piwonice	0.079	2.03	0.079	2.02	0.079	2.12	0.079	2.2	0.079	2.12
Boguslaw	0.066	1.66	0.065	1.64	0.067	1.73	0.066	1.82	0.065	1.74

SE_v, in m sec^{-1}; SE_q, in m^3 sec^{-1}.

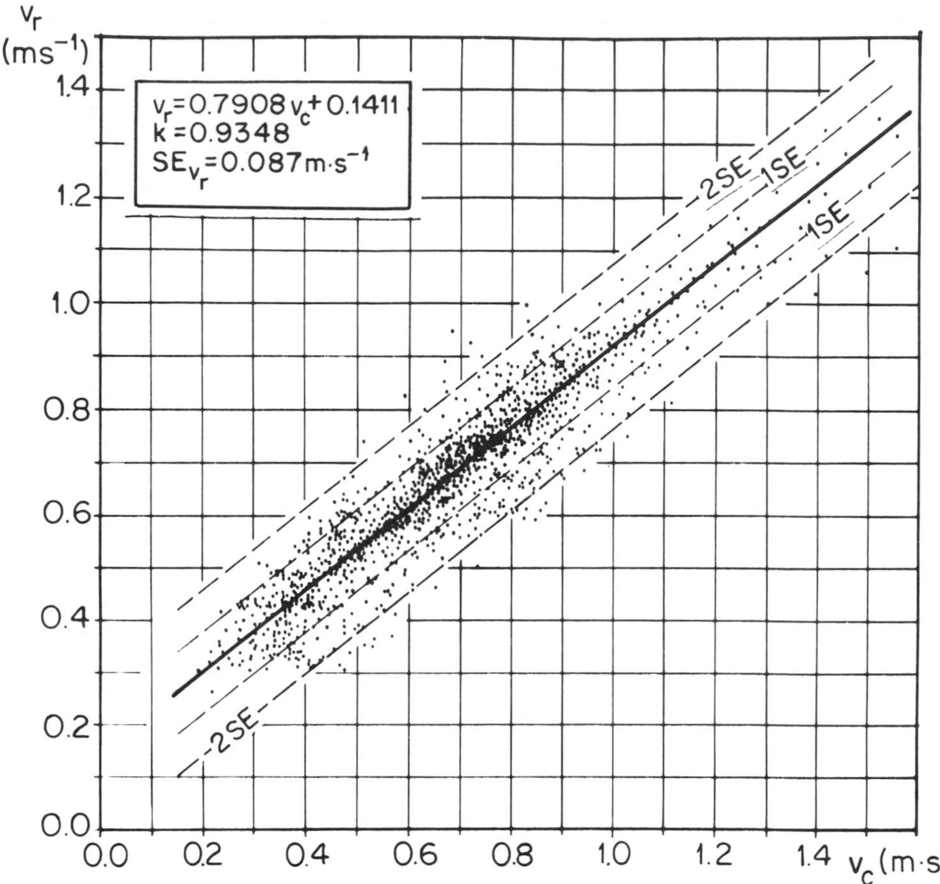

FIGURE 14.6 RELATION BETWEEN MEAN FLOW VELOCITY CALCULATED BY CHÉZY-MANNING FORMULA AND REAL FLOW VELOCITY ON THE BASIS OF 1352 PAIRS OF DATA FOR THE ODRA DRAINAGE BASIN

and its full form with the application of the Chézy–Manning formula is as follows (Figure 14.6):

$$V_{r_{(Ma)}} = \frac{0.7908}{N} R^{1/6}\sqrt{RS} + 0.1411 \tag{30}$$

where $k = +0.9348$ and $SE = 11.85$ per cent.

The universal form of the mathematical formula for the computation of real discharge on the basis of calculated discharge is as follows (Figure 14.7):

$$Q_r = 0.9208\,Q_c + 2.3616 \tag{31}$$

or, with the application of the Chézy–Manning formula:

$$Q_{r_{(Ma)}} = \frac{0.9208}{N} A_c R^{1/6}\sqrt{RS} + 2.3616 \tag{32}$$

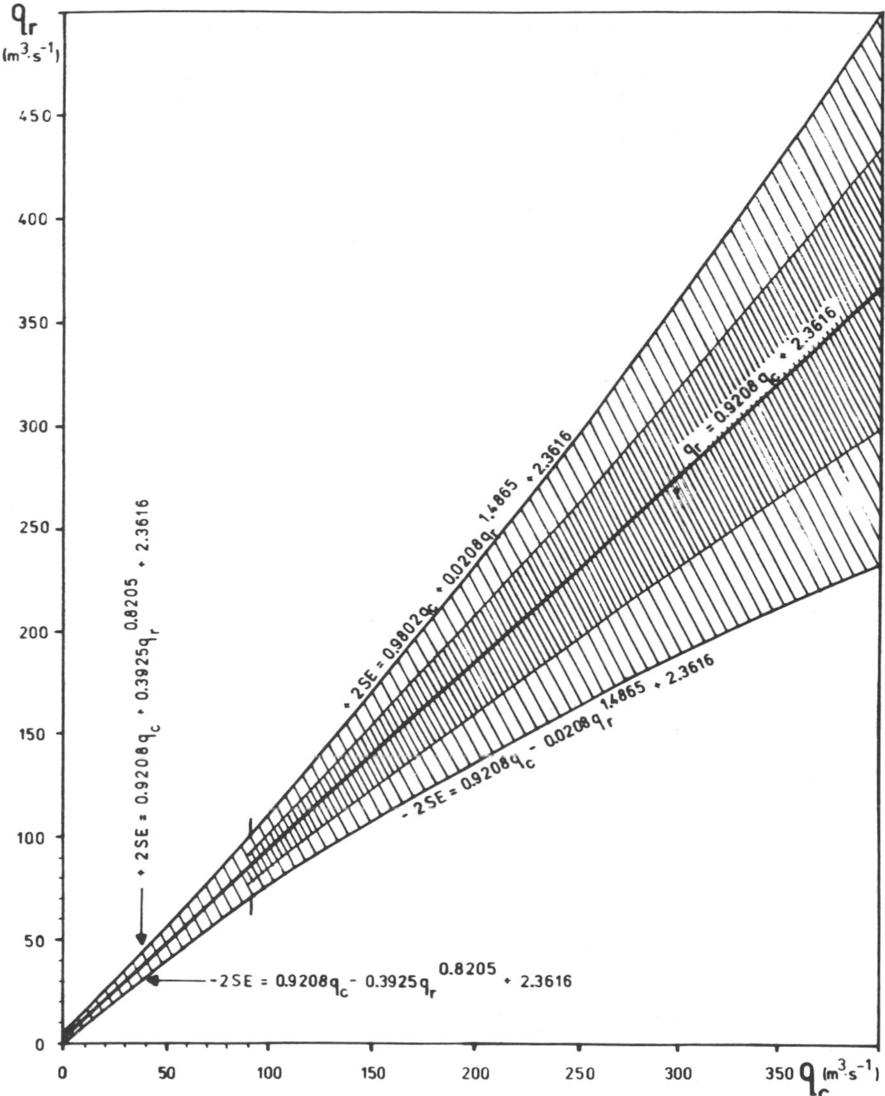

FIGURE 14.7 RELATION BETWEEN REAL DISCHARGE AND THAT CALCULATED BY CHÉZY-MANNING FORMULA FOR 1352 PAIRS OF DATA FROM THE ODRA DRAINAGE BASIN, APPROXIMATE BANDS OF ONE AND TWO STANDARD ERRORS ARE MARKED

where $k = +0.9857$. The standard error of this formula is variable and depends on the discharge value; $2SE$ for real discharges ranging from 0 to 85 m^3 sec^{-1} are:

$$2SE_{Qr_{(Ma)}} = 0.3925\, Q_{r_{(Ma)}}^{0.8208} \tag{33}$$

and for real discharges from 85 to 1,000 m^3 sec^{-1}:

$$2SE_{Qr_{(Ma)}} = 0.0208\, Q_{r_{(Ma)}}^{1.4865} \tag{34}$$

The curves delimiting the $Q_{r_{(Ma)}} \pm 2SE$ band are given by the following functions (Figure 14.7):

for real discharge from 0 to 85 m^3 sec^{-1}:

$$\pm 2SE = 0.9208\, Q_c \pm 0.3925\, Q_{r_{(Ma)}}^{0.8205} + 2.3616 \tag{35}$$

for real discharges ranging from 85 to 1,000 m^3 sec^{-1}:

$$\pm 2SE = 0.9208\, Q_c \pm 0.0208\, Q_{r_{(Ma)}}^{1.4865} + 2.3616 \tag{36}$$

The curves delimiting the $Q_r \pm 2SE$ band are given by approximate functions. For this reason, standard error values read from them, or calculated with the help of equations (35) and (36), are slightly higher than real values of the error calculated by means of partial regression (Table 14.4).

TABLE 14.4 STANDARD ERROR OF THE RETRODICTION OF FORMER DISCHARGE COMPUTED BY MEANS OF EQUATION (32)

Discharge (Q_r) (m^3 sec^{-1})	One *SE* (m^3 sec^{-1})	One *SE* (%)
3.5	0.8094	22.48
7.0	1.1926	15.28
10.0	1.4211	14.21
20.0	2.7781	13.87
30.0	3.3480	10.99
40.0	3.6014	9.00
50.0	3.8972	7.75
60.0	4.2765	7.14
70.0	5.5074	7.87
80.0	6.5989	8.25
90.0	7.8446	8.71
100.0	9.1510	9.15
150.0	15.2481	10.16
250.0	36.6510	14.70
500.0	106.9200	21.38
750.0	195.3490	26.04
Schumm's equations:	$l_m = 438\,(q_b^{0.43}/m^{0.47})$	40.2 average
	$l_m = 1\,890\,(q_m^{0.34}/m^{0.74})$	37.7 average
	$w = 37\,(q_m^{0.38}/m^{0.39})$	32.8 average

The value of the error of equation (32) depending on the value of real discharge is shown in Table 14.4. As can be seen, the best results in real discharge determination are obtained for the values of Q_b ranging from 15 m^3

sec^{-1} to 250 m^3 sec^{-1}. These errors are considerably smaller than those of the Schumm (1968, 1969) or Carlston (1965) formulae.

The regression equations (30) and (32) show that real mean flow velocities and real discharges are very significantly related to the calculated ones. Both formulae therefore can be model mathematical formulae for the computation of former discharges (Q_r) and mean flow velocities (V_r) on the basis of cutoff meander channels. At the present stage of modelling of past discharges and mean flow velocities, the retrodiction of these variables should be limited to lowland rivers with sandy or gravelly bottoms of the moderate climatic zone, displaying the following features: rain–snow replenishment; channel slope: 0.00001–0.01; mean flow velocity: 0.1–1.3 m sec^{-1}; bankfull discharge: 5–500 m^3 sec^{-1}; and roughness coefficient: 0.010–0.060.

Geological aspects of the method

The calculation of former discharges of meandering rivers by means of the Chézy–Manning formula or of other semi-empirical formulae concerning steady uniform flow requires use of the following data concerning a given palaeochannel of a cutoff meander: the value of hydraulic radius (R_h); roughness coefficient (N); channel slope (S); and cross-sectional area (A_c).

If we assume that for the time being we calculate former discharges for channels found on floodplains only (Figure 14.8), then channel slope (S) can be determined with the help of a precise topographic map at the scale of 1:10,000, assuming that the water-level surface in a palaeochannel was parallel to the floodplain in the examined section of the valley bottom.

In order to establish the hydraulic radius, roughness coefficient and cross-sectional area, it is necessary to identify the spatial distribution of palaeomeanders by means of a precise topographic map and aerial photographs (Figure 14.8). This stage is necessary to establish the relative chronology of the activity of particular meander channels in the past, as well as to choose the palaeomeanders best suited for former discharge investigations. Then the geological and geodesic identification of the palaeomeander consists of establishing its actual and precise course and the morphology of its bottom (Figure 14.9). Very often an additional criterion for the correct determination of the channel bottom is provided by the spatial variability of the grain-size distribution characteristic of meandering channels (Figures 14.10 and 14.11). Finally, the identification of the bottom topography makes it possible to find the preserved banktop and to establish the cross-sectional area of the channel and other parameters needed for bankfull discharge determination (Figures 14.12 and 14.13). The morphology of the channel bottom, its variability and grain-size distribution are very helpful in the estimation of roughness coefficient n by means of Cowan's (1956) procedure. For a satisfactory identification of the morphology of the palaeochannel of a cutoff meander, it is

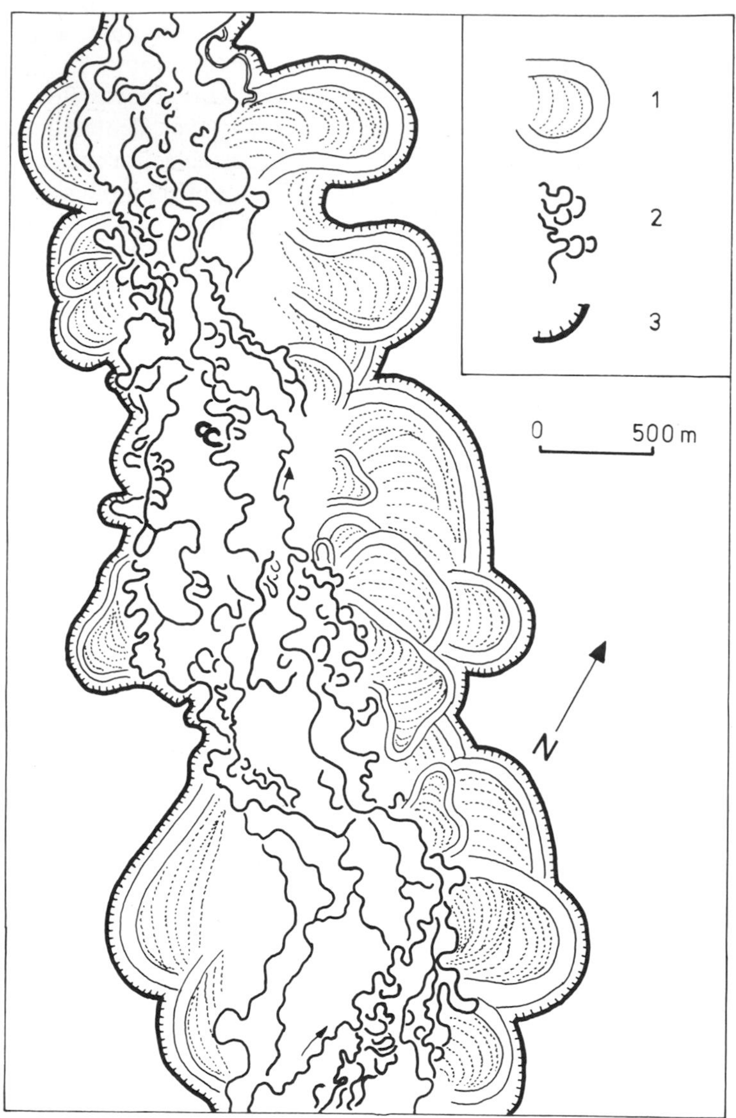

FIGURE 14.8 CARTOGRAPHIC IDENTIFICATION OF PALAEOMEANDERS BASED ON PHOTO-INTERPRETATION
1: preserved fragments of large palaeomeanders; 2: small palaeochannels and recent channel; 3: edge of older terraces

necessary to perform dozens of geological borings, preferably with a core auger or a driving rod of the 'Instorf' type. Determining the channel bottom of a palaeomeander by means of geological and sedimentological methods gives a

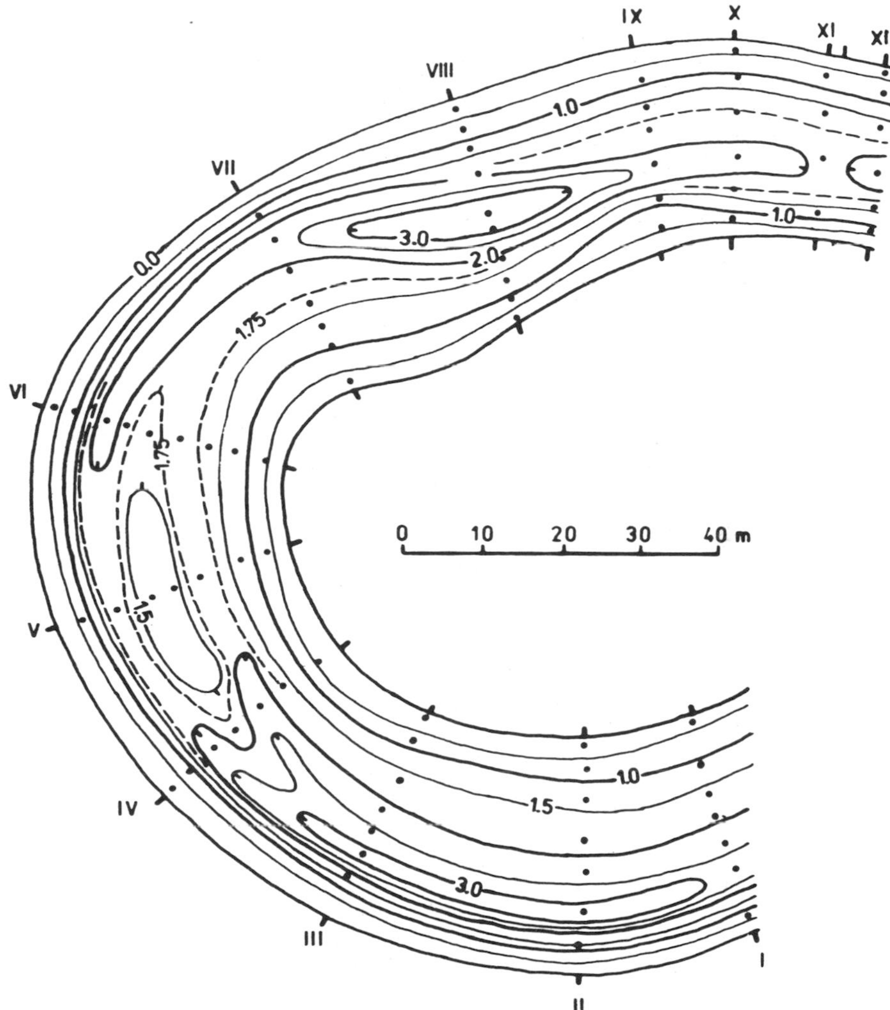

FIGURE 14.9 BATHYMETRY OF PALAEOMEANDER IN METRES; DOTS INDICATE BORINGS

good insight into the stratigraphy of deposits filling the palaeomeander, which permits a choice of samples best suited for the determination of the age of the palaeomeander with the help of radiocarbon dating or palaeobotanic analysis.

The first results of calculating mean flow velocity and discharge with the help of equations (30) and (32) for two palaeomeanders and for the present channel of the Prosna river near the gauging station Mirków in the southeast of Wielkopolska, Poland are given in Table 14.5. Mean annual discharge was reckoned on the basis of the present relation of bankfull discharge to mean annual discharge, which is exactly 5:1 in the Mirków cross-section.

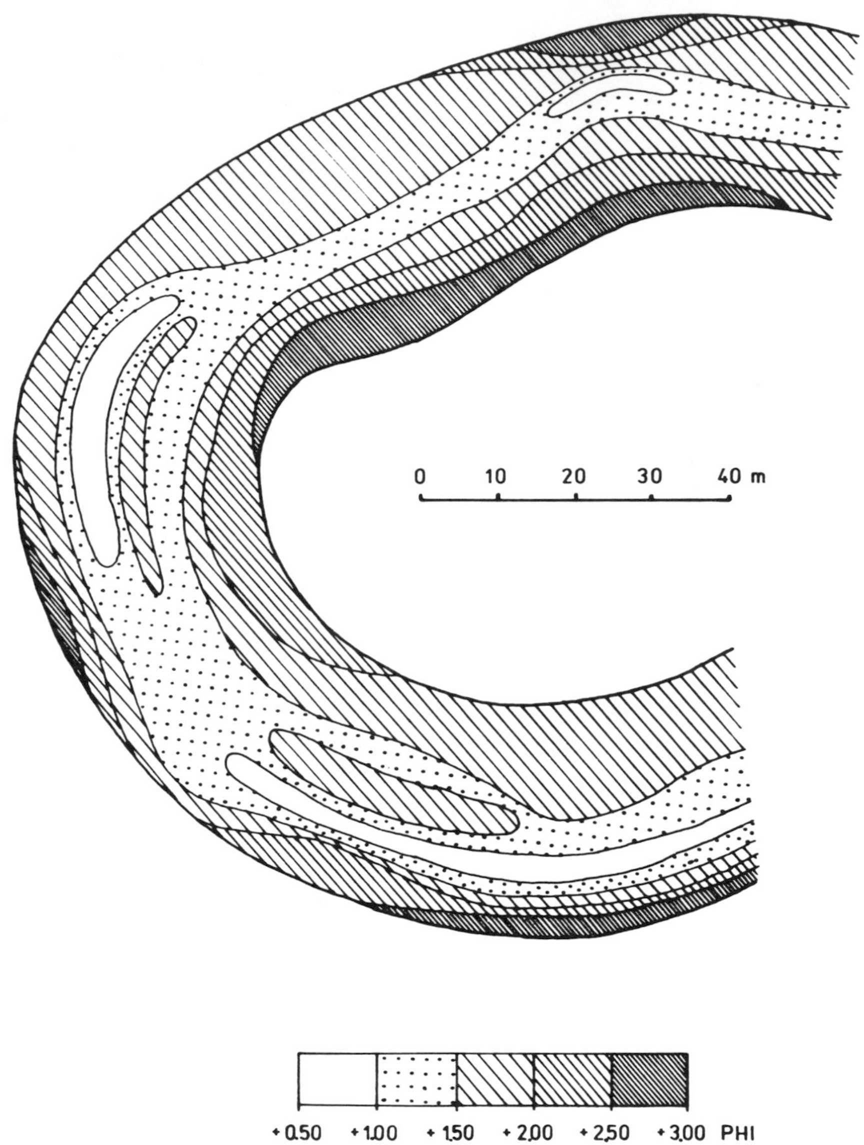

FIGURE 14.10 SPATIAL VARIABILITY OF THE MEAN DIAMETER OF THE MATERIAL CONSTITUTING THE BOTTOM OF A PALAEOMEANDER

CONCLUSIONS

Preliminary results of the application of semi-empirical formulae, especially the Chézy-Manning formula, to modelling the past discharges of meandering

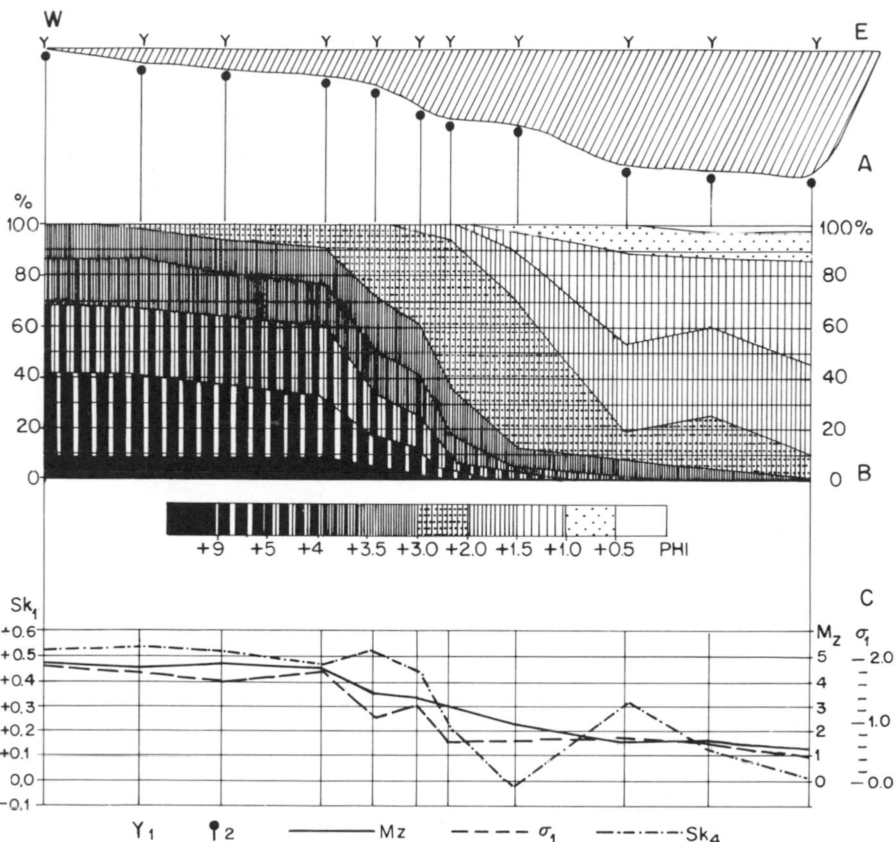

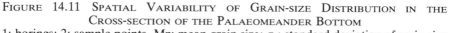

FIGURE 14.11 SPATIAL VARIABILITY OF GRAIN-SIZE DISTRIBUTION IN THE CROSS-SECTION OF THE PALAEOMEANDER BOTTOM
1: borings; 2: sample points. Mz: mean grain size; σ_1: standard deviation of grain size distribution; Sk_1: skewness of grain size distribution

rivers are promising and encourage further studies. In the near future the first phase will be completed of the studies of the history of quantitative changes of discharge in the Late Vistulian and Holocene on the North Polish Plain related to the factors that conditioned those changes. In future, one should check in a systematic way the application of steady uniform flow formulae for the modelling of past discharges of rivers of other climatic zones, as well as those of a wider range of such parameters as channel slope (s), roughness coefficient (n) and bankfull discharge (q_b).

It must be remembered that further improvement in modelling past river discharges depends on the state of our knowledge about laws governing modern stream flow and the controlling factors.

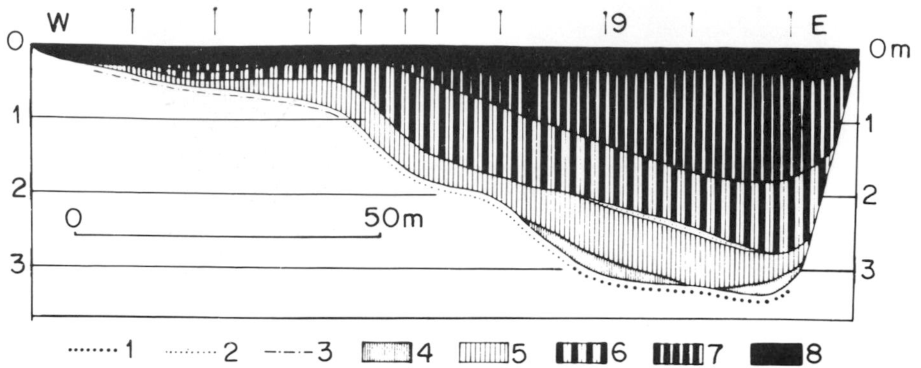

FIGURE 14.12 CROSS-SECTION OF A PALAEOMEANDER AND THE INFILL DEPOSITS
1: coarse-grained sand; 2: fine-grained sand; 3: very fine-grained sand and silt; 4: grey silt; 5: beige silt; 6: reed-sedge peat; 7: wood peat; 8: well decayed peat

ACKNOWLEDGEMENTS

The present research has been carried out within the IGCP Project No. 158 with the financial support of Adam Mickiewicz University in Poznań, and for the last two years also from the Institute of Geography of the Polish Academy of Sciences. I would like to express warm thanks to Professor K. J. Gregory of the University of Southampton and Professor L. Starkel of the Polish Academy of Sciences in Cracow, who inspired the present chapter, for their kind interest in the author's research in the field of quantitative palaeohydrology. Numerous discussions with Professor L. Starkel on palaeohydrologic issues in the last few years were a constant encouragement for me and I think that these discussions have been reflected in the present chapter. Special thanks are also due to my wife Jadwiga, whose extensive hydrologic knowledge was a constant and great help and a corrective of my thinking in terms of palaeohydrology. She assisted me in all the tedious computation work. I am also grateful to my younger colleagues: Dr. K. Borówka, Dr. Z. Młynarczyk, A. Wojciechowski (MSc), M. Borówka (MSc), W. Jaszkowski (MSc), the driver M. Janowicz, and my students, for the years of help in the difficult field examinations of about 20 palaeomeanders.

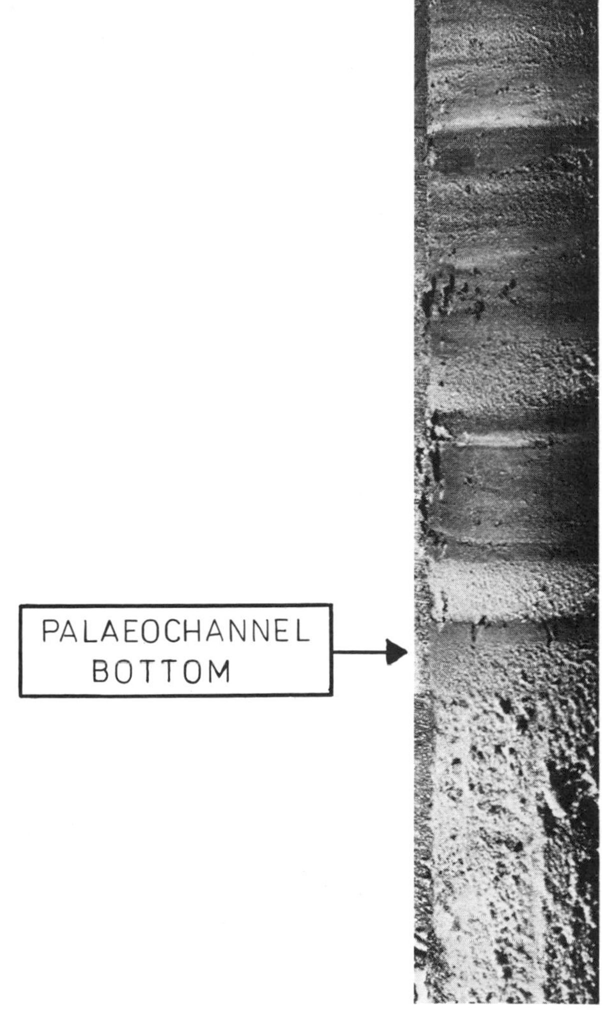

FIGURE 14.13 PALAEOMEANDER BOTTOM CORE

TABLE 14.5 DISCHARGE VALUES FOR MEANDERING CHANNELS OF VARIOUS AGES CALCULATED USING EQUATION (32)

	Palaeomeander 1	Palaeomeander 2	Meander 3	
Age	Younger Dryas	4 500–5 000 years B.P.	1980 A.D.	
Radius of meander curvature	270 m	25 m	30 m	
Channel width at bankfull discharge	129.4 m	17.0 m	18.5 m	
Hydraulic radius	1.98 m	1.573 m	1.450 m	
Cross-sectional area at bankfull discharge	258.48 m^2	29.83 m^2	32.2 m^2	
Roughness coefficient	0.020	0.035	0.045	
Channel slope	0.000377	0.00020	0.0004431	
			Measured	Computed
Mean flow velocity computed by means of equation (30)	0.525 m sec^{-1}	0.572 m sec^{-1}	0.699 m sec^{-1}	0.616 m sec^{-1}
Bankfull discharge computed by means of equation (32)	117.79 m^3 sec^{-1}	17.356 m^3 sec^{-1}	22.50 m^3 sec^{-1}	20.60 m^3 sec^{-1}
Mean annual discharge	23.56 m^3 sec^{-1}	3.47 m^3 sec^{-1}	4.50 m^3 sec^{-1}	4.12 m^3 sec^{-1}

APPENDIX LIST OF SYMBOLS USED IN THE CHAPTER

Constants: *c*; exponents: *z*

Subscripts: measured, observed, real = $_r$; computed = $_c$; computed by Chézy–Manning formula = $_{\text{Ma}}$

Present streams	Characteristics	Former streams
w	channel width	W
d	channel depths	D
r_h	hydraulic radius	R_h
p_e	wetted perimeter	P_e
f	width/depth ratio	F
n	roughness coefficient	N
γ	roughness coefficient in Bazin's formula	Γ
s	channel slope	S
v	mean flow velocity	V
a_c	cross-sectional area	A_c
l_m	meander wavelength	L_m
w_m	meander belt width	W_m
r_m	radius of meander curvature	R_m
ω	channel sinuosity	Ω
q_m	mean annual discharge	Q_m
q_b	bankfull discharge	Q_b
q_{maf}	mean annual flood	Q_{maf}
q_{max}	maximum flood discharge	Q_{max}
m	channel silt–clay percentage	M
m_c	silt–clay percentage in channel bottom	M_c
m_b	silt–clay percentage in channel bank	M_b
k	correlation coefficient	k
SE	standard error	SE
δ_n	standard deviation	

REFERENCES

Allen, J. R. L. 1977. Changeable rivers: some aspects of their mechanics and sedimentation. In Gregory, K. J. (Ed.), *River Channel Changes*, Wiley, Chichester, 15–45.

Baker, V. R. 1981. Palaeoflood hydrologic analysis from slackwater deposits. Symposium *Paleohydrology of the Temperate Zone, Abstracts of Papers*, Adam Mickiewicz University, 11–12.

Biswas, A. K. 1970. *History of Hydrology*, North-Holland Publishing Company.

Carlston, C. W. 1965. The relation of free meander geometry to stream discharge and its geomorphic implications. *American Journal of Science,* **263**, 864–885.

Chow, V. T. 1959. *Open-channel Hydraulics*, McGraw-Hill, New York, 655 pp.

Cowan, W. L. 1956. Estimating hydraulic roughness coefficients. *Agricultural Engineering,* **37**, 473–475.

Daniel, J. F. 1971. Channel movement of meandering Indiana streams. *United States Geological Survey Professional Paper 732–A,* Washington.
Dury, G. H. 1952. The alluvial fill of the valley of the Warwickshire Itchen near Bishop's Itchington. *Proceedings Coventry District Natural History and Scientific Society,* **2**, 180–185.
Dury, G. H. 1964a. Principles of underfit streams. *United States Geological Survey Professional Paper 452–A*, Washington.
Dury, G. H. 1964b. Subsurface exploration and chronology of underfit streams. *United States Geological Survey Professional Paper 452–B*, Washington.
Dury, G. H. 1965. Theoretical implications of underfit streams. *United States Geological Survey Professional Paper 452–C,* Washington.
Dury, G. H. 1976. Discharge prediction, present and former, from channel dimensions. *Journal of Hydrology,* **30**, 219–245.
Dury, G. H. 1977. Underfit streams: retrospect, perspect, and prospect. In Gregory, K. J. (Ed.), *River Channel Changes*, Chichester, 281–293.
Falkowski, E. 1965. History of Holocene and prediction of evolution of Middle Vistula valley from Zawichost to Solec. *Materiały Sympozjum: Geologiczne problemy zagospodarowania Wisły środkowej od Sandomierza do Puław*, SIT Górn. Zarząd Główny, Katowice, 45–61.
Falkowski, E. 1967. Evolution of the Holocene Vistula from Zawichost to Solec with an engineering–geological prediction of further development (Summary). *Biuletyn Instytutu Geologicznego,* **198**, 57–150.
Falkowski, E. 1970. History and prognosis for the development of bed configurations of selected sections of Polish Lowland Rivers (Summary). *Biuletyn Geologiczny,* **12**, 5–110.
Falkowski, E. 1972. Regularities in development of lowland rivers and changes in river bottoms in the Holocene. *Excursion Guidebook, Symposium INQUA Comm. Stud. of the Holocene,* **2**, PAN Poland, 3–30.
Gregory, K. J. 1981. Human activity and palaeohydrology. *Symposium 'Palaeohydrology of the temperate zone', Abstracts of Papers,* Adam Mickiewicz University, 25–26.
Hjulström, F. 1935. *Studies of the Morphological Activity of Rivers as Illustrated by the River Fyris*, Uppsala, 527 pp.
Inglis, C. C. 1941. Digest of answers to the Central Board of Irrigation questionnaire on meandering of rivers with comments on factors controlling meandering and suggestions for future action. In *Annual Report Technical*, 1939–1940, India, Central Board of Irrigation, Pub. 24, 100–114.
Inglis, C. C. 1949. *The Behaviour and Control of Rivers and Canals with the Aid of Models*, Poona, India, Central Waterpower Irrig. Navigation Research Sta., Research Pub. 13, pt. 1, 298 pp.
Jefferson, M. S. W. 1902. Limiting width of meander belts. *National Geographic Magazine,* **13**, 373–384.
Joppling, A. V. 1966. Some principles and techniques used in reconstructing the hydraulic parameters of a paleo-flow regime. *Journal of Sedimentary Petrology,* **36**, 5–49.
Kaczmarek, Z. 1970. *Statistical Methods in Hydrology and Meteorology* (in Polish), Warszawa, 312 pp.
Klimek, K., and Starkel, L. 1974. History and actual tendency of flood-plain development at the border of the Polish Carpathians. In *Geomorphologische Prozesse und Prozesskombinationen in der Gegenwart unter verschieden Klimabedingungen. Abhandlungen der Akademie der Wissenschaften in Göttingen*, 185–196.

Klimek, K., and Starkel, L. 1981. Some palaeohydrological reconstructions. In Starkel, L. (Ed.), *The Evolution of the Wisłoka Valley Near Dębica During the Late Glacial and Holocene*, Ossolineum, Kraków, 78–83.
Kochel, R. C., and Baker, V. R. 1982. Paleoflood hydrology. *Science,* **215**, 353–361.
Kozarski, S. 1981. River channel changes in the Warta valley to the south of Poznań. *Symposium 'Paleohydrology of the temperate zone', Guidebook of Excursions*, Adam Mickiewicz University, Poznań, 6–23.
Kozarski, S., and Rotnicki, K. 1977. Valley floors and changes of river channel patterns in the North Polish Plain during the Late Würm and Holocene. *Quaestiones Geographicae,* **4**, Adam Mickiewicz University, Poznań, 51–93.
Lambor, J. 1971. *Engineering Hydrology* (in Polish), Arkady, Warszawa, 362 pp.
Langbein, W. B. *et al.* 1949. Annual runoff in the United States. *United States Geological Survey Circular,* **52**, 14 pp.
Leopold, L. B., and Maddock, T. Jr. 1953. The hydraulic geometry of stream channels and some physiographic implications. *United States Geological Survey Professional Paper 252*, Washington.
Leopold, L. B., and Wolman, M. G. 1957. River channel patterns; braided, meandering and straight. *United States Geological Survey Professional Paper 282–B*, Washington, 39–85.
Leopold, L. B., Wolman, M. G., and Miller, J. P. 1964. *Fluvial Processes in Geomorphology*, Freeman, San Francisco, 522 pp.
Mycielska-Dowgiałło, E. 1969. An attempt at reconstructing the palaeohydrodynamics of a river, based on sedimentological studies in the Vistula Valley near Tarnobrzeg (Summary). *Przegląd Geograficzny*, **41**, 409–429.
Mycielska-Dowgiałło, E. 1972. Stages of Holocene evolution of the Vistula valley on the background of its older history, in the light of investigations carried out near Tarnobrzeg. *Excursion Guidebook. Symposium INQUA Comm. Stud. of the Holocene,* **2**, PAN, Poland, 69–82.
Mycielska-Dowgiałło, E. 1977. Channel pattern changes during the last glaciation and Holocene, in the northern part of the Sandomierz Basin and the middle part of the Vistula valley, Poland. In Gregory, K. J. (Ed.), *River Channel Changes*, Wiley, Chichester, 75–87.
Padgett, G. V., and Ehrlich, R. 1976. Palaeohydrologic analysis of a late Carboniferous fluvial system, southern Morocco. *Geological Society of America Bulletin,* **87**, 1101–1104.
Richards, K. S. 1973. Hydraulic geometry and channel roughness — a non-linear system. *American Journal of Science,* **273**, 877–896.
Rose, J., Turner, G., Coope, G. R., and Bryan, M. D. 1980. Channel changes in a lowland river catchment over the last 13,000 years. In Cullingford R. A., Davidson, D. A., and Lewin, L. (Eds.), *Timescales in Geomorphology*, Wiley, Chichester, 159–175.
Rotnicki, K. 1982. The method of retrodiction of former meandering river discharge and its significance for the investigation of river hydrology changes during the Holocene. *Abstracts,* **2**, *XI INQUA Congress*, Moscow, 269.
Schumm, S. A. 1960. The shape of alluvial channels in relation to sediment type. *United States Geological Survey Professional Paper, 352–B*, Washington.
Schumm, S. A. 1963. Sinuosity of alluvial rivers on the Great Plains. *Geological Society of America Bulletin,* **74**, 1089–1100.
Schumm, S. A. 1965. Quaternary palaeohydrology. In Wright, H. E., and Frey, D. G. (Eds.), *Quaternary of the United States*, Princeton University Press, 783–794.
Schumm, S. A. 1968. River adjustment to altered hydrologic regimen—Murrumbidgee River and palaeochannels. Australia. *United States Geological Survey Professional Paper 598*, Washington.

Schumm, S. A. 1969. River metamorphosis. Journal of the Hydraulics Division, **95**, *Proceedings of the American Society of Civil Engineers,* 251–273.
Schumm, S. A. 1971. Fluvial geomorphology. In Shen, H. W. (Ed.), *River Mechanics*, Fort Collins, Colorado.
Schumm, S. A. 1972. Fluvial palaeochannels. In Rigby, J. K., and Hamblin, W. K. (Eds.), *Recognition of Ancient Sedimentary Environments*, Society Economic Palaeontologists and Mineralogists, Special Publication, **16**, 98–107.
Starkel, L. (Ed.) 1981. The evolution of the Wisloka Valley near Dębica during the late glacial and Holocene. *Folia Quaternaria,* **53**, Ossolineum, Kraków, 91 pp.
Starkel, L., and Thornes, J. B. 1981. Palaeohydrology of river basins. *Technical Bulletin No. 28, British Geomorphological Research Group*, 107 pp.
Sundborg, A. 1967. Some aspects on fluvial sediments and fluvial morphology. I. General views and graphic methods. *Geografiska Annaler,* **49A**, 333–343.
Szumański, A. 1972. Changes in the development of the lower San's channel pattern in the Late Pleistocene and Holocene. *Excursion Guidebook. Symposium INQUA Comm. Stud. of the Holocene,* **2**, PAN Poland, 55–69.
Szumański, A. 1981. The evolution of the lower San river during the late glacial and the Holocene. *Symposium 'Paleohydrology of the temperate zone', Abstracts of Papers*, Adam Mickiewicz University, 81–82.
Thornes, J. B. 1977. Hydraulic geometry and channel change. In Gregory, K. J. (Ed.), *River Channel Changes*, Wiley, Chichester, 91–100.
Wisler, C. O., and Brater, E. F. 1959. *Hydrology*, Wiley, New York, 408 pp.

Background to Palaeohydrology
Edited by K. J. Gregory

15

River channel adjustment to climatic change in west central Poland

STEFAN KOZARSKI

Adam Mickiewicz University, Quaternary Research Institute, Poznań, Poland

In west central Poland called the Great Polish Lowland (Figure 15.1) most of the alluvial valley floors reveal palaeochannel scars which can be easily identified on aerial photographs (Kozarski, 1974a, 1976), as well as in the field where they are rather fresh forms in the valley floor topography. Palaeochannels in alluvial valley floors are a common feature of the North Polish Plain (Falkowski, 1971; Kozarski and Rotnicki, 1977), of the Sub-Carpathian Basins (Szumański, 1972; Trafas, 1975; Mycielska-Dowgiałło, 1977; Alexandrowicz *et al.*, 1981), and of the uplands (Harasimiuk and Henkiel, 1980).

Study of aerial photographs shows (Kozarski, 1974a, 1976) the presence of some regularities in the images of alluvial valley floors in west central Poland. Firstly, contrary to the flat middle and high terraces that sometimes exhibit braided palaeochannel scars and, as a rule, more or less linear erosional scarps, the low terraces and floodplains display clear signs of higher level of water flow organization in the form of meandering palaeochannels accompanied by well preserved point-bar ridges. Secondly, sets of palaeomeanders are of different sizes and at least two generations can be distinguished amongst them. Thirdly, freely-meandering palaeochannels and wide meander belts are characteristic features of alluvial valley floors with low gradients.

These regularities are extremely well manifested in the valley floor of the Warta River to the south of Poznań (Figure 15.1). Therefore detailed studies of this valley reach have been undertaken since 1967 and developed into the Warta Valley Research Project (Kozarski, 1981a). In 1977 it was included in the IGCP Project No. 158 'Palaeohydrological changes of the temperate zone in the last 15,000 years Sub-project A. Fluvial environment'.

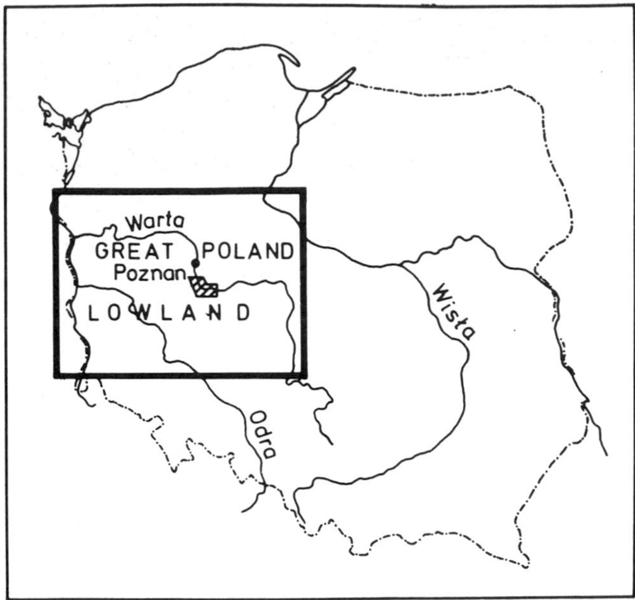

FIGURE 15.1 LOCATION MAP OF THE CASE STUDY AREA

Investigations of the case study area (Figure 15.1) have so far been concentrated on channel pattern changes and their age, as well as on vegetation history which has been studied by Dr. K. Tobolski and his co-workers. The first results of investigations have been summarized in a joint paper (Kozarski and Rotnicki, 1977) referring to the North Polish Plain. Results of the advanced Warta Valley Research Project have been presented for discussion during the international symposium 'Palaeohydrology of the temperate zone', Poznań, Poland 1981, and these provide a good background for the palaeohydrological approach.

STUDY AREA

The case study area covers a long reach of the Warta River valley (Figure 15.1) in its middle course between Nowe Miasto and Mosina, over a distance of 36 km. It lies to the south of Poznań in a major Pleistocene drainage track called the Warsaw–Berlin Pradolina, which divides west central Poland or the region of the Great Polish Lowland into two longitudinal parts. The location of the case study area offers some important advantages which include: firstly, position between two major extent lines of the last ice sheet with the maximum extent line in the south (Figure 15.2A); secondly, varied valley floor morphology with many palaeochannel scars; and thirdly, rich palaeochannel fills containing organic sediments suitable for palaeobotanical and radiocarbon dating.

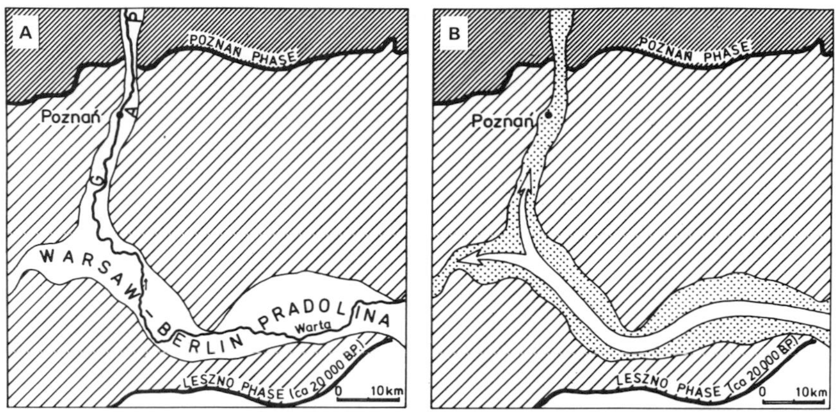

FIGURE 15.2 A– MAJOR EXTENT LINES OF THE LAST ICE SHEET AND THE WARSAW — BERLIN PRADOLINA IN THE SURROUNDINGS OF POZNAN B– WATER BIFURCATION IN THE WARSAW — BERLIN PRADOLINA (ca 17 000 to ca 13 000 years B.P.)

Geomorphology, ice sheet recession, and drainage system organization

The vast geomorphic background of the case study area belongs to the longitudinal landscape realm in the centre of the Great Polish Lowland, which consists of several geomorphic sub-units. The longitudinal extension of both the realm and geomorphic sub-units have been affected by events associated with the last Scandinavian glaciation called the 'Baltic Glaciation' in Poland. This was the most drastic phase in the entire Vistulian cold period. Events are well marked in the relief as subsequent ice front extent positions are represented by marginal forms with a parallel drainage track, the Warsaw-Berlin Pradolina between them (Figure 15.2A).

Moving from the south to north, the following three geomorphic sub-units can be distinguished (Figure 15.2A):

1. The marginal zone of the Leszno Phase which marks the maximum extent of the last ice sheet. Its age is estimated as *ca* 20,000 radiocarbon years B.P. (Kozarski, 1980a; Pazdur *et al.*, 1981). To the north of the marginal zone the area is predominantly covered with ground moraine. The surface of the ground moraine is flat or slightly undulating. In many places it is dissected by subglacial channels.
2. The Warsaw–Berlin Pradolina which is genetically and chronologically connected with the extent of the last ice sheet called the Poznań Phase. The Pradolina was the main drainage track during this phase.
3. The marginal zone of the Poznań Phase manifested in the relief as thrust end moraines and/or as moraine-like ridges and hillocks built of fluvioglacial deposits and ablation tills. The marginal zone was formed during a mostly steady state position of the retreating ice front after the maximum extent of the ice sheet. In many places outwash plains stretch

southwards from the marginal zone to the Warsaw–Berlin Pradolina and they represent the proglacial drainage system. Flat and undulating ground moraine surfaces are another common feature in the front of the Poznań Phase marginal zone. Ground moraine areas with numerous minor end moraine ridges and outwash plains prevail to the north (Kozarski, 1962). They were the basic forms used for the reconstruction of the post-imposed drainage system and the formation of the Warta River gap (Kozarski, 1962 and earlier literature quoted in this book; Witt, 1975).

The first organized drainage system coincides chronologically with the ice sheet position during the Poznań Phase which in the stratigraphical sequence belongs to the Upper Pleni Vistulian (Kozarski, 1980a). According to new investigations (Witt, 1975 and older literature quoted in this paper) at that time proglacial waters gathered along the drainage track of the Warsaw–Berlin Pradolina were oriented more or less parallel to the ice sheet margin. Shortly after the Poznań Phase, the Warta River cut through end moraines to the north due to ice sheet recession and formed a meridional gap (Figure 15.2A). This event took place during the Chodzież Sub-phase (Witt, 1975) and was recognized by studies on river terraces and related outwash plains.

In the evolution of the Upper Pleni Vistulian drainage system during the Chodzież Sub-phase an important process occurred, namely bifurcation of the pradolina waters. One part of the water flowed westward following the pradolina track, and the other flowed northward through the gap (Figure 15.2B). This peculiar situation of bifurcating waters lasted till the beginning of the Late Vistulian. Studies of palaeochannel fills indicate that the westward oriented pradolina drainage track was abandoned presumably during the Bølling Interstadial.

Valley floor morphology

The valley floor in the case study area consists of three low levels (Bartkowski, 1957; Kozarski, 1974b, 1981a; Kozarski and Rotnicki, 1977) namely:

1. Bifurcation terrace 66–65 m a.s.l. which comprises the abandoned bottom of the Warsaw–Berlin Pradolina and terrace remnants in the gap (Bartkowski, 1957).
2. Transitional terrace sloping down the valley from 66 to 61 m a.s.l.
3. Floodplain sloping downstream from 60 to 58 m a.s.l.

No distinct boundary between the bifurcation terrace and the transitional one can be observed in the field. Instead, an extremely gentle slope becomes evident only from contour lines on topographic maps. On the other hand, a low scarp 2–4 m high separates the transitional terrace from the floodplain. The bifurcation terrace is the oldest morphological level on which scars of palaeochannels were identified by means of aerial photographs. They reveal a

FIGURE 15.3 WARTA RIVER BRAIDED AND MEANDERING PALAEOCHANNEL PATTERNS

NW oriented braided pattern (Figure 15.3) typical of the Upper Pleni Vistulian drainage in pradolinas generated by periglacial climatic conditions.

The Warta River

According to Polish hydrological classifications (Mikulski, 1963) the Warta River is a river with a second order drainage basin. Its length (L) is 808 km and the drainage area (A) covers 53,710 km^2. The headwater parts of the Warta River are situated in the Jurassic Kraków–Częstochowa Upland, but the dominant part of the drainage basin occupies the lowlands and therefore, in hydrological descriptions (Mikulski, 1963, Dynowska, 1971) the Warta River is regarded as a lowland river. It is the main drainage line of west central Poland.

The Warta River has a regionally varying hydrologic regime. In the hydrologic typology of Polish rivers (Dynowska, 1971) the Warta River has a uniform regime in the headwater parts, but only a broadly uniform regime with specific, very low summer discharges and spring peak discharges, prevails in the middle and lower course. More information on the Warta River regime close to the case study area is provided by data from the gauging station in Poznań. It is situated not far from the so-called basin centre of gravity (Chorley, 1973) because nearly half of the drainage basin (25,053 km^2) lies upstream. For the years 1951–1965 the mean annual discharge (Q) was 82.1 m^3 s^{-1}, the maximum discharge observed (Q_{max}) reached 706 m^3 s^{-1}, and the minimum discharge (Q_{min}) only 12.3 m^3 s^{-1}. The computed mean annual, maximum and minimum runoff values attain: 3.27 l s^{-1} km^2, 28.1 l s^{-1} km^2 and 0.49 l s^{-1} km^2, respectively. Peak discharges and floods are mostly associated with snow melt. In a long-term analysis for the years 1850–1969 for the cross-section of Srem which is the centre of the case study area, Michałowska (1973) estimated the highest frequency (68 per cent) of peak discharges and floods for the three months, February, March, and April. On the other hand, very low discharges were recorded for September and October. Present-day flood discharges of the Warta River appear effective in the vertical accretion of the floodplain (Michałowska, 1973).

LATE VISTULIAN AND HOLOCENE CHANNEL PATTERNS

As mentioned above, the recent stage of investigations has been concentrated on changes of channel pattern and upon their age. For this purpose a standard investigation procedure was used which involved three stages. Firstly, aerial photograph interpretation including measurements of geometrical parameters (channel width and curvature radius of meandering palaeochannels), as well as a comparison of detected palaeochannels with their images on topographic maps at scales 1:25,000 and 1:10,000. Secondly, geomorphological mapping and detailed stratigraphical investigation, especially of the palaeochannel fills. The latter were based on excavations in the braided palaeochannels and on borings in palaeomeander fills. Ten to twenty boreholes were made along several cross-section lines with a modified 'Instorf' core-sampler in each of the selected palaeomeanders. For age determinations of palaeomeander fills complete cores were taken from the deepest parts which were pools, and samples were collected only from organic matter containing layers of braided palaeochannel fills as available in excavations. Thirdly, laboratory investigation of sediment samples taken from palaeochannel fills included grain-size analysis of mineral deposits, determination of calcium carbonate content, loss-on-ignition of organic matter, and also age determinations by palaeobotanical methods and radiocarbon measurements.

Braided palaeochannels

Braided palaeochannel scars occur only on the bifurcation terrace. From a detailed analysis by Antczak (1981a) it follows that they are present in the topography as elongated hollows and low bars. The average heights of the bars range within 0.5–0.8 m above the channel bottom. The mean sinuosity coefficient for major palaeochannels and the coefficient of braiding after Brice (1964) were estimated (Antczak, 1981a) as 1.06 and 8.84, respectively.

Palaeochannel fills and bars consist largely of coarse sands and gravels. The mean particle size for the fills was determined as 0.8 mm and such deposits must be interpreted as bed-load material. This can also be inferred to a large extent from analysis of the structure of deposits in pits (Antczak, 1981a). Sedimentary units occurring there belong to the large-scale trough cross-stratification and tabular cross-stratification which is typical of a braided river depositional environment. According to Antczak (1981a) dip orientations measured within these sedimentary units are placed in the NW sector, and show a general flow direction which tends to follow the general orientation of palaeochannel scars. Antczak (1981a) reports also that palaeochannel sediments reveal characteristics of cyclothems which can be related to flow conditions in a braided river. Gravels and coarse sands occur at the bottom of each cyclothem whereas silty deposits complete it at the top. The structure and particle size of palaeochannel bars are similar to those of the fills. Perhaps the most striking difference is a greater admixture of fine particles in the top strata.

Among typical channel fill deposits of braided patterns grey silts containing sporomorphes and plant macroremains are also known (Antczak, 1981b). These were deposited in the stagnant water of an abandoned channel. The palaeobotanical analysis carried out by Dr. K. Tobolski is not yet finished but the absolute dating of macroremains offers promising results. The age given by two samples is of significance: 12,770 ± 190 years B.P. (Gd–924) and 12,630 ± 160 years B.P. (Gd–945), because it means that the grey silts are a deposit which belongs to the Bølling Interstadial (Table 15.3). At that time the braided channel system was already abandoned by the river. Thus, the system is older than the Bølling Interstadial and was functioning until the Oldest Dryas cold period.

Meandering palaeochannels

In the early stages of research there were several geomorphological reasons for dividing the meandering palaeochannel pattern into two groups (Kozarski, 1974a, 1976; Kozarski and Rotnicki, 1977). These reasons included the fact that a glance at aerial photographs reveals differences in the size of palaeomeanders. Large palaeomeanders occur farther from the Warta River, and smaller ones closer to the river. Furthermore the palaeomeanders exist at two levels: the larger are characteristic features of the transitional terrace, whereas the smaller remain typical for the floodplain. In addition a low erosional scarp separates the levels and both groups of palaeomeanders. For these reasons the two groups of meandering channel were called generations

TABLE 15.1 GEOMETRICAL PARAMETERS OF PALAEOMEANDERS (in m)

Generation	r_m	b	d	F	Cross-section area in m^2
Older	247.1	58.9	3.64	16.2	107.2
Younger	140.6	45.0	2.17	20.7	48.8

(Kozarski, 1974a, 1976; Kozarski and Rotnicki, 1977), and differences in their age have been presumed.

Primarily (Kozarski, 1974a; Kozarski and Rotnicki, 1977) two geometrical parameters were measured and computed to describe the palaeomeanders, namely mean radius of curvature (r_m) and mean channel width (b). When the study palaeomeander fills developed, an increased number of borings allowed the estimation of a third important parameter, which was mean channel depth (d). The currently corrected geometrical parameters in comparison with older values (Kozarski, 1974b; Kozarski and Rotnicki, 1977) together with some new values are given in Table 15.1. The search for mean channel depth (d) is a very time-consuming procedure as it requires hundreds of borings to be made in palaeomeander fills. Therefore the parameter d must be regarded as a first estimate, the accuracy of which will develop with a further increase in the number of borings available. The varying mean depth values will of course affect the width/depth ratio (F) and the cross-section area, and thus the palaeohydrological reconstruction of discharges. Results already obtained show that there are major differences in the size of palaeomeanders which belong to the initially distinguished two generations. Thus, there must have been substantial differences in the discharges which sculptured palaeochannels during successive periods of valley floor formation.

From detailed stratigraphical study of palaeomeander fills conclusions can be drawn about the periods mentioned. Fills in the deepest parts (pools) of older palaeomeanders reveal a more or less constant general sequence of sediments (Figure 15.4) varying in thickness. A layer of fine, medium and/or coarse sand, the former bed-load (Gonera, 1981), lies at the bottom. It is overlain by silts from several centimetres to as much as 2 m thick. The silts are covered with a layer of different kinds of gyttja (calcareous gyttja, coarse gyttja, sporadically black gyttja) which also ranges in thickness from a few centimetres up to 3 m. The topmost layer is represented by peats including Carex peat and wood peat. Detailed examination of cores indicates that the sequence in many boreholes includes sand lenses with organic material or silts and sandy silts. A good example of a sedimentary rich core was analysed in an earlier study (Kozarski and Rotnicki, 1977, Figure 5). According to present-day observations of processes filling oxbow lakes, thin and discontinuous sand and silt intercalations in palaeomeander fills can be

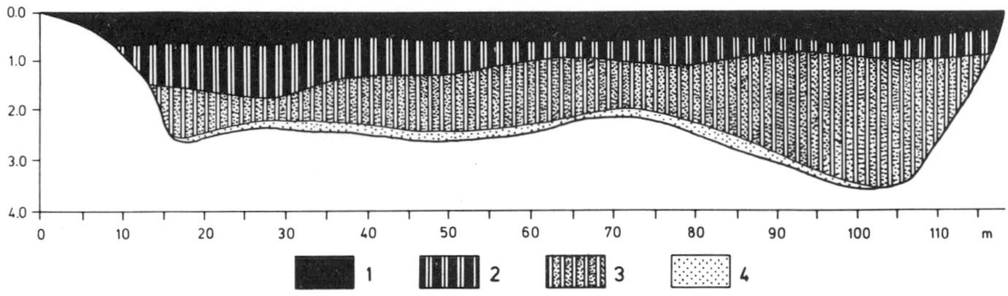

FIGURE 15.4 GENERAL SEQUENCE OF FILLS IN PALAEOMEANDERS OF OLDER GENERATION 1—peat, 2—gyttja, 3—silt, 4—sand

interpreted as the depositional consequences of floods extending into former oxbow lakes.

For consideration of channel changes in time it is most important to have age determinations of the bottom parts of the cores taken from the palaeomeander fills. For age determinations the palaeobotanical record was preferred with complementary radiocarbon measurements because of the possibility of accidental contamination (Okuniewska and Tobolski, 1982; Kozarski, 1980b, 1981b). The dated fills of selected older palaeomeanders contain in their bottom parts sediments deposited during different periods of the Late Vistulian (Table 15.3, logs 2 to 4), i.e. the Older Dryas in site Jaszkowo (Tobolski, 1981), at least the Allerød Interstadial in site Mechlin (Okuniewska and Tobolski, 1981) and the Younger Dryas in site Zbrudzewo I (Czerniak *et al.*, 1981). Therefore the older generation of palaeomeanders belongs to a channel system which was functioning during the Late Vistulian from the Bølling Interstadial till the end of the Allerød Interstadial.

The younger generation of the palaeomeanders differs in the stratigraphy of fills from the older one. At the bottom there is a layer of fine, medium and/or coarse sand which can be interpreted (Kozarski, 1981a) as stream-bed material. On average, larger particles occur in the pool or close to it, whereas smaller ones cover the rising channel bed to the convex bank. The basal layer is usually overlain by a sandy peat veneer several centimetres thick, but sporadically it is covered with a thin gyttja-like deposit. Above the basal layer and the sandy peat veneer, the dominant deposit in cross-sections (Figure 15.5A and B) is peat interbedded with sandy silt and silt. The entire thickness of this kind of fill in pools ranges between 2.2 and 4.0 m.

The palaeobotanical record (Czerniak *et al.*, 1981; Okuniewska and Tobolski, 1981) and radiocarbon measurements (Kozarski, 1980b), with one exception due to contamination, demonstrate a much younger age of palaeomeander fills on the floodplain. The data presented in Table 15.3 (logs 5 to 7) show that the basal parts of the fills correspond with the Boreal Period and the Sub-Boreal Period. Thus, the younger generation of palaeomeanders

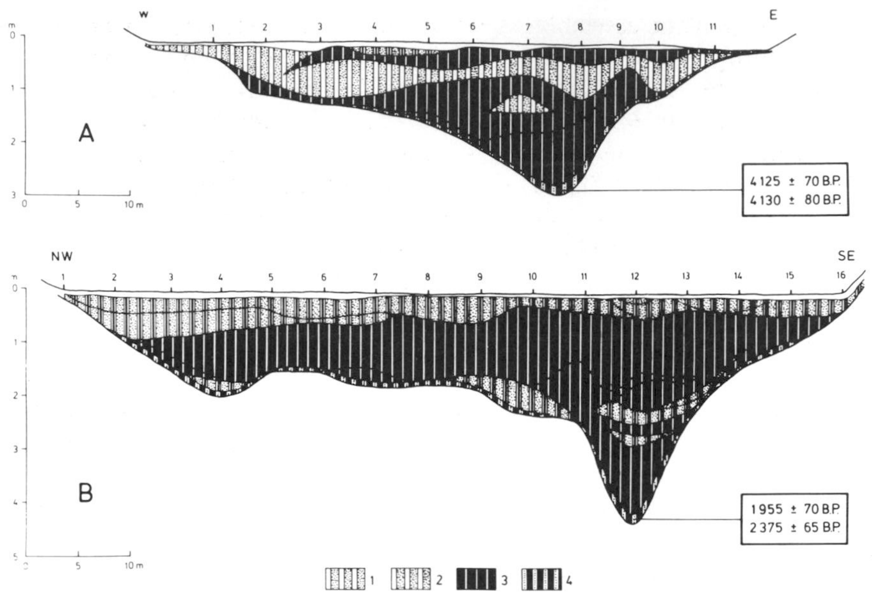

FIGURE 15.5 SEQUENCE OF FILLS IN HOLOCENE PALAEOMEANDERS AT SITE CZMONIEC
(after M. Otolski, partly modified) A– basal layer deposited during the middle part of the Sub-Boreal Period (SB_2); B– basal layer deposited during the upper part of the Sub-Boreal Period (SB_3) 1—silt, 2—sandy silt, 3—peat, 4—sandy peat

represents a channel pattern developed during the Holocene (Kozarski, 1980b, 1981a).

CHANNEL ADJUSTMENT TO CLIMATE CHANGES

Vegetation and climate changes

From the compiled (Kozarski, 1980a, 1981c) palaeobotanical record it can be concluded that after the recession of the last ice sheet prior to the Late Vistulian there was a pre-vegetation period in west central Poland. This belongs to the Upper Pleni Vistulian (Table 15.3) and at that time permafrost reappeared in areas abandoned by the last ice sheet. This is documented by already known (Kozarski, 1971, 1974c, 1980a, 1981) and new, as yet unpublished, findings of ice-wedge casts and fossil frost cracks.

The first immigration of plants into west central Poland corresponds with the Oldest Dryas (Tobolski, 1966; Borówko-Dłuzakowa, 1969). In general the vegetation cover developed as a shrub tundra with 50 per cent of herbaceous plants among plant assemblages on average. The subsequent warming of climate during the Bølling Interstadial was a short-term but very important climate change. Pollen diagrams (Tobolski, 1966; Borówko–Dłuzakowa, 1969) record the first immigration of trees, birch and pine, for this interstadial. The vegetation cover became a park tundra with first forest patches. Frenzel (1980) concludes that during the European Late Glacial Period there were two important transitions from cold to warm climatic conditions; the first one from Oldest Dryas to Bølling and the other from Younger Dryas to Holocene. In both cases the climatic alteration was of rather short duration since it occupied 300 or 400 years.

Frenzel's (1980) opinion is well supported for the entire northwest of Poland by radiocarbon measurements of organic deposits derived from the Bølling Interstadial (Table 15.2). The expansion of vegetation was really a very rapid process in such a vast area stretching from the central realm of the country to the Baltic Coastal Region and obviously this influenced the runoff ratio very significantly. Park tundra assemblages also dominated during cold periods of the Older Dryas and Younger Dryas (Tobolski, 1966; Borówko–Dłuzakowa, 1969). Climatic conditions throughout these stadials must have been severe because the Late Vistulian ice-wedge casts and fossil frost cracks, both epigenetic and intraformational, found in northwest Poland (Kozarski, 1971, 1974c) prove the occurrence of permafrost. On the other hand, during the warmest Late Vistulian interstadial, i.e. the Allerød, climatic amelioration promoted the development of birch–pine forests (Tobolski, 1966; Borówko–Dłuzakowa, 1969; Szafrański, 1973). This was the first compact forest cover after the glaciation. Subsequently, vegetation density decreased during the Younger Dryas cold period but forests reappeared quickly in Early Holocene times.

Although Frenzel (1980) is correct to distinguish two important transitions from cold to warm periods at the turn of Upper Pleni Vistulian/Late Vistulian and Late Vistulian/Holocene, the first appearance of forests during the Allerød Interstadial cannot be overlooked in a discussion of river behaviour and its response to climate change. The case study area provides evidence for the assumption that altered conditions of runoff and sediment yield, especially bed load transport, were reflected in changed palaeochannel geometry at that time. Anticipating the forthcoming discussion it must be stressed that climate and type of vegetation change during the Allerød Interstadial were more significant for river behaviour than analogous changes at the Late Vistulian/Holocene boundary.

Summarizing the above information on plant cover changes, two events in Late Vistulian vegetation development must be underlined for further consideration. Firstly, the development of park tundra during the Bølling Interstadial, and secondly the first appearance of forests during the Allerød Interstadial. The third event in vegetation development which is important for

TABLE 15.2 RADIOCARBON AGE OF BØLLING INTERSTADIAL DEPOSITS IN NW POLAND

Site	Location	^{14}C age B.P.	References
Konin	E of the case study area	12 980 ± 130	Barówko-Dłużakowa, 1969
Żabinko	Case study area	12 630 ± 160	Antczak, 1982 (unpubl.)
Niechorze	Baltic coast	12 920 ± 330	Kopczyńska-Lamparska, 1976

river behaviour is of Holocene age, and is the occupation of the land by dense forests.

River metamorphosis and channel changes

Although the geomorphological, geological and palaeobotanical record so far obtained is mostly qualitative in character, it seems to be sufficient for a discussion of two basic, closely interrelated, problems in the area of the case study. Firstly, in the question of river metamorphosis and channel changes in response to climate fluctuations, and secondly the general palaeohydrological trend from the Upper Pleni Vistulian till the Holocene.

Geomorphic facts and age determinations of palaeochannel fills show that river metamorphosis and channel changes varied with time. The following sequence of changes in channel organization was reconstructed (Table 15.3):

TABLE 15.3 AGE OF CHANNEL CHANGES RELATED TO VEGETATION TYPES AND CHRONOSTRATIGRAPHICAL UNITS

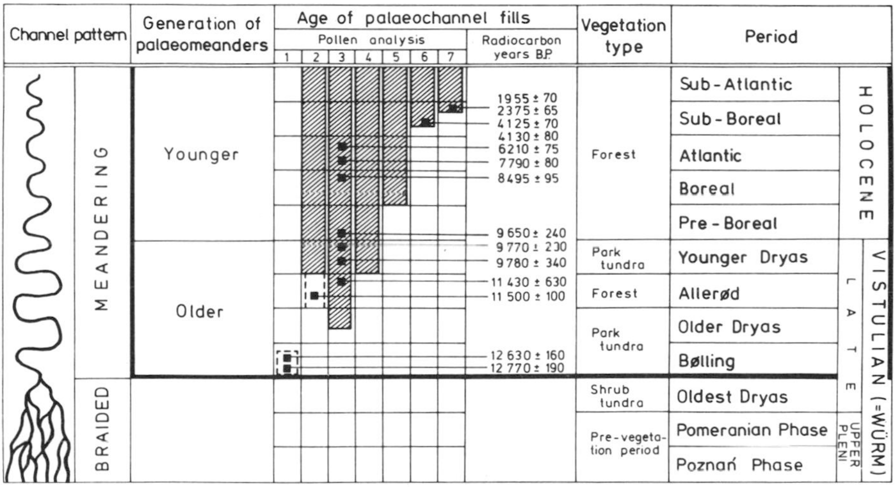

Channel pattern	Generation of palaeomeanders	Age of palaeochannel fills: Radiocarbon years B.P.	Vegetation type	Period		
MEANDERING	Younger	1955 ± 70 2375 ± 65	Forest	Sub-Atlantic	HOLOCENE	
		4125 ± 70 4130 ± 80		Sub-Boreal		
		6210 ± 75 7790 ± 80		Atlantic		
		8495 ± 95		Boreal		
		9650 ± 240		Pre-Boreal		
	Older	9770 ± 230 9780 ± 340	Park tundra	Younger Dryas	LATE	VISTULIAN (=WÜRM)
		11 430 ± 630 11 500 ± 100	Forest	Allerød		
			Park tundra	Older Dryas		
		12 630 ± 160 12 770 ± 190		Bølling		
BRAIDED			Shrub tundra	Oldest Dryas		
			Pre-vegetation period	Pomeranian Phase	UPPER PLENI	
				Poznań Phase		

Pollen analysis columns: 1, 2, 3, 4, 5, 6, 7.

Cores and/or samples taken at sites/location in Figure 15.3/:
1—Żabinko, 2—Mechlin, 3—Jaszkowo, 4—Zbrudzewo I,
5—Zbrudzewo II, 6—Czmoniec A, 7—Czmoniec B

Upper Pleni Vistulian — braided channel
Late Vistulian — meandering channel (large meanders)
Holocene — meandering channel (small meanders)

Boundaries between these periods of different channel pattern occurrence are clearly determined by palaeobotanical analyses and/or radiocarbon measurements. In the chronostratigraphical scheme of the Great Polish Lowland (Kozarski, 1980a, 1981c) the first boundary lies between the Oldest Dryas and the Bølling Interstadial. It is well demonstrated by the fact that in the braided palaeochannel silty fill, organic matter is contained which was deposited during the Bølling Interstadial (Tables 15.2 and 15.3, log 1). Obviously the braided channels must have been abandoned at the time when the deposition of the silty fill with organic remains took place. The silts are deposits of stagnant water. This permits the conclusion that the braided channel system was active during the Oldest Dryas and prior to it. From the age of the basal layers of large and small palaeomeander fills assigned to the Younger Dryas and the Boreal Period, respectively (Table 15.3, logs 4 and 5) an inference can be made about the other boundary, which lies at the turn of Late Vistulian to Holocene.

The above boundaries coincide with major periods of vegetation change. It is not an accidental but a causal coincidence, since in the hierachy of variables (Schumm, 1977) vegetation (type and density) remains the first variable among the dependent variables. This variable as an environmental factor reflects climatic and edaphic conditions in a complex form. In palaeoenvironmental reconstructions it seems sufficient to use this factor as a basis for inferences about river behaviour and channel pattern change instead of more precise estimation of palaeoclimatic parameters like temperature, precipitation, transpiration, and speculations on runoff/precipitation ratios. Difficulties in this respect are discussed by Frenzel (1980).

During the pre-vegetation period and the domination of shrub tundra assemblages during the Oldest Dryas, braided channels were a typical pattern in the area under discussion. Runoff conditions, discharges and sediment yield were strongly controlled by permafrost which enabled only overland flow and throughflow in the active layer with no infiltration at all. The overland flow favoured intensive removal of sediments from bare or nearly bare slopes and their yield to the rivers. But most of the sediment yield was derived from the unconsolidated glacial deposits due to lateral erosion of the major braided river. It is reported from periglacial regions (Church, 1972) that periglacial rivers carry sometimes over 90 per cent of their total load as bedload sediments, that is coarser material. As has been mentioned before such material, gravel and coarse sand, is characteristic of the braided system in the study area. Braided bedload channels have very great width/depth ratios (Schumm, 1968) as manifested in the Warsaw–Berlin Pradolina by shallow and wide palaeochannel scars, as well as by the vast bifurcation terrace. The latter is the result of prevailing lateral erosion and aggradation of the Upper Pleni Vistulian braided river.

Abrupt vegetation changes after the Oldest Dryas promoted the main impulses which induced allogenic (Lewin, 1977) channel alterations. The transition from shrub tundra to park tundra must have produced threshold conditions (Brunsden and Thornes, 1979) within the system under which an extrinsic threshold (Schumm, 1977, 1979) was crossed and river metamorphosis (Schumm, 1971, 1977) from braided to meandering took place. It was the most dramatic river and channel change during the entire timespan of climatic changes and channel adjustment discussed in this paper. After the Bølling Interstadial the river channel remained meandering.

The previously discussed geometric parameters and geomorphic features were the basis for the distinction of two generations among meandering palaeochannels. Dated sequences, especially the palaeobotanical record (Table 15.3) show that threshold conditions within the meandering system appeared twice: firstly, during the Allerød Interstadial when there was the first forest expansion after the last glaciation; and secondly during the Pre-Boreal Period when there was the beginning of Holocene forest development after park tundra ceased to dominate at the very end of the Late Vistulian (Younger Dryas). When the first threshold conditions were crossed the river responded with incision into the transitional terrace. This is evidenced by the existence of an erosional scarp between the transitional terrace and the floodplain, as well as by the Younger Dryas age of the basal layer of the palaeomeander fill in one of the large palaeomeanders (Table 15.3, log 4). The river responded to the second threshold condition by decreasing the size of its channel (Table 15.1).

It has been clearly pointed out by Schumm (1973) that changes of river channel morphology depend on the magnitude of changes in discharge and sediment load. The decreasing size of palaeomeanders in the Warta River valley is a measurable fact because characteristic geometric parameters of younger palaeomeanders, i.e. channel width (b), channel depth (d), and mean radius of curvature (r_m) are smaller. Radius of curvature in regression equations shows (Leopold and Wolman, 1960) a linear relation to meander wavelength (λ). This allows the use of geometrical record in terms of Schumm's (1977) equations. The decreasing values of channel width (b^-), channel depth (d^-), meander wavelength (λ^-), and channel cross-section area of smaller meanders related to larger ones in the Warta River valley (Table 15.1) must have resulted from a situation in which average magnitude of both discharge and sediment load decreased due to an increased vegetation density and decreased denudation in the river basin. The only discrepancy in reconstructed and compared parameters is revealed in the width/depth ratio (F) but it will be calculated again after continued borings in palaeomeander fills are finished. The nature of meandering channel change in the Warta River valley at the transition from the Late Vistulian to the Holocene can be expressed by one of the Schumm's (1977, p. 136) equations:

$$Q^- Q_s^- \simeq b^-, d^-, \lambda^-, S^\pm P^+, F^-$$

with the restriction established above in reference to the width/depth ratio (F). The gradient (S) of the floodplain in comparison with the transitional terrace is decreased in the Warta River valley. On the other hand the sinuosity index (P) seems to be increased because of the decreased mean radius of curvature (r_m).

The river metamorphosis and channel changes discussed permit the conclusion that a general hydrological trend occurred in the middle segment of the Warta River valley from the Upper Pleni Vistulian until the younger part of the Holocene and decreasing average quantities of water passed through the river channel as it was adjusting to climate changes. The Warta River was not an exception. A similar trend can be observed in many Polish lowland rivers (Kozarski and Rotnicki, 1977). Although this conclusion has a qualitative character it seems to provide an important background for more advanced palaeohydrological reconstruction because of the well differentiated and dated channel changes in the Warta River valley.

CONCLUSIONS

The case study in the Warta River valley provides an interesting example of how dated channel changes related to the evaluation of palaeobotanical record can be explained as system responses to climate fluctuations. They also offer the possibility of discovering threshold conditions which occurred in relatively short timespans and which induced channel changes. Detailed investigations and dating of palaeochannel fills, together with the evaluated palaeobotanical record which has been evaluated, can at level 1 enrich the palaeohydrological "think back model" proposed by Baker (1978). It has been proved in this study that 'discontinuities between stable periods are not to a large degree related to relatively abrupt climate changes', as pointed out by Knox (1975), but they are directly related to such changes, and what is more, in shorter time scales than Holocene. Dated sequences in the Warta River valley reveal two changes during 7 000 years.

Concentration on the dating of palaeochannel fills at an early stage of research gives a chance to discover and use better defined timespans in channel change reconstruction. It seems to be a very valuable part of the research procedure because it eliminates an unnecessarily general treatment of fluvial episodes and period determinations like 'late Pleistocene' or 'late Quaternary' which are climatically heterogeneous; errors in sampling of mineral deposits for advanced reconstructions of past hydrologic regimes; and premature construction of curves of erosion and deposition which usually display incorrect correlations with known time scales.

In age determinations of palaeochannel fills a combined procedure of palaeobotanical analysis and radiocarbon dating should be preferred (Kozarski, 1980b, 1981). Due to contamination there are examples of strong deviations from results of pollen analysis in the radiocarbon age of organic deposits (Okuniewska and Tobolski, 1980) in the case study area. Therefore dating of palaeochannel fills cannot be exclusively based on radiocarbon measurements.

REFERENCES

Alexandrowicz, S. W., Klimek, K., Kowalkowski, A., Mamakowa, K., Niedziałkowska, E., Pazdur, M., and Starkel, L. 1981. The evolution of the Wisłoka valley near Dębica during the late glacial and Holocene. *Folia Quaternaria*, **53**, 91 pp.

Antczak, B. 1981a. Morphology and deposits within the braided palaeochannel pattern, site 2: Zabno. In Kozarski, S., and Tobolski, K. (Eds.), *Symposium 'Palaeohydrology of the temperate zone', Poznań, Poland'81, Guide-book of excursions*, 26–28.

Antczak, B. 1981b. Deposits and braided channel scars in the bifurcation terrace, site 1: Zabinko. In Kozarski, S., and Tobolski, K. (Eds.), *Symposium 'Palaeohydrology of the temperate zone', Poznań, Poland '81, Guide-book of excursions*, 24–25.

Baker, V. R. 1978. Adjustment of fluvial systems to climate and source terrain in tropical and subtropical environments. In Miall, A. D. (Ed.), *Fluvial Sediment ology*, Canadian Society of Petroleum Geologists Memoir 5, 211–230.

Bartkowski, T. 1957. Rozwój polodowcowej sieci hydrograficznej w Wielkopolsce środkowej (Zfs.: Die Entwicklung des post-glazialen Entwässerungssystems in Mittleren Grosspolen). *Zeszyty Naukowe Uniwersytetu im. A. Mickiewicza, Geografia 1*, Poznań.

Brice, J. C. 1964. Channel patterns and terraces of the Loup River in Nebraska. *US Geological Survey Professional Paper 422–D*.

Brunsden, D., and Thornes, J. B. 1979. Landscape sensitivity and change. *Transactions, Institute of British Geographers, New Series*, **4**, 463–484.

Borówko–Dłuzakowa, Z. 1969. Palynological investigations of late glacial and Holocene deposits at Konin. *Geographia Polonica*, **17**, 267–281.

Chorley, R. J. 1973. The drainage basin as the fundamental geomorphic unit. In Chorley, R. J. (Ed.), *Introduction to Fluvial Processes*, 30–52.

Church, M. 1972. Baffin Island sandurs; a study of Arctic fluvial processes. *Geological Survey Canada Bulletin*, **216**, 208 pp.

Czerniak, A., Okuniewska, I., and Tobolski, K. 1981. Palaeobotanical investigation of palaeomeander fills at Zbrudzewo. In Kozarski, S., and Tobolski, K. (Eds.), *Symposium 'Palaeohydrology of the temperate zone', Poznań, Poland '81, Guide-book of excursions*, 44–49.

Dynowska, I. 1971. Typy reżimów rzecznych w Polsce (Summary: Types of river regimes in Poland). *Zeszyty Naukowe Uniwersytetu Jagiellońskiego, Prace Geograficzne*, **28**, 155 pp.

Falkowski, E. 1971. Historia i prognoza rozwoju układu koryta wybranych odcinków rzek nizinnych Polski (Summary: History and prognosis for the development of bed configurations of selected sections of Polish lowland rivers). *Biuletyn Geologiczny*, **12**, 5–121.

Frenzel, B. 1980. Klima der Letzten Eiszeit und der Nacheiszeit in Europa. *Veröffentlichungen der Joachim Jungius-Gesellschaft für Wissenschaften in Hamburg*, **44**, 9–46.

Gonera, P. 1981. Morphodynamic interpretation of channel deposits from palaeomeanders of the Warta River near Mechlin. In Kozarski, S., and Tobolski, K. (Eds.), *Symposium 'Palaeohydrology of the temperate zone', Poznań, Poland '81, Guide-book of excursions*, 36–38.

Harasimiuk, M., and Henkiel, A. 1980. The influence of neotectonics upon valley floor development: a case study from the Wieprz valley, Lublin Upland, *Quaestiones Geographicae*, **6**, 35–53.

Knox, J. C. 1975. Concept of the graded stream. In Melhorn, W. N., and Flemal, R. C. (Eds.), *Theories of Landform Development*, Publications in Geomorphology, State University of New York, Binghamton, 169–198.

Kopczyńska-lamparska, K. 1976. Radiocarbon datings of the late glacial and Holocene deposits of western Pomerania. *Acta Geologica Polonica*, **26**, No. 3, 413–417.

Kozarski, S. 1962. Recesja ostatniego lądolodu z północnej części Wysoczyzny Gnieźnieńskiej a kształtowanie się pradoliny Noteci-Warty (Summary: Recession of the last ice sheet from the northern part of Gniezno Pleistocene plateau and formation of the ice-marginal valley of rivers Noteć-Warta). *Poznańskie Towarzystwo Przyjaciół Nauk, Prace Komisji Geograficzno-Geologicznej*, 2, 3, Poznań, 154 pp.

Kozarski, S. 1971. Ślady działalności późnowürmskich procesów peryglacjalnych w regionie dolnej Odry i dolnej Warty (Summary: Traces of the activity of Late Würm periglacial processes in the Lower Odra and Lower Warta Region). *Badania fizjograficzne nad Polska zachodnia*, **24**, Ser. A. Geogr. fiz., 97–177.

Kozarski, S. 1974a. Późnoglacjalne i holoceńskie zmiany w ukladzie koryt rzecznych niżowej części dorzecza Odry (Late glacial and Holocene changes of river channel patterns in the lowland part of the Odra drainage basin). *Krajowe Sympozjum 'Rozwój den dolinnych...etc.', Wrocław–Poznań, Streszczenia referatów i komunikatów*, 17–19.

Kozarski, S. 1974b. Stanowisko Jaszkowo koło Śremu. Migracje koryta Warty na południe od Poznania w późnym glacjale i holocenie — Generacje meandrów (Site Jaszkowo near Srem. Migrations of the Warta channel to the south of Poznań during Late glacial and Holocene — Generations of meanders). *Krajowe Sympozjum 'Rozwój den dolinnych...etc.', Wrocław–Poznań, Przewodnik wycieczki*, 46–49.

Kozarski, S. 1974c. Evidence of Late Würm permafrost occurrence in northwest Poland. *Quaestiones Geographicae*, **1**, 65–85.

Kozarski, S. 1976. Air-photo interpretation in the geomorphological survey of valley floors in the North Polish Plain. *Československa Akademia Ved, Geografický Ustav Brno, Studia Geographica*, **55**, 79–81.

Kozarski, S. 1980a. An outline of Vistulian stratigraphy and chronology of the Great Poland Lowland. *Quaternary Studies in Poland*, **2**, 21–35.

Kozarski, S. 1980b. The Holocene generation of palaeomeanders in the Warta valley, Great Poland Lowland. *Geologisches Jahrbuch*, Ser.A, Hannover (in press).

Kozarski, S. 1981a. River channel changes in the Warta valley to the south of Poznań. In Kozarski, S., and Tobolski, K. (Eds.), *Symposium 'Palaeohydrology of the temperate zone', Poznań, Poland '81, Guide-book of excursions*, 6–23.

Kozarski, S. 1981b. Generations of palaeomeanders, site 5: Zbrudzewo. In Kozarski, S., and Tobolski, K. (Eds.), *Symposium 'Palaeohydrology of the temperate zone', Poznań, Poland '81, Guide-book of excursions*, 41–43.

Kozarski, S. 1981c. *Stratygrafia i chronologia Vistulianu Niziny Wielkopolskiej* (Summary: Vistulian stratigraphy and chronology of the Great Poland Lowland). Polska Akademia Nauk Oddział w Poznaniu, Ser.: Geografia, 6, Państwowe Wydawnictwo Naukowe Warszawa–Poznań, 44 pp.

Kozarski, S., and Rotnicki, K. 1977. Valley floors and changes of river channel patterns in the North Polish Plain during the Late Würm and Holocene. *Quaestiones Geographicae*, **4**, 51–93.

Leopold, L. B., and Wolman, M. G. 1960. River meanders. *Bulletin of the Geological Society of America*, **71**, 769–794.

Lewin, J. 1977. Channel pattern changes. In Gregory, K. J. (Ed.), *River Channel Changes*, J. Wiley and Sons, 167–184.

Michałowska, E. 1973. Współczesne procesy akumuacyjne na terasie zalewowej Warty koło Sremu (Present-day aggradation processes on the Warta River floodplain near Srem). Unpubl. MS thesis.

Mikulski, Z. 1963. *Hydrografia Polski*, Państwowe Wydawnictwo Naukowe, Warszawa, 287 pp.

Mycielska-Dowgiałło, E. 1977. Channel pattern changes during the last glaciation and Holocene, in the northern part of the Sandomierz basin and the middle part of the Vistula Valley, Poland. In Gregory, K. J. (Ed.), *River Channel Changes*, J. Wiley and Sons, 75–87.

Okuniewska, I., and Tobolski, K. 1982. Wstępne wyniki badań palaeobotanicznych osadów z dwóch paleomeandrów w dolinie Warty koło Poznania (Summary: Preliminary results of palaeobotanical investigations of deposits from two palaeomeanders in the Warta Valley near Poznań). *Badania fizjograficzne nad Polska zachodnia*, **34**, ser.A.Geogr.fiz. 149–160.

Okuniewska, I., and Tobolski, K. 1981. Palaeobotanical investigations of deposits from the palaeomeander of the younger generation at Czmoniec. In Kozarski, S., and Tobolski, K. (Eds.), *Symposium 'Palaeohydrology of the temperate zone', Poznań, Poland '81, Guide-book of excursions*, 52–56.

Pazdur, M. F., Stankowski, W., and Tobolski, K. 1981. Litologiczna i stratygraficzna charakterystyka profilu z kopalynymi utworami organogenicznymi w Malińcu koło Konina (Summary: Lithological and stratigraphical description of a profile with organic deposits at Maliniec near Konin). *Badania fizjograficzne nad Polska zachodnia*, **33**, ser.A.Geogr.fiz., 79–80.

Szafrański, F. 1973. Roślinność Wielkopolskiego Parku Narodowego w późnym glacjale i holocenie w świetle badań palynologicznych nad osadami Jeziora Budzyńskiego (Summary: Vegetation of the Great Poland National Park during late glacial and Holocene in the light of palynological investigations of deposits from the Budzyńskie Lake). *Folia Quaternaria*, **42**, 1–36.

Schumm, S. A. 1968. River adjustment to altered hydrologic regimen — Murrumbidgee River and palaeochannels, Australia. *US Geological Survey Professional Paper*, **598**, 65 pp.

Schumm, S. A. 1971. Fluvial geomorphology: channel adjustment and river metamorphosis. In Shen, H. W. (Ed.), *River Mechanics*, **I**, Fort Collins, Colorado, 5.1–5.22.

Schumm, S. A. 1973. Geomorphic implications of climatic changes. In Chorley, R. J. (Ed.), *Introduction to Fluvial Processes*, 202–211.

Schumm, S. A. 1977. *The Fluvial System*, J. Wiley and Sons, 338 pp.

Schumm, S. A. 1979. Geomorphic thresholds: the concept and its applications. *Transactions Institute of British Geographers*, New Series 4, No. 4, 485–515.

Szumański, A. 1972. Changes in the development of the Lower San channel pattern in the Late Pleistocene and Holocene. *Symposium of the INQUA Commission on Studies of the Holocene 'Changes in the palaeogeography of valley floors of the Vistula drainage basin during Holocene', Guide-book of excursion, 2nd part-The Polish Lowland*, 55–58.

Tobolski, K. 1966. Późnoglacjalna i holoceńska historia roślinnośc na obszarze wydmowym w dolinie środkowej Prosny (Summary: Late glacial and Holocene history of vegetation in the dune area of the Middle Prosna Valley). *Poznańskie Towarzystwo Przyjaciół Nauk, Prace Komisji Biologicznej*, **22(1)** 69 pp.

Trafas, K. 1975. Zmiany Koryta Wisły na wschód od Krakowa w świetle map archiwalnych i fotointerpretacji (Summary: Changes of the Vistula river bed east of Cracow in the light of archival maps and photointerpretation). *Zeszyty Naukowe Uniwersytetu Jagiellońskiego, Prace Geograficzne 40*, 80 pp.

Witt, A. 1975. Rekonstrukcja kierunku odpływu wód w poziomie najwyższej terasy przełomowego odcinka Warty pod Poznaniem (Summary: Reconstruction of direction of water outflow in the highest terrace level of gap of the Warta near Poznań). *Badania fizjograficzne nad polska zachodnia*, **27**, ser.A.Geogr.fiz., 179–208.

Background to Palaeohydrology
Edited by K. J. Gregory

16

Floodplain deposits and accelerated sedimentation in the lower Severn basin

A. G. Brown

Geography Department Southampton University

Studies of the geomorphological history of floodplains can help to bridge the temporal gap between the backward extrapolation of the results gained from studies of geomorphological processes, and traditional studies of Quaternary landforms and sediments. This may allow contemporary processes and rates to be evaluated on a chronological scale of several millennia. The integrated study of the fill of a large river valley such as the Severn can also help to bridge the gap between process studies from small basins and studies of Quaternary landform development at the regional scale. The sites described in this chapter were investigated as part of a study of the floodplain sediments of the Lower Severn Valley, conducted under the auspices of the International Geological Correlation Programme (IGCP Sub-Project 158A: Palaeohydrological Changes in the Temperate Zone in the Last 15 000 Years: Fluvial Environments). However, the approach taken sought to exploit the overlap between palaeoecological studies of recent palaeoenvironments, as described and standardized by Berglund (1979–1982) in the IGCP 158B guides (Lake and Mire Environments), and standard geomorphological methodology. This interdisciplinary approach was particularly applicable because of the generally fine nature of the floodplain sediments under investigation. The aim of the studies described here was to produce inferential palaeohydrological information as distinct from empirically-based palaeohydrological estimates, from palaeohydraulics of palaeochannel geometry, or estimates deduced from palaeohydrological modelling as typified by Lockwood's (1979) attempt to compute a water balance equation for the Atlantic and Devensian.

METHODOLOGY

The techniques used to investigate the valley fill sediments included the construction of a borehole databank, field investigation, and sampling of sediments using a variety of coring and augering devices. The databank contained information from 96 boreholes located on the floodplain between Bewdley (GR SO 787754) and Longney (GR SO 760140) extending over a valley distance of just under 100 km. One line or computer card was used for each borehole with various geomorphological and sedimentary parameters being used as variables. The approach taken was similar to that used by Rhind and Sissons (1971) for the databanking of drift borehole records in the Edinburgh area. The first six variables described the surface location and the other seventeen described the depths and thicknesses of the fill types. In order to keep the matrix as small as possible the fill sediments had to be classified into eight types: sand and gravel, sand, silt, organics (usually peat), semi-organic silt/clay, and bedrock. The predominant sediment type of each unit noted in the borehole log was used as the sediment type.

Field methods included the use of an extension auger, adapted for use with Russian corer rods, and bank sections, as well as a variety of chambered coring devices. The description and diagrammatic representation of sediments in the field followed a modified version of the Troels–Smith method, as advocated by Barber (1976) and Aaby (1979). For the detailed sampling of organic sediments, Russian fixed chamber corers were used where possible, often in conjunction with a pit dug through superficial inorganic sediments. Samples were routinely subject to loss-on-ignition, estimation of the $CaCO_3$ content, and to grain size analysis using dry sieving and a coulter counter. Standard palaeoecological methods were used (Berglund, 1979–1982) with minor modifications (Brown, 1983).

THE MAIN VALLEY FILL

Sections published by Beckinsale and Richardson (1964) suggested a relatively stable channel throughout the later Holocene with the predominant deposition of silts, clays, and peat. From boreholes abstracted from the borehole databank six additional valley cross-sections were constructed (Brown, 1983), two of which are shown in Figure 16.1. At the base of the Worcester Link Road section a broken and weathered zone of Keuper Marl was frequently encountered, often grading into a mixed unit of silts and clays with shattered lithorelics of Keuper Marl material. Above this lies a dense to loose gravel with sand which is laterally variable and which occasionally was found to contain silt, sand and clay lenses. Above this occurs a mixed silt with frequent organic lenses and shells, overlain by a red/brown silty clay with few organic remains. Horizontal laminations were commonly recorded in both the grey silts and the

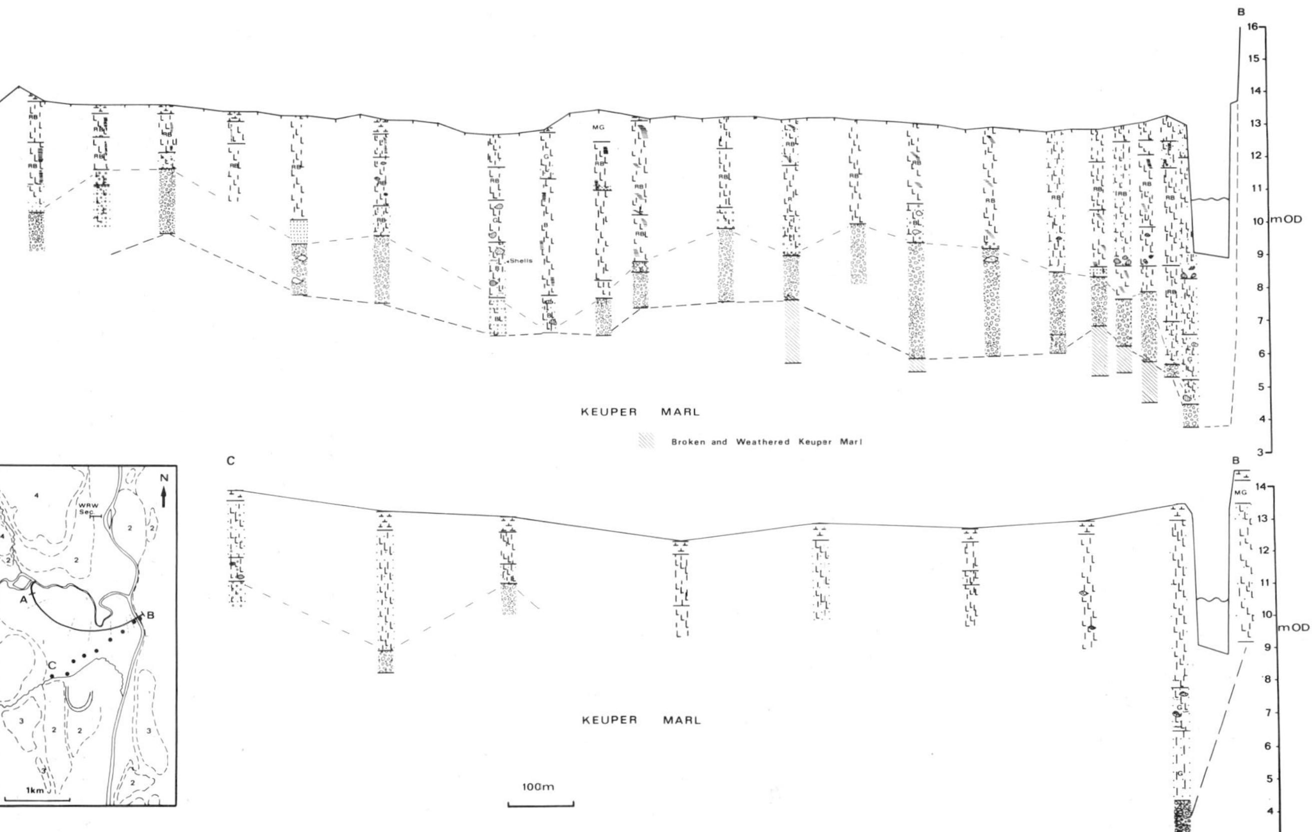

Figure 16.1 Worcester Link Road (A–B) and Foul Sewer (C–B) Sections

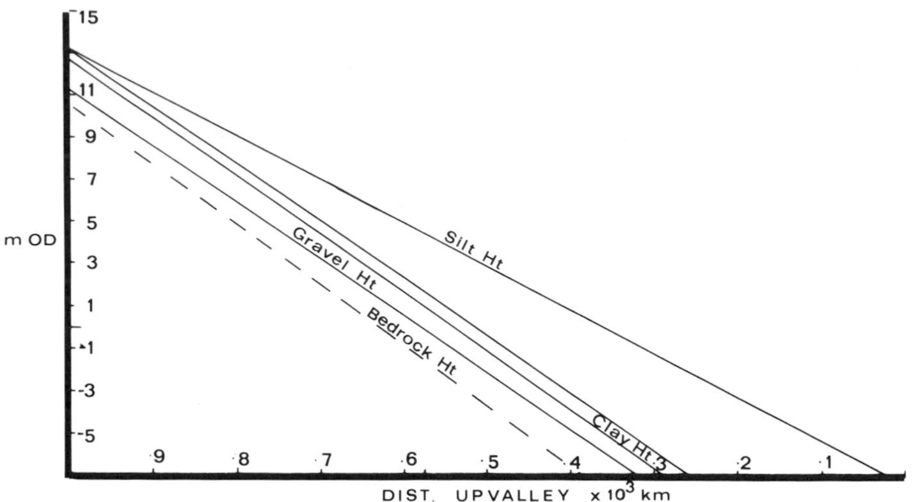

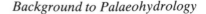

FIGURE 16.2 FLOODPLAIN SEDIMENT TYPE, STATISTICAL SLOPES

red brown silty clay. From a cross-section to the north at Worcester, Beckinsale and Richardson (1964) described a basal gravel as "probable Power House or Power Station gravels". However, it is relevant that Wills (1938) did not regard the Power Station Terrace as one unit and united under this term all the very low terrace occurring above Worcester. Therefore dates relating to Power Station Terrace from the Stourport area (Wilden; Coope, and Shotton, 1981) are not necessarily applicable to basal gravel units within the fill to the south of Worcester.

Using the borehole databank and bivariate regression a crude model of the longitudinal stratigraphic sequence was constructed and is depicted in Figure 16.2. Despite low and variable R^2 values, and the undoubtedly non-linear upstream variations, the diagram does give a simplified picture of the major unit slopes employing all the available data. The sand unit slope is not surprisingly similar to that of the generally underlining gravels. The clay unit slope is also indistinguishable from the sand and gravel, and sand unit slopes. However the silt slope is considerably less, and it is suggested that this is due to rapid overbank deposition in response to an altered hydrological situation with an increase in the importance of more local sedimentary sources. Preliminary Markov analysis of the borehole data suggests a strongly deterministic element within the reach stratigraphy. The procedure used, as described by Davis (1973), involved the combining of the silt and clay categories and the exclusion of bedrock. Miall (1973) has shown the applicability of Markov chain analysis to both coal measure cyclothems and

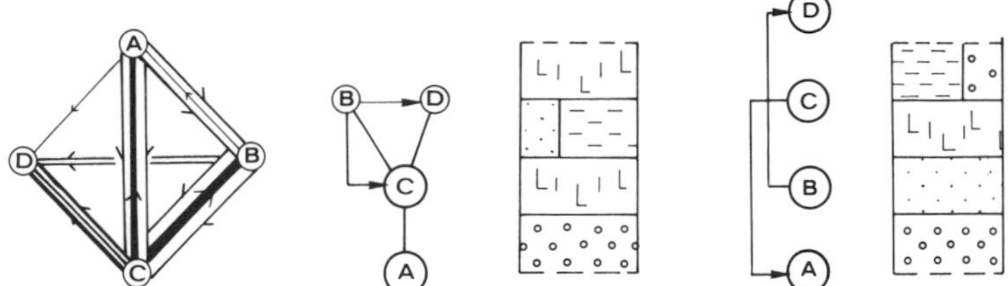

FIGURE 16.3 DIAGRAMMATIC REPRESENTATION OF TRANSITION PROBABILITIES AND RESULTANT FLOODPLAIN SEQUENCES

fluvial fining-upward cycles. The analysis revealed two preferred pathways (Figure 16.3): a typical fining-upward sequence culminating in a peat, and more commonly a sequence omitting the sand unit and progressing from sand and gravel to silts and clays, to peat, and then to silts and clays again.

From the cross-sections and the borehole data it is suggested that sand and gravel deposition carried on into the early Holocene, after which clays were deposited which were often semi-organic and of varying thickness. The loss-on-ignition of these sediments varies between 5–10 per cent, and they are interpreted as the deposits of backswamps and semi-lacustrine facies. Within this unit are occasional sand lenses, probably caused by channel migration or crevasse splays. The commonly occurring middle (fine) unit is more organic, and is often a peat, with loss-on-ignition figures varying from below 17 per cent to 65 per cent and with obvious, and in some cases identifiable, plant remains. Peat within this macro-unit is often interstratified with silty clays. The environment of deposition is interpreted as backswamp marsh and, as revealed by pollen and macrofossil analysis, it was covered by alder carr and reed beds. The top macro-unit is in most cases an inorganic silt with often a fine sand component. The unit is typically a relatively well sorted silt with a mean grain size of between 5 and 7ϕ, but it is coarser in more marginal locations or in tributary valleys. The unit has been referred to as the buff red silty clay by Shotton (1978) who postulates prehistoric land use change as the cause of its rapid sedimentation, which began around 2600 B.P. Shotton (1978) also suggested that the major source area for the sediment was the local Keuper and Bunter lowlands. The vegetation history of this southerly and easterly portion of the basin is poorly known because the majority of the pollen diagrams within and adjacent to the basin are restricted to its western and upland fringes (Figure 16.4). Therefore in order to investigate the palaeoenvironmental history pertinent to changes in valley deposition primary sites were located close to the floodplain edges or in tributary valleys within the lowland zone. The predominant type of site used was a peat filled palaeochannel.

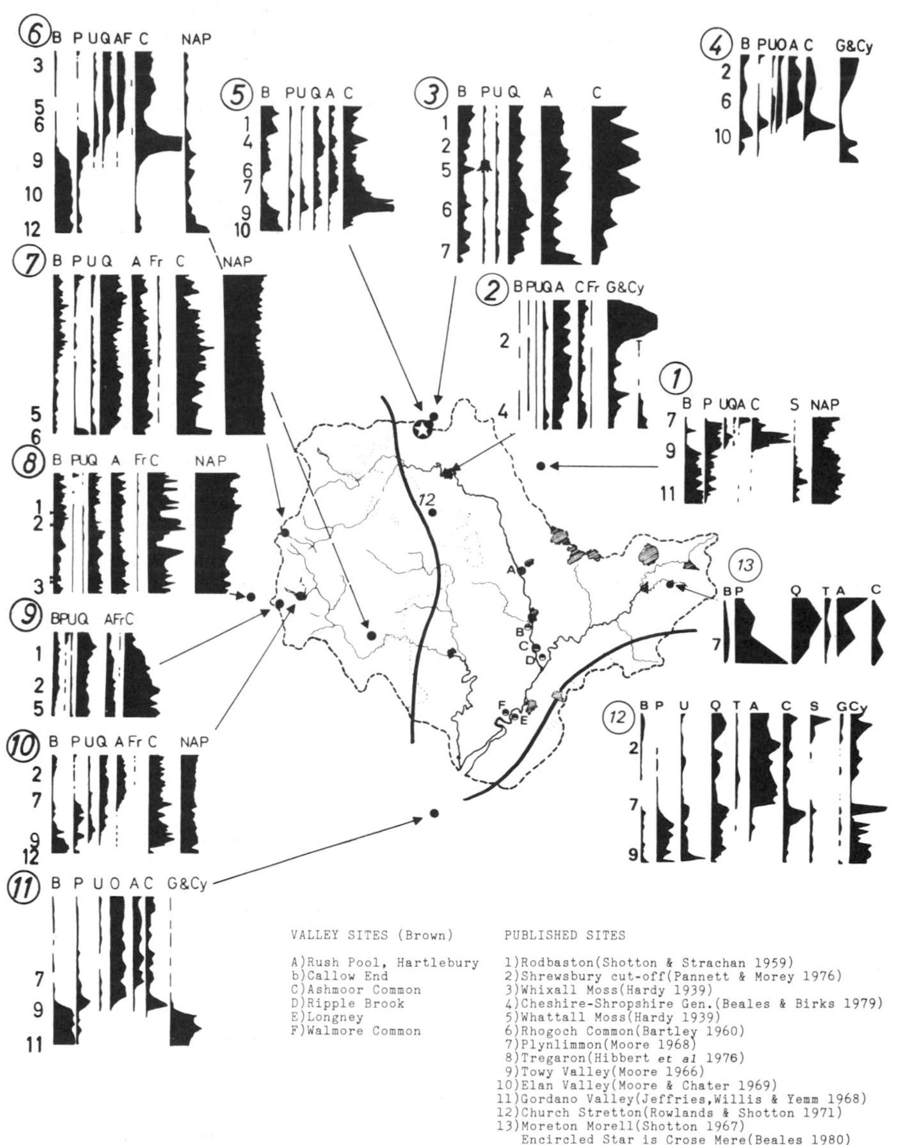

FIGURE 16.4 SEVERN BASIN, PRIMARY SITE LOCATIONS AND SIMPLIFIED PUBLISHED POLLEN DIAGRAMS

THE TERRACE AND LOWLAND ENVIRONMENT

Five sites will be discussed here, and the palaeoecological evidence of three of the sites has also been briefly discussed by Brown (1982), while further details of all the sites may be found in Brown (1983).

At Hartlebury Common near Stourport a peat-filled terrace palaeochannel yielded a pollen diagram primarily reflecting the terrace environment, which is under 10 m from the core. The bog is located on the First Terrace (Wills, 1926) and it is situated 600 m from the River Severn, although the terrace depressions were most likely formed by an ancient course of the River Stour. A ^{14}C date from a pine log just above the peat base gave an age of 9710 ± 130 B.P. (HAR 4842) which suggests that peat growth began during the Pre-Boreal period. Although the typical sequence of Holocene pollen fluctuations can be seen (Figure 16.5), a typical Atlantic phase, often dominated by alder and oak and below the appearance of cereals, seems to be poorly developed. A hiatus in peat deposition is possible and further ^{14}C dating is under way; however, there is no stratigraphic evidence for a hiatus and neither are there abrupt changes in the pollen curves. The diagram suggests that the terrace vegetation was dominated first by pine and birch and then hazel with a subsequent rise in grasses. The diagram shows how recent vegetation change on the terrace such as the expansion of ericaceous species was accompanied by an inwash of sand. The site is surrounded by a blown sand complex (Wills, 1926) from which the basal sand within the lower sandy peat was derived, and probably deposited by wind. However, grain size studies of the upper sandy peat revealed a less well sorted grain size distribution with a considerable fine tail, suggestive of inwashing from the surrounding soils. This inwash may have been associated with destruction of the vegetation cover (especially trees and scrub) caused by 18th century sand and gravel exploitation. Erosion of the dry sandy soils of the Common is a serious problem today, with large areas with no vegetation cover at all. Further to the south, the palaeoecological sequence upon the terraces alters as lime becomes the dominant taxon of the mid-Holocene.

At Ashmoor Common (7 km south of Worcester) a relict tract of floodplain has filled with autochthonous peat over the last 6000 years (basal date 5930 ± 70 B.P., HAR 4350). As can be seen from the stratigraphy (Figure 16.6) organic deposition may have started at approximately the same time as the main fill, but due to its protected situation this deposition has continued up to the surface which is now at a level above the main floodplain. From this site a pollen diagram with four ^{14}C dates (Brown, 1983) shows the deforestation of the surrounding terrace woodland at about 3600 B.P. and a much later two stage clearance of the floodplain alder woodland. This two stage, or stepped, alder clearance is clearly seen at two other floodplain sites, Ripple Brook and Callow End, and it is suggestive of some form of management, probably

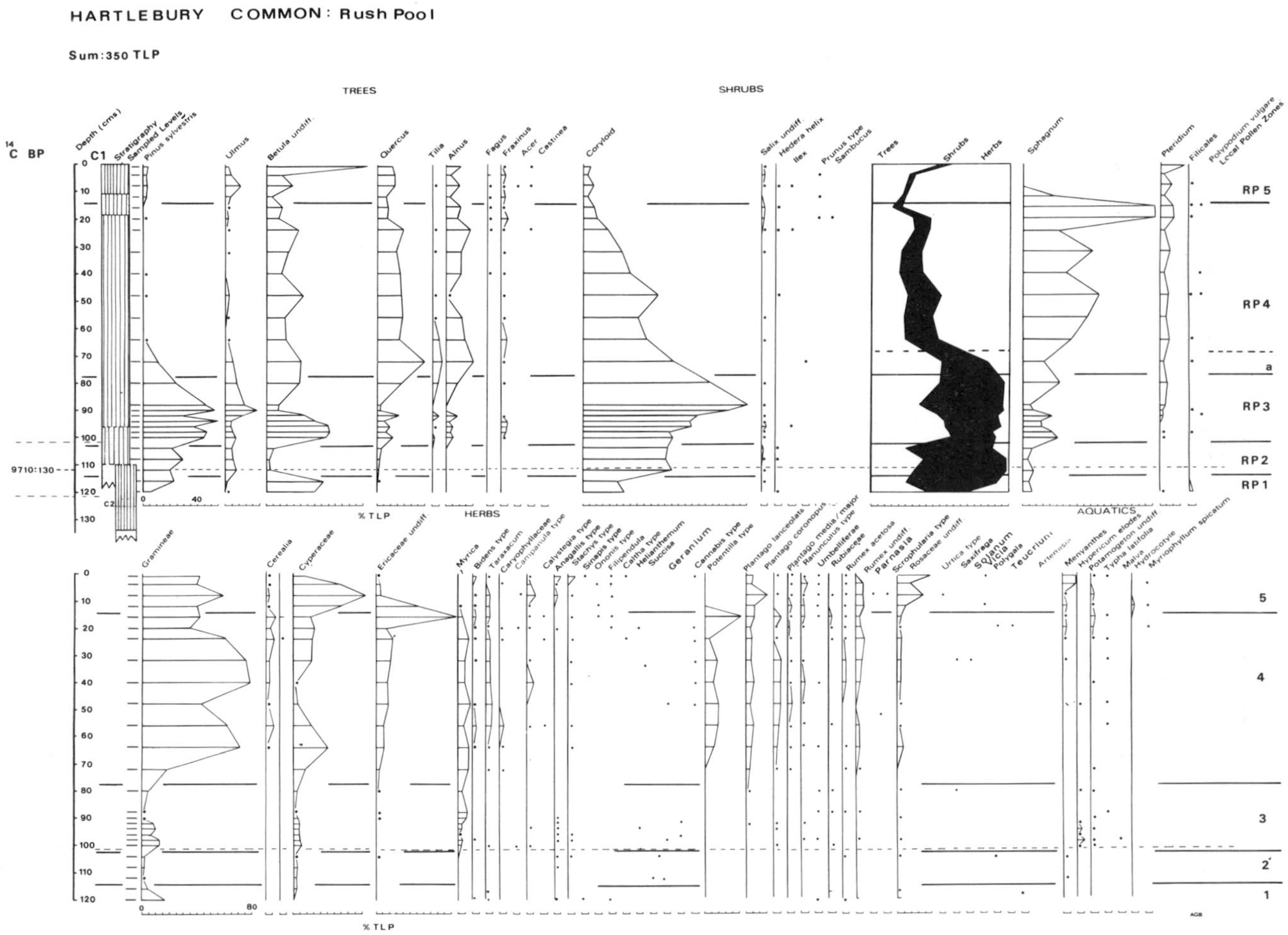

FIGURE 16.5 HARTLEBURY COMMON (RUSH POOL) POLLEN DIAGRAM

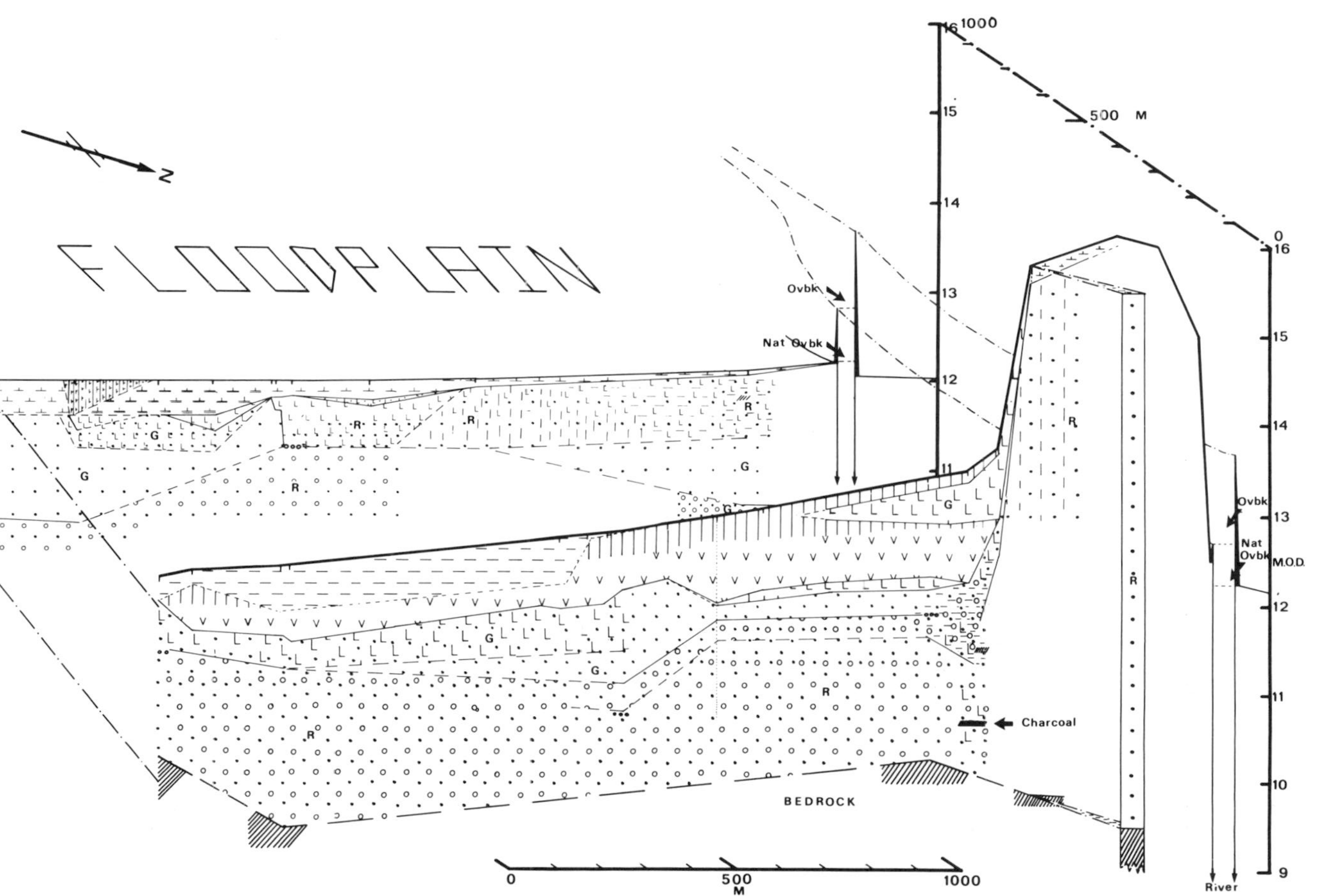

FIGURE 16.6 ASHMOOR COMMON LONG-SECTION STRATIGRAPHY

coppicing (Brown, 1982). A date of around 2000 B.P. can be estimated for the clay inwash of terrace soil at the head of the bog on the basis of pollen correlation with the main core.

At Ripple Brook (4 km north of Tewkesbury) a similar sedimentary sequence to that of the main floodplain has accumulated in an overdeepened terrace channel. It now drains a small low relief catchment of 19 km^2 in area which is situated adjacent to the Severn's floodplain. The catchment's underlying lithologies are Keuper Marl, Keuper Sandstone, and terrace deposits. The channel feature varies between 45 m and 150 m in width, 2.5 m and 5 m in sedimentary depth, and is 4 km long. The sedimentary sequence is complex, both laterally and vertically, but three macro-units can be distinguished (Figure 16.7). At the base (C) is a variable sand and gravel with some silts and occasionally included wood fragments. Above it is a unit (B) which is less variable and consists of either silty peats or organic rich silts. The peat often grades from a silty wood peat at the base to herbaceous peat above. The superficial unit (A) is a clay containing some sand and occasionally stones of terrace origin. The terrace is under 30 m from the main core which was located at the northern end of the palaeochannel. By correlation with the main sequence into which it grades, it seems likely that both macro-units B and C equate with the middle (fine) Severn unit. Because of the small basin size (the bog width) it can be realistically assumed that both the majority of the incoming pollen and the sediments deposited originated from a common source area. The pollen sequence of just over one metre, and with four ^{14}C dates, spans only 500 years giving a rapid accumulation rate of 1 cm every five years. This overall accumulation rate is similar to that of the grey clays at Pilgrims Lock investigated by Shotton (1978). The vegetation sequence shows a decline in dry land tree pollen, especially lime and hazel, at about 2800 B.P. (Figure 16.8; for full pollen diagram see Brown, 1982, 1983). The sequence indicates a considerable lag before inorganic sediments containing terrace-derived material were deposited. The Main Terrace flanks the site on both sides. This lag suggests either a delayed erosion response, spatially restricted deforestation or upstream sediment stores. The Ripple Brook water was alkaline throughout the period of peat deposition, and possibly reached as high as 8.5 pH as it contained an abundant alkaliphilous diatom flora and 5 per cent alkalibiontic species. The present pH of the Ripple Brook water fluctuates about 8.2 pH. Most of the species encountered were suited to eutrophic conditions, however the hydrological conditions were probably not static. This is reflected by changes in species composition and relative abundance within the diagram. An example of these fluctuations is a peak in *Synedra Ulna* near the top of the diagram (Figure 16.8), a species which is particularly tolerant of high suspended sediment concentrations (Gemeinhardt, 1926).

Unfortunately mollusca were not recovered from positions throughout the organic unit, but a bulk sample from the top of the unit yielded a limited

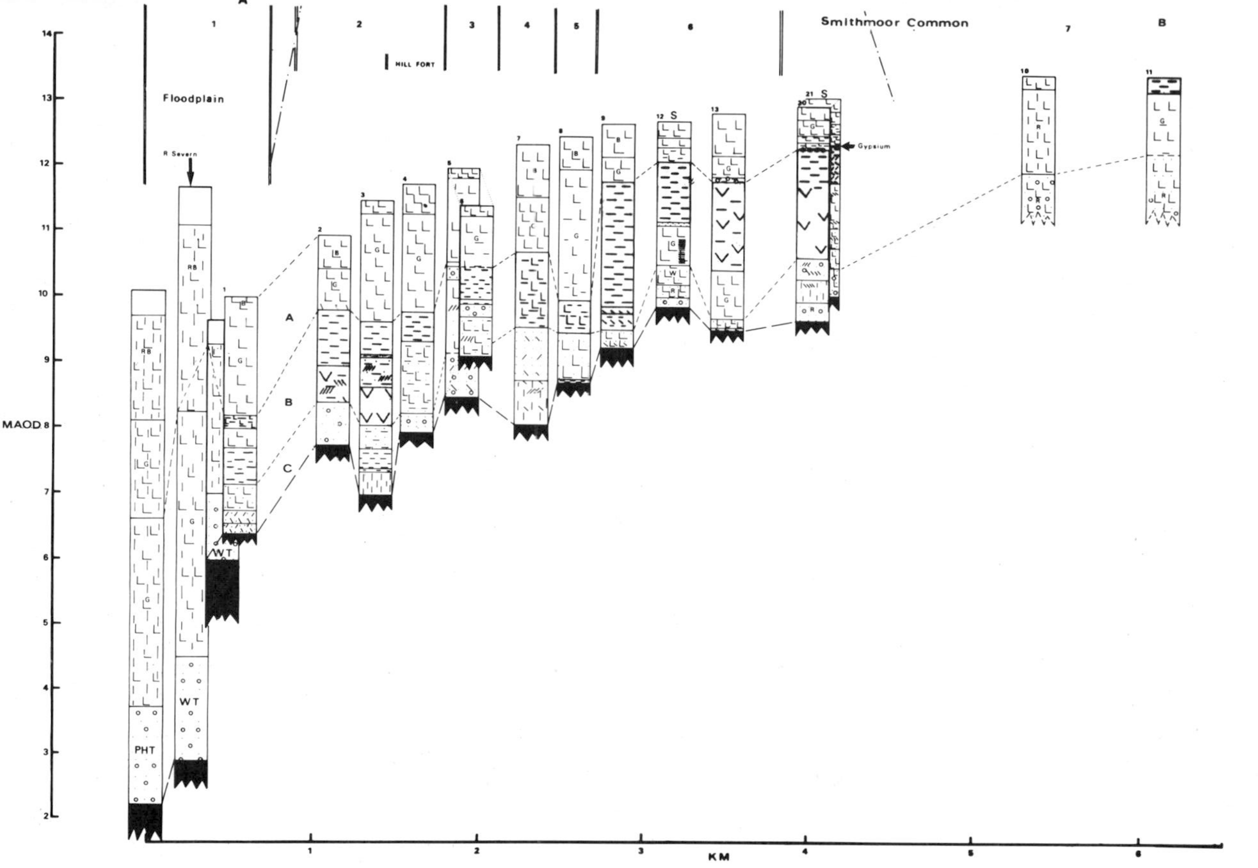

FIGURE 16.7 RIPPLE BROOK LONG-SECTION STRATIGRAPHY

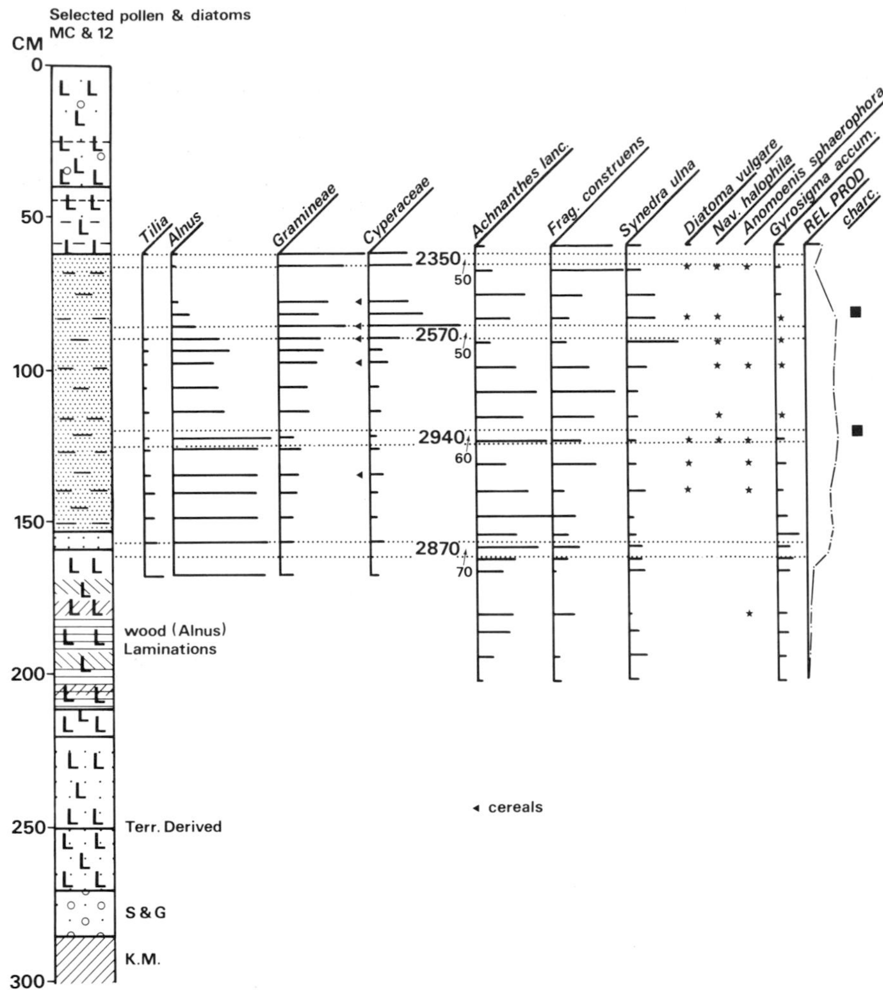

FIGURE 16.8 RIPPLE BROOK: SELECTED POLLEN AND DIATOM TAXA From pollen correlated cores

spectrum characteristic of slow sluggish riverine conditions (Shotton anal., 1981) similar to the top of the spectra at Pilgrim Lock described by Shotton (1978).

If the inorganic peat fraction (which is 60 per cent by weight) and the accretion rate are considered together it becomes clear that there is an immediate increase in the inorganic accretion rate, which slows down as the bog surface becomes drier and is covered by emergent aquatics, reeds, and sedges. This inorganic fraction is composed of fine silts and clay, and the likely sources are the local Keuper Marl soils and also Smithmoor Common which is a large low-lying and once peat-filled depression upstream of the site. The age inversion of the third and basal ^{14}C samples was probably caused by the inwashing of old pedogenic carbon from the surrounding topsoils. The accretion rate suggests high suspended sediment output from the catchment, much of which would have been transported directly to the main floodplain. After about 2300 B.P. coarser sediments of proximal terrace origin (probably including a colluvial component) were deposited. The superficial unit also contains a thin intermittent band of gypsum probably derived from the Keuper Marl. Cope (1973) has reported secondary gypsum from grey floodplain-surface clays at Ashleworth, Tirley, and Norton. It is also noteworthy that the two upstream cores (Figure 16.7) have a thin (2–4 cms) gravel band dividing the peat from the superficial unit, which is indicative of a hydrological change, and may have been caused by a single event.

A similar stratigraphic sequence can be seen on the main floodplain at Callow End, which is located just to the south of the Severn–Teme junction. Adjacent to the Worcester terrace a large cut-off meander bend was preserved as a backswamp marsh until it was filled-in in 1982 (Figure 16.9). Five cross-sections revealed an infill of clay with occasional wood fragments, an organic unit and, at the top, a silty clay which graded laterally into the red silty clay of the main fill (Figure 16.10). The silty peat has, in contrast to the other sites, few macro-remains, and the stratigraphy in cross-sections 3, 4, and 5 are suggestive of an inwashed origin. However, the remains of reeds and the relatively undisturbed pollen profile suggest an autochthonous as well as an allochtonous input. The pollen diagram which covers only 54 cms in vertical extent spans a striking decline of lime from 27 per cent TLP to under 1 per cent (Brown, 1982). This feature of the diagram is probably so well developed because of the close proximity of the terrace edge to the core (19 m), and it suggests, along with the rest of the diagram, lime dominant terrace woodlands followed by clearance and arable cultivation. Some of the organic and probably much of the inorganic fraction of the peat was derived from soil erosion caused by the destruction of the woodland cover and subsequent cultivation. A thin band of sand within the peat probably reflects an intensive runoff event which caused erosion of the sandy terrace subsoil. The change to entirely inorganic sedimentation may have been caused by an increase in overbank sedimentation by the Teme and Severn and/or by further land use changes on the terraces such as intensified cultivation.

Another core, at Kempsey, close to the floodplain edge, revealed further evidence of floodplain and terrace deforestation. From the village of Kempsey (GR SO 851493) to the Severn–Teme junction the edge of the floodplain is sharply defined by a steep slope of Keuper Marl with Main Terrace capping it at 26 m OD (Figure 16.9). The slope is covered by mixed deciduous woodland with a rich ground flora, and the floodplain is under arable cultivation adjacent to the channel but under wet grassland in other areas. The main core was sampled using a Hiller corer and it was situated 25 m from the slope base and

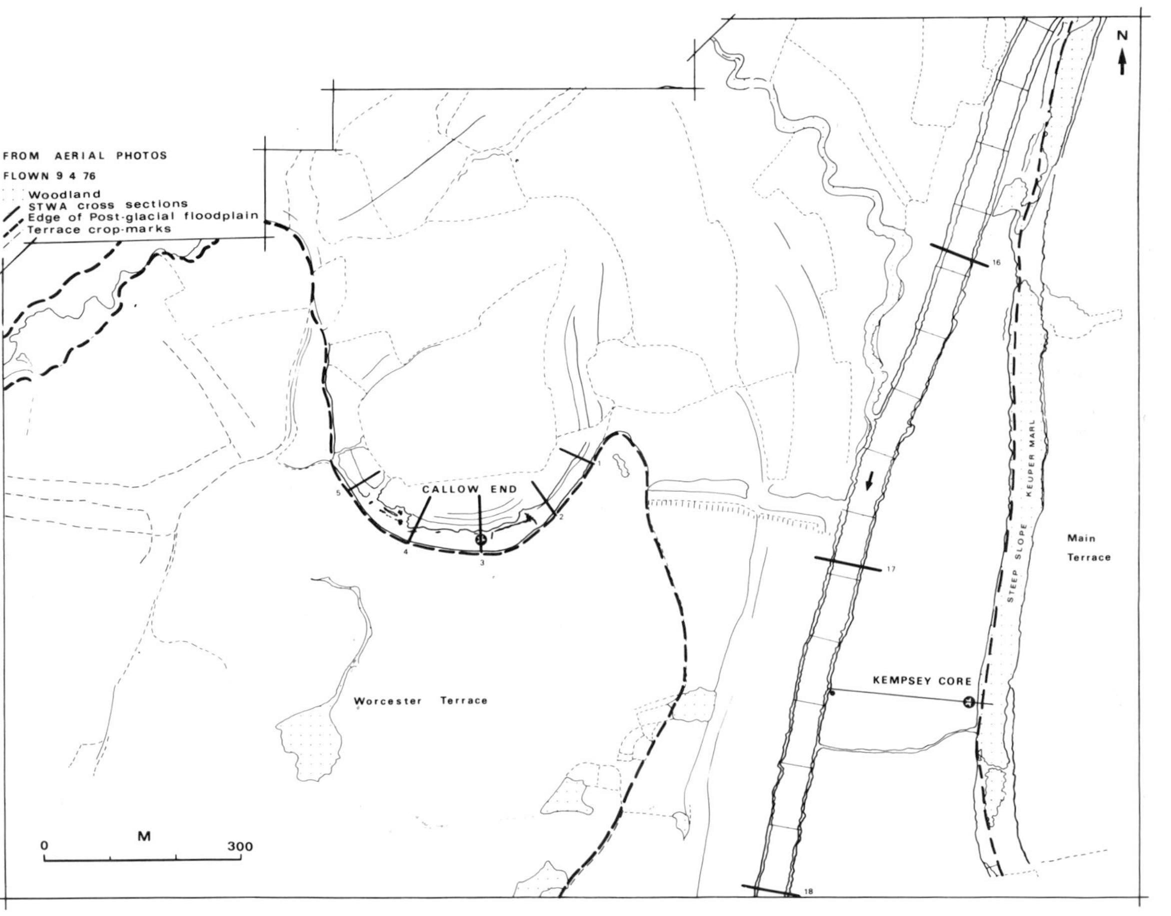

FIGURE 16.9 MAP OF CALLOW END AND KEMPSEY SITES

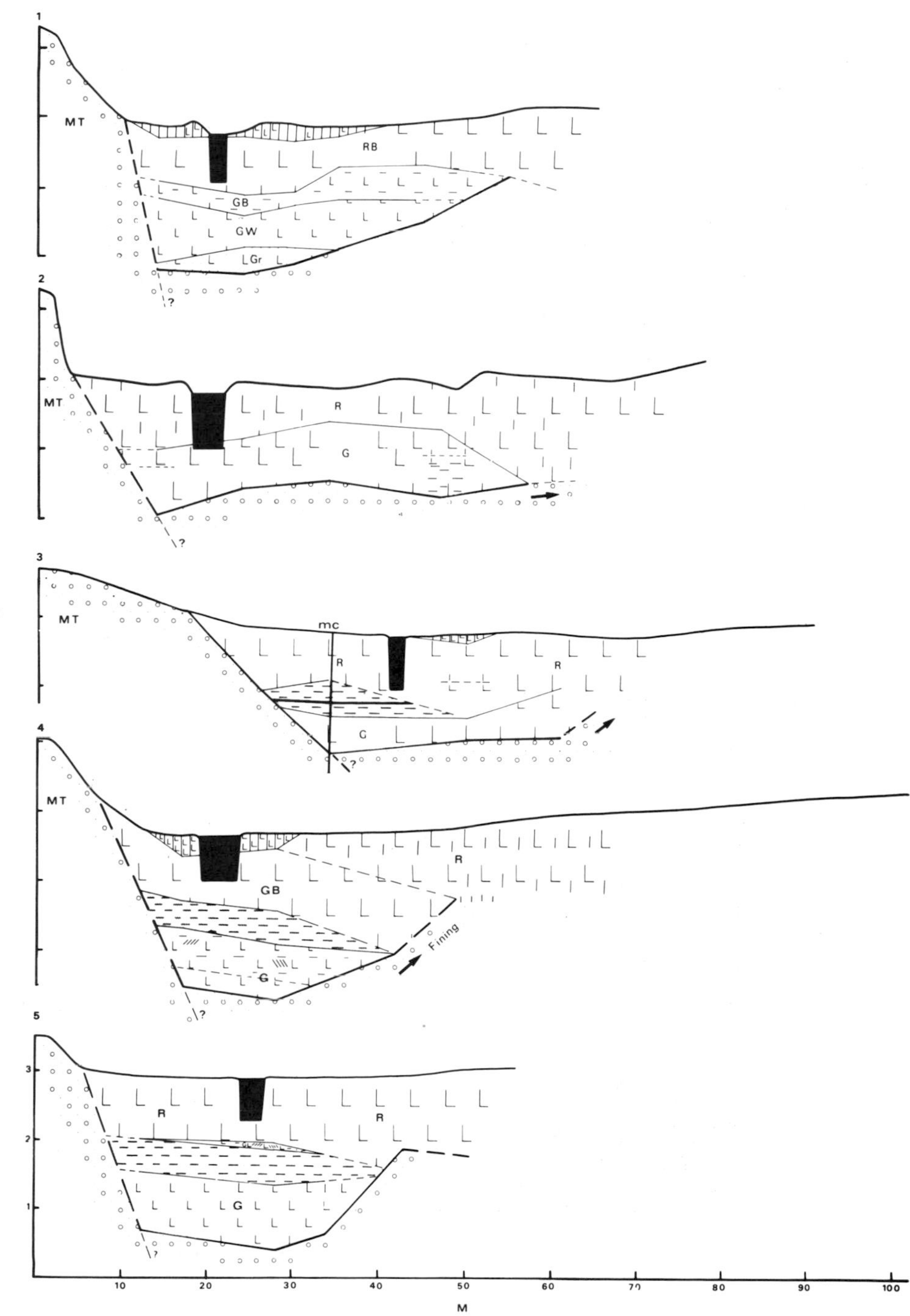

Figure 16.10 Callow End Stratigraphic Cross-sections

240 m from the River Severn. The stratigraphy of this core is summarized in Table 16.1. By using large quantities of sediment, and the cold hydrofluoric acid method (Brown, 1983), six pollen counts were possible between 200 cm and 475 cm, despite the largely inorganic nature of the sediments. Due to low concentrations the bottom two counts had a reduced sum of 150–200 TLP; however, the good condition of the pollen grains and coherence of the diagram suggests that differential pollen destruction had not occurred. Alder dominates the diagram (Figure 16.11) throughout but with willow rising to 10 per cent TLP at the top. The diagram also shows a decline of lime from over 15 per cent TLP to under 1 per cent TLP, and it is during this decline that cereals appear in substantial quantities, as do the weeds indicative of arable cultivation. It is sugested that this reflects the deforestation and subsequent agricultural exploitation of the adjacent Main Terrace. The palynological impact of these land-use changes was probably lessened by the buffering effect of the woodland on the intervening slope which was probably not deforested because of its precipitous gradient and its susceptibility to landslipping. The pollen diagram additionally indicates that the River Severn channel has not migrated to this side of the floodplain at this location during at least the last 4000 years.

FLOODPLAIN AND CHANNEL DEVELOPMENT

Closer to the present channel the thickness of the top inorganic unit is greater than at the sites previously described, and at the Severn–Teme junction it is over 4 m in thickness. This red silty clay is well sorted and generally homogeneous except at the top of the unit where sand/silt banding occurs, and also at the base above the underlying gravels (Figure 16.12). The grain size distribution of the sand bands is similar to that of recent flood sands. At this location (Powick Common GR SO 832523) the river has not been embanked and it is suggested that this upward increase in grain size is caused by the increased heights now required before flooding occurs and the concomitant increased intensity required of the flood event. This has been caused by relative channel incision within the last 3000 years. The floodplain which is now almost entirely deforested may have lost some sediment during the late Middle Ages when map evidence and palynological studies of a historical cutoff indicate intensive strip farming. Both the studies at Powick and the dated rise in the floodplain surface at Ripple Brook generally conform with the hypothetical overbank sedimentation model of floodplain development discussed by Wolman and Leopold in 1957. This model would require a disequilibrium between the floodplain accretion rate and the channel bed level, and this is independently suggested by the palaeochannels studied. For example the maximum palaeochannel depth at Callow End is 3.5 m, as compared to contemporary maximum Teme depths in unembanked reaches of over 5 m, despite similar channel capacities and virtually the same bankfull discharges

TABLE 16.1 KEMPSEY—SIMPLIFIED STRATIGRAPHY

Sediment	Depth (m)	Thickness (m)
Grey brown clay with wood fragments	0–0.58	0.58
Grey clay with wood fragments	0.58–1.29	0.71
Grey clay	1.29–1.89	0.60
Brown silty peat with clay highly humified (H4)	1.89–2.08	0.19
Grey silty clay with wood and other organic inclusions	2.08–4.25	2.17
Grey silty clay with organic material	4.25–4.90	0.65
Grey sand with fine gravel	4.90–5.10	0.20
Impenetrable sand and gravel	5.10–	—

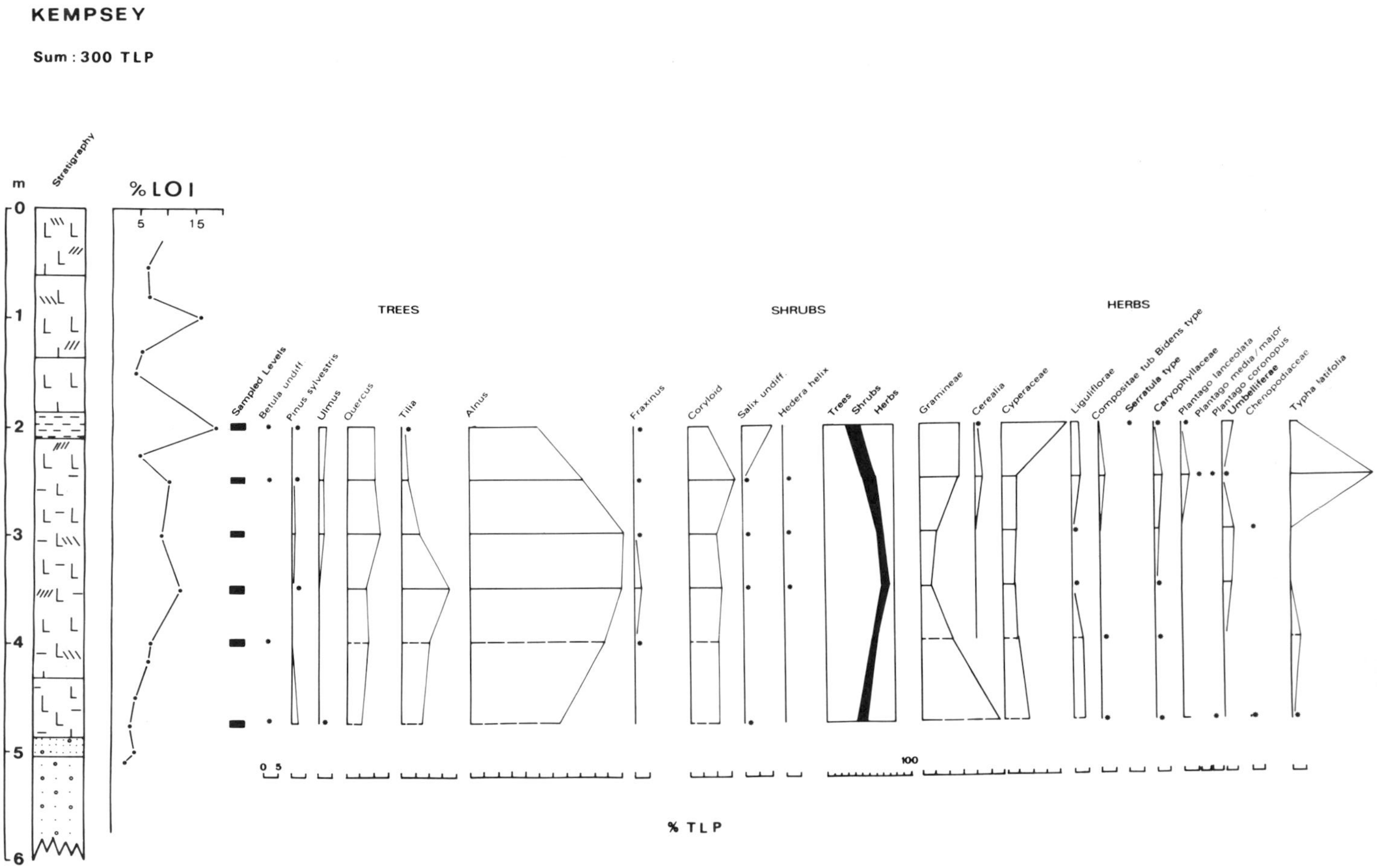

KEMPSEY

Sum : 300 TLP

m
0
1
2
3
4
5
6

Stratigraphy

% LOI

5
15

Sampled Levels

TREES

Betula undiff.

Pinus sylvestris

Ulmus

Quercus

Tilia

Alnus

Fraxinus

SHRUBS

Coryloid

Salix undiff.

Hedera helix

Trees

Shrubs

Herbs

HERBS

Gramineae

Cerealia

Cyperaceae

Liguliflorae

Compositae tub Bidens type

Serratula type

Caryophyllaceae

Plantago lanceolata

Plantago media / major

Plantago coronopus

Umbelliferae

Chenopodiaceae

Typha latifolia

0 5

100

% TLP

FIGURE 16.11 KEMPSEY POLLEN DIAGRAM

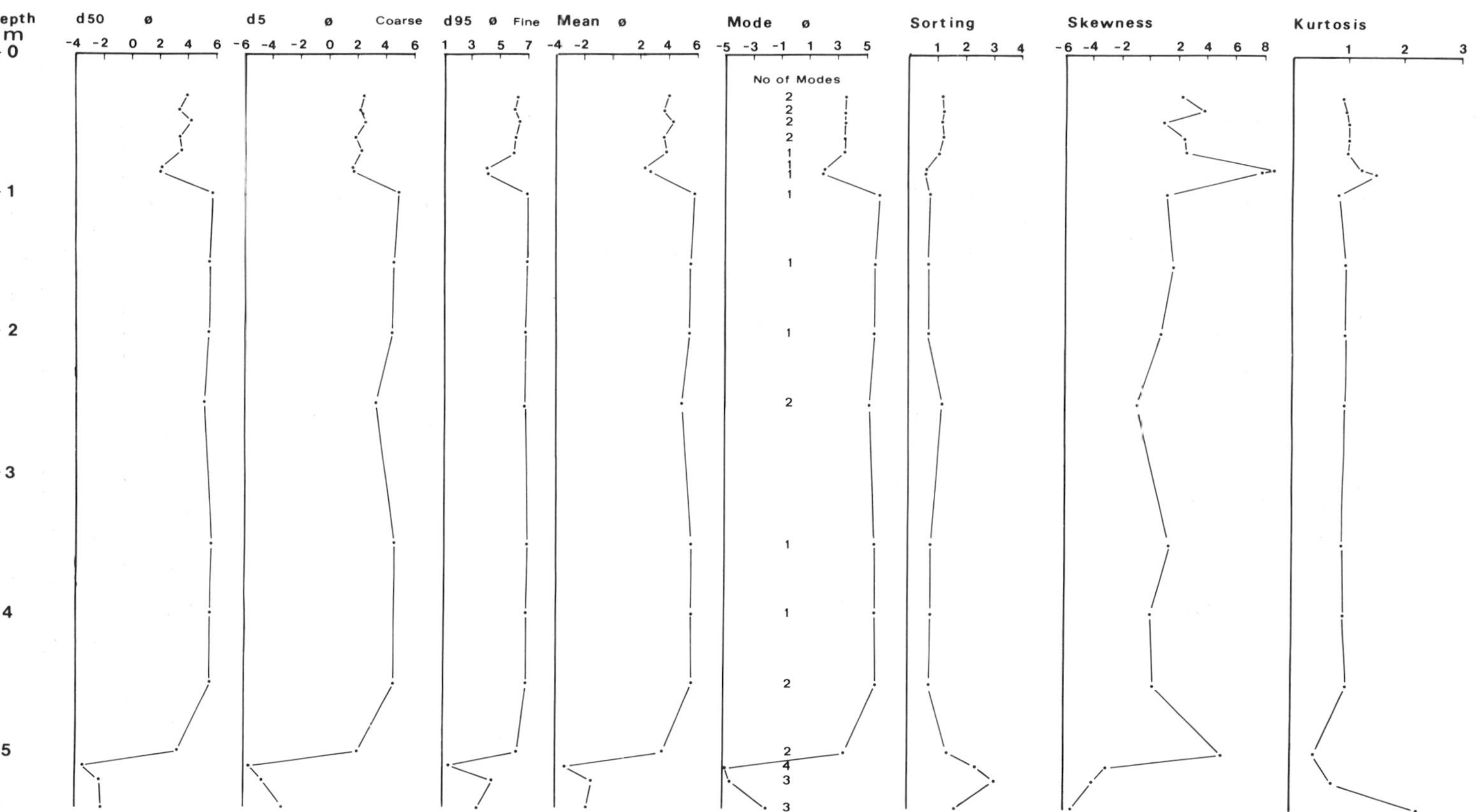

FIGURE 16.12 POWICK, BANK SECTION, GRAIN SIZE STATISTICS

(estimated using Manning's equation for Callow and hydrological data for the Teme). However, if the semi-exponential type of curve of floodplain rise, as predicted by the model, is transposed to the early Holocene when fine overbank deposition probably began, it becomes clear that it does not reflect the full accretion rate trend and that a threshold, or reinitiation, of the curve is required. Independent evidence of the resultant sigmoidal curve can be seen in the floodplain level curves for some river valleys in northwestern France produced by Morzadec-Kerfourn (1974).

The sites discussed suggest the importance of relatively local sources for some recent floodplain sediments. The sedimentary development of the Lower Severn over the last 3 000 years has been dominated by material from the lowland, central and easterly portions of the basin. The temporal relationship between vegetation change (however induced) and sedimentation is complicated both by antecedent soil conditions and by subtle fluctuations in climate. These factors and the distance of the sources from the channels will lead to the diachrony of correlatable sedimentary features even in response to synchronous vegetation changes.

The existence of a superficial inorganic unit overlying organic sediments is common in lowland Britain, and examples include the Tame (Shillitoe, 1961), and the Thames at Farmoor (Lambrick and Robinson, 1979), and Thatcham (Churchill, 1962). Culleton and Mitchell (1976), Evans (1975), and Bell (1982) have pointed to evidence of increased inorganic sedimentation from the late Neolithic onwards due to deforestation and agriculture. One of the few specific studies of floodplain alluviation in Britain was a study of the Windrush by Hazeldine and Jarvis (1979). They recognized two macro-units comprising a sandy gravel overlain by clays. Radiocarbon dates from the gravel top sugested that the clay was deposited after about 2600 B.P., and considering the modern channel type, they concluded that this was an allogenic change caused by a rapid increase in local forest clearance and in the amount of ploughed land. The processes operating in the late Bronze Age and early Iron Age in Britain, as suggested by this and other studies, are similar to those invoked as the effects of European settlement in parts of the USA by Knox (1972) and Trimble (1979). These included increased floodplain accretion, changes in sediment character and the production of lag effects by slope-based sediment stores.

Without detailed examination of floodplain organics it is not safe to assume that they represent floodplain halt, or standstill, phases. The peak of terrestrial snails reported by Shotton from the grey clays at Pilgrim Lock and the changes in the diatom flora at Ripple Brook, show that palaeohydrological changes may not always be reflected in physical sedimentary characteristics. Although some palaeoenvironmental analysis of British floodplain sequences has been performed, especially in relation to sea level changes, a wider application of its methods may help in the identification and estimation of the effects of land use changes on Holocene floodplain development. The sites described in this chapter and studies of the stratigraphy of the Lower Severn Valley suggest that the major palaeohydrological event during the Holocene was an increase in fine sediment output from the basin, which caused floodplain aggradation and may have been largely responsible for the planform stability of the channel over the last two to three thousand years.

ACKNOWLEDGEMENTS

This work was carried out while the author held a NERC research studentship. The author is grateful to Professor K. J. Gregory and Dr. K. E. Barber for their help and advice. Thanks are also due to Hereford and Worcester County Council for borehole records and local landowners for access to the sites mentioned.

REFERENCES

Aaby, B. 1979. Characterisation of peat and lake deposits. In Berglund, B. (Ed.), *Palaeohydrological Changes in the Temperate Zone in the Last 15,000 Years*, IGCP 158 Sub-Project B Guide, Lund, Vol.. 1, 77–99.

Barber, K. E. 1976. History of vegetation. In Chapman, S. V. (Ed.), *Methods in Plant Ecology*, Blackwell, Oxford, 5–84.

Bartley, D. D. 1960. Rhosgoch Common, Radnorshire: stratigraphy and pollen analysis. *New Phytologist,* **59**, 238–262.

Beales, P. W. 1980. The Late Devensian and Flandrian vegetational history of Crose Mere, Shropshire. *New Phytologist,* **85**, 133–161.

Beales, P. W., and Birks, H. J. B. 1979. Palaeolimnological studies in N Shropshire and Cheshire. *Bulletin of the Shropshire Conservation Trust,* **35**, 12–15.

Beckinsale, R. P., and Richardson, L. 1964. Recent findings on the physical development of the Lower Severn Valley. *Geographical Journal,* **130**, 87–105.

Bell, M. 1982. The effect of land-use and climate on valley sedimentation. In Harding, A. F. (Ed.), *Climatic Change in Later Prehistory*, Edinburgh, University Press, 127–142.

Berglund, B. 1979–1982. *Palaeohydrological Changes in the Temperate Zone in the Last 15,000 Years*, IGCP 158B Lake and Mire Environments, Vols. 1–3, Lund.

Brown, A. G. 1982. Human impact on the former floodplain woodlands of the Severn. In Bell, M. (Ed.), *Archaeological Aspects of Woodland Ecology*, British Archaeological Reports, International Series 116, 93–105.

Brown, A. G. 1983. *Late Quaternary Palaeohydrology, Palaeoecology and Floodplain Development of the Lower River Severn*, unpublished Ph.D. Thesis, University of Southampton.

Churchill, D. M. 1962. The stratigraphy of the Mesolithic sites III and V at Thatcham, Berkshire, England. *Proceedings of the Prehistorical Society,* **28**, 362–370.

Coope, G. R., and Shotton, F. W. 1981. The late glacial domain: the Devensian terraces of the Severn below Ironbridge and of the Warwickshire Avon. *IGCP 158A Progress Report* (unpublished).

Cope, D. W. 1973. Soils in Gloucestershire 1 (Norton). *Soil Survey Records,* **13** (Sheet No. 5082).

Culleton, E. B., and Mitchell, G. F. 1976. Soil erosion following deforestation in the early Christian period in southern Wexford. *Journal of the Society of Antiquaries, Ireland,* **106**, 120–123.
Davis, J. C. 1973. *Statistics and Data Analysis in Geology*, Wiley, Chichester.
Evans, J. G. 1975. *The Environment of Early Man in the British Isles*, Elek, London.
Gemeinhardt, K. 1926. Die gattung in systematischer, zytologischer und okoligischer Bezienhung. (The systematics, cytology and ecology of the genus *Synedra*). *Pflanzenforschung,* **6**, 1–88.
Hardy, E. M. 1939. Studies in the post-glacial history of British vegetation. V. The Shropshire and Flint Maelor mosses. *New Phytologist,* **38**, 364–396.
Hazeldine, J., and Jarvis, M. G. 1979. Age and significance of alluvium in the Windrush valley, Oxfordshire. *Nature,* **282**, 291–292.
Hibbert, F. A., and Switsur, V. R. 1976. Radiocarbon dating of Flandrian pollen zones in Wales and northern England. *New Phytologist,* **77**, 793–807.
Jeffries, R. L., Willis, A. J., and Yemm, E. W. 1968,. The late- and post-glacial history of the Gordano valley, north Somerset. *New Phytologist,* **67**, 335–348.
Knox, J. C. 1972. Valley alluviation in south-western Wisconsin. *Annals of the Association of American Geographers,* **62**, 401–410.
Lambrick, H. T., and Robinson, M. 1979. *Iron Age and Roman Riverside Settlements at Farmoor, Oxfordshire*, Council for British Archaeology Research Report 32.
Lockwood, J. G. 1979. Water balance of Britain, 50 000 B.P. to the present. *Quaternary Research,* **12**, 297–310.
Miall, A. D. 1973. Markov chain analysis applied to an ancient alluvial plain succession. *Sedimentology,* **20**, 347–364.
Moore, P. D. 1966. *Stratigraphical and Palynological Investigations of Upland Peats in Central Wales*, unpublished Ph.D. Thesis, University of Wales.
Moore, P. D. 1968. Human influence upon vegetational history in north Cardiganshire. *Nature,* **217**, 1006–1009.
Moore, P. D., and Chater, E. H. 1969. The changing vegetation of west-central Wales in the light of human history. *Journal of Ecology,* **57**, 361–379.
Morzadec-Kerfourn, M. T. 1974. Variations de la ligne de rivage Armoricaine au Quaternaire. *Memoire de la Societé Geologique et Mineralogique de la Bretagne*, 17.
Pannet, D. J., and Morey, C. 1976. The origin of the old river bed at Shrewsbury. *Bulletin of the Shropshire Conservation Trust,* **35**, 7–12.
Rhind, D. W., and Sissons, J. B. 1971. Databanking of drift borehole records for the Edinburgh area. *Institute of Geological Sciences Report,* **71/15**.
Rowlands, P. H., and Shotton, F. W. 1971. Environmental change and palaeobotany. Pleistocene deposits of Church Stretton, Shropshire and its neighbourhood. *Quarterly Journal of the Geological Society,* **127**, 599–622.
Shillitoe, J. S. 1961. Borings in the first terrace and alluvium of the river Tame at Tamworth. *Proceedings of the Coventry and District Natural History and Scientific Society,* **3**, 133–138.
Shotton, F. W. 1967. Investigation of an old peat moor at Moreton Morrell, Warwickshire. *Proceedings of the Coventry and District Natural History and Scientific Society,* **4**, 13–16.
Shotton, F. W. 1978. Archaeological inferences from the study of alluvium in the lower Severn–Avon valleys. In Limbrey, S., Evans, I. G. (Eds.), *Man's Effect on the Landscape: The Lowland Zone*, Council for British Archaeology Research Report 21, 27–32.
Shotton, F. W., And Strachan, I. 1956. The investigation of a peat moor at Rodbaston, Staffordshire. *Quarterly Journal of the Geological Society,* **115**, 1–15.

Trimble, S. W. 1979. *Sedimentation in Coon Creek Wisconsin*, unpublished manuscript, 1–12.

Wills, L. J. 1926. Geology and soils of Hartlebury Common. *Proceedings of the Birmingham Natural History Society,* **15**,. 95–101.

Wills, L. J. 1938. The Pleistocene development of the Severn from Bridgnorth to the sea. *Quarterly Journal of the Geological Society,* **94**, 161–242.

Wolman, M. G., and Leopold, L. B. 1957. River floodplains: some observations on their formation. *United States Geological Survey Professional Paper 282C.*

Background to Palaeohydrology
Edited by K. J. Gregory

17

Osage-type underfitness on the River Severn near Shrewsbury, Shropshire, England

G. H. Dury

Sidney Sussex College, Cambridge

The Osage-type underfit stream is one that, although greatly diminished in channel dimensions, and displaying a pool-and-riffle sequence on its existing bed, has failed to develop stream meanders. If such a stream is contained in a meandering valley, as is frequently the case, then the apparent l/w ratio, where l is meander wavelength and w is bedwidth between banktops, will run high, say at about 40/1, as opposed to the approximate 11/1 expectable from meandering streams in general. In actuality, the observed ratio is L/w, the ratio between the wavelength of former meanders and the bedwidth of the present stream.

Discussion of the apparently anomalous wavelength/width ratio of meandering streams incised into bedrock began with Jefferson (1902), who was followed by Bates (1939). The discussion appears to have been confused through the agency of poor mapping. Thus, Bates uses the Kickapoo River in southwest Wisconsin as an example of the combination of great meander wavelength with modest bedwidth; but, in actuality, that river also possesses stream meanders appropriate in wavelength to its existing channel (Dury, 1964a, Figure 20 and accompanying text, 1964b, Figure 30 and accompanying text). That is to say, the Kickapoo, combining valley meanders with stream meanders, is by definition a manifest underfit. Numerous other examples could be adduced, where valley meanders have been mapped but stream meanders have not (see for example Dury, 1964a, Figures 13 and 14a).

Confusion has also arisen from time to time, in relation to the fact that bedrock structure can influence meander plan geometry. The most abundant set of measurements on valley meanders ingrown into strongly-jointed bedrock with which I am acquainted, is that of R.W. Young (1974); as can be shown, when the former wavelengths and former bedwidths are averaged for meander trains, the wavelength/width ratio L/W turns out to be 10.43, well within the range observed for alluvial meanders (Dury, 1977).

At least eight kinds of underfit stream can be imagined (Dury, 1964a, Figure 4). Up to now, only two kinds have been studied — the manifest underfit, which combines valley meanders with stream meanders, and the Osage-type, which combines valley meanders with a non-meandering present channel that displays pool-and-riffle. Manifest underfits can be recognized on sight from air photographs or from accurate maps. The test for Osage-type underfitness demands profiling of the bed.

Bed profiles need not always be determined by field survey: the pool-and-riffle sequence on the type river Osage is clearly visible on air photographs at very low stages (Dury, 1964a, Figures 43 and 44, and accompanying text). Field survey was however employed in the first critical experiment on the Colo River in NSW, Australia (Dury, 1966) when the term *Osage-type underfit* is first established. Unlike the Osage, the Colo does not reveal its riffles at low stages. The Severn near Shrewsbury also covers its riffles at the stage of ordinary low flow, but the riffles near Montford show up well on air photos taken in colour, on account of the aquatic plants that colonize them.

THE FIELD AREA AND BASIC DATA

The Severn near Shrewsbury displays trains of valley meanders, including a cutoff at Shrewsbury itself. The existing channels of the Severn and the major tributary Vyrnwy, upstream of the confluence, display stream meanders; but these are either very poorly developed or are absent downstream of the confluence, until one reaches the short series of loops near Leighton, a short distance upstream of the Ironbridge Gorge (Figure 17.1). The existing channel does however possess a well defined sequence of pool and riffle.

As has been repeatedly demonstrated in flume experiments, the conversion of a straight to a meandering channel begins with deformation of the bed. Because in such experiments time is greatly scaled down, some early reports had bed deformation followed immediately by side-to-side swing. Later work, however, has shown that deformation of the bed need not necessarily be followed by deformation of the banks (O'Brien, 1979).

It is necessary to infer that the same holds for the Severn near Shrewsbury. Shrinkage of the stream from the size as measured by channel dimensions and/or by channel-forming discharge, appropriate to the valley meanders, to the size appropriate to the present channel, has been accompanied by deformation of the bed but by little or no deformation of the banks. A preliminary statement on this matter (Dury, Sinker, and Pannett, 1972) provoked, much to the surprise of the original authors, considerable debate (Ferguson, 1973; Kennedy, 1972; Kirkby, 1972, Richards, 1972; Tinkler, 1972; and Cogley, 1973; for a rejoinder, see Dury, 1973). In the subsequent

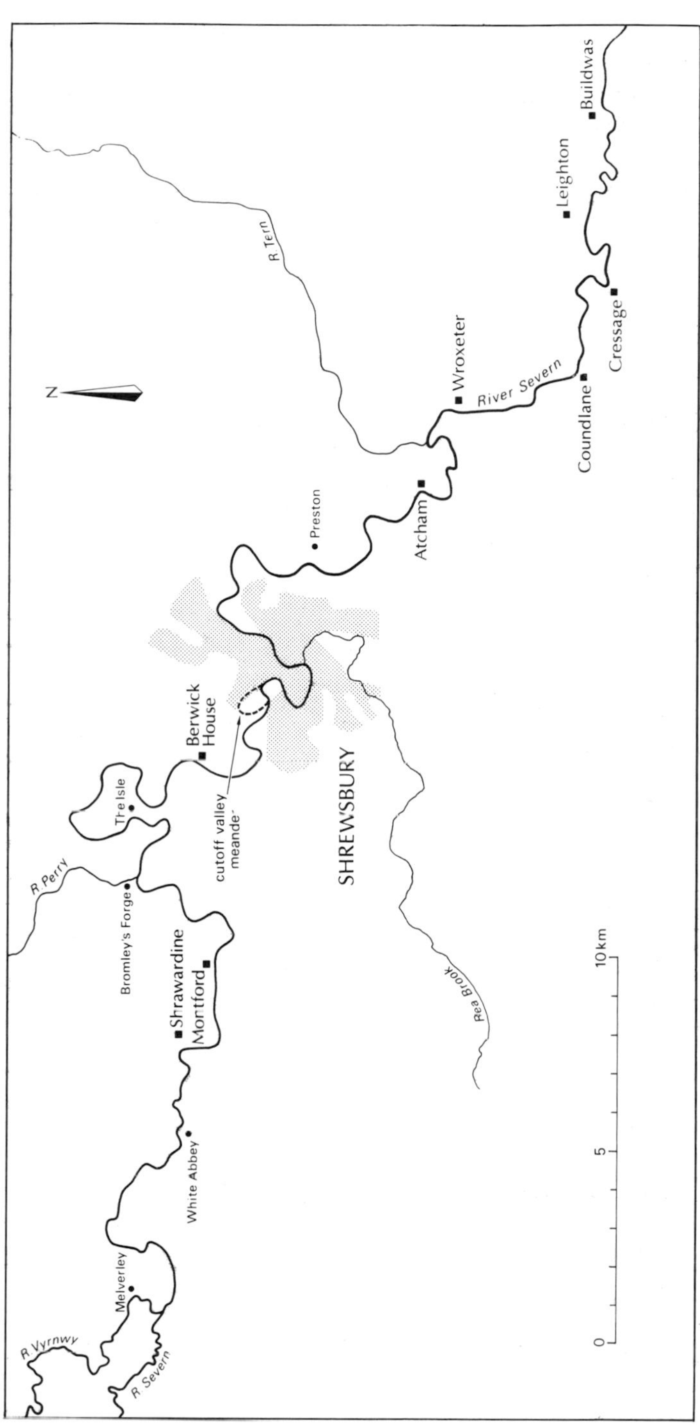

FIGURE 17.1 LOCATION MAP

decade, additional soundings and cross-sectional measurements have been made on the Severn, with the results that the relationship of the pool-and-riffle sequence to bedwidth and to drainage area can now be defined for some 30 km of channel length. As will appear, it has become possible to offer a provisional explanation of why the former stream could meander, whereas the present stream cannot.

The bedwidth values to be employed in the following discussion are the arithmetic means of measurements taken on air photographs or on the 1:10.560 map.

Air photos were used for reaches upstream of Shrewsbury, where the photo scale made it convenient to measure width at intervals of 113.5 m ground distance along the channel, and where widths were determined to the nearest 0.5 m. For reaches downstream of Shrewsbury, map measurements were made at intervals of 100 m of ground distance and to the nearest 1.0 m. Banktop is very well defined in wholly alluvial reaches. Where terrace materials or bedrock abut on one side of the channel, the typically very steep slopes ensure that errors in the determination of bedwidth between banktops are minimized.

As I propose to show elsewhere, bedwidth on the study reaches of the Severn undergoes rapid step-functional variation in the downstream direction. For any given reach, for example with unusually high or unusually low average bedwidth, the observed widths obtained from the intense sample produce a Poisson frequency distribution. Arithmetic means may not therefore be the best measure of representative width for a given reach, but since such means have been widely employed in the past, they will be retained here.

The reaches surveyed are located in Figure 17.1. Basic data are contained in Table 17.1. As there indicated, pool-to-pool or riffle-to-riffle spacing has been determined for 65 cases, to which can be added the four determinations made from the meanders near Leighton. When the average spacing per reach is doubled, to provide potential meander wavelength l', and when l' is regressed on mean bedwidth, the result is that shown in Figure 17.2. The mean wavelength for the Leighton meanders falls squarely on the line for the regression equation, which differs very little indeed from the arithmetic relationship

$$l' = 11.32w$$

very close indeed to the value of $11w$ obtained for the actual meanders of a very large sample of streams (Dury, 1976).

The identification of the Severn near Shrewsbury as an Osage-type underfit is abundantly confirmed. The channel has shrunk far below the dimensions appropriate to the former meanders: it has developed pool-and-riffle, but in very large part has failed, on the existing channel, to develop side-to-side

Table 17.1 Spacing of Pools or Riffles, and Bedwidth, on the Present Channel of the Severn near Shrewsbury

Reach	i channel distance m	ii number of pools/ riffles	iii mean spacing m	iv 2 × mean spacing m	v mean bedwidth m	vi ratio cols. iv/v
Shrawardine to gauge	1290	6	215	430	36.6	11.75
Montford eyot to Parry confluence	3185	13	245	490	46.9	10.45
Near Berwick House to Shrewsbury	5930	24	247	494	45.75	10.78
Atcham to Wroxeter	4200	14	300	600	52.8	11.36
Upstream of Coundlane	2600	8	326	652	53.5	12.19
					weighted mean	11.07

Values in col. i relate to the numbers of spacings indicated in col. ii

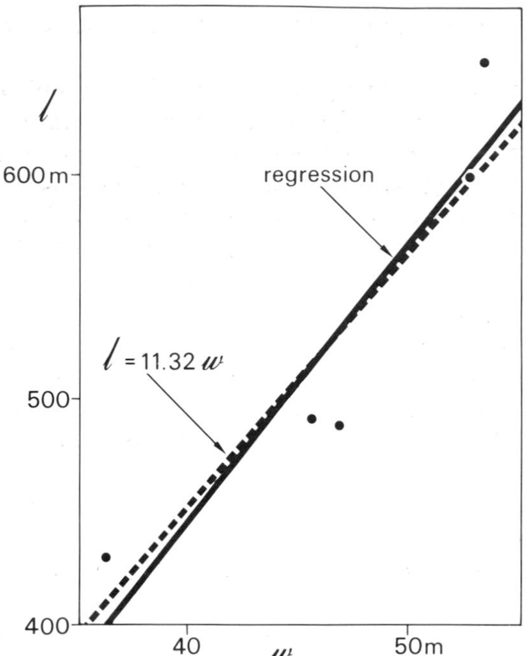

FIGURE 17.2 DOUBLE SPACING OF POOLS OR RIFFLES AGAINST BEDWIDTH
Data as in Table 17.1, cols. iv and v. Regression gives $r = +0.929$, $r^2 = 0.863$, $0.05 > P$

swing. Reasons for this set of circumstances can be suggested, but only as yet in the strictly limited context of the study reaches. As will be made clear, there exist some grave contingent problems.

HYDRAULICS AND PALAEOHYDRAULICS

The study channel traverses the deposits of the Irish Sea Glacier, including proglacial sediments (for environmental reconstructions of the sedimentary record, see Shaw, 1972). In part, the channel is shaped in lacustrine sediment, other parts are shaped in till, outwash, and terrace gravels. The general downvalley slope, determined essentially by deposition as opposed to fluvial erosion, is slight.

Both the existing and the former, larger, channel may be compared to the stereotype meandering stream (Dury, 1976), in respect not only of dimensional characteristics but also of hydraulic characteristics. The data employed to obtain the specifications of the stereotype stream, however, took no account of the possible difference of channel-forming discharge, and therefore of recurrence interval, between pool and riffle sections. Moreover, they made no allowance for the apparently typical step-functional variation in flood regimes, such as has only very recently come to light (Dury and Kohchi, 1982; Dury, 1982).

As I propose to show elsewhere, the equation for bankfull discharge derived by Williams (1978) predicts for pool sections on the Severn near Montford a discharge very close to the 257 cumecs of the 1.58-year flood on the total flood regime. For riffle sections, Williams' equation predicts a discharge very close to the 212 cumecs of known channel capacity at bankfull. This value is that of the 1.15-year flood on the total regime, but corresponds to the 1.58-year flood on the regime of low peaks. I have accordingly made separate comparative calculations for the two discharges (Table 17.2, cols. i–iv). When the averages of the pool values and the riffle values are employed, we work with a bankfull discharge of 234.7 cumecs ((Table 17.2, cols. v–vi): entry of the mean cross-sectional area and the observed slope into Williams's equation results in a predicted discharge of 234.5 cumecs. For the Atcham reach, hard information is as yet available only on riffle sections (Table 17.2, cols. vii–viii).

As shown, the observed slopes are about half the stereotype slopes. In consequence, although the observed cross-sectional areas and the observed velocities through the cross-sections closely resemble the stereotype values, the slope-dependent hydraulic characteristics of stream power, shear stress at the boundary as a whole, and frictional shearing velocity at the boundary as a whole, run low on the real stream. In strong contrast, quantities calculated for the bed, even though they involve the roughness coefficient which is itself in part slope-dependent, turn out to be in the main very similar to stereotype values. Shear stress at the bed, in fact, is quite high on average and is distinctly high in pool sections.

An hypothesis may now be offered, to the effect that the unduly low slope on the study reaches so reduces stream power, shear stress on the banks, and frictional shearing velocity on the banks, that little bank deformation can occur. Some contribution is undoubtedly made by bank strength, which in places runs high, but in view of the results of the hydraulic calculations, the main influence of bank strength could be thought to be the influence exerted on the form ratio. Part of the right bank on the Montford reach is cut in bedrock, but most is in till and/or alluvium that yields, if it yields at all, by slumping at high angles. Analyses of grab samples of bank sediment, collected during a far from intensive sampling programme, indicate for the Montford reach averages of D_{50} = 0.115 mm, D_{84} = 0.153 mm, and for the Atcham reach D_{50} = 0.098 mm, D_{84} = 0.153 mm. As Table 17.2 shows, the bedwidths on these two reaches run well below the stereotype values. By contrast with what happens, or fails to happen, to the banks, the bed has undergone marked deformation, that in part of the Montford reach includes the excavation of pools in bedrock.

The stated hypothesis, needless to say, requires many tests on additional reaches, including reaches of other streams, before it can be considered as a potential general explanation of Osage-type underfitness. One might intuitively

TABLE 17.2 DIMENSIONAL AND HYDRAULIC CHARACTERISTICS OF THE EXISTING CHANNEL

	i Stereotype stream	ii Montford riffles (mean of 6)	iii Stereotype stream	iv Montford pools (mean of 4)	Means of cols. i & iii	Means of cols. ii & iv	Stereotype stream	Atchamm riffles (mean of 3)
q	212	212	257	257	234.5	234.5	230	230
w	56.9	40.4	63.26	43.75	60.08	42.08	59.5	50.3
d	3.054	4.101	3.272	4.731	3.163	4.416	3.144	3.425
A_c	173.7	165.7	207.0	207.0	190.35	186.35	187.0	172.3
v	1.219	1.279	1.2415	1.2415	1.230	1.260	1.230	1.335
R	2.757	3.409	2.965	3.890	2.861	3.650	2.842	3.015
S	7.5^{-4}	3.2^{-4}	6.5^{-4}	3.2^{-4}	7.0^{-4}	3.2^{-4}	7.1^{-4}	3.7^{-4}
n	0.0442	0.0317	0.0424	0.0356	0.0433	0.0337	0.0435	0.0302
Fr	0.2227	0.2016	0.2191	0.1822	0.2209	0.1919	0.2214	0.2303
ω	163.3	67.84	167.05	82.24	165.2	75.04	163.3	80.5
τ	2.018	1.091	1.927	1.245	1.973	1.168	2.018	1.055
μ^*	0.141	0.102	0.137	0.110	0.139	0.106	0.141	0.101
μ^*_b	0.065	0.062	0.064	0.075	0.065	0.069	0.065	0.066
f_b	0.023	0.019	0.021	0.029	0.022	0.024	0.022	0.029
τ_b	4.225	3.844	4.123	5.587	4.174	4.716	4.225	4.322

q discharge in $m^3 s^{-1}$
w bedwidth in m
d mean depth in m
A_c channel cross sectional area in m^2
v mean flow velocity in $m\ s^{-1}$

R hydraulic radius in m
S slope
n Manning roughness coefficient
Fr Froude number
ω stream power calculated as 1000 Sq

τ shear stress at the boundary in Kgm^{-2} calculated as 1000 SR
μ^* frictional shearing velocity at the boundary in ms^{-1} calculated as $\sqrt{gRS}$
μ^*_b frictional shearing velocity at the bed in $m\ s^{-1}$ calculated as $v/7{\cdot}75 \log (d/n) + 8{\cdot}5$
f_b friction factor at the bed calculated as $8\,(\mu^*_{b/v})^2$ in $m\ s^{-1}$
τ_b shear stress at the bed in Kgm^{-2} calculated as $125 f_b (v)^2$

expect such a general explanation, if one should ever be forthcoming, to resemble that suggested for the study reaches of the Severn simply because Osage-type underfits deform their beds, but their banks little or not at all. Two contexts of likely difficulty, however, at once suggest themselves. The first is that of manifest underfits, which deform their banks despite a reduction of former slope by a factor equal to present sinuosity. There is admittedly some evidence that the width/depth ratios of the former, larger, streams were higher than are those of the existing streams. A reduction of width/depth ratio is however to be expected as an accompaniment of significant reduction of channel-forming discharge; the available evidence permits no statement about whether or not the observed reduction is greater than would be expected from the reduction in discharge. The second context is that of meandering tidal creeks, where a typical mean velocity through the cross-section, at maximum discharge, seems to be about 0.5 m s^{-1}. I have been unable to locate enough information on channel slope for such creeks to allow me to calculate values of hydraulic characteristics, but an obvious possibility seems to be that such values could run distinctly low. If that should prove to be the case, then it would become necessary to search for hydraulic thresholds between the non-meandering and the meandering habits, additional to those considered here.

Scrutiny of the former Severn, that is of the stream which shaped the valley meanders, involves reconstruction of former discharge at the stage of channel-forming flow. Such reconstruction perforce depends on measurements of former channel dimensions. The most easily measurable of these dimensions is of course valley-meander wavelength, determined as the average for a train. Measurements of former wavelength on this part of the Severn are however not at all easy, since most trains are short, and the pattern next upstream of Shrewsbury is highly distorted. Nevertheless, when measurements on tributaries are also taken into account, power-function regression of wavelength on drainage area can be shown to indicate for the Montford reach a former wavelength of some 1360 m, and for the Atcham reach one of some 1380 m, values that agree quite well with observed averages. Entry of these values into the relevant equations for the stereotype stream retrodicts former bedwidths of about 124 and 125 m, former cross-sectional areas of 628 and 644 m^2, and former bankfull discharges of 870 and 895 cumecs.

These reconstructed values of discharge may, however, be too low. As Table 17.2 shows, the existing channel is distinctly narrower than the stereotype channel. Since meander wavelength depends directly on bedwidth and only indirectly on discharge, and since the former channel was shaped in the same kinds of material that border the present channel, we can infer that the former channel also possessed an unusually low width/depth ratio, and therefore an unusually low wavelength/discharge ratio. The converse being, needless to say, that the discharge/wavelength ratio was unusually high.

My only information on former channel dimensions, apart from wavelength, relates to the cutoff valley meander at Shrewsbury, where the alluvial fill has been drilled across at intervals of 15 m (Pannett and Morey, 1976). Most of the fill was logged as clay, but part of the section, 45–50 m in width and extending from top to base, consists of peat. This part would seem to represent a former course of the present channel; if that is so, then the cutoff was not effected until the Severn had shrunk to its existing size. Here, as is usual, identification of the former banktops is difficult to impossible. Depending on where former banktop level is assumed to have been, within the range of reason, former bedwidth on this section was between 150 and 180 m, with former cross-sectional area between 565 and 825 m^2. Because no comparative sections are available from nearby, it is impossible to judge if this is a pool, a riffle, or an intermediate section, although this last would seemed to be excluded by the observed cross-profile of the bed. The location of the line of drilling, through the grid reference point SJ 495150, would suggest an intermediate section. The average of the reconstructed former discharges, is calculated for the two reaches as 1200 cumecs, and retrodicts for the cross-section mean velocities of 2.124 to 1.454 m s^{-1}. As will be seen, something in the lower part of this range seems the most likely, the implication being that the line of drilling did in fact traverse part of a pool, although not the maximum depth, identifying a channel where the effective cross-sectional area was of the order of 750 m^2, and the effective bedwidth in the approximate range 120–135 m.

The outcrop of bedrock that constricts the existing channel on part of the Montford reach can be assumed also to have affected the palaeochannel. Let us therefore take the Atcham data, which relate to a wholly sedimentary reach, as fairly representing the effects of bank strength, both on the present and on the former channel. When the observed means of valley-meander wavelength are adjusted by the factor obtained by division of observed into stereotype bedwidth (i.e., on the present channel), adjusted values are obtained of 1610 and 1633 m, corresponding to former bedwidths of about 124 and 126 m, and to former stereotype discharges of 1185 and 1215 cumecs through cross-sectional areas of about 767 and 784 m^2, and former velocities of about 1.55 m s^{-1}. With these reconstructed values, the interpretation of the cross-section at Shrewsbury agrees well.

Confirmation that the former bedwidth on the Montford reach was of the order calculated comes from measurements of the width of the existing floodplain. The floodplain alluvium is here inset into till and a few patches of No. 2 terrace, the surficial materials resting on Triassic bedrock. On account of the steepness of the channel banks (present: observed; former: inferred) we may assume that any errors resulting from the non-coincidence of the floodplain surface with former water level are not great. The average floodplain width obtained from 15 regularly spaced measurements on the straight stretch near Montford is 128.5 m, reassuringly close to the calculated value of 123 m.

Table 17.3 lists, for the reconstructed discharges, stereotype values of channel and hydraulic characteristics (cols. i, iv). It also lists stereotype values related to observed slopes (cols. ii, v), comparing these with values reconstructed for the former stream (cols. iii, vi). Confirmation that we seem to be working in the right general area comes from the entry of the stereotype values of cross-sectional area and slope into Williams' equation. The bankfull discharges predicted by this equation are some 7.5 per cent in excess of the reconstructed values, probably not a serious discrepancy in view of the considerable noise in the data from which the stereotype was originally derived and of the uncertainties involved in any exercise of reconstruction such as the present.

As Table 17.3 shows, the values calculated for the former stream run close to those calculated for the stereotype stream, when observed slopes are used. The former stream can be identified as hydraulically capable of deforming both its bed and its banks, that is of shaping the meanders that are actually observed.

OUTSTANDING SPECIFIC PROBLEMS

The general problems likely to be posed by manifest underfits and by meandering tidal creeks, in relation to hydraulic explanations of Osage-type underfitness, have been mentioned earlier. Outstanding specific problems posed by the Severn near Shrewsbury relate to the well-developed stream meanders on the Severn and the Vyrnwy upstream of the confluence, and to the short meander train on the trunk channel near Leighton.

If the two channels upstream of the confluence turn out to have unduly low channel slopes in relation to discharge, as seems possible, then slope alone will not suffice to account for the difference between the meandering and the non-meandering habits. Bankfull discharge at Leighton can be taken as sensibly identical to the 250 cumecs known for Buildwas, a short distance downstream. The stereotype slope for this discharge is 0.00066, whereas the observed slope on the Leighton reach is 0.000355. The per cent discrepancy between observed and stereotype slope is the same for the Leighton reach, which does meander, as for the Atcham reach, which does not.

Grab samples of bank sediment taken at 1 m below the floodplain surface, at various points from the confluence through the Atcham reach, and on and near the Leighton reach, display significant differences of means (the data exclude samples of lacustrine sediment, 96 per cent of which passes the 0.142-mm mesh). The cutoff values for the Leighton reach are nearly half as large again as the average elsewhere. For all that, all the sampled material in the range D_{50}-D_{84} falls within the fine sand grade (US Bureau of Soils classification).

It cannot possibly be claimed that the sampling programme was in any way adequate: I foresee that the results of far more intensive sampling will be presented. Nevertheless, the results of the analyses actually made (Table 17.4), in association with the fact that part of the bank in the Leighton reach is shaped

TABLE 17.3 DIMENSIONAL AND HYDRAULIC CHARACTERISTICS OF THE FORMER CHANNEL

	Montford reach			Atcham reach		
	i Stereotype values	ii Stereotype values up to *R* inclusive	iii Reconstructed values throughout	iv Stereotype values	v Stereotype values up to *R* inclusive	vi Reconstructed values throughout
q	1185	1185	1185	1215	1215	1215
w	146.6	146.6	124	148.7	148.7	125.7
d	5.675	5.675	6.18	5.723	5.723	6.237
A_c	832	832	766.5	851	851	784
v	1.424	1.424	1.546	1.428	1.428	1.550
R	5.267	5.267	5.621	5.314	5.314	5.674
S	2.1^{-4}	3.2^{-4}	3.2^{-4}	2.07^{-4}	3.1^{-4}	3.1^{-4}
n	0.0308	0.0380	0.0366	0.0307	0.0375	0.0361
Fr	0.1908	0.1908	0.1985	0.1905	0.1905	0.1981
ω	249	379	379	252	377	377
τ	1.106	1.685	1.799	1.100	1.647	1.759
μ^*	0.1039	0.1282	0.1325	0.1036	0.1268	0.1310
μ^*_b	0.0662	0.0698	0.0726	0.0662	0.0678	0.0725
f_b	0.0173	0.0181	0.0176	0.0172	0.0180	0.0175
τ_b	4.385	4.588	5.258	4.384	4.588	5.255

Key to symbols as in Table 17.2

TABLE 17.4 BANK SEDIMENT SAMPLES, 1 m BELOW FLOODPLAIN SURFACE

Location	Grid reference (km only)	Bank	D_{50} mm	D_{84} mm
Near Melverley	344163	left	0.092	0.140
Near White Abbey	376153	right	0.113	0.150
Near Montford†	410145	left	0.115	0.153
Near Bromley's Forge	446164	left	0.115	0.162
Near The Isle	464165	right	0.130	0.180
Near Preston	523120	left	0.110	0.180
U/s of Atcham Bridge	537095	left	0.098	0.160
Opposite Coundmoor Brook confluence	567063	left	0.087	0.150
At Coundlane	570050	left	0.118	0.175
Near Cressage	595046	left	0.112	1.148
		means	0.109	0.135
Near Leighton	620045	left	0.140	0.180
U/s of Buildwas Bridge	640045	right	0.130	0.213
		means	0.1598	0.1965

The sieves ranged from 20 through 100 mesh/inch. I have determined the values of D_{50} and D_{84} by probability plotting. The values† reported for the sample near Montford are averages for samples at 0.75 and 1.25 m below the floodplain surface

in terrace gravel, can be read as suggesting an hypothesis that would deserve testing. This is, that at least on some reaches of some streams, the transition from a non-meandering to a meandering habit relates primarily or even entirely to bank strength, in the same way that the transition from single-channel flow to braiding can, in certain cases, be shown to do.

ACKNOWLEDGEMENTS

Most of my fieldwork on the Severn was conducted from the Preston Montford Field Centre. People attached to the Centre who assisted in survey and in the provision of equipment included Charles Sinker, David Pannett, Peter Robinson, Tony Evans, and Adrian Bayley. Thanks are also due to the Centre's anonymous (to me) laboratory analyst who sieved my sediment samples. My wife, Muriel Dury, acted as rod-runner and field recorder. The work was in part supported by a grant from the Graduate School, The University of Wisconsin–Madison (Project No. 181441).

REFERENCES

Bates, R. E. 1939. Geomorphic history of the Kickapoo region. *Bulletin, Geological Society of America*, **50**, 819–880.

Cogley, J. G. 1973. On runoff at the time of glaciation. *Area*, **5**, 33–37.
Dury, G. H. 1964a. Principles of underfit streams. *United States Geological Survey Professional Paper 452–A*, 67 pp.
Dury, G. H. 1964b. Subsurface exploration and chronology of underfit streams. *United States Geological Survey Professional Paper 452–B*, 56 pp.
Dury, G. H. 1966. Incised valley meanders on the lower Colo River, New South Wales. *Australian Geographer*, **10**, 17–25.
Dury, G. H. 1973. Channel habit, underfit streams, and the conduct of technical debate. *Area*, **5**, 152–154.
Dury, G. H. 1976. Discharge prediction, present and former, from channel dimensions. *Journal of Hydrology*, **30**, 219–245.
Dury, G. H. 1977. Underfit streams: retrospect, perspect, and prospect. In Gregory, K. J. (Ed.), *River Channel Changes*, Wiley, Chichester 280–293.
Dury, G. H. 1982. Step-functional analysis of long records of streamflow. *Catena*, **9**, 379–396.
Dury, G. H., and Kohchi, N. 1982. Step-functional change in annual peak flow illustrated for the Ishikari River at Hashimoto-chō, Hokkaido. *The Japanese Journal of Limnology*, **43**, 147–156.
Dury, G. H., Sinker, C. A., and Pannett, D. J. 1972. Climatic change and arrested meander development on the River Severn. *Area*, **4**, 81–85.
Ferguson, R. I. 1973. Channel pattern and sediment type. *Area*, **5**, 38–41.
Jefferson, M. S. W., 1902. Limiting widths of meander-belts. *National Geographic Magazine*, **13**, 373–384.
Kennedy, B. A. 1972. 'Bankfull' discharge and meander forms. *Area*, **4**, 284–288.
Kirkby, M. J. 1972. Alluvial and non-alluvial meanders. *Area*, **4**, 284–288.
O'Brien, R. J. 1979. *A study of the hydrodynamics of meander initiation*, unpublished M.S. thesis, The University of Wisconsin-Madison.
Pannett, D. J. and Morey, C. 1976. The origin of the old river bed at Shrewsbury. *Shropshire Conservation Trust Bulletin*, **35**, 7–12.
Richards, K. S. 1972. Meanders and valley slope. *Area*, **4**, 288–290.
Shaw, J. 1972. The Irish Sea glaciation of north Shropshire — some environmental reconstructions. *Field Studies*, **3**, 603–631.
Tinkler, K. J. 1972. The superimposition hypothesis — a general rejection and specific test. *Area*, **4**, 86–61.
Williams, G. P. 1978. Bankfull discharge of rivers. *Water Resources Research*, **14**, 1141–1154.
Young, R. W. 1974. *The Meandering Valleys of the Shoalhaven River System: a Study of Stream Adjustment to Structure and Changed Hydrologic Regimen*, unpublished Ph.D. thesis, The University of Sydney.

Background to Palaeohydrology
Edited by K. J. Gregory

18

The utilization of Arctic river analogue studies in the interpretation of periglacial river sediments from southern Britain

IAN D. BRYANT

Department of Geology, University of Nottingham

River terraces in southern Britain are usually underlain by variable thicknesses of sand and gravel. Many of these sequences have been shown to contain fossil assemblages indicative of deposition in a cold, open country environment, whilst evidence of *contemporaneous* permafrost is often provided by the presence of ice wedge casts *within* the sediments (e.g. Briggs *et al.*, 1975). Recently, it has been argued that all such successions resulted from deposition by braided river systems (Briggs and Gilbertson, 1980; Castleden, 1980) although frequently a braided river origin is only inferred on the basis of the coarse grade of the sediments and the absence of 'point-bar sequences' (Allen, 1964) (e.g. Clarke and Dixon, 1981). Braided palaeochannel patterns have been recognized on the surfaces of low terraces in the Thames (Hare, 1947) and Kennet (Cheetham, 1980) valleys but it is suggested that such features may only represent the final phase of the rivers' occupation, rather than characterizing the succession as a whole (Maizels, 1983). By contrast, Clayton (1977) has suggested that many such gravel aggradations in southern Britain were deposited by meandering channel systems. This possibility is supported by recent work on gravel-bed meandering rivers (McGowen and Garner, 1970; Gustavson, 1978; Bluck, 1971, 1979) which has shown that these river systems may deposit sequences closely analogous to those deposited by gravelly braided river systems (Bluck, 1979). Furthermore, a wide variety of river types exist in modern periglacial environments and it seems reasonable to expect that a similar diversity may have existed in lowland Britain during cold stages of the Pleistocene.

It is of considerable importance to the palaeohydrologist to establish whether the sediments under investigation are of either multiple channel, low sinuosity or single channel, high sinuosity origin (Cheetham, 1976; Starkel and Thornes, 1981) since this will affect the proportion of the rivers discharge carried by palaeochannels, whilst for a given slope braided channels have a higher discharge than meandering channels (Leopold and Wolman, 1957). The value of a sedimentological approach to this problem is illustrated by analyses of deposits in the Kennet Valley which enabled Cheetham (1976, 1980) to establish a braided river model upon which to base palaeodischarge estimates.

Detailed investigation of modern sedimentary environments has enabled various workers to establish facies models (Reading, 1978; Walker, 1979) which have then been applied to interpretation of ancient vertical facies sequences. This approach has been particularly successful in the field of fluvial sedimentology (Smith, 1970; Miall, 1977, 1978; Cant, 1978) and is applied in this article to the interpretation of Pleistocene periglacial fluvial sediments from southern Britain. Unfortunately, the term 'braided river' is potentially confusing since braiding may occur in single channel rivers whilst meandering may occur in multiple channel systems (Rust, 1978). In this account the term 'braided river' is taken to imply a multiple channel, low sinuosity river system.

MODERN PERIGLACIAL RIVER SYSTEMS

Modern periglacial rivers adopt a variety of channel forms and transport sediments of widely varying calibre (Bryant, 1982a). Such rivers have been classified according to the degree of seasonality exhibited by their annual discharge regimes, into 'Arctic nival' and 'Arctic proglacial' rivers (Church, 1974). Intermediate types exist between these two extremes (Marsh and Woo, 1981) but it is useful to consider these two river regimes as 'end members' of such a continuum in the following discussion.

Arctic nival discharge regimes are marked by only one major flood per annum, usually associated with the spring break-up flood when rapid melt of the accumulated winter's precipitation (snow) results in a high magnitude discharge event. Subsequent summer flows are derived from slow meltout of late lying snow patches or from occasional summer rainfall. In the extreme case (which we will consider here) these flows are never of equal magnitude to the spring break-up flood.

Arctic proglacial discharge régimes are likewise subject to a high magnitude flood at the commencement of the arctic summer but, due to a glacially derived component of discharge, flows remain high throughout the summer. In some cases discharges of equivalent or greater magnitude than the break-up flood may occur.

Both river types experience no flow or very low flows during the winter months. Clear division of drainage basins into one or other of these categories is complicated by the fact that either delayed snowmelt and/or intense summer storms may result in rivers of non-glacierized catchments possessing discharge characteristics similar to those of 'Arctic proglacial' rivers (March and Woo, 1981). However, field investigations in Spitsbergen and on Banks Island, N.W.T., Canada (Bryant, 1982a) suggest that this subdivision provides a useful framework for study of sedimentation in periglacial river systems.

The high discharges associated with the break-up flood may be equally well accommodated in low sinuosity, multiple channel or high sinuosity, single channel river systems. Indeed the Saganavirktok River, Alaska, possesses both of these channel types in different reaches (Scott, 1978) as does the Sachs River, Banks Island (Bryant, 1982a). The lowest reaches of the Sachs River predominantly deposit sand grade sediments and are considered more fully elsewhere (Bryant, 1982a; Good and Bryant, 1982). Attention is instead focused on those reaches of the Sachs, and other rivers, which predominantly deposit sands and gravels since these appear to be of wider application in interpretation of Pleistocene fluvial successions in southern Britain.

Miall (1977, 1978) has classified coarse-grained sediments into a series of lithofacies-types each of which may be abbreviated to a lithofacies code. In the following discussion these lithofacies will initially be described but thereafter lithofacies codes will be employed for brevity.

Sedimentation in Arctic proglacial discharge regimes

Rivers of this type are well known (see Miall, 1977 for summary; also Boothroyd and Nummedal, 1978; Shaw and Healey, 1981) and gravelly rivers of this type have been summarized in the Scott- and Donjek-type braided river facies models of Miall (1977, 1978).

The Adventelva, Vestspitsbergen is an arctic proglacial river which will be used to illustrate the variability of lithofacies sequences which are deposited by such rivers (Figure 18.1). Sedimentation in the upper reaches of the river and in tributary streams is primarily in the form of longitudinal bars, comprising massive or crudely horizontally bedded gravels (facies Gm) (*cf.* Boothroyd, 1972; Hein and Walker, 1977). Erosional contacts between units of Gm are frequently marked by gravel lags (Figure 18.1a), whilst conformable contacts may be marked by horizons of strongly imbricated, framework-supported clasts (Figure 18.1b) and pockets of scour-fill sands (facies Ss) (Figure 18.1a). In these steep channels, however, sediment tends to occur as a thin veneer over the bedrock and is subjected to frequent reworking, resulting in erosional contacts (Figure 18.1a) and a rapid downstream decrease of maximum clast size (Boothroyd, 1972). Bedforms such as transverse ribs (McDonald and

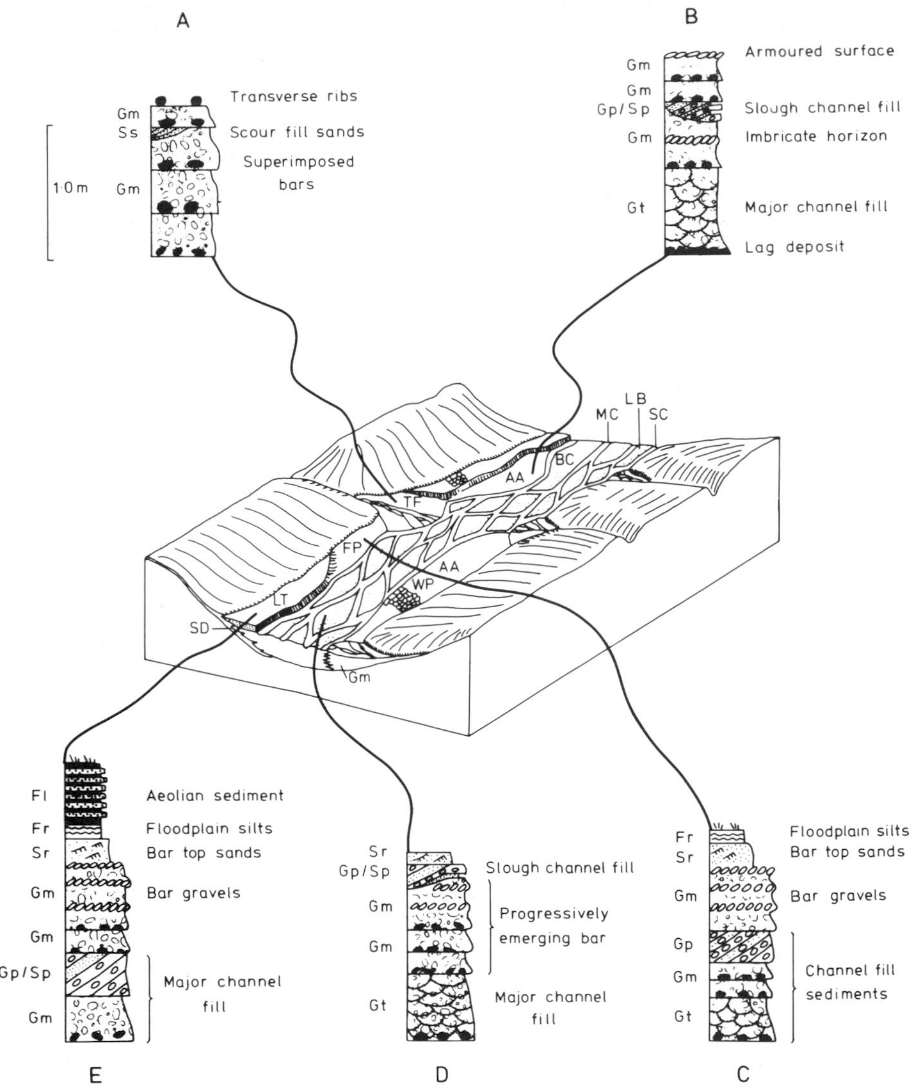

FIGURE 18.1 SEDIMENTATION IN A RIVER OF ARCTIC PROGLACIAL CHARACTER
A schematic block diagram to illustrate the spatial variability of lithofacies vertical sequences deposited in a river with an arctic proglacial annual discharge régime. (LT = low terrace; FP = floodplain; TF = tributary fan; AA = abandoned alluvium; MC = main channel; SC = slough channel; LB = lateral bar; WP = wedge polygons; SD = solifluction deposit. Lithofacies codes are fully explained in the text)

Banerjee, 1971) may occur in these reaches and may be used to infer various palaeohydraulic parameters (Koster, 1978), although they are difficult to recognize in vertical sections of ancient deposits. These sequences in tributary valleys (Figure 18.1a) and valley margin fans closely resemble the Scott-type sequences of Miall (1977, 1978).

By contrast sedimentary sequences in the middle reaches of the axial valley result from accumulation in a diverse milieu of subenvironments. Various elements of the alluvium rise to differing elevations such that a series of 'topographic levels' exists (Williams and Rust, 1969). As sediment accumulates at progressively higher levels so mean grain size decreases. The coarsest sediment floors the main channels of the principal sediment dispersal routes (Level 1 of Williams and Rust) as a series of trough cross-bedded or massive gravels (facies Gt and Gm respectively). Lateral expansion of these channels appears to lead to reduced competence levels and the growth of within channel bars (Cheetham, 1979) which become progressively emergent (Figure 18.1d). Thereafter supra-bar sedimentation of rippled sands (facies Sr) occurs as near bankfull flows are diverted across the bar to produce lateral bar migration (Bluck, 1974, 1976, 1979). At the margins of the active channel belt thicker sequences of bar top sands (Sr) may be capped by vertical accretion floodplain sediments (facies Fr) (Figure 18.1e). (N.B. Facies Fr is equivalent to Miall's facies Fl; the facies code Fl in this account is reserved for description of laminated aeolian sediments.) In more central areas of the alluvial tract, fine grained bar top sediments that are not stabilized by vegetation are subject to aeolian deflation at low stage (Nickling, 1978) to leave an armoured bar surface (Figure 18.1b). This aeolian sediment may accumulate at the valley margins as low terraces (Péw, 1968; Pissart *et al.*, 1977; Bryant, 1982b) or as dune fields (Boothroyd and Nummedal, 1978). In Adventdalen these sediments are interbedded with buried organic horizons to produce laminated fine grained accumulations (facies Fl) (Figure 18.1e).

Careful examination of sedimentary structures on similar bars of Icelandic sandar has identified the importance of lateral bar migration (Bluck, 1974). Each bar unit consist of a lower *bar core* overlain by a *bar platform*. The lower part of the bar core seems to consist of either massive gravels or festoon bedded gravels (Walker, 1979) generated within major channels. As the bar accretes to become emergent so successive increments of low angle gravel sets (facies Gm) are added to the bar surface until the bar reaches a near bankfull elevation. Subsequent aggradation is primarily of washover sands (Sr) and other *supra-platform* sediments. At this point subtle fluctuations in stage may cause the bar to be emergent with an armoured surface or submerged to a shallow depth. Whilst the bar is submerged, flow in the main channel is deflected across the bar surface to transport either sand or (at higher flow strengths) gravel into the slough or inner channel (Figure 18.2). As a result such channels are frequently infilled by planar foreset bedded sands (facies Sp)

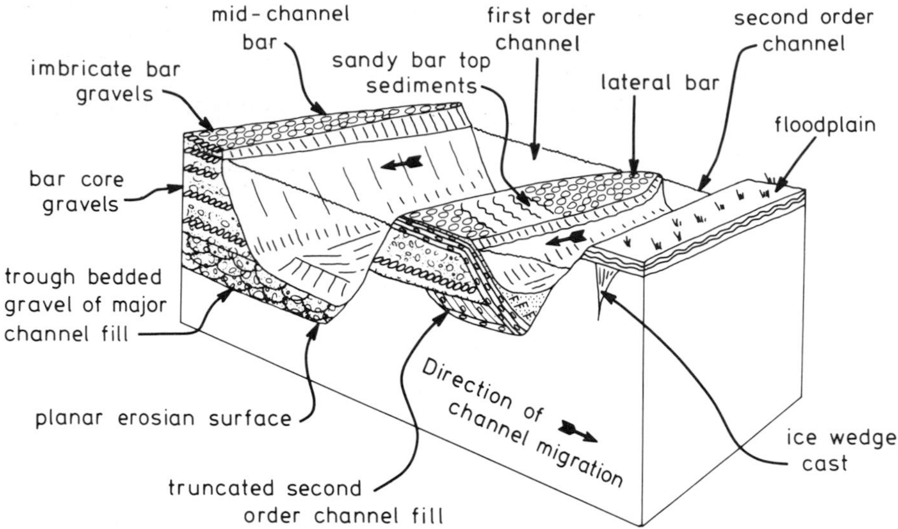

FIGURE 18.2 TRUNCATION OF FACIES SEQUENCES BY LATERAL CHANNEL MIGRATION
A schematic block diagram to illustrate: (a) the relationship between geomorphological units and lithofacies units and, (b) the truncation of complete vertical facies sequences by lateral channel shift of the main (first order) and slough (second order) channels

or gravels (facies Gp) (Figure 18.2). Flow through slough channels is usually inhibited by riffles at the head and mouth of the channel composed of sediment washed out of the main channel. However, resurgences of flow strength in the channel due to breaching of the riffle or rapid stage changes may cut reactivation surfaces in the foresets (Fahnestock and Bradley, 1973) and deposit trough cross-bedded (St) or ripple bedded (Sr) sands. These complex slough channel fills (Figure 18.1b and d) seem to characterize the sediments of low sinuosity, multiple channel river systems (Bluck, 1979) but *only* where frequent fluctuations of stage at or near bankfull discharge occur in order to overtop the bars and rework the high stage (bar platform) deposits into the adjacent slough channel.

Larger sets of foreset bedded sands and gravels are also generated at the downstream apex of bars (Fahnestock and Bradley, 1973; Bluck, 1979) but these are usually thicker, orientated parallel to the main channel axis and interfinger with horizontal gravel sheets of the bar platform (Figure 18.3).

Sedimentation in Arctic nival discharge regimes

Rivers which possess an Arctic nival discharge regime are less well described than those of Arctic proglacial character and the observations reported herein

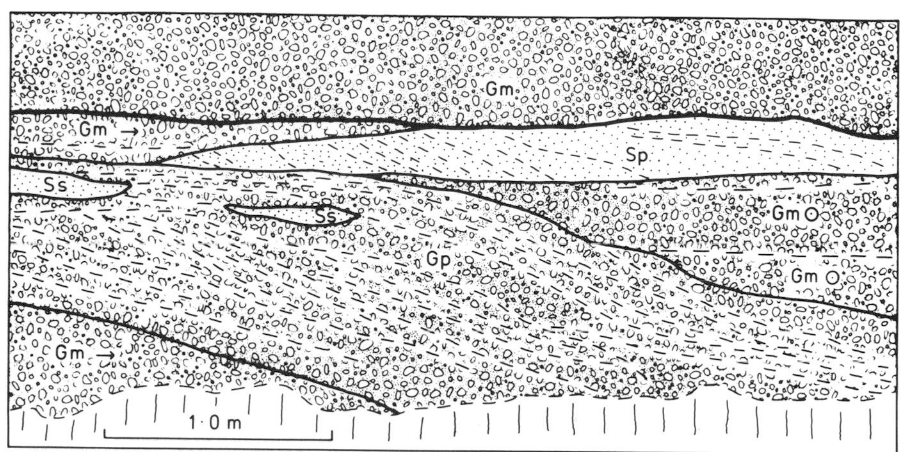

FIGURE 18.3 FORESET BEDDED GRAVELS DEPOSITED AT THE DOWNSTREAM APEX OF A BAR

A true scale diagram to illustrate the transition from bar core sediments (Gm) to planar, tabular foreset bedding of gravels (Gp) at the downstream apex of a bar. The channel into which the foresets were building (right, centre) was subsequently infilled by massive gravels (Gm) deposited by axial flow. (Drawn from a photograph of a quarry exposure at Tattershall Thorpe, Lincolnshire.) (Palaeoflow in the plane of the page is designated by arrows; through the page by dotted circles)

relate to investigation of only three rivers. Two short, steep streams i.e. the Schetigelva and Kvadehukselva from Vestspitsbergen and part of the larger, lower gradient Sachs River from Banks Island (Bryant, 1982a). These rivers possess a low sinuosity, multiple channel habit at peak discharge but at low stage they adopt a more sinuous frequently single channel course flowing in the deeper channels cut by the high stage course (Figure 18.4).

Sedimentary sequences similar to those generated in Arctic proglacial discharge regimes are also produced in these river systems. Steep tributary streams and the headwaters of the main river are dominated by thin sequences of superimposed bar sediments (Figure 18.4a) which may thicken downstream into alluvial fans (Figure 18.4b). Floodplain sequences (Figure 18.4c), sequences in the central part of the alluvial tract (Figure 18.4d), and beneath marginal terraces of aeolian sediments (Figure 18.4e), closely resemble those produced in analogous situations within Arctic proglacial river systems. However floodplain silts (Fr) appear to be less abundant whilst very rarely do slough channels become infilled by planar foresets (Figure 18.4e). Slough channel fills, where they do occur, tend to relate to one phase of infill rather than repeated episodes. This phenomenon is due to the absence of frequent discharge events during the summer of sufficient magnitude to overtop bars or rework high-stage sediment and this effect is often compounded by strong imbrication of the bar and channel floor sediments during flow recession from

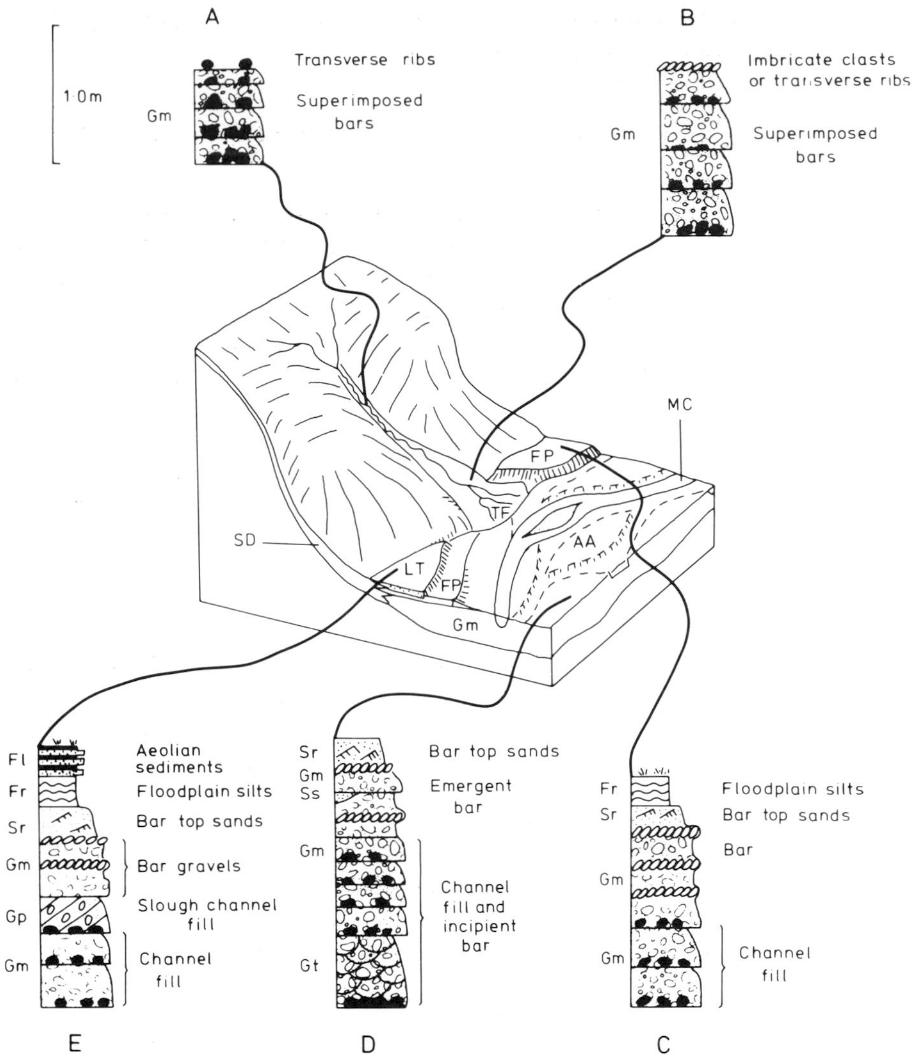

Figure 18.4 Sedimentation in a River of Arctic Nival Character
A schematic block diagram to illustrate the spatial variability of lithofacies vertical sequences deposited in a river with an arctic nival annual discharge régime. (See Figure 18.1 for key to the abbreviations used)

the break-up flood. As a result sequences deposited by Arctic nival river systems tend to comprise predominantly high stage facies types (Gm with subsidiary Gt) with low proportions of facies deposited during moderate to low stage reworking (i.e. facies Gp, Sp, St or thick units of Sr). Most foreset bedding in such successions is of the downstream bar apex type (Figure 18.3) whilst most sandy sediment represents thin supra-bar accumulations (Sr) rather than slough channel infills (Sp, St).

PERIGLACIAL PLEISTOCENE RIVER SEDIMENTS IN SOUTHERN BRITAIN

Very little detailed facies analysis of fluvial successions of known periglacial character has been reported from southern Britain (Cheetham, 1980; Rose *et al.*, 1980) yet these two studies indicate considerable variability of river habit within the Last Glacial Stage (Devensian). This work has been extended (Bryant, 1982a) to encompass other sites in southern Britain (Figure 18.5). Sequences containing a high proportion of Jurassic limestone clasts were found to be particularly informative since they exhibited strong imbrication and were subject to early cementation, thereby preserving the integrity of individual facies units. Two groups of lithofacies sequences were identified which were readily interpretable in terms of facies models outlined in the preceding section.

Group 1

Facies sequences were recorded from quarry exposures at Linch Hill, near Oxford, and at Tattershall Thorpe, near Boston, Lincolnshire (Figure 18.5). Both of these successions contain fossil assemblages of periglacial affinity and intraformational ice wedge casts (Bryant, 1983), thereby indicating their deposition, at least in part, under permafrost conditions. The vertical sequences of facies (Figure 18.6) have been analysed by facies transition matrices (Harms *et al.*, 1975) to deduce facies relationship diagrams (Table 18.1). These indicate that facies Gt consistently grades upwards into facies Gm; to be overlain by facies characteristic of bar top or channel fill sedimentation (Sr, Sp, Gp, Ss). The spatial relationship of the various lithofacies is also consistent with the model based on the Adventelva (Figure 18.1), e.g. bar core gravels (Gm) are characterized by imbricate framework supported horizons which can be traced laterally into foreset bedded sands (Sp) and gravels (Gp) indicative of bar tail deposition (Figure 18.3).

The vertical facies sequences (Figure 18.6), especially the abundance of slough channel fills (Figure 18.7), suggests accumulation in a braided river environment. The truncation of ice wedge casts and slough channel fills (Figure 18.7) further suggests that substantial lateral channel shift during aggradation has resulted in only partial preservation of each depositional cycle (Bryant, 1983). These facies sequences closely resemble those recorded from proximal proglacial outwash sequences (Eynon and Walker, 1974; Fraser, 1982) which is surprising since conventional wisdom suggests that the

FIGURE 18.5 LOCATION OF STUDY SITES IN SOUTHERN BRITAIN
The Devensian maximum ice limit is shown (after Bowen, 1978) as a heavy black line. Tattershall Thorpe (TT); Linch Hill (LH); Brimpton (B) and Farnham (F) are marked by dots

successions aggraded in drainage basins beyond the accepted contemporary limits of glaciation (Figure 18.5). It seems reasonable to expect that the similarity of the sediments at Linch hill and Tattershall Thorpe to sediments from glacial outwash sequences results from a common depositional process, namely the reworking of bar core sediments (high stage) into channels by lower stage discharges. Thus at these two localities such stage fluctuations were not the result of diurnal variations in ablation rates of glaciers within the catchment (*cf*. Fahnestock, 1963; Church, 1972) but were of non-glacial origin. Two likely mechanisms to account for this phenomenon may be suggested on the basis of investigation of similar non-glacierized catchments in the Arctic (Marsh and Woo, 1981) i.e. either delayed snowmelt and/or heavy summer rainfall. Little evidence is currently available to distinguish the relative

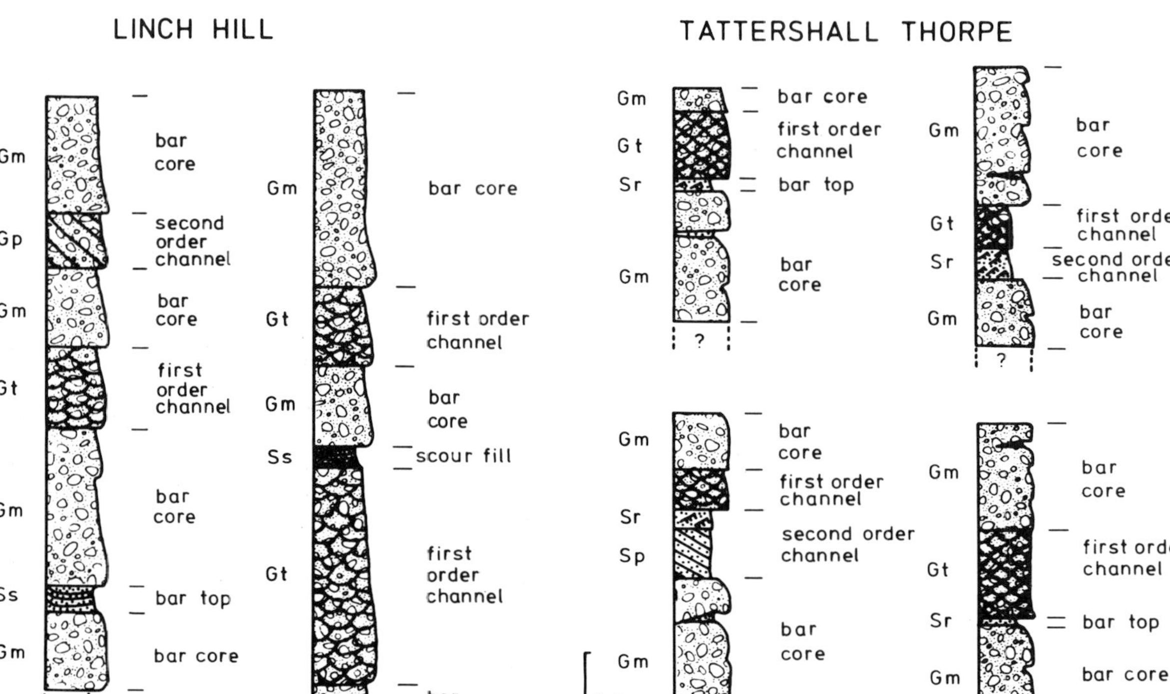

FIGURE 18.6 INTERPRETATION OF LITHOFACIES SEQUENCES RECORDED FROM QUARRIES IN SOUTHERN ENGLAND
Representative partial stratigraphic logs from two quarries in southern England and their interpretation, as suggested by analogy with the lithofacies sequences shown in Figure 18.1

TABLE 18.1 SIMPLIFIED FACIES RELATIONSHIPS
Reproduced from Bryant, 1983 by permission of Blackwell Scientific Publications Ltd.

Linch Hill:	Gt ⇌ Gm → Sp ↘ ↙ Gm
Tattershall Thorpe:	Gt → Gm → Sr, Sp, Ss/Sp

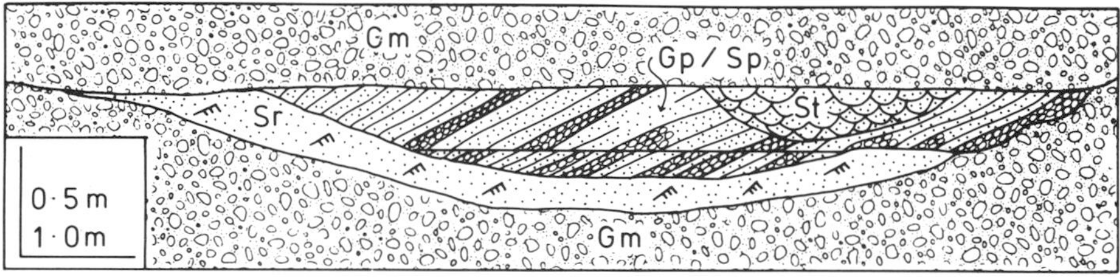

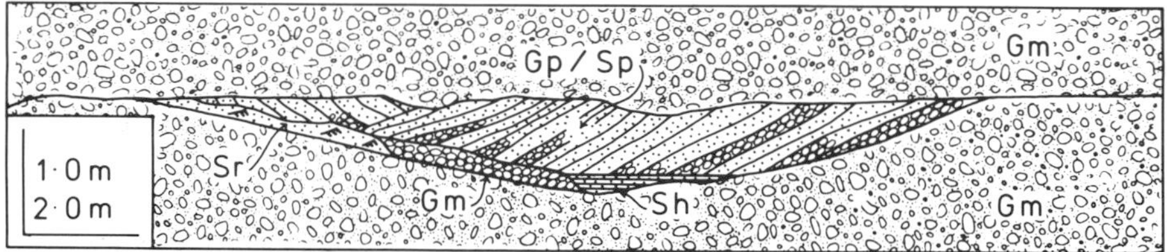

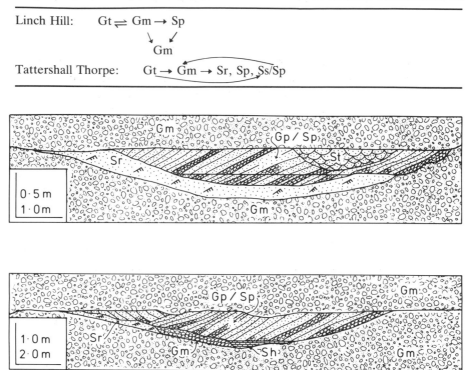

FIGURE 18.7 TRUNCATED SLOUGH CHANNEL FILLS
Diagrammatic representations of two slough channel fills. Drawn from quarry exposures at Linch Hill, Oxfordshire. (Note the vertical exaggeration indicated by the vertical and horizontal scales in each diagram) Reproduced from Bryant, 1983 by permission of Blackwell Scientific Publications Ltd.

importance of these two potential sources of discharge although the probably continental Devensian palaeoclimate (Coope, 1975) suggests that the former mechanism may have been more important. This also seems probable on the basis of *a priori* considerations regarding latitudinal effects. In the Arctic the summer sun maintains a high angle throughout the summer thereby reducing the likelihood of snowpatches remaining shaded throughout the day and increasing the relative importance of the break-up flood. By contrast, in middle latitudes the sun always shines from the south and thus steep north-facing slopes may have retained substantial snow covers well into the summer (although it is difficult to conclusively demonstrate the importance of this mechanism). This investigation does, however, demonstrate firstly, a braided river origin for the sediments; secondly, a preferential preservation of the lowest part of each depositional cycle; and thirdly, the existence of an annual discharge regime akin to that of modern Arctic proglacial river systems.

Group 2

Facies sequences recorded from Farnham, Surrey (Bryant, *et. al.* in press) and Brimpton, Berkshire (Bryant, Holyoak and Mosley, in press) (Figure 18.5) differed markedly from those described above. For the most part they consisted of monotonous sequences of bar gravels (Gm), occasionally separated by sandy pockets (Ss) or bar top sands (Sr). No foreset bedded slough channel fills were recorded at either locality and abandoned channels were usually infilled by silts, clays, and organic debris indicative of a slack water environment. Trough cross-bedding (Gt) was rarely recorded from the sequences but was consistently overlain by bar gravels (Gm). Foreset bedded gravels were also infrequently encountered and were invariably of downstream bar apex origin (Figure 18.3).

The low facies diversity of these sediments suggests that they may have originated in a sedimentary environment similar to that illustrated in Figure 18.4 by the superposition bars which were only active at high stage. It should also be noted, however, that similar Gm dominated sequences might be expected to result from preferential preservation of the lowest parts of sedimentary cycles deposited by meandering gravel bed streams (Bluck, 1971, 1979; Blacknell, 1982). Facies sequences of this type are therefore less environmentally diagnostic but (whatever their precise depositional environment) they appear to represent a hydrologic regime dominated by superimposed flood cycles with very low amounts of falling or lower stage reworking.

Several possible explanations of the differences between the two groups of sediments might be considered and either location within the palaeofloodplain (compare the various sequences in Figures 18.1 and 18.4); or deposition under differing climatic regimes due either to geographical variation or non-time equivalence of the sedimentary successions are possibilities which remain profitable areas for further research.

IMPLICATIONS FOR PALAEOHYDROLOGICAL INVESTIGATION

Palaeodischarge estimates frequently rely upon employment of values for palaeochannel slope, palaeochannel width and depth, and/or some assessment of sediment particle size (usually mean or maximum particle diameter). The implications of the foregoing discussion to determination of each of these parameters will now be separately considered.

1. Palaeochannel slope is often considered to be related to the slope of the river terrace (Cheetham, 1980) or to the slope of the lower bounding surface of the fluvial sediments beneath the terrace surface (Clarke and Dixon,

1981). The relationship of channel slope to either terrace or rock-head gradient is dependent upon the sinuosity of the river channel (i.e. the ratio of channel length to valley length) and hence channel slope only approximates to the terrace slope for sinuosities of *ca* 1.0. Facies analysis of the sediments may result in recognition of sequences similar to those in Group 1; such that a braided river origin is inferred and a sinuosity value typical of braided river systems may be adopted such that (if uniform flow is assumed) terrace slope may approximate to hydraulic slope. Sedimentary sequences of Group 2 character have also been ascribed to a braided river origin (Clarke and Dixon, 1981) but this assumption must be regarded with some caution.

2. Palaeochannel width is difficult to determine in most exposures due to the obliquity of the section. This difficulty has been overcome by measurement of palaeochannel width from aerial photographs (Cheetham, 1980) but this is not always possible and may not be representative of channel widths represented by the underlying sedimentary sequence. Similarly, channel fills within sedimentary sequences are almost invariably truncated (Figure 18.7) and so represent minimum channel depths. An estimate of former channel depth from such channel fills may be obtained by projection of truncated foreset surfaces upward to an intersection point with the channel margin, although this too is likely to underestimate true channel depth since slough channels are frequently shallower than major channels (Bluck, 1979; Bryant, 1982a). In this context foresets which are associated with supra-platform bar topsets and appear to have accumulated at the downstream apex of bars (Figure 18.3) may provide the best estimate of maximum channel depth. Even then channel depth may not be equivalent to flow depth since much deposition in braided channels occurs in association with overbank flood events (Maizels, this volume pp.101).
3. Many modern periglacial braided rivers are characterized by pronounced downvalley sediment sorting (e.g. Boothroyd, 1972; Maizels, 1979) and particular downvalley sorting patterns have been claimed to be diagnostic of braided river transport (Frostick and Reid, 1980). However, similar analyses of Pleistocene sediments in southern Britain (Clarke and Dixon, 1981; Bryant, 1982a) reveal no significant downvalley trends. Two explanations of this phenomenon may be considered.

The first is that many large, low gradient, braided river systems do not appear to transport significant quantities of coarse sediment *downstream* e.g. the Adventelva is characterized in its lower course by sandy, rather than gravelly, bars (Bryant, 1982a), whilst its delta is predominantly composed of sands and silts (Prior *et al.*, 1981). A similar situation is reported for the Knik River, Alaska (Fahnestock and Bradley, 1973). Thus, gravelly sediment, which has been well sorted by transport in steeper tributary streams and upstream reaches of the river (e.g. Maizels, 1979) may subsequently enter prolonged storage with the lateral bars in the middle reaches of the axial valley (*cf.* 'storage bars' of Church and Jones, 1982) without further downvalley sorting.

An alternative explanation of the absence of significant sorting is the potentially limited availability of coarser clasts. However, coarse sarsen stones occur frequently in upstream reaches of the Kennet Valley (Clark *et al.*, 1967) and occasionally in periglacial fluvial sediments of the middle reaches, thereby suggesting that coarse sediment was available to the river system. The infrequent occurrence and anomalously large diameter of sarsen clasts in the periglacial river sediments downstream may however reflect ice-rafting of sediment (*cf.* Walker, 1973) rather than a competence control resulting in downvalley grading.

Clast size in bar core gravels (Gm) may thus severely underestimate (due to the unavailability of large clasts) or overestimate (due to the presence of ice-rafted debris) peak discharge. However some sedimentary units (e.g. gravel foresets) can show extremely good sorting and have been used to give relatively precise palaeocurrent velocities for various elements of the channel system (Eynon and Walker, 1974).

SUMMARY

Periglacial river systems in the modern arctic are divisible into Arctic nival and Arctic periglacial rivers on the basis of the degree of seasonality exhibited by their annual discharge regimes. The deposits of Arctic proglacial river systems are distinguishable from those of Arctic nival river systems by the greater abundance of lithofacies types attributable to reworking of bar sediments during recurrent high summer flow events.

Sequences of Pleistocene periglacial fluvial sediments in southern Britain are interpretable by recourse to facies models derived from study of modern periglacial rivers provided that account is taken of preferential preservation of only the lower part of each depositional cycle. Late Pleistocene periglacial fluvial systems in southern Britain seem to have possessed discharge regimes of both 'Arctic nival' and 'Arctic proglacial' character. Some of the sediments examined are clearly ascribable to a braided river origin, but where slough channel fills are absent such successions may be ascribed to braided or meandering river deposition. Determination of the channel type and origin of each lithological unit has important implications for estimation of palaeodischarge, whilst the absence of significant downvalley decrease in mean grain size suggests that the use of grain size parameters in estimation of peak discharge may produce misleading results.

ACKNOWLEDGEMENTS

The author is grateful to Mr. J. Rose, Dr. J. K. Maizels, and Dr. P. Worsley for their helpful comments on an early draft of the manuscript.

REFERENCES

Allen, J. R. L. 1964. Studies in fluvial sedimentation: six cyclothems from the Lower Old Red Sandstone, Anglo-Welsh Basin. *Sedimentology,* **3**, 163–198.

Blacknell, C. 1982. Morphology and surface sedimentary features of point bars in Welsh gravel-bed rivers. *Geological Magazine,* **119**, 181–192.

Bluck, B. J. 1971. Sedimentation in the meandering River Endrick. *Scottish Journal of Geology,* **7**, 93–138.

Bluck, B. J. 1974. Structural and directional properties of some valley sandur deposits in southern Iceland. *Sedimentology,* **21**, 533–554.

Bluck, B. J. 1976. Sedimentation in some Scottish rivers of low sinuosity. *Transactions of the Royal Society of Edinburgh,* **69**, 425–456.

Bluck, B. J. 1979. Structure of coarse grained braided stream alluvium. *Transactions of the Royal Society of Edinburgh,* **70**, 181–221.

Bluck, B. J. 1980. Structure, generation and preservation of upward fining braided stream cycles in the Old Red Sandstone of Scotland. *Transactions of the Royal Society of Edinburgh,* **71**, 29–46.

Boothroyd, J. C. 1972. *Coarse Grained Sedimentation on a Braided Outwash Fan, Northeast Gulf of Alaska*, Technical Report No. 6, Coastal Research Division, University of South Carolina, Columbia, 127 pp.

Boothroyd, J. C., and Nummedal, D. 1978. Proglacial braided outwash: a model for humid alluvial-fan deposits. In Miall, A. D. (Ed), *Fluvial Sedimentology, Memoir of the Canadian Society of Petroleum Geologists,* **5**, 597–604.

Bowen, D. Q. 1978. *Quaternary Geology: a Stratigraphic Framework for Multi-disciplinary Work*, Pergamon, Oxford 221 pp.

Briggs, D. J., Coope, G. R., and Gilbertson, D. D. 1975. Late Pleistocene terrace deposits at Beckford, Worcestershire, England. *Geological Journal,* **10**, 1–15.

Briggs, D. J., and Gilbertson, D. D. 1980. Quaternary processes and environments in the upper Thames Valley. *Transactions of the Institute of British Geographers*, N.S. **5**, 53–65.

Bryant, I. D. 1982a. *Periglacial River Systems: Ancient and Modern*, unpublished Ph.D. Thesis, University of Reading.

Bryant, I. D. 1982b. Loess deposits in Lower Adventdalen, Vestspitsbergen. *Polar Research,*

Bryant, I. D. 1983. Facies sequences associated with some braided river deposits of late Pleistocene age from southern Britain. In Collinson, J. D. and Lewin, J. (Eds.), *Modern and Ancient Fluvial Systems: Sedimentology and Processes, Special Publication of the International Association of Sedimentologists,* **6**, 267–275.

Bryant, I. D., Gibbard, P. L., Holyoak, D. T., Switsur, V. R. and Winkle, A.G. in press. Stratigraphy and palaeontology of Pleistocene cold stage deposits at Alton Road Quarry, Farnham, Surrey, England. *Geological Magazine.*

Bryant, I. D., Holyoak, D. T., and Moseley, K. A. in press. Devensian deposits at Brimpton, Berkshire, England, *Proceedings of the Geologists Association.*

Cant, D. J. 1978. Development of a facies model for sandy braided river sedimentation: comparison of the South Saskatchewan River and the Battery Point Formation. In Miall, A. D. (Ed.), *Fluvial Sedimentology, Memoir of the Canadian Society of Petroleum Geologists,* **5**, 597–604.

Castleden, R. 1980. Fluvioperiglacial pedimentation: a general theory of fluvial valley development in cool temperate lands, illustrated from western and central Europe. *Catena,* **7**, 135–152.

Cheetham, G. H. 1976. Palaeohydrological investigations of river terrace gravels. In Davidson, D. A., and Shackley, M. L. (Eds.), *Geoarchaeology: Earth Science and the Past*, Duckworth, London, 335–344.

Cheetham, G. H. 1979. Flow competence in relation to stream channel form and braiding. *Bulletin of the Geological Society of America,* **90**, 877–886.

Cheetham, G. H. 1980. Late Quaternary palaeohydrology: the Kennet Valley case-study. In Jones, D. K. C. (Ed.), *The Shaping of Southern England,* Institute of British Geographers Special Publication, **11**, 203–223.

Church, M. 1972. Baffin Island sandurs; a study of Arctic fluvial processes. *Geological Survey of Canada Bulletin,* **216**, 208 pp.

Church, M. 1974. Hydrology and permafrost with reference to northern North America. In *Permafrost Hydrology*, Proceedings of Workshop Seminar, Canadian National Committee, 7–20.

Church, M., and Jones, D. 1982. Channel bars in gravel-bed rivers. In Hey, R. D., Bathurst, J. C., and Thorne, C. R. (Eds.), *Gravel-bed Rivers*, Wiley, London, 291–338.

Clark, M. J., Lewin, J., and Small, R. J. 1967. The sarsen stones of the Marlborough Downs and their geomorphological implications. *University of Southampton, Research Series in Geography,* **4**, 3–40.

Clarke, M. R., and Dixon, A. J. 1981. The Pleistocene braided river deposits in the Blackwater Valley area of Berkshire and Hampshire, England. *Proceedings of the Geologists Association,* **92**, 139–158.

Clayton, K. M. 1977. River terraces. In Shotton, F. W. (Ed.), *British Quaternary Studies: Recent Advances*, Clarendon, Oxford, 153–167.

Coope, G. R. 1975. Climatic fluctuations in northwest Europe since the last interglacial indicated by fossil assemblages of Coleoptera. In Wright, A. E., and Moseley, F. (Eds.), *Ice Ages: Ancient and Modern, Geological Journal Special Issue,* **6**, 153–168.

Eynon, G., and Walker, R. G. 1974. Facies relationships in Pleistocene outwash gravels, southern Ontario: a model for bar growth in braided rivers. *Sedimentology,* **21**, 43–70.

Fahnestock, R. K. 1963. Morphology and hydrology of a glacial stream — White River, Mount Rainier, Washington. *United States Geological Survey Professional Paper, 442–A*, 70 pp.

Fahnestock, R. K., and Bradley, W. C. 1973. Knik and Matanuska Rivers, Alaska: a contrast in braiding. In Morisawa, M. (Ed.), *Fluvial Geomorphology, Proceedings of 4th Geomorphology Symposium, State University, Binghampton, New York*, 220–250.

Fraser, J. Z. 1982. Derivation of a summary facies sequence based on Markov chain analysis of the Caledon outwash: a Pleistocene braided glacial fluvial deposit. In Davidson-Arnott, R., Nickling, W., and Fahey, B. D. (Eds.), *Research in glacial, glacio-fluvial and glacio-lacustine systems, Proceedings of 6th Guelph Symposium*, Geobooks, Norwich, 175–202.

Frostick L. E., and Reid, I. 1980. Sorting mechanisms in coarse-grained alluvial sediments: fresh evidence from a basalt plateau gravel, Kenya. *Journal of the Geological Society of London,* **137**, 431–441.

Good, T. R., and Bryant, I. D. 1982. A study of modern fluvial and eolian processes, Lower Sachs River, Banks Island, N.W.T., Canada. *Abstracts: International Association of Sedimentologists Congress, 1982, Hamilton, Ontario,* 147.

Gustavson, T. C. 1978. Bedforms and stratification types of modern gravel meander lobes, Nuecas River, Texas. *Sedimentology,* **25**, 401–426.

Hare, F. K. 1947. The geomorphology of a part of the middle Thames. *Proceedings of the Geologists Association,* **58**, 294–339.

Harms, J. C., Southard, J. B., Spearing, D. J., and Walker, R. G. 1975. Depositional environments as interpreted from primary sedimentary structures and stratification sequences. *Society of Economic Palaeontologists and Mineralogists Short Course Notes,* **2**, Dallas, 69–73.

Hein, F. J., and Walker, R. G. 1977. Bar evolution and development of stratification in the gravelly, braided, Kicking Horse River, British Columbia. *Canadian Journal of Earth Sciences,* **14**, 562–570.

Koster, E. H. 1978. Transverse ribs: their characteristics, origin and palaeohydraulic significance. In Miall, A. D. (Ed.), *Fluvial Sedimentology, Canadian Society of Petroleum Geologists Memoir,* **5**, 161–186.

Leopold, L. B., and Wolman, M. G. 1957. River channel patterns — braided, meandering and straight. *United States Geological Survey Professional Paper, 282–B*, 39–85.

Maizels, J. K. 1979. Proglacial aggradation and changes in braided channel patterns during a period of glacier advance: an alpine example., *Geografiska Annaler,* **61A**, 87–101.

Maizels, J. K. 1983. Proglacial channel systems: change and thresholds for change over long, intermediate and short timescales. In Collinson, J. D., and Lewin, J. (Eds.), *Modern and Ancient Fluvial Systems: Sedimentology and Processes, Special Publication of the International Association of Sedimentologists,* **6**, 251–266.

Marsh, P., and Woo, M.-K. 1981. Snowmelt, glacier melt and high arctic stream-flow regimes. *Canadian Journal of Earth Services,* **18**, 1380–1384.

McDonald, B. C., and Banerjee, I. 1971. Sediments and bedforms on a braided outwash plain. *Canadian Journal of Earth Sciences,* **8**, 1282–1301.

McGowen, J. M., and Garner, L. E. 1970. Physiographic features and stratification types of coarse-grained point bars: modern and ancient examples. *Sedimentology,* **14**, 77–111.

Miall, A. D. 1977. A review of the braided-river depositional environment. *Earth Science Reviews,* **13**, 1–62.

Miall, A. D. 1978. Lithofacies types and vertical profile models in braided river deposits: a summary. In Miall, A. D. (Ed.), *Fluvial Sedimentology, Memoir of the Canadian Society of Petroleum Geologists,* **5**, 597–604.

Nickling, W. G. 1978. Eolian sediment transport during dust storms: Slims River Valley, Yukon Territory. *Canadian Journal of Earth Sciences,* **15**, 1069–1084.

Péwé, T. L. 1968. Loess deposits of Alaska. *Proceedings of 23rd International Geological Congress,* 1968, **8**, 297–309.

Pissart, A., Vincent, J.-S., and Edlund, S. A. 1977. Dépôts et phénomènes éoliens sur l'île de Banks, Territories du Nord-Ouest, Canada. *Canadian Journal of Earth Sciences,* **14**, 2462–2480.

Prior, D. B., Wiseman, W. J., and Bryant, W. R. 1981. Submarine chutes on the slopes of fjord deltas. *Nature, London,* **290**, 326–328.

Reading, H. G. (Ed.) 1978. *Sedimentary Environments and Facies*, Blackwell Scientific, Oxford, 557 pp.

Rose, J., Turner, C., Coope, G. R., and Bryan, M. D. 1980. Channel changes in a lowland river catchment over the last 13 000 years. In Cullingford, R. A., Davidson, D. A., and Lewin, J. (Eds.), *Timescales in Geomorphology*, Wiley, London. 159–175.

Rust, B. R. 1978. A classification of alluvial channel systems. In Miall, A. D. (Ed.), *Fluvial Sedimentology, Memoir of the Canadian Society of Petroleum Geologists,* **5**, 187–198.

Scott, K. M. 1978. Effects of permafrost on stream channel behaviour in arctic Alaska. *United States Geological Survey Professional Paper, 1068*, 19 pp.

Shaw, J., and Healy, T. R. 1981. Morphology of the Onyx River System, McMurdo Sound Region, Antarctica. *New Zealand Journal of Geology and Geophysics,* **23**, 223–238.

Smith, N. D. 1970. The braided stream depositional environment: comparison of the Platte River and some Silurian clastic rocks, North–Central Appalachians. *Geological Society of America Bulletin,* **81**, 2993–3014.

Starkel, L., and Thornes, J. B. 1981. Palaeohydrology of river basins. *British Geomorphological Research Group Technical Bulletin,* **28**, 106 pp, Geobooks, Norwich.

Walker, H. J. 1973. Morphology of the North Slope. In Britton, M. (Ed.), *Alaskan Arctic Tundra*, Arctic Institute of North America, Technical Paper No. 25, 49–92.

Walker, R. G. (Ed.) 1979. *Facies Models*, Geoscience Canada, Reprint Series, **1**, 211 pp.

Williams, P. F., and Rust, B. R. 1969. The sedimentology of a braided river. *Journal of Sedimentary Petrology,* **39**, 649–679.

Background to Palaeohydrology
Edited by K. J. Gregory

19

Palaeovegetation and palaeohydrology in upland Britain

P. E. J. WILTSHIRE and P. D. MOORE

Department of Plant Sciences, King's College, University of London

Pollen analysis has become a standard technique for the investigation of past vegetation, and the deposits most frequently used for such botanical reconstructions are usually lacustrine sediments or peats, both of which may also be expected to register concurrent changes in palaeohydrological regimes. In a lake sediment, an increased discharge from the catchment results in sediment changes of a textural and a chemical nature. Often the rate of sedimentation increases, as in the lakes of the English Lake District (Pennington, 1978a), as a result of greater erosion from the catchment accompanying hydrological and land use changes. In the case of peat deposits, the nature of the evidence for hydrological modifications varies with the physiography of the mire; fens and valley, and basin mires which are rheotrophic in their hydrological regime (*sensu* Moore and Bellamy, 1974) may register increased inwash by a raised proportion of inorganic material in the sediment, derived from eroded soils in the catchment. They may also respond by increasing their rate of peat accumulation as a consequence of impaired decomposition and/or changes in the litter-forming, contemporary vegetation. Ombrotrophic, rain-fed mires, on the other hand, will respond to increased water input only by alterations in the rate of peat accretion, again resulting from lower decomposition rates following waterlogging or floristic changes, or both.

In all these cases it is possible to use the sediments which register the changes as a source of pollen, by means of which one can investigate the contemporaneous vegetation of the catchment and of the surrounding regions. In the case of lake sites, as much as 85 per cent of the pollen contained within such sediments may have been derived from inwash rather than direct fallout and rainwash from the atmosphere (Bonny, 1978), and this complicates the picture in that some of the pollen is likely to be secondarily derived from older sediments and soils. The same will apply to rheotrophic mire peats, though the precise proportion of pollen entering the system in this way has never been quantified and is likely to be very variable from site to site.

The input of allochthonous material into lake basins and rheotrophic mires during periods of increased water discharge from the catchment leads to a number of complications in the interpretation of pollen stratigraphy. Firstly, eroded soils contain older pollen and do not, therefore, necessarily reflect accurately the contemporaneous vegetation of the catchment. A change in pollen proportions at such stratigraphic horizons should not automatically be regarded as representative of changes in surrounding vegetation. Secondly, radiocarbon dating of such horizons can lead to anomalous dates (Pennington, 1975, 1978b), in which case it is difficult to determine accurately the overall rate of sedimentation and particularly difficult to ascertain the short-term changes in sedimentation rate directly associated with phases of soil erosion. Thirdly, when the problems of precise dating of these sediments containing older carbon are coupled with a consideration of the innate errors of the dates (expressed usually as only one standard deviation whereas two would be far preferable statistically), the use of absolute pollen influx data within such phases of hydrological disturbance is clearly fraught with difficulty. There are, nevertheless, examples of the use of such techniques which have provided valuable information, e.g. Pennington (1975). Even if good, absolute data are obtained, Pennington (1973) has shown that pollen deposition rates are more strongly influenced by the efficiency of the lake as a pollen trap than by the local pollen productivity of the vegetation. Fourthly, these considerations all argue for caution when trying to differentiate between hydrological changes caused by human agencies and those of a climatic nature.

In the case of the ombrotrophic mires, such as raised bogs and blanket mires, there is effectively no catchment from which drainage water is received, therefore climatic influences can be registered directly. In fact, this is something of an oversimplification because the surfaces of such mires are not perfectly uniform and the existence of a complex microtopography of hummocks, pools, and drainage channels means that certain parts of such mires do possess a local catchment, though normally it is only of the order of a few square metres. Certain parts of blanket mires which lie in depressions, either in valleys or plateau basins, will of course have rather larger catchment areas. Even so, the existence of such catchments does not normally lead to an input of transported material to a site of sedimentation during a period of increased ground wetness, unles there is a physical erosion and redistribution of peat within that catchment.

The subject of climatically-controlled stratigraphic changes in ombrotrophic peats has recently been reviewed by Barber (1981), who concludes that extension of pool systems and general increases in peat growth rates over a bog surface can be related to the crossing of climatic thresholds at that particular site. The thresholds involved, however, may vary from site to site, which makes regional climatic reconstruction on the basis of such stratigraphic data a difficult task.

The outcome of all these considerations is that if we are to elucidate the hydrological history of an area we must be content to restrict our conclusions in a geographical sense and to confine them to the region under study, and also seek to combine the information provided by rheotrophic (water receiving) and ombrotrophic (water shedding) sites within the study area. Under such circumstances, a combination of stratigraphic analysis (providing evidence of hydrological change and local vegetation changes) with pollen analysis (providing data relating to contemporaneous vegetation changes over a wider region) may permit the differentiation of climatic and human contributions to any particular palaeohydrological event.

As an example of this course of reasoning, we present here some results obtained from blanket mire and valley mire studies in upland mid-Wales. This part of Britain is of particular interest in that it has provided much of the data upon which the proposal that blanket mire initiation was induced by prehistoric human activity has been based (Moore, 1975).

A CASE STUDY FROM CENTRAL WALES

The study area consists of a part of the main watershed of the Cambrian Mountains, west of the small town of Rhayader in mid-Wales. Here a number of peat sites from a variety of topographic locations has been subjected to palynological and stratigraphic analysis (PEJW, unpublished). Two of these sites will be described here (Figure 19.1) in an attempt to illustrate how vegetation changes and palaeohydrological changes are linked and how they can be interpreted in terms of changing prehistoric land use and changing climate.

The area is one of relatively uniform geology, consisting largely of Silurian shales and mudstones (Pringle and George, 1948). The area was glaciated in the Devensian and the general physiography is one of rounded hills and plateaux with 'U' shaped valleys, further incised by post-glacial stream action. The area is extensively peat-covered, mostly by blanket peats up to 3 m deep on plateau sites, and underlain in some depressions by late-Devensian lake deposits and early Flandrian peats (e.g. Moore and Chater, 1969). Land use in the area currently consists of sheep grazing, forestry, and water storage, which has involved the construction of a series of reservoirs (Figure 19.1).

Two sites only will be considered here, Pwll-nant-ddu and Esgair Nantybed-dau. Pwll-nant-ddu (SH 866675) lies at an altitude of approximately 520 m on a watershed plateau. It is typical of the deep peat blanket mires of this region, having a vegetation cover dominated by *Sphagnum capillifolium, Calluna vulgaris, Eriophorum angustifolium, Andromeda polifolia*, etc.

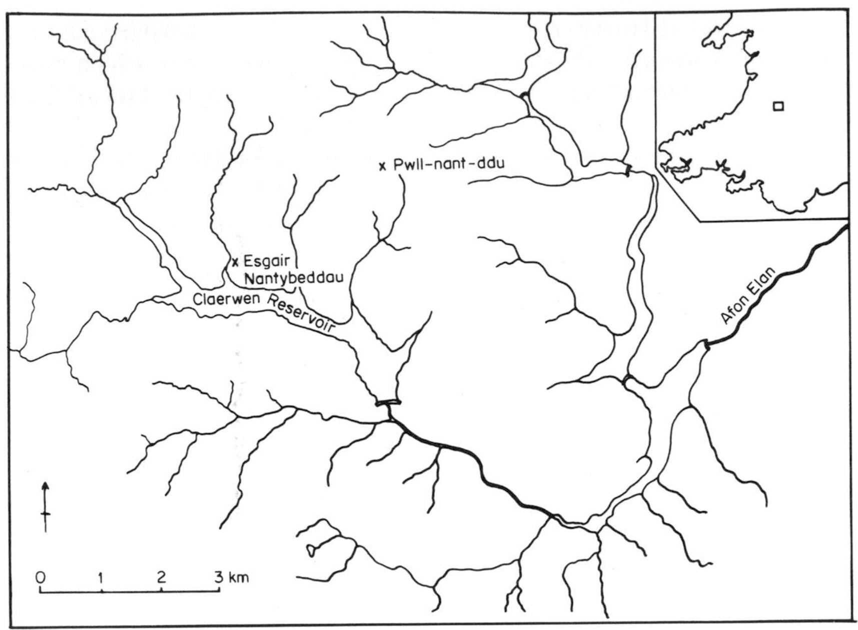

FIGURE 19.1 SKETCH MAP OF STUDY AREA
The sites from which peat cores have been collected and analysed are indicated by an X. Inset shows a map of Wales with the study area indicated

A stratigraphic profile and pollen diagram from this site is shown in Figures 19.2 to 19.4. The diagram and its zonation will not be described in detail here, but the points relevant to an understanding of the palaeohydrology of the site will be enumerated.

The peat is developed over a gleyed soil profile which is underlain by till. The transition from mineral soil to peat is accompanied by a considerable fall in the pollen proportions of *Pinus, Ulmus*, and Coryloid pollen and is accompanied by a rise in Gramineae, *Potentilla* type, and *Alnus*. Comparison with other diagrams in the area suggests that this horizon represents the 'elm decline' of Neolithic times, dated at about 5000 years ago (Smith and Pilcher, 1973). The decline of *Pinus* at the same time as *Ulmus* is a feature of many blanket mires, such as that of Plynlimmon (Moore and Chater, 1969b) and several Lake District sites (Pennington, 1975) where it is interpreted as a consequence of high altitude forest destruction by fire on the part of Neolithic people.

Within the basal peats there is an abundance of charcoal which lends credence to the idea of a high incidence of fire, though there is some dispute concerning the possible origin of charcoal in peats of this kind (Boyd, 1982a, 1982b; Moore, 1982). These charcoal layers are accompanied by a steep increase in the pollen of Ericaceae, largely comprised of *Calluna vulgaris*, and

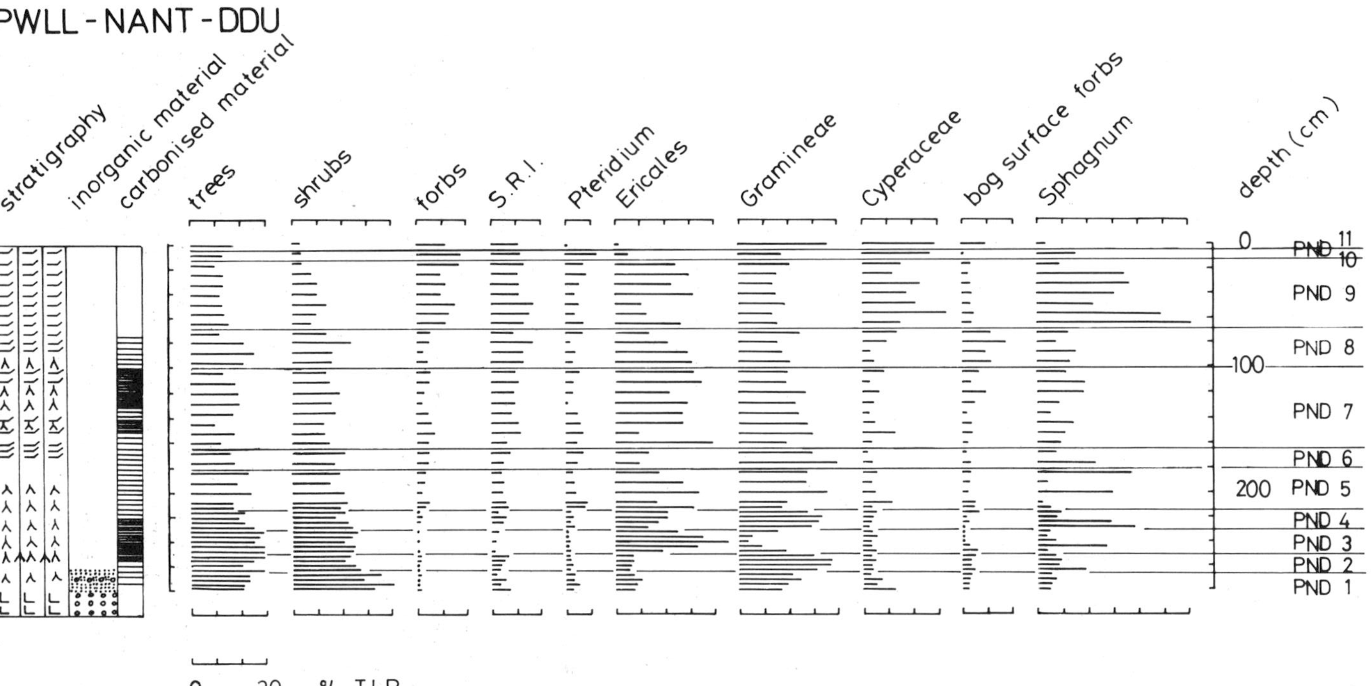

FIGURES 19.2–19.4 PARTIAL POLLEN DIAGRAM FROM PWLL-NANT-DDU

Stratigraphic symbols: Horizontal dashes—*Sphagnum*; inverted V = Ericaceous wood; V = wood; L = clay; dots in fourth column = inwashed mineral material; vertical lines = monocot remains; horizontal lines in fifth column = charcoal (density of lines indicates abundance of charcoal)

SRI column is the species richness index

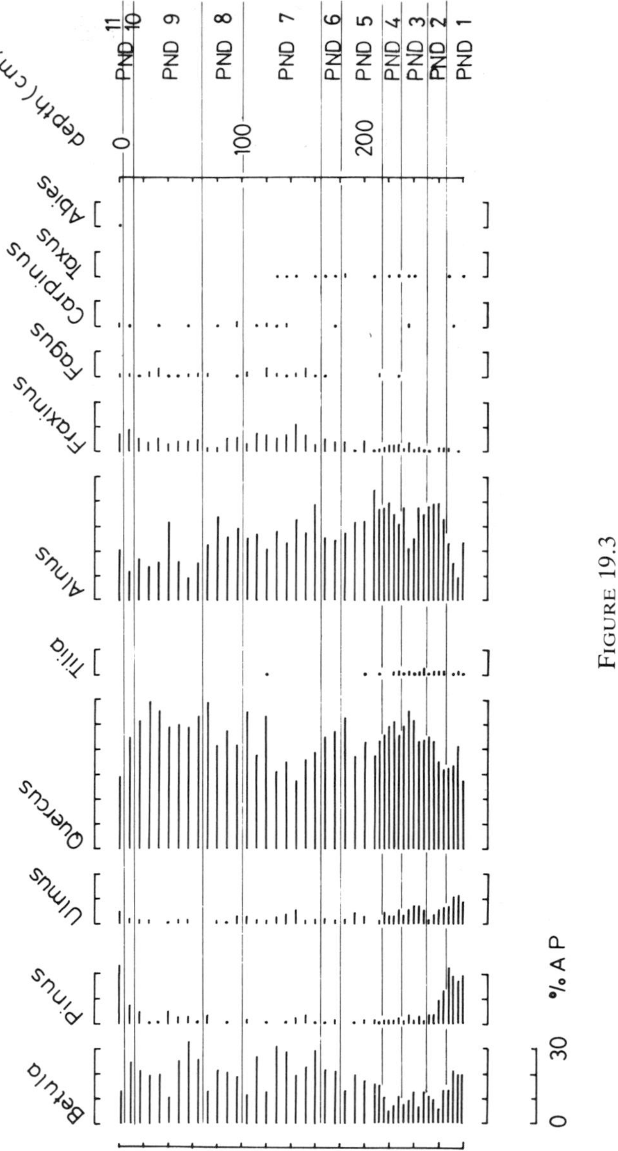
Betula
Pinus
Ulmus
Quercus
Tilia
Alnus
Fraxinus
Fagus
Carpinus
Taxus
Abies
depth (cm)
0
100
200
PND 11
PND 10
PND 9
PND 8
PND 7
PND 6
PND 5
PND 4
PND 3
PND 2
PND 1
0
30
% A P

Figure 19.3

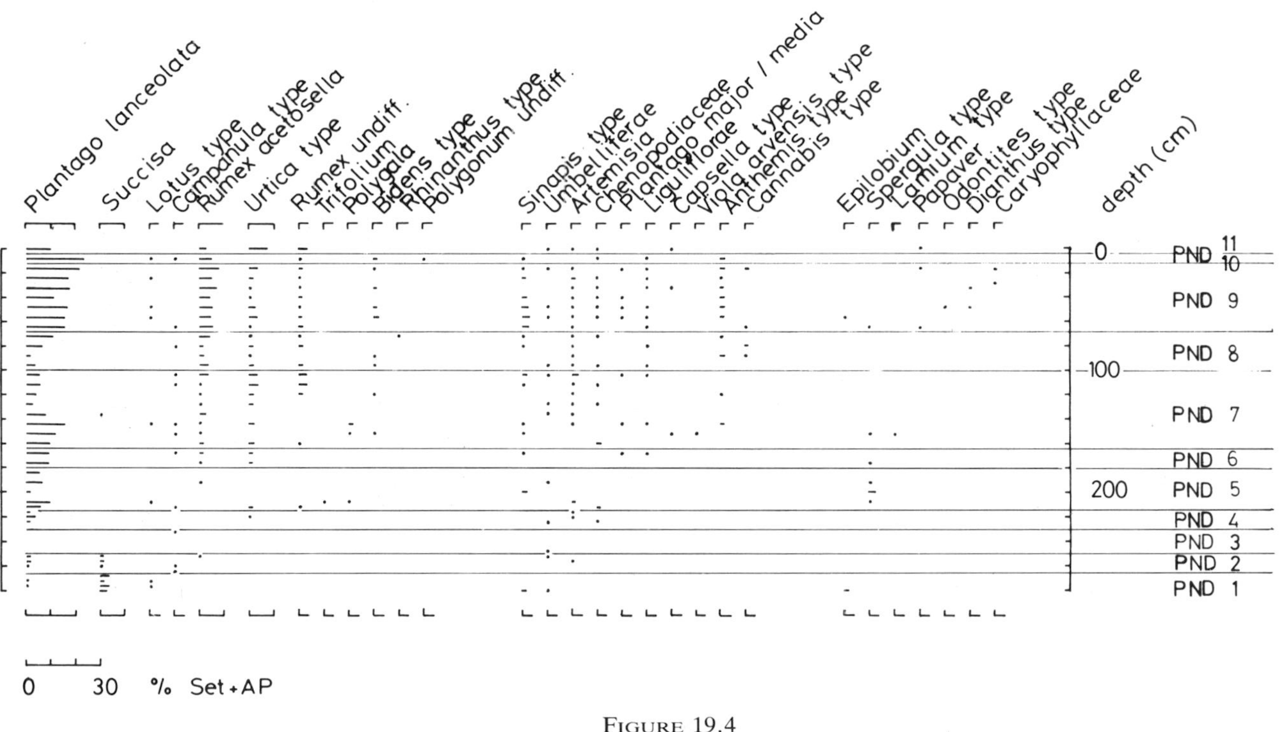

Plantago lanceolata
Succisa
Lotus type
Campanula type
Rumex acetosella
Urtica type
Rumex undiff.
Trifolium
Polygala
Bidens type
Rhinanthus type
Polygonum undiff.
Sinapis type
Umbelliferae
Artemisia
Chenopodiaceae
Plantago major / media
Liguliflorae
Capsella type
Viola arvensis type
Anthemis type
Cannabis type
Epilobium
Spergula type
Lamium type
Papaver
Odontites type
Dianthus type
Caryophyllaceae
depth (cm)
0
100
200
PND 11
PND 10
PND 9
PND 8
PND 7
PND 6
PND 5
PND 4
PND 3
PND 2
PND 1
0 30 % Set+AP

FIGURE 19.4

woody, ericaceous material is also the dominant feature of the peat stratigraphy. The basal peats are also rich in the pollen of *Succisa* and contain some *Filipendula* and *Chrysosplenium*, all of which indicate a degree of nutrient flushing in the wet, marshy soils of the period. This is in keeping with the concept of an open environment being disturbed by burning. Bryophyte analysis of the basal peat layers has not been attempted in detail here, but similar peat sites in the Berwyn Mountains of North Wales have been analysed in this way by Bostock (1980) who shows that the wet, nutrient-enriched phase at the commencement of blanket peat initiation is often characterized by a bryophyte assemblage in which *Drepanocladus* features strongly.

As far as the palaeohydrology of the site is concerned, the very fact that the mineral soil begins to accumulate a peat, or at least a mor humus, surface layer is indicative of increased wetness.

The next period in which increased wetness is implied by the stratigraphic changes is at a depth of 200 cm, where the peat rich in ericaceous materials becomes a pure *Sphagnum* peat. Total tree and shrub pollen has been declining steadily up to this event and the pastoral indicator *Plantago lanceolata* has exhibited at least two peaks, recording clearance phases. But at about this time *Plantago* rises to higher values than previously and these are maintained for 50 cm of the peat profile. At other sites in the region (Moore, Merryfield, and Price, in press) this very distinctive feature has been radiocarbon dated to about 2500 b.p., corresponding with the Iron Age.

Within this band of peat, many other pollen types suggest profound vegetation modification on the part of contemporaneous human cultures. Gramineae pollen become more abundant, including grains of larger dimensions (50–70 μm), which is suggestive both of extension in pastoral areas and of cereal growing, presumably in the valleys (Moore, 1980). *Pteridium* reaches a series of peaks, its spread perhaps being a consequence of continued burning in the uplands, to the ill effects of which its deep rhizome renders it resistant. There are also peaks in *Potentilla* type and in *Melampyrum*. The former can best be interpreted as indicating increasing grazing pressures, allowing this straggling species to flower more profusely, but the *Melampyrum* is more difficult to interpret. It is often associated with disturbed habitats, such as burnt moorland and clearings in woodland. The increase in *Narthecium* from the Iron Age onwards suggests a local wetness, possibly resulting from this particular site receiving some flow from local drainage water.

A further, and very conspicuous, hydrological change can be seen commencing in late zone PND 8 and continuing into PND 9. This shows itself stratigraphically as an increase in *Sphagnum* component of the peat and palynologically as an increase in *Sphagnum* spores, together with Cyperaceae. *Narthecium* declines at this point, as does *Calluna* pollen, so there is evidently a local vegetation change taking place which would seem to involve the replacement of *Calluna* and *Narthecium* by *Sphagnum* and, probably,

Eriophorum sp. This could, perhaps, result from increased climatic wetness, but it is more likely a result of changes in local land use. Other indications in the pollen diagram, such as the decreasing shrubs and trees (especially *Alnus*) and increasing *Plantago lanceolata, Pteridium*, and weed species, point to a general intensification of forest clearance and agricultural (including pastoral) activity in the region. Comparison with other sites would indicate that this corresponds with Monastic and Mediaeval times (12th century onwards).

Esgair Nantybeddau (SH 854660) is situated in a tributary valley of the Claerwen, at an altitude of approximately 400 m. The mire surface is dominated by tussocky *Molinia caerulea*, which is characteristic of valley floors and flushed, wet slopes in this area. A stratigraphic profile of this site is shown in Figure 19.5 and the pollen diagram in Figures 19.5 to 19.7. The basal peats are rich in wood remains, indicating that forest grew directly on the underlying clay soils. The pollen evidence suggests that these early woods were rich in *Corylus* and that *Alnus* subsequently became an important component but, given the nature of the depositional site under such conditions, it is probable that such data is representative of a very limited pollen source area (Jacobson and Bradshaw, 1981), so one cannot assume that the entire valley floor exhibited this sequence.

The sediments lie below the elm decline which is a clear feature of this pollen diagram, commencing at 380 cm and then accelerating at about 360 cm depth. This second stage in the elm decline is marked in the stratigraphy both by charcoal and by an inwash of mineral material, which suggests fire and soil instability in the catchment leading to some erosion. At this site there is again an evident *Pinus* decline near the base, as at Pwll-nant-ddu, but here it clearly precedes the elm decline horizon, as does the decline in Coryloid pollen. These features are not therefore, associated with the disturbance which accompanies the elm decline.

The rheophilous nature of the mire site in its early stages of development is more evident than in the blanket peat basal layers. *Succisa* figures prominently, and *Filipendula* is particularly well represented. Various other pollen taxa, such as *Ranunculus, Potentilla, Hydrocotyle*, Rubiaceae, *Valeriana*, etc. also denote a flushed, relatively nutrient-rich marsh environment. The quantities of *Melampyrum* pollen are very high both in the basal layers of peat and above. The difficulties of interpreting such a feature have already been touched upon, but it could be taken as indicative of disturbance and its periods of expansion correlate well with the charcoal content of the peat.

Despite these indications of disturbance within the valley catchment, the local valley carr woodland remains intact, though there is a brief phase (320–300 cm) where wood peat is replaced by *Sphagnum*. Since there is only a small decrease in the overall tree pollen component of the pollen sum, however, this is likely to represent only a local clearing in the carr. The same is probably true of the *Sphagnum* band at about 200 cm.

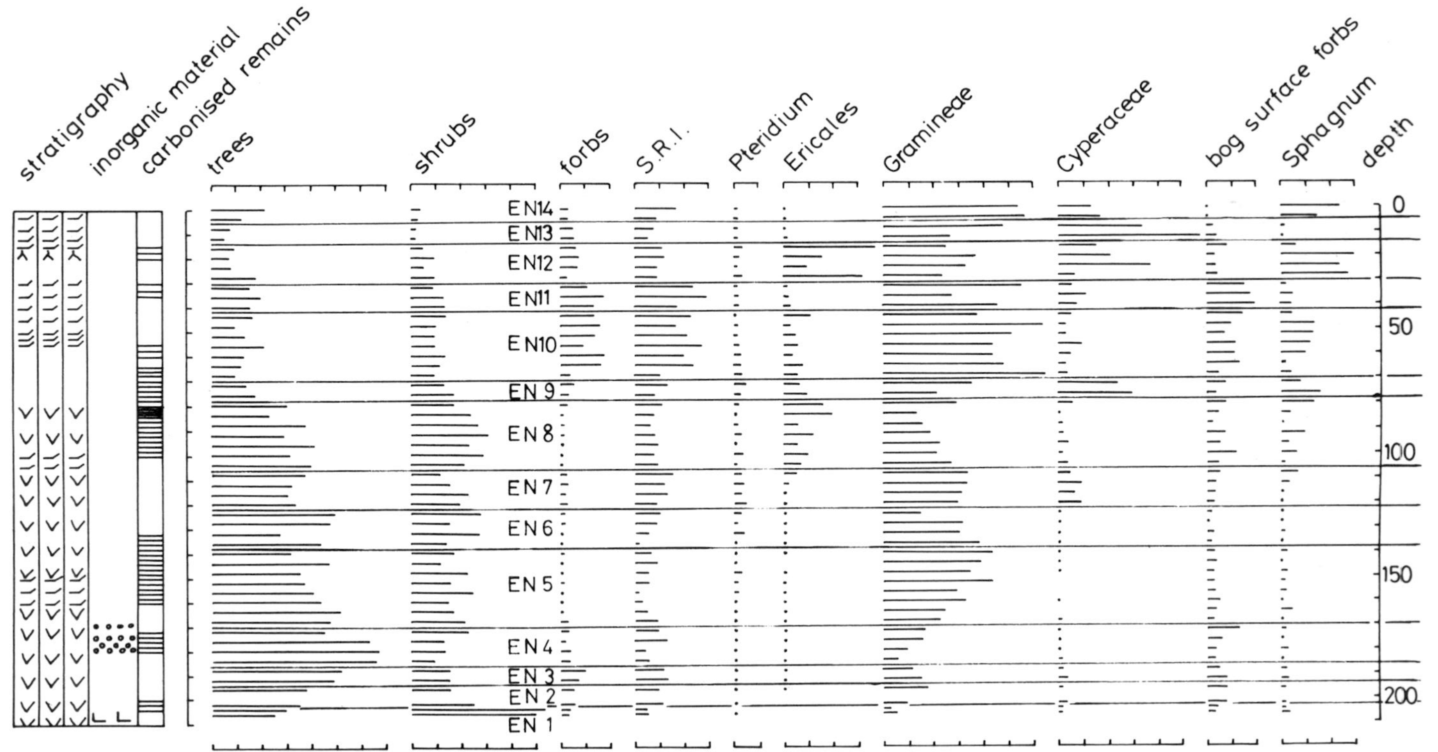

Figures 19.5–19.7 Partial Pollen Diagram from Esgair Nantybeddau
Stratigraphic and other symbols as in Figures 19.2–19.4

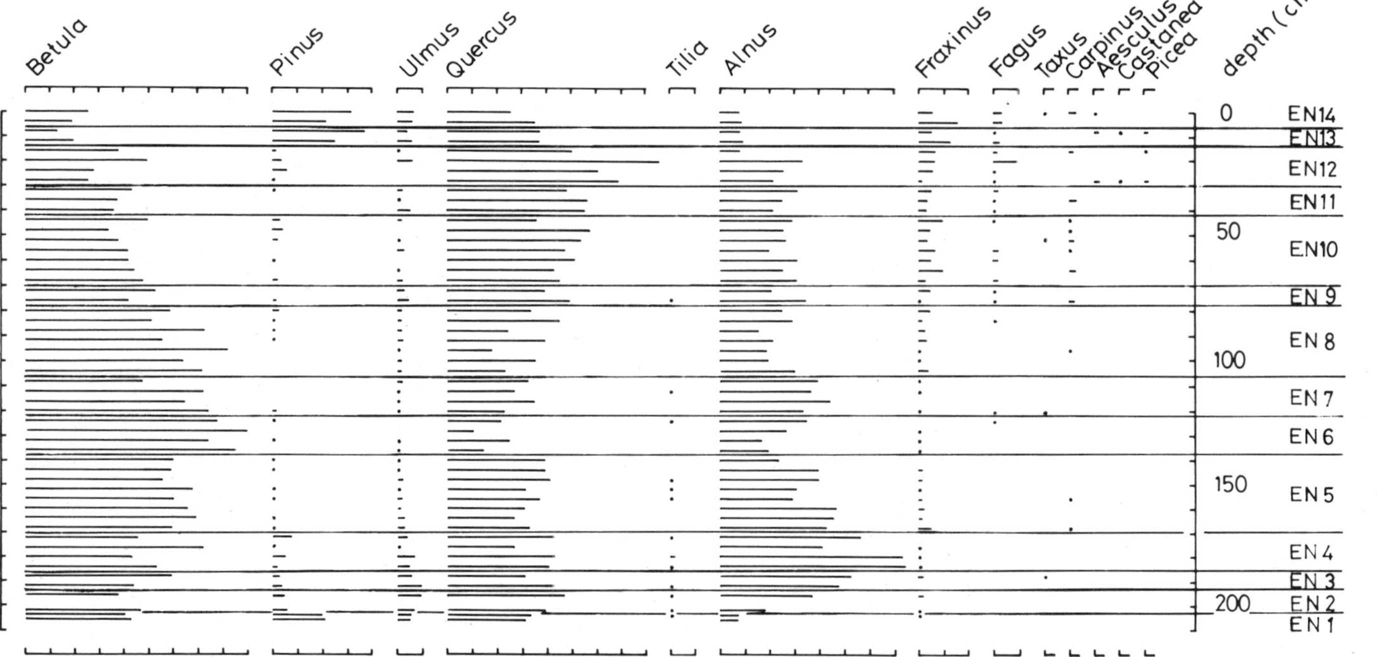
Betula
Pinus
Ulmus
Quercus
Tilia
Alnus
Fraxinus
Fagus
Taxus
Carpinus
Aesculus
Castanea
Picea
depth (cm)
0
50
100
150
200
EN14
EN13
EN12
EN11
EN10
EN 9
EN 8
EN 7
EN 6
EN 5
EN 4
EN 3
EN 2
EN 1
0
30
% AP – excluding Alnus

FIGURE 19.6

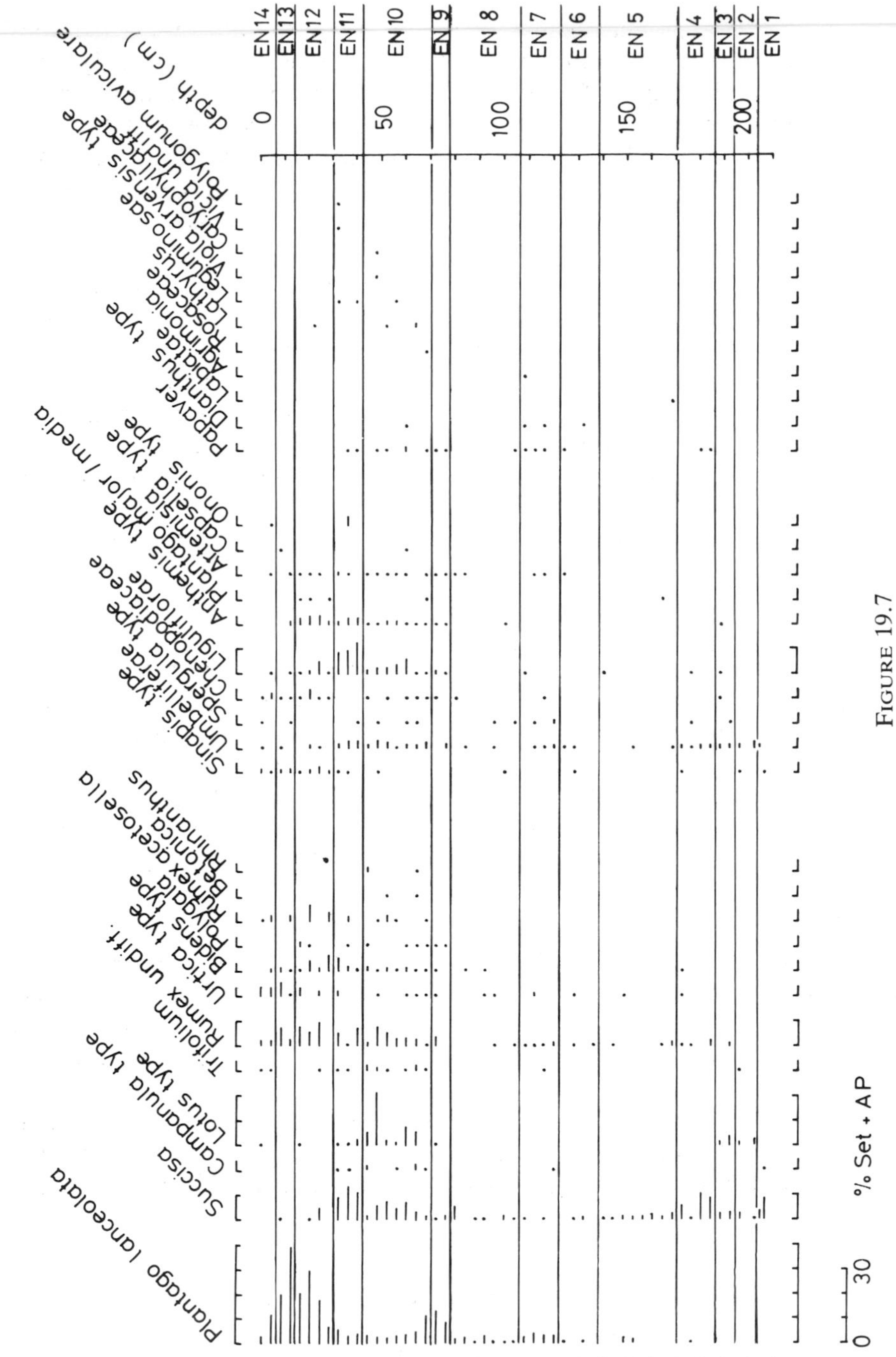

depth (cm)
Plantago lanceolata
Succisa
Campanula type
Lotus type
Trifolium
Rumex
Urtica undiff.
Bidens type
Polygala
Rumex acetosella
Veronica
Rhinanthus
Sinapis type
Umbelliferae
Spergula type
Chenopodiaceae
Liguliflorae
Anthemis type
Plantago major / media
Artemisia
Capsella type
Ononis type
Papaver
Dianthus type
Lapiatae
Agrimonia
Rosaceae
Lathyrus
Leguminosae
Viola arvensis type
Caryophyllaceae
Vicia undiff.
Polygonum aviculare
EN14
EN13
EN12
EN11
EN10
EN 9
EN 8
EN 7
EN 6
EN 5
EN 4
EN 3
EN 2
EN 1
0
50
100
150
200
0
30
% Set + AP

FIGURE 19.7

A major stratigraphic change takes place at 160 cm, where the wood content of the peat decreases dramatically and gives way to monocot and *Sphagnum* peat, thus indicating the final demise of the valley carr as it became part of the general blanket peat cover of the area (compare the loss of forest at Carneddau Hengwm, Moore, 1972, and the stratigraphic profiles of Coom Rigg Moss, Chapman, 1964). The event is accompanied by very high concentrations of charcoal in the peat, suggesting that the final loss of woodland resulted from fire.

In the pollen diagram, the features leading up to this event are of considerable interest. Tree and shrub pollen are declining even before the local tree cover is lost, and ericaceous (largely *Calluna*) pollen builds up from the preceding *Sphagnum* band at 200 cm, as does the charcoal in the peat. Gramineae and Cyperaceae form peaks at the time of stratigraphic change and general bog taxa (especially *Narthecium*) show a steady increase. The loss of the carr habitat is shown most strongly be the decline in *Betula* and in fern spores.

Plantago lanceolata attains a series of high values at the time of the stratigraphic change and, largely as a result of the behaviour of this species, the event is considered to have taken place in Iron Age/Roman Times. There is some difficulty in deciding the precise point in the stratigraphic profile where the Iron Age rise in *Plantago* actually begins. From as low as 240 cm, *Plantago lanceolata* has a steady input of pollen to the site and, considering the high local tree pollen production within the carr, one must regard these values as artificially depressed. It is possible that the final loss of woodland at this site occurred quite late in Roman times. Additional palynological evidence for intensive land use in the area during these times is afforded by such taxa as *Pteridium*, large Gramineae pollen, *Ranunculus, Urtica, Potentilla* type, Liguliflorae, etc.

The further loss of shrubs and trees, which is seen at a depth of about 60 cm, is accompanied by very high values of *Plantago lanceolata* and *Pteridium* and probably corresponds to the period of history spanning the Monastic development (12th century) up to the Napoleonic wars around the turn of the 19th century, where *Plantago* reaches its highest values throughout this region (Moore and Chater, 1969b). In terms of local vegetation at this site, the period is marked by early peaks in *Calluna* pollen, a build up in Cyperaceae (probably derived largely from *Eriophorum*) and peaks in *Sphagnum* spores. The last two of these features were also found at the plateau blanket peat site of Pwll-nant-ddu.

PALAEOHYDROLOGY OF THE SITES

When the two sites of Pwll-Nant-Ddu and Esgair Nantybeddau are compared, it is found that both exhibit three episodes of clearly recognizable palynological change, that is the Neolithic elm decline, the Iron Age/Roman episode, and the onset of Mediaeval land use changes. Of these, the first is accompanied by strong evidence for hydrological change at Pwll-nant-ddu, leading to peat accumulation on this plateau site, whereas at the Esgair Nantybeddau valley site the period was marked by evidence for erosion but no strong local changes in hydrology. The Iron Age phase shows increased wetness at the upland site, but any such change in the valley is masked by the destruction of local woodland. The third phase, at the opening of the Middle Ages, again shows evidence for increased wetness at the upland site, which does not register strongly in the valley, save for certain changes in local vegetation.

A further comparison of these two sites and the vegetational and hydrological changes which have taken place during the course of their development can be attempted using ordination techniques. Principal components analysis (PCA) was first used by Birks (1974) as one means for erecting a zonation scheme for pollen diagrams using objective, numerical techniques, and was subsequently used as a means of comparing data from pollen and chemical analysis of lake sediments by Pennington and Sackin (1975). It can also be employed to trace the course of vegetation development, as recorded in a pollen diagram, and to compare the sequence described by two or more such diagrams (Birks and Berglund, 1979). It is in this way that the technique will be applied here.

The analysis was performed on all taxa with a representation of greater than 2 per cent in the pollen data. Two separate analyses were conducted for each site, one of which considered all taxa and the second only trees. Mean values along each principle axis were obtained for each zone, and these points are plotted in Figures 19.8 and 19.9 and consecutive zones are linked by lines. Also marked on these diagrams are locations of the particular horizons discussed in the previous section. Since the distance between points provides an index of pollen changes between zones, it can now be observed whether there is evidence for major environmental shifts at any of these horizons and whether the same pattern is displayed at both sites. Considering all taxa (Figure 19.8), there is little evidence for major vegetation changes at either site at the elm decline horizon, though both sites (and Pwll-nant-ddu in particular) show considerable changes in the tree pollen at this time (Figure 19.9). Evidently, the elm decline was felt more in terms of forest composition than in major vegetation changes.

In what are regarded as Bronze Age times, there are considerable movements recorded in the all-taxa tracks of both sites, but most especially in Pwll-nant-ddu. In terms of tree pollen, the changes are more marked at Esgair Nantybeddau. This culminates at Esgair with the opening of the Iron Age, when there are large shifts in both the all-taxa and the tree pollen tracks and these are in such a direction that they tend to converge with those of Pwll. This is undoubtedly a consequence of the loss of local woodland at Esgair and the

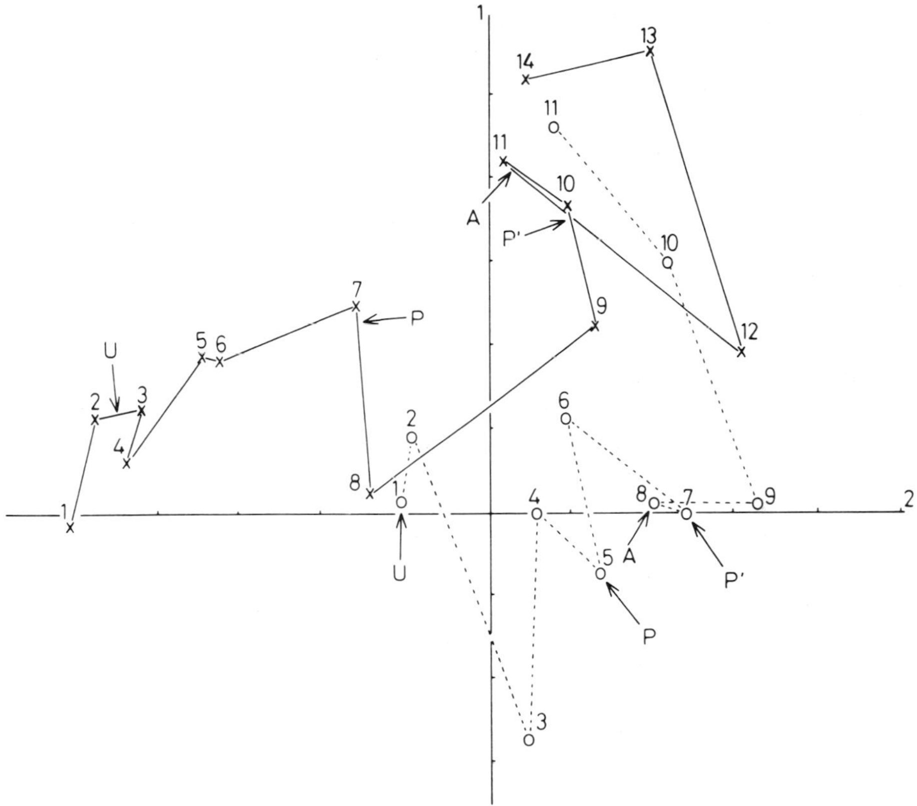

FIGURE 19.8 PCA OF ZONE MEANS CONSIDERING ALL TAXA
Axis 1 is the First Principal Component axis. Axis 2 is the Second Principal Component axis. Numbers refer to pollen assemblage zones. U = *Ulmus* decline position. P = first major *Plantago lanceolata* rise (commencement of Iron Age). P′ = decline in *Plantago* (end of Iron Age/Roman period). A = *Alnus* decline (Mediaeval).
O ------ O = Pwll-nant-ddu X ——— X = Esgair Nantybeddau

establishment of open 'blanket mire' vegetation. At Pwll-nant-ddu the Iron Age changes in both tracks are far smaller than those at Esgair Nantybeddau.

At the end of the Iron Age, both sites show a period of stability, both in their all-taxa and in their tree pollen PCA tracks. Acceleration of change occurs once more with the arrival of Mediaeval times, which shows some of the most pronounced changes detected, both in all-taxa and trees.

CONCLUSIONS

This case study illustrates the problems encountered when attempting to differentiate between climatic and man-induced vegetation changes in the

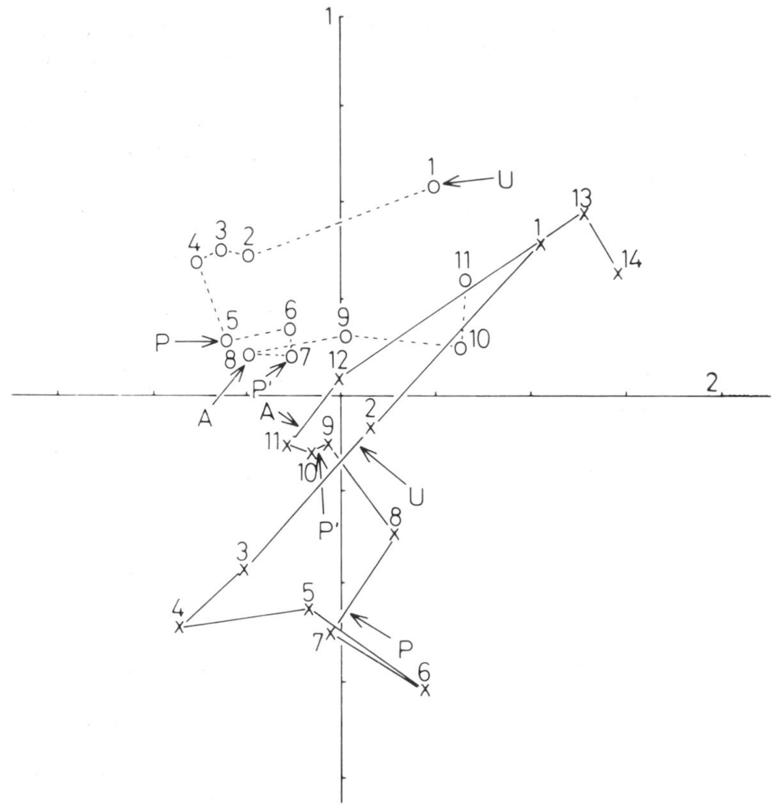

FIGURE 19.9 PCA OF ZONE MEANS CONSIDERING TREE TAXA (EXCLUDING *Corylus*)
Symbols are as in Figure 19.8

development of palaeohydrological events. The PCA plots indicate that at the elm decline horizon there is little overall change in vegetation (as reflected by the pollen of all taxa) and such change as occurs is probably to be accounted for in the internal changes in the tree taxa plus *Corylus*. These alterations in forest composition and, to a certain degree, extent are almost invariably accompanied by evidence for human acitivity and their nature is such that they are more satisfactorily explained in terms of selective exploitation of certain species, such as *Ulmus, Corylus*, and, perhaps, *Pinus*, than by any climatic alternative.

The elm decline event is, however, often also accompanied by evidence for hydrological change as in the commencement of peat formation at Pwll-nant-ddu and the soil erosion phase at Esgair Nantybeddau. The question arises whether the relatively small modifications in the vegetation of the upland environment shown in the all-taxa PCA plots are sufficient alone to account for the palaeohydrological changes. There is evidence (see Lamb, 1974) which is suggestive of climatic deterioration at this time (about 5000 b.p.) involving raised precipitation:evaporation ratios and it is possible that the human impact on the forests represented the final factor needed to cross that environmental threshold which prevented tree regeneration on the high plateaux and permitted the establishment of blanket bog. Similarly, the human disturbance of hillside and valley forest, although not sufficient to result in widespread destruction, could have resulted in the soil erosion episode recorded at Esgair Nantybeddau, especially under increasingly wet climatic conditions.

Major vegetation changes are indicated in the Bronze Age at Pwll-nant-ddu by the large, erratic movements of the PCA track, and these continue into the Iron Age. The Bronze Age changes are more marked at the upper site than in the sheltered valley at Esgair, which displays major shifts in the Iron Age. In hydrological terms, it is the Iron Age period which shows the most marked changes at both sites, with shifts in the stratigraphic profile indicative of increased wetness leading to *Sphagnum* domination. The evidence at Esgair Nantybeddau, in the form of dense charcoal at the level where trees are lost, would strongly favour the idea that these hydrological changes were man induced. Yet, once again, the period under consideration (the first millenium BC) was a time of extreme climatic instability (Lamb, 1974) and this makes it difficult to predict what the outcome of the intensified human land use would have been in the absence of concurrent climatic changes. Evidence from these sites, and many others in the area, indicates that this was a time when considerable inroads were being made into the forests of mid-Wales by late Bronze Age and Iron Age peoples, but the hydrological consequences of their activities were undoubtedly aggravated by increased climatic wetness, perhaps of a fairly sudden nature.

A further, major shift in vegetation is indicated in what are considered to be Mediaeval times, which involves the further loss of trees from the area, especially *Alnus, Quercus*, and *Betula*. There can be no doubt that these changes resulted from human acitivity, particularly the spread and intensification of grazing in the area as a result of the establishment of the Cistercian Monastery of Strata Florida. There is little evidence of major hydrological change in either site at this time, however, but this may not be surprising since both sites had by then lost their local tree cover and were part of an extensive complex of blanket mires, valley mires, and slope flushes. The forest clearance observed will have been located further down the valleys and hence their hydrological impact would not be registered in these upland sites.

The overall conclusion reached, therefore, is that peat sites, as well as lake sites, provide useful evidence of palaeohydrological events and by choosing critically-sited study areas, such as those described here in adjacent water shedding and water receiving areas, one can observe the consequences of such events in some detail. Pollen analysis adds to the stratigraphic evidence, on which the inferences concerning palaeohydrology are based, in that it provides a time dimension and correlation. It also provides evidence of concurrent regional and local vegetation changes and may indicate the involvement of man in them. The use of principal components analysis in the sorting of pollen data further permits the identification of periods of vegetation instability.

This particular study shows that the hydrological change found at the elm decline horizon is not accompanied by general vegetation change, but the Bronze Age/Iron Age hydrological event is. In both cases the impact of man must be considered as a factor which serves to amplify the effects of the unstable periods in the climate and places the vegetation under a stress which results in the crossing of thresholds in respect to the regenerative capacity of certain trees. Whether such thresholds would have been crossed in the absence of the human impact cannot be tested and is thus purely conjectural. The Mediaeval vegetation changes are of little hydrological consequence in the upper reaches of the valleys and on the hill plateaux simply because these areas were already devoid of trees.

ACKNOWLEDGEMENTS

We are indebted to the Natural Environmental Research Council for their support in the form of a studentship held by one author (PEJW). We are also grateful to Dr. I. C. Prentice of the Geography Department of Southampton University for his valued assistance with the numerical analysis of data.

REFERENCES

Barber, K. E. 1981. *Peat Stratigraphy and Climatic Change*, A. A. Balkema, Rotterdam, 219 pp.

Birks, H. J. B. 1974. Numerical zonation of Flandrian pollen data. *New Phytologist,* **73**, 351–358.

Birks, H. J. B., and Berglund, B. E. 1979. Holocene pollen stratigraphy of southern Sweden: a reappraisal using numerical methods. *Boreas,* **8**, 257–279.

Bonny, A. P. 1978. The effect of pollen recruitment processes on pollen distribution over a sediment surface of a small lake in Cumbria, *Journal of Ecology,* **66**, 385–416.

Bostock, J. L. 1980. *The History of the Vegetation of the Berwyn Mountains, North Wales, with Emphasis on the Development of the Blanket Mire*, unpublished Ph.D. thesis, University of Manchester.

Boyd, W. E. 1982a. Sub-surface formation of charcoal and its possible relevance to the interpretation of charcoal remains in peat. *Quaternary Newsletter No. 37*, 6–8.

Boyd, W. E. 1982b. Sub-surface formation of charcoal: an unexplained event in peat. *Quaternary Newsletter No. 38*, 15–16.

Chapman, S. B. 1964. The ecology of Coom Rigg Moss, Northumberland. I. Stratigraphy and present vegetation. *Journal of Ecology,* **52**, 299–313.

Jacobson, G. L., and Bradshaw, R. H. W. 1981. The selection of sites for palaeovegetational studies. *Quaternary Research,* **16**, 80–96.

Lamb, H. H. 1974. Climate, vegetation and forest limits in early civilized times. *Philosophical Transactions of the Royal Society of London,* **A 276**, 195–230.

Moore, P. D. 1972. The influence of prehistoric cultures upon the initiation and spread of blanket bog in upland Wales. *Nature,* **241**, 350–353.

Moore, P. D. 1975. Origin of blanket mires. *Nature,* **256**, 267–269.

Moore, P. D. 1980. Resolution limits of pollen analysis as applied to archaeology. *Museum Applied Science Centre for Archaeology (MASCA) Journal* (University of Pennsylvania), **1** (4) 118–120.

Moore, P. D. 1982. Sub-surface formation of charcoal: an unlikely event in peat. *Quaternary Newsletter No. 38*, 13–14.

Moore, P. D., and Bellamy, D. J. 1974. *Peatlands*, Elek Science, London 221 pp.

Moore, P. D., and Chater, E. H. 1969a. Studies in the vegetational history of mid-Wales. I. The post-glacial period in Cardiganshire. *New Phytologist,* **68**, 183–196.

Moore, P. D., and Chater, E. H. 1969b. The changing vegetation of west-central Wales in the light of human history. *Journal of Ecology,* **57**, 361–379.

Moore, P. D., Merryfield, D. L., and Price, M. D. R. (in press). The vegetation and development of blanket mires. In Moore, P. D. (Ed.), *European Mires*, Academic Press, London.

Pennington, W. 1973. Absolute pollen frequencies in the sediments of lakes of different morphometry. In Birks, H. J. B., and West, R. G. (Eds.), *Quaternary Plant Ecology*, Blackwell Scientific Publications, Oxford.

Pennington, W. 1975. The effect of Neolithic man on the environment in north-west England: the use of absolute pollen diagrams. In Evans, J. G., Limbrey, S., and Cleere, H. (Eds.), *The Effect of Man on the Landscape: The Highland Zone*, Council for British Archaeology, Research Report No. 11, 74–86.

Pennington, W. 1978a. Responses of some British lakes to past changes in land use on their catchments. *Verhandlugen der internationalen Vereinigung für theoretische und angewandte Limnologie,* **20**, 636–641.

Pennington, W. 1978b. The impact of man on some English lakes: rates of change. *Polskie Archiwum Hydrobiologii,* **25**, 429–437.

Pennington, W., and Sackin, M. J. 1975. An application of principal components analysis to the zonation of two late-Devensian profiles. I. Numerical analysis. *New Phytologist,* **75**, 419–440.

Pringle, J., and George, T. N. (1948) *British Regional Geology South Wales*, H.M.S.O., 101 pp.

Smith, A. G., and Pilcher, J. R. 1973. Radiocarbon dates and vegetational history of the British Isles. *New Phytologist,* **72**, 903–914.

Background to Palaeohydrology
Edited by K. J. Gregory

20

Large-scale fluvial palaeohydrology

VICTOR R. BAKER

Department of Geosciences, University of Arizona, Tucson, Arizona 85721 U.S.A.

Hydrology is the science dealing with continental waters, their occurrence, distribution and movements through the entire cycle of precipitation, evapotranspiration, atmospheric circulation, surface flow, and subsurface flow. Palaeohydrology is the study of past occurrences, distributions, and movements of continental waters. Palaeohydrology is the interdisciplinary effort that links scientific hydrology to sciences concerned with earth history and past environments (Schumm, 1967). The linkage extends in both directions: modern hydrologic data is useful in creating models of past environments, and data from past environmental change can be used to calibrate modern hydrologic models in order to predict environmental change.

Palaeohydrology encompasses several subdisciplines, including palaeofluminology — the study of ancient stream systems; palaeolimnology — the study of conditions and processes of ancient lakes; hydrologic aspects of palaeoclimatology — the study of past climates; palaeohydraulics — application of engineering mechanics to calculating past water flows; and palaeohydrogeology — the study of the ancient occurrence and movement of subsurface waters. This chapter will concentrate on the large-scale aspects of palaeofluminology, or fluvial palaeohydrology.

The time span of palaeohydrologic concern can be operationally defined as extending beyond the range of conventional monitoring techniques including meteorological stations, stream gauges, monitored wells, and evaporation pans. Techniques for palaeohydrology (see Chapter 1, pp. 10) include studies of the hydrologic implications of the following: 1. Ancient soils (palaeopedology), 2. Tree rings, 3. Pollen and spores (palynology), 4. Ancient cultural evidence (archaeology); 5. Written and oral records; 6. Sediments (sedimentology and stratigraphy); 7. Landforms (geomorphology and Quaternary geology); 8. Ancient floral and faunal distributions; 9. Caves and ancient karst topography; and 10. Isotopic chemistry of ancient waters. While many of the above also constitute the tools of palaeoclimatology, the palaeohydrologic use of these tools is directed toward reconstructing conditions of the ancient continental hydrosphere.

Numerous problems of hydrologic concern involve the need for long-range projection of present or past trends. Water-supply planning, hazardous waste isolation, the chance of catastrophic flooding, and the chance of prolonged drought are a few of these concerns. As society becomes increasingly more complex, interdependent, and interconnected, low probability vicissitudes of nature can exert devastating impacts on key transportation, communication, and economic networks. The lack of long-term hydrologic data is a true impediment to progress in understanding both rare and long-acting hydrological processes. Through the interdisciplinary efforts of geologists, hydrologists, palaeoecologists, climatologists, and related specialists new and exciting techniques can be developed for extending hydrologic records.

On the academic side of the earth and environmental sciences, long-term hydrologic data are essential for answering several fundamental scientific questions. What is the role of rare, high magnitude events (storms, floods, debris flows) in the shaping of the landscape? What hydrologic changes accompanied the great shifts of human, plant, and animal populations that occurred during the glacial–interglacial cycles of the last few million years, the part of earth history in which civilization emerged? What do the discoveries of aqueous processes on other planets and satellites in the solar system mean for the origin of our own water-rich environment and the life that thrives in it? How does the earth's great climatic engine relate to the hydrologic cycle? How did sediment yields, soil moistures, and streamflow vary in the remote geologic past when the physical and ecological environments were very different from those today?

Neither the contributions of micro-scale hydrology nor the concentration of recent research on small catchments as fundamental geomorphic units need be reviewed here. These concerns propelled the advancement of hydro geomorphology through the middle twentieth century. Nevertheless, there is a growing realization that contemporary process studies are of little worth in evaluating long-term landscape evolution (Church, 1980). Moreover, the micro-scale view provides no clue to the broader problems of changed inputs to drainage systems and the general feedback that numerous changed drainage basins provide to the systems (structural and climatic) that induce those inputs. Indeed, the important hydrogeomorphic questions for the next century seem to be at the macro-scale. Although some upward extrapolation is possible (Church and Mark, 1980), a balanced approach to these problems demands research on expanded spatial and temporal scales.

This chapter proposes to explore change in fluvial systems on the largest possible temporal and spatial scales. The proposal is far from new, since it is merely the restatement of what was well known to the great geomorphologists of the last century. The parent science of geology has long emphasized large-scale features in its central discipline of tectonics. Although early proponents of large-scale crustal mobilism, such as Alfred Wegener, were decidedly renounced by the mainstream scientific community, their ideas provided the stimulus for work that eventually transformed the earth sciences. The plate tectonic model that emerged in the late 1960's was but a quantitative geophysical confirmation of the elegant hypothesis developed by careful attention to large-scale structural patterns on the earth's surface. Similarly, during the 1970's megasedimentology, the study of sedimentary basins and larger areas, became the most rapidly expanding subdivision of sedimentary geology (Potter, 1974). Even more recently elements of the hydrologic community have realized that new problems of water-resource management required a global approach to hydrology and climate (Eagleson, 1982). Expansion of the time scale of observation is an important component of these changing scientific concerns.

TEMPORAL AND SPATIAL SCALING

The importance of size and scale in geomorphology has been reviewed by Church and Mark (1980). Because the most significant changes in landscapes cannot be observed during the short timescales or in the limited study areas of process geomorphological field studies, Church and Mark (1980) advocate the extension of the existing body of empirical proportionality relations to develop key questions about system behaviour. Examples of such relations include the well known correlations of alluvial fan area to drainage basin area (Bull, 1964), and of mainstream length to drainage area (Hack, 1957). Such scaling relationships are fundamental to all science, so this introduction must be necessarily brief.

The operative temporal and spatial scales of geologic phenomena span an immense range (Figure 20.1). Note that fluvial phenomena (floods) occupy an intermediate position in this scaling. The various phenomena all follow the general equation

$$S = aT^b \quad (1)$$

where S is the size of the feature, T is the time (duration), and a and b are constants. The constant b is generally a scaling factor (equal to about 1.0) showing that big phenomena tend to last longer. The constant a seems to relate to the intensity of the process, that is how rapidly it expends energy per unit area.

Geologic phenomena can also be scaled in dynamic terms, and for this purpose the concept of power is quite useful. Although the connection is complex between the geomorphically significant parameter (rate of erosion) and the physical concept (power), the comparison of various geomorphic process agents

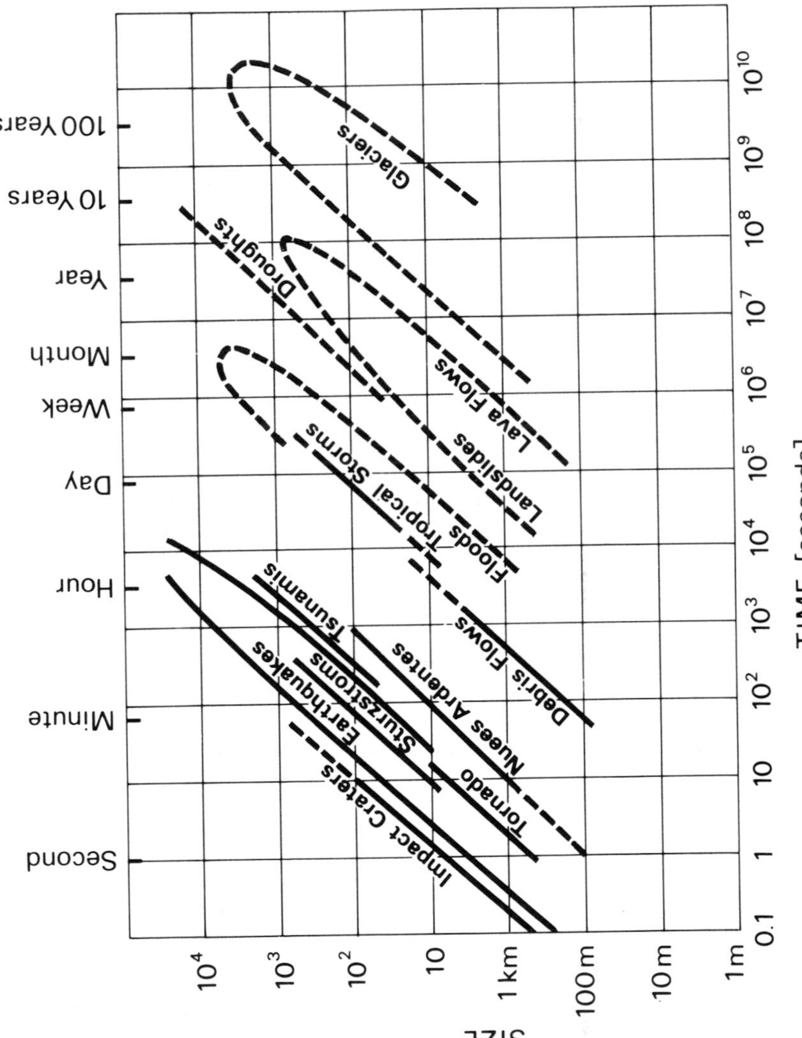

FIGURE 20.1 SCALE RELATIONSHIPS FOR VARIOUS GEOLOGIC AND GEOMORPHIC PROCESSES
The relationships are illustrative and based on the work of Carey (1962)

TABLE 20.1 COMPARATIVE FLOW DYNAMICS OF VARIOUS RIVERS IN FLOOD

River	Slope	Depth (m)	Shear stress (dynes cm^{-2})	Velocity (m s^{-1})	Power per unit area (ergs $cm^{-2}s^{-1}$)	Discharge (m^3 s^{-1})
Amazon	0.00001	60	59	2	1×10^4	3×10^5
Mississippi	0.00005	12	59	2	1×10^4	3×10^4
East Fork (Wyoming)	0.001	1.2	100	1.2	1×10^4	23
Missoula Flood	0.003	70	2×10^4	10	2×10^7	1×10^7
Elm Creek	0.0045	7	3×10^3	6.4	2×10^6	1×10^3

is facilitated by power theory. Because total power scales with the size of the process, it is most appropriate to consider power per unit area of bed ω calculated as

$$\omega = \frac{\gamma QS}{w} = \gamma dS\bar{V} = \tau\bar{V}, \tag{2}$$

where w is the bed width, Q is the discharge, S is the energy slope, γ is the specific weight of the fluid, d is the flow depth, $\bar{V}$ is the mean flow velocity, and τ is the bed shear stress.

Comparisons of power per unit area generated by various rivers in flood reveal an interesting result (Table 20.1). It is not the discharge of a river that dictates its power per unit area. Rather, an ideal combination of gradient, depth, and velocity can generate remarkable power for an unremarkable discharge. The small Texas stream of Elm Creek in its flood of May 11, 1972 (Baker, 1977) generated 200 times the power unit area of either the Amazon or the Mississippi in flood. Clearly the scaling of phenomena must give due consideration to intensities of action as well as to absolute size.

NEW SOURCES OF PALAEOHYDROLOGIC DATA

The first cargo mission of the space shuttle was flown in November 1981. The orbiter Columbia carried among other scientific experiments a shuttle imaging radar (SIR–A). The radar system acquired a variety of earth-surface images (Elachi *et al.*, 1982), but the most spectacular were of the dry Selima Sand Sheet of the eastern Sahara (McCauley *et al.*, 1982). The radar penetrated the sand cover to reveal sand- and alluvium-filled valleys (Figure 20.2). McCauley *et al.* (1982) propose that these relict valleys and drainage systems were carved at times when this hyperarid region was subject

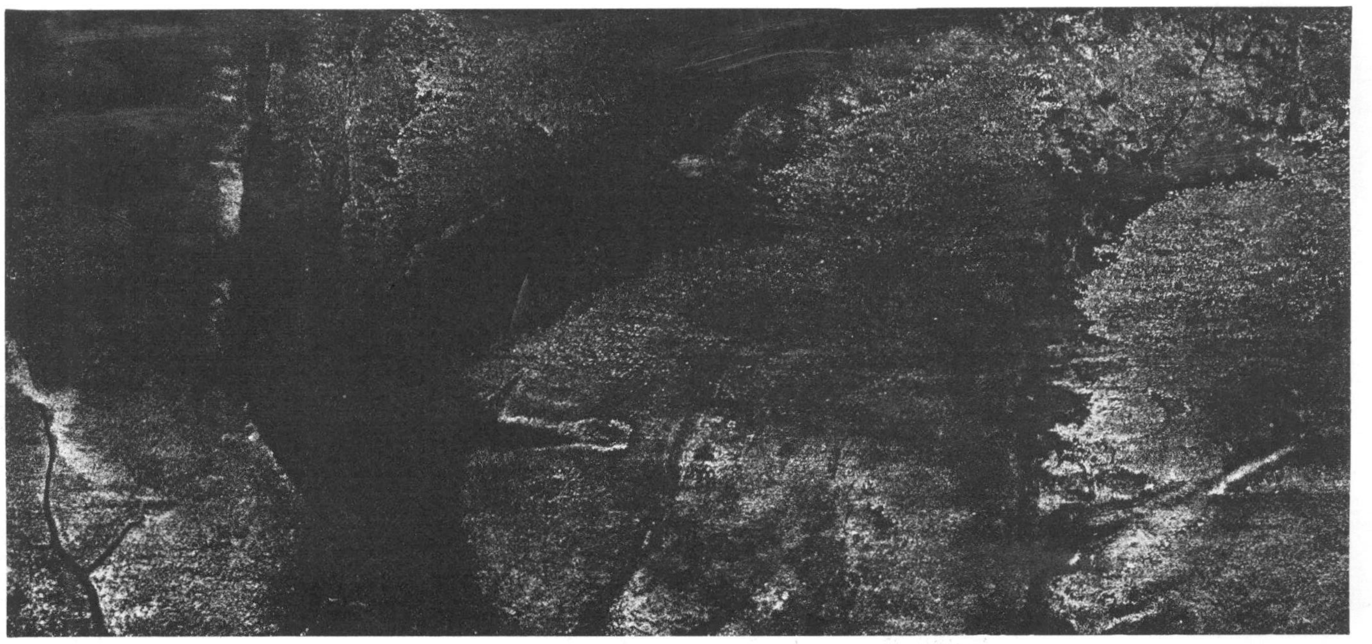

FIGURE 20.2 RADAR IMAGE SHOWING PALAEODRAINAGE FEATURES OBSCURED BY SAND IN THE EASTERN SAHARA
This picture was generated in November 1981 by the Shuttle Imaging Radar (SIR–A) experiment and shows fluvial valleys (dark areas) concealed beneath dunes and drift sand of the Selima Sand Sheet in northeastern Sudan. Radio penetration of the extremely dry sand is at least 5 m (McCauley *et al.*, 1982), revealing broad, shallow valleys of a regional south-flowing drainage system. McCauley *et al.* (1982) interpret the large valleys as Tertiary in age, and note that small wadis were superimposed on them during Quaternary pluvials. The present desert is hyperarid and lacks fluvial features. This picture shows a scene 90 × 40 km

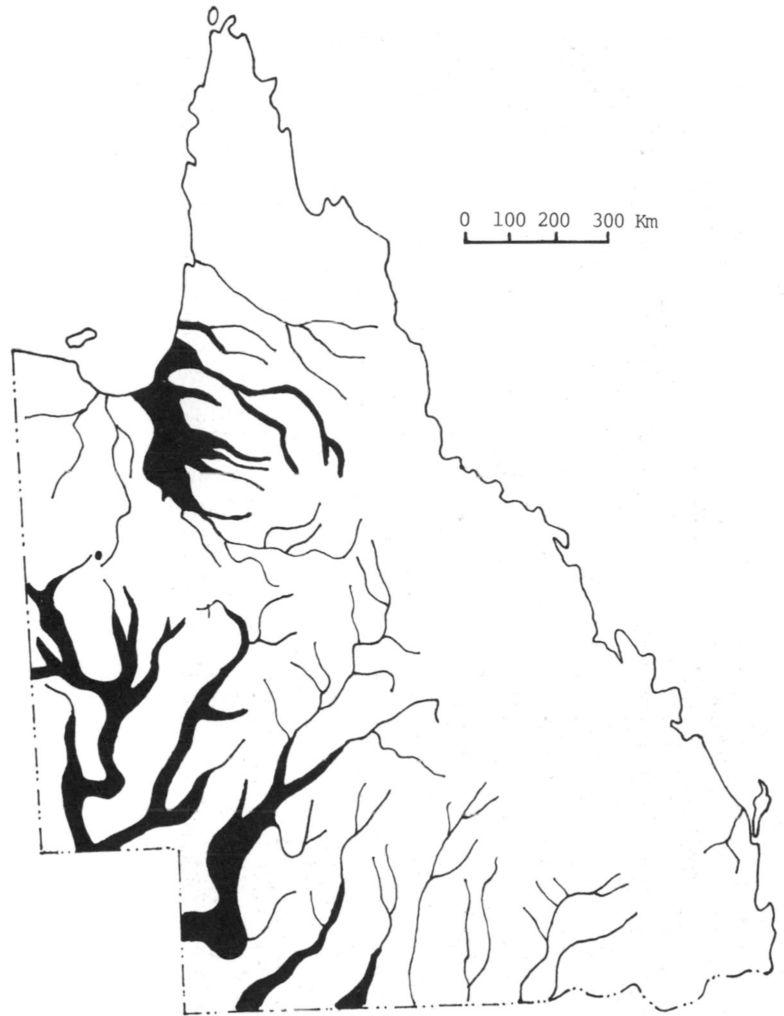

FIGURE 20.3 EXTENT OF MONSOONAL FLOODING IN QUEENSLAND, AUSTRALIA, IN JANUARY 1974
The flooded areas were mapped from LANDSAT images

to extensive erosion by running water, probably during pluvial episodes of the Quaternary.

The profound changes possible in desert hydrology are also graphically illustrated by LANDSAT images of immense regional flooding in the Australian state of Queensland (Figure 20.3). During January 1974 nearly a metre of rainfall occurred in the headwater areas of western Queensland rivers

(Heatherwick, 1974). Associated with a north Australian monsoon of unprecedented strength and southward penetration, this rainfall was as much as four to six times the annual average.

Rare, large hydrologic phenomena are now being documented through new global observation systems. Global change has become a new focus of concern in the American space program (N.A.S.A., 1982). The purpose of the National Aeronautics and Space Administration research initiative is to document, understand, and (if possible) predict long-term (5–50 years) global changes that can affect the habitability of the earth. Part of this program will be the development of even more advanced remote sensing systems. A recent improvement has been made in the LANDSAT system, which generated its first digital multispectral images in 1972. The early LANDSATS relied on a multispectral scanner system that generated data in four spectral bands at 80 m resolution. A fourth LANDSAT, launched in 1982, contains the Thematic Mapper, which generates data in seven spectral bands at 30 m resolution. The variety and quality of the new data systems are facilitating a global view of landscapes and genetic processes that was not available to the large-scale geomorphologists of the last century.

Advances in the resolution of temporal scales through geochronology are opening a variety of new cross-disciplines in the earth sciences. Palaeohydrologic research can gain much from new dating tools such as tandem accelerator mass spectrometers (Currie, 1982). Such instruments can directly measure $^{14}C/^{13}C$, resulting in analytical time of minutes for radiocarbon samples as small as a few milligrams.

TROPICAL CONCERNS

For fluvial geomorphology an irony of recent advances in the field is that the world's largest rivers remain poorly understood. Consider the largest of all, the Amazon, whose total drainage area is about six million square kilometres (Gibbs, 1967). Much of this area is covered by dense broadleaf, evergreen tropical rainforest. The mean annual precipitation in the Amazon Basin is 2400 mm (Gibbs, 1967), but the annual precipitation varies from less than 1000 mm to more than 5000 mm (Prance, 1973). Mean annual runoff from the basin is equivalent to only about 800 mm of annual precipitation. Most of the rainfall in the basin is derived from rainforest evapotranspiration within the basin (Friedman, 1977). Nevertheless, the total fresh water discharge from this basin is enormous. The mean annual flow has been estimated at 160 000 $m^3 s^{-1}$. The yearly volume of the Amazon amounts to approximately 15 per cent of all the fresh water added to the world's oceans each year (Sternberg, 1975). The mean annual sediment discharged to the sea is approximately 8 to 9 $\times 10^8$ tonnes (Meade *et al.*, 1979).

Such impressive statistics indicate that the great basin of the Amazon experiences meteorological, hydrological, and botanical events that may effect the Earth's environment on a global scale. Consider, for example, what would happen to the global heat balance and general world climatic conditions if the Amazon rainforest were cleared for the expansion of cultivation. Suppose large amounts of water were dammed or held in the interior, or directed to other drainage basins. Would this change the atmospheric circulation coming from the equatorial regions which provides heat exchange between the hot tropical zone and cold polar regions? If the forest were cleared from the Amazon Basin (a major part of the equatorial area), and the land converted to farmland with subsequent control of water level and flow, this change could make the climate of the basin more arid with less cloud cover. Would this situation change the global climate through complex linkages in the Earth's atmosphere? No one yet knows the answer to these questions. It is clear that major modifications of the environment of the Amazon Basin could trigger subtle and/or dramatic changes in the total global ecosystem (Newell, 1971).

A surrogate for man-induced environmental change, which has only begun to influence the Amazon Basin, is Quaternary climatic change, which affected the region in the past. During the Quaternary the Amazon experienced climatic change induced by dislocation of latitudinally-distributed atmospheric circulation zones. In times of glaciation for polar and alpine regions, the westerlies were displaced toward the equator. The intertropical convergence zone became narrow, reducing the equatorial rainforest climatic zone to a much narrower band than at present. The evidence for climatic change includes palynological data indicative of drier and colder climates within and around the Amazon Basin (Van der Hammen, 1974); arid climate sediments in basin fills and river terraces, and on the continental shelves of northeastern South America (Damuth and Fairbridge, 1970; Klammer, 1971); aeolian dunes in areas presently stabilized by tropical vegetation (Tricart, 1974); arid climate landforms, including pediments (Mabesoone *et al.*, 1977), and possible relict drainage (Tricart, 1974, 1975, 1977); calcrete, ferricrete, and complete laterite soils in tropical settings where these horizons cannot presently form (Klammer, 1971); and peculiar patterns of speciation in tropical lizards, butterflies, birds, and trees, that indicate periods of evolution in formerly isolated refugees that are separated by no ecological barriers today (Haffer, 1969; Prance, 1973).

In the Amazon Basin, the most important environmental thresholds affecting morphogenesis are probably phytoclimatic. These include the transition from forest to savana, and from continuous to broken cover grassland. Presently, the rainforest in many areas may exist in precarious stability (Friedman, 1977). The rainforest maintains and is, in turn, maintained by highly leached soils. The luxuriance of the vegetation derives from the ideal climatic conditions for vegetative growth, rather than from the fertility of the soils. Moreover, the climate of the rainforest is determined by extensive meteorologic recycling of moisture initiated by transpiration from the forest

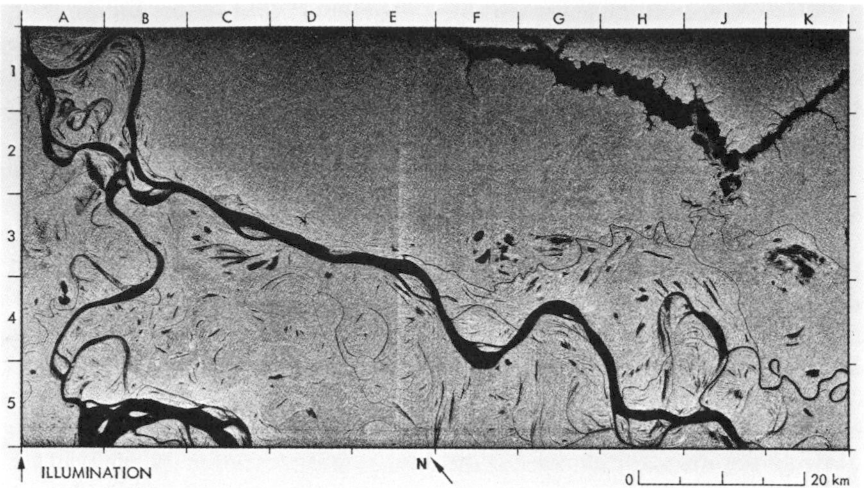

FIGURE 20.4 RADAR IMAGE SHOWING THE RIO JAPURA (CENTRE) NEAR ITS JUNCTION WITH THE RIO SOLIMÕES (BOTTOM, LEFT)
These rivers in the western Amazon Basin of Brazil were studied during the National Aeronautics and Space Administration SIR–A experiment (see also Figure 20.2). Pictures can be obtained from the National Space Science Data Center, Code 601, NASA–Goddard Space Flight Center, Greenbelt, Maryland 20771, U.S.A

itself (Friedman, 1977). Where sufficiently large areas of forest are cleared, savanna becomes established (Zonneveld, 1968). Presently juxtaposed rainforest and savanna environments on the Island of Maranjo at the mouth of the Amazon River experience dramatically different climates (Friedman, 1977). The savanna is subjected to a long, severe dry season, while the rainforest receives almost daily precipitation throughout the year. Thus, minor climatic change external to the Amazon Basin can produce a major climatic response within the basin if savanna replaces forest. Massive deforestation of the Amazon would have the immediate consequence of decreasing precipitation in the western, downwind portion of the basin (Lettau *et al.*, 1979).

Orbital remote sensing imagery of western Amazon Basin rivers (Figure 20.4) reveals a remarkable diversity of fluvial patterns. Preliminary analyses (Baker, 1978a; Holz *et al.*, 1979; Holz and Baker, 1981) show that rivers of the western Amazon display pronounced deviations from established sediment size–sinuosity relationships. All the rivers lack appreciable bedload, yet sinuosities (P) were found to vary from nearly straight ($P = 1.1-1.2$) to tortuous ($P = 3.0$). Many of the complexities probably result from the relative abilities of different rivers to rework coarse, relict alluvium that was deposited during relatively arid full-glacial phases of the Pleistocene. Additional factors include firstly, the role of tropical vegetation on bank stability (Savat, 1975), and secondly, river size and flow variability. Rivers directly draining the

Andes, for example, have seasonal flows and pronounced floods that can overcome the bank resistance provided by the vegetation. In contrast tropical lowland rivers have moderate discharges that remain relatively uniform throughout the year. These streams have the greatest sinuosity and are most dominated by vegetative effects.

Clearly, the rivers, soils, vegetation, and climate of the Amazon comprise an extremely complex, interrelated system operating on an enormous scale. Although the system contains many feedback elements that are only partially understood, the accelerating pace of man-induced environmental change in this region will require evaluation of possible impacts. Studies of palaeohydrologic change in the region offer an excellent analogic model for this environmental change, since the full-glacial climate of the region seems to have been a tropical savanna type, which is exactly what may be induced anthropogenically.

RIVER METAMORPHOSIS

River systems can be remarkably sensitive to changes in environment, especially when certain thresholds for change are exceeded (Schumm, 1979). Although relict fluvial features are a tempting source of palaeohydrologic information, the current state-of-the-art in the reconstruction of late Quaternary fluvial palaeohydraulics and palaeohydrology reveals both promise and problems. The problems derive from the chain of processes and responses that connects palaeoclimatic and palaeohydrologic events to discernible field evidence. As discussed more fully by Baker (1978a), the various linkages in a palaeohydrologic 'think back' are complicated by information losses and feedback effects.

The promise of large-scale fluvial palaeohydrology lies in the immense diversity of late Quaternary palaeochannels that are preserved in various climatic and tectonic settings. Many of the rivers responsible for these palaeochannels have experienced immense adjustments of discharge and sediment load, major drainage diversions, and/or episodes of cataclysmic flooding. The pattern changes represent a phenomenon that Schumm (1969, 1977) terms 'river metamorphosis'. The directions and, in some cases, the magnitudes of change can be deduced from various empirical relations and classifications recently summarized by Schumm (1977). The most spectacular examples of river metamorphosis occur on broad alluvial plains, such as the Riverine Plain of New South Wales, Australia (Schumm, 1968; Butler *et al.*, 1973; Bowler, 1978); the Gulf Coastal Plain of the United States (Baker and Penteado-Orellana, 1977, 1978); and the north-European fluvial plains of rivers such as the Vistula (Starkel, 1982).

The palaeohydrologic implications of various river metamorphosis examples proceed from the assumption that progressive changes in channel morphology and sedimentology reflect changes in the sediment load and runoff of the catchment. A large number of empirical relationships are available for calculating the flow characteristics of ancient rivers from their preserved morphology and sediments (e.g. Schumm, 1969, 1972; Dury, 1965, 1976). However, these relationships are limited by a number of severe constraints on their use (Ethridge and Schumm, 1978). For example, although Schumm's empirical relationships between sinuosity, channel shape, and sediment type work quite well for single-channel rivers in temperate regions, the relationships do not explain the western Amazon River variations noted above (Baker, 1978a; Holz *et al.*, 1979). Nevertheless, as more diverse examples of metamorphosis are encountered and explained, the relevant genetic theory can be generalized to encompass the full range of natural variability.

An even broader perspective comes when attempting to interpret the metamorphosis of ancient rivers from purely geologic criteria. For example, the ubiquitous presence of braided stream deposits, with a paucity of fine-grained sediment, carries important information about pre-Devonian (Schumm, 1977) and even Pre-cambrian (Long, 1978) fluvial systems. The theme of very large rivers in earth history has fascinating opportunities. For example, Potter (1978) found that rivers such as the Amazon, Zaire, Mississippi, Nile, and Brahmaputra have been remarkably persistent landscape elements. Their relationships to tectonic and sedimentation histories indicate a prolonged response of megageomorphology on a continental scale, rather than to local elements of regimen.

HYDROCLIMATOLOGY

The global hydrologic cycle is intimately linked to the atmospheric circulation through exchanges of sensible and latent heat at the earth's surface. Thus, climatology requires an understanding of water and its various phase transformations (vapour, liquid, solid) in order to better characterize the general atmospheric circulation. The global interactions of the land surface, atmosphere, and oceans require global analytical models as advocated by the WMO–I CSU Global Atmospheric Research Program (GARP, 1975). Recently general circulation models (GCMs) have been developed to simulate the global features of the atmosphere (Washington and Williamson, 1977). These models have already revealed some profound relationships. For example, high soil moisture conditions in the U.S. (providing maximum evapotranspiration) appear to be self-enhancing, yielding twice as much summer precipitation as do relatively dry conditions (Shukla and Mintz, 1982). Anthropogenic changes in albedo (Sagan *et al.*, 1979), vegetative cover (Charney, 1975), and carbon dioxide (Hansen *et al.*, 1981) will all lead to profound hydrologic change in the coming decades.

The possibility that river changes can affect climate has only been realized with the advent of macro-engineering projects and proposals. Drainage of the White Nile swamps, flooding of the Qattara depression, and deforestation of the

Amazon have all been considered. In the most spectacular example, the Soviet Union plans major alterations in the drainages of three of the world's largest rivers, the Ob, the Yenisey, and the Lena. Water that usually flows northward to the Arctic Ocean would be diverted southward to irrigate new croplands in the arid Aral–Caspian basin region. In addition to the regional effects on the impacted river regimens (Micklin, 1978), this project could have a global climatic influence. Removal of fresh water from the Kara Sea might promote a system instability that would increase the ice cover (Micklin, 1981). The effect on northern hemisphere climate would then be major, since sea ice plays a key role in the ocean–atmosphere mass/energy exchange that dictates atmospheric pressure and circulation patterns (Lamb and Morth, 1978). Only a coupled hydrologic–climatic model can adequately deal with such a complex interactive system.

The new GCMs are still simplistic, generally expressing geomorphic and hydrologic variables as static boundary conditions. Their improvement requires better modelling of the critical physical processes at dynamically adequate (but computationally economical) time and space intervals (Eagleson, 1982). These models must couple systems that operate on widely disparate temporal and spatial scales. Most important, the models need to be verified with global data sets. While such data sets exist for the current land-surface state (i.e., soils, vegetation, land use, water surface, snow cover), they do not exist for past or future states. Moreover, the key element of these models is their ability to predict change. Ideally, palaeohydrologic information can play an important role here in calibrating the models to explain past changes, hopefully thereby enabling the prediction of future change.

A useful reference point for evaluating the palaeohydrologic/palaeoclimatic linkages of the late Pleistocene is provided by the CLIMAP (1976) reconstruction of the 18 000 yr B.P. global climate from a compilation of the earth's ocean surface temperatures. A preliminary extension of the analysis to continental areas revealed numerous problems of local variability (Peterson *et al.*, 1979). Such comparisons between general atmospheric conditions, as modelled by GCMs, and palaeohydrological reconstructions hold great promise for understanding large-scale climate-water interactions. The goal here is to generate a self-enhancing spiral of understanding, with models pointing to key palaeohydrologic questions and palaeohydrologic data refining the models.

The need for a long-term hydrologic–climatologic perspective prompted Kilmartin (1980) to call for a much needed new cross-discipline, hydroclimatology. He proposed a multidisciplinary emphasis on understanding long period events in hydrology with specific reference to the consideration of past and/or future climatic changes. The greatest need for this approach is in the analysis of so-called 'Noah events' (Mandelbrot and Wallis, 1968), that is, prolonged occurrences of abnormally high or low precipitation or streamflow. Such events are not understood by a basin hydrology approach, since they result from anomalies in atmospheric circulation patterns on the near hemispherical scale. Indeed conventional water resource planning has generally ignored long-term trends in historical data series and avoided the troublesome question of future extrapolation of such trends. The period of such short sightedness is classically illustrated by water allocation in the Colorado River Basin of the United States, which was based on a record (1907–1932) of anomalously high, persistent runoff, demonstrated by tree ring analysis to be the greatest and longest such anomaly in the last 450 years (Stockton, 1975).

Another type of process that has confounded hydrologists is the so-called 'Noah event' (Mandelbrot and Wallis, 1968), that is, the occurrence of extreme precipitation or streamflow (superfloods). These events have traditionally been analysed by two methods: 1. Hydrometeorological correlation of synoptic scale atmospheric processes to basin hydrology; and 2. Statistical analysis of the tails of hydrologic data distributions to determine the recurrence intervals of the events. Although 'Noah events' are of short duration, they are noteworthy for their long-term causes and effects. The synoptic pattern producing a 'Noah event' has a climatologic history of its own, generally related to a broadscale atmospheric circulation anomaly of wide extent and relatively long duration. The results of a 'superflood' can, in the appropriate geomorphic settings, be preserved for long periods as elements of the landscape (Baker, 1977).

As with 'Joseph events', the water resource manager confronting the problem of 'Noah events' has generally found his basin-scale, short-term methodologies wanting. 'Noah events' are viewed according to two myths (Willeke, 1980), as follows: firstly, the myth of infinitesimal probability, which holds that the probability of occurrence of predicted probable maximum (or minimum) events is infinitesimal, and secondly, the myth of impossibility, which holds that hydrometeorologic estimates can be made of events so large that they cannot or will not occur. Unfortunately, standard hydrologic procedures do not prove whether or not the atmosphere is (or was) physically capable of generating a series of events that behave according to these myths. The resolution of this difficulty lies in the analysis of palaeohydrologic and palaeoclimatic records.

The interpretation of long-term river flood records from slack-water sediments provides a promising new approach to evaluating 'Noah events' and various changes in the magnitude and frequency of streamflow (Kochel and Baker, 1982; Baker *et al.*, in press). Important stratigraphic sites occur along confined bedrock canyons where major floods transport sand and silt in suspension. The thickest slack-water sequences develop at protected localities where current velocities are reduced, such as back-flooded tributary mouths, at abrupt channel expansions, downstream from bedrock spurs, in bedrock caves and niches along canyon walls, and upstream from constrictions that are capable of hydraulically damming a major flood. Where these slack-water sedimentation sites are protected from various post-flood erosion processes, the sites may continue to accumulate additional sediment from subsequent floods of similar or greater magnitude.

Multiple slack-water sites along a given river system allow the cross-correlation of partial flood records that may occur at any one site. The primary correlation is by time stratigraphy. Datable organic materials intercalated in slack-water sediments include wood, charcoal, buried soil organics, and fine-grained organic litter transported by the individual floods. The Pecos River of western Texas preserves a remarkable 10 000–year sequence of such flood deposits (Kochel *et al.*, 1982; Patton and Dibble, 1982). These long hydrologic records contain the results of hydroclimatic adjustment throughout the late Quaternary. When enough such records are analysed in enough localities, they will constitute an immense resource of hydroclimatic information.

LARGE-SCALE ANOMALIES

It can be argued that the greatest advances in science occur not by reaction to present concerns, embellishing models established in the current paradigm. Rather, major advancement comes from a concentration upon the anomalies in the present scientific understanding, the points at which models fail. Modern hydrology is weakest for predicting extreme long-term and large-scaled phenomena. It is therefore upon such anomalies that effort should be expended, with the goal of advancing understanding relations among hydrologic phenomena (Baker, 1982a).

Palaeofluminology of Mars

The scale concerns of palaeohydrologic research took a quantum leap in 1971, when the Mariner 9 spacecraft returned the first pictures of channels and valleys on Mars. The genesis of these features by the action of flowing water has now been established by studies of high resolution pictures provided by the orbiting Viking spacecraft in 1976–1980 (Baker, 1982b). The planet-wide distribution of valleys and channels on another planet invites a comparison to the distribution of such features on our own. Yet, where is there a global analysis of fluvial phenomena on Earth?

Martian channels (Figure 20.5) are immense features, as much as 100 km wide and 2000 km in length. The gradients of channel floors range from about 1 to 2.5 m km^{-1}. A suite of bedforms on the channel floors indicate that large-scale fluid flows were the primary agents of channel genesis (Baker, 1978b). Most channels show evidence that the fluid flows emanated from complex collapse zones known as 'chaotic terrain'. Such channels are termed

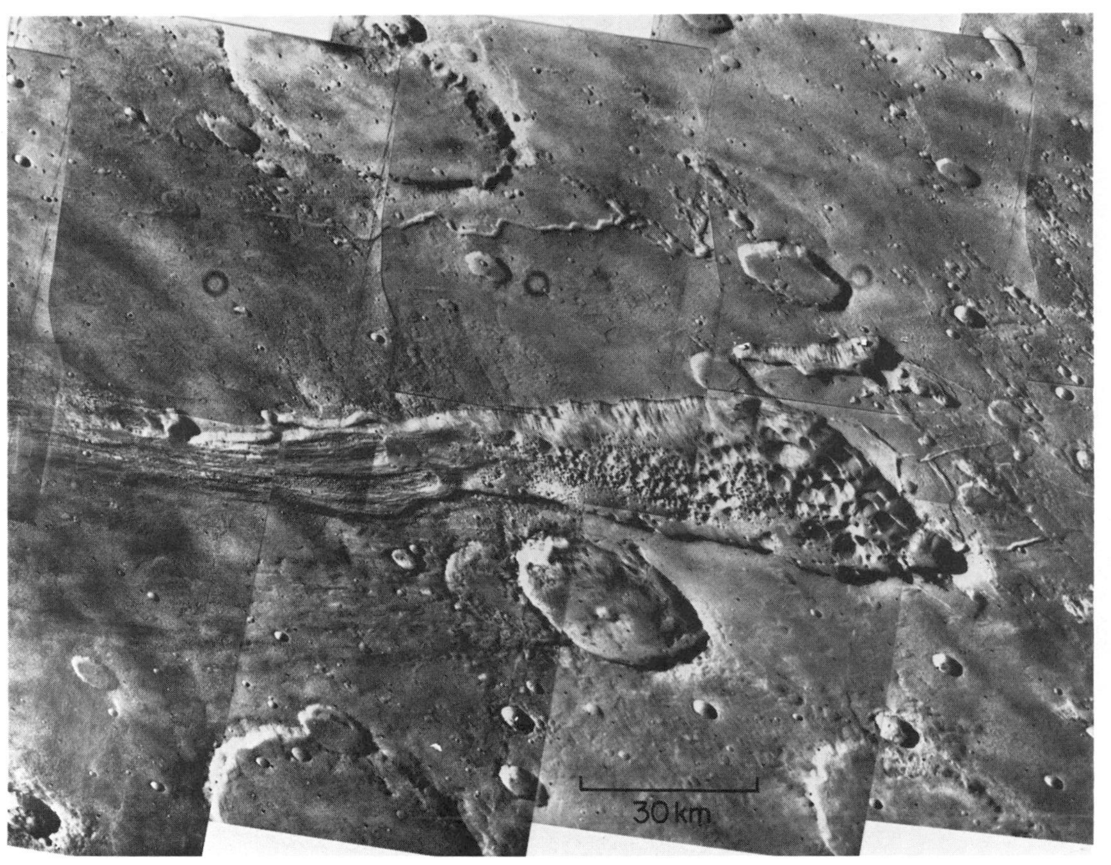

Figure 20.5 Viking Spacecraft Images of a Martian Outflow Channel
This is a mosaic of olique pictures and shows the chaotic collapse zone (right) from which large-scale fluid flows emanated to form Ravi Vallis (left)

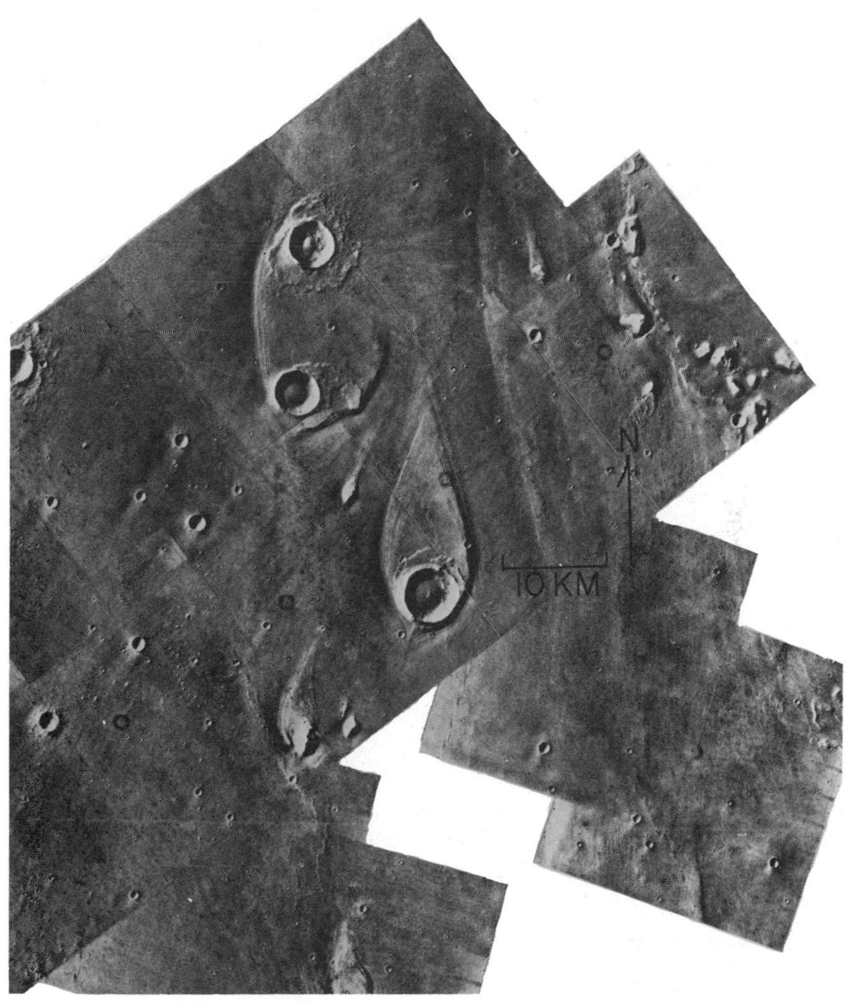

FIGURE 20.6 VIKING SPACECRAFT IMAGES OF STREAMLINED HILLS IN A MARTIAN OUTFLOW CHANNEL
The raised crater rims acted as barriers to the fluid flows coming from the south

'outflow' and probably formed by a headward extension of the collapse zones and concomitant erosion of downstream troughs by the channel-forming fluids. Morphological relationships in the channels show that the eroding fluid had a free upper surface, demonstrated by its ponding upstream of flow constrictions. It eroded scour holes, deposited bar-like sediment accumulations, spilled over low divides, and shaped magnificent streamlined hills (Figure 20.6). Among cosmically abundant substances water seems best suited to satisfy these and other constraints on the natural primary agent of channel genesis.

The outflow channel geometry and bedforms require aqueous erosion on an immense scale, a scale achieved only in some terrestrial examples of catastrophic glacial outburst flooding. After an initial phase of immense flooding, the outflow channels experienced extensive modification of their floors and walls by processes that include cratering, ground ice melting, aeolian erosion and deposition, landsliding, debris flowage, and rilling. Even though polygenetic and highly modified features abound, the analogic reasoning process with terrestrial counterparts requires that water was a necessary ingredient in channel formation.

In addition to the flow morphologies, the large Martian channels invite a comparison of planetary flow dynamics. Thus, various investigators have evaluated channel formation in terms of sediment transport theory (Komar, 1979), ice-covered rivers (Wallace and Sagan, 1979), streamlining (Baker and Kochel, 1978), cavitation, and macroturbulence (Baker, 1979).

The term 'valley network' applies to Martian trough systems, which appear to form by fluid flow, but which, unlike the outflow channels, lack a suite of bedforms on their floors. The Martian valleys of greatest interest consist of interconnected, digitate networks that dissect extensive areas of heavily cratered uplands on the planet (Figure 20.7). Some networks may have formed by surface runoff processes requiring the concentration of rainfall, but most show low drainage densities, theatre-headed valleys, and short tributaries which indicate that rainfall need not have been a direct cause of the patterns. Rather, a headward sapping process appears best able to account for the observed morphology. Ground water, derived from an ice-rich permafrost, probably played a key role in the genesis of many valley networks. The interplanetary comparison of these features has reawakened an interest in sapping, piping, and headward erosion as processes in terrestrial drainage development.

The ages of Martian channels and valleys are determined by cross-cutting relationships and by the densities of impact craters on the terrain. Age interpretation is complicated by assumptions concerning cratering rates and the resurfacing of cratered areas. Given this limitation, most researchers conclude that channels and valleys on Mars are extremely ancient, many as much as 4 billion years old. Some channels, or at least the depositional mantles that cover their floors, may be relatively young by Martian standards, less than 1 billion years old. In any case, this time scale is phenomenal by terrestrial standards, where the oldest 'palaeoforms' date to the Mesozoic (Twidale, 1976). Although still quite speculative, one exciting prospect of interplanetary comparative palaeofluminology is that the present surface of Mars may be a window to the hydrology of the early Precambrian Earth.

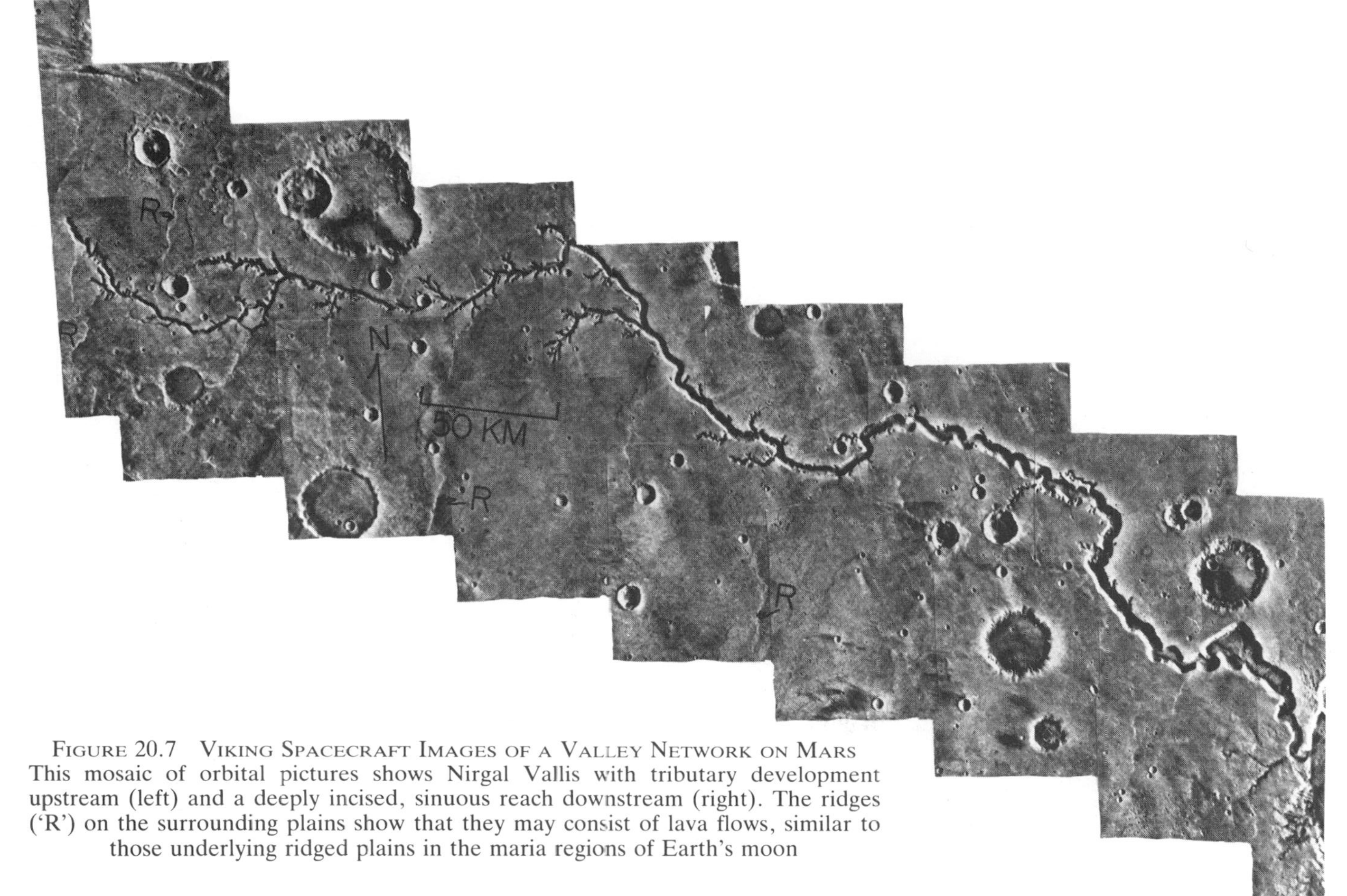

FIGURE 20.7 VIKING SPACECRAFT IMAGES OF A VALLEY NETWORK ON MARS
This mosaic of orbital pictures shows Nirgal Vallis with tributary development upstream (left) and a deeply incised, sinuous reach downstream (right). The ridges ('R') on the surrounding plains show that they may consist of lava flows, similar to those underlying ridged plains in the maria regions of Earth's moon

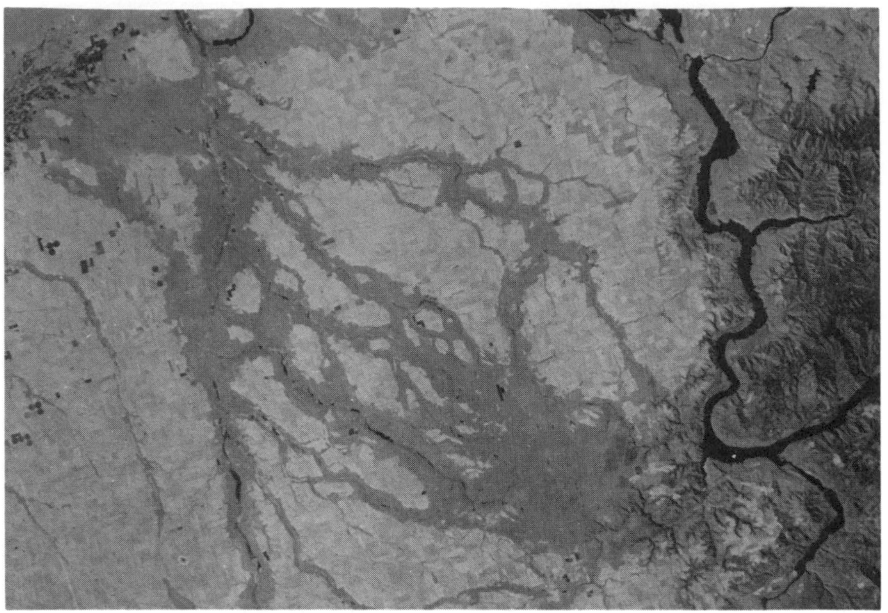

FIGURE 20.8 LANDSAT PICTURE OF A PORTION OF THE CHANNELED SCABLAND
Note the anastomosing pattern of flood channels carved into basalt. The picture is 70 × 150 km. North is to the right of the picture and flood flows moved right to left

Cataclysmic Pleistocene flooding of the Columbia River system

The Martian outflow channels invited immediate comparison to a flood-scarred plexus of channelways in eastern Washington State (Baker and Milton, 1974), a region which J Harlen Bretz named the 'Channeled Scabland'. In a fascinating scientific debate during the 1920's, Bretz encountered vehement opposition to his contention that the Channeled Scabland was the product of catastrophic flooding (Baker, 1978c, 1981). He insisted that great late Pleistocene flood flows had filled pre-flood stream valleys to overspilling, thereby producing divide crossings that converged to comprise an anastomosing complex of channelways scoured in rock and loess (Figure 20.8).

Bretz was able to convince the critics of his theory by careful documentation of the overwhelming evidence for flood-produced landforms in the Channeled Scabland. These features were subsequently shown to be consistent with the flow physics of cataclysmic flooding (Baker, 1973, 1978d). Immense bars of boulders and gravel formed wherever large flow separations were generated by various flow obstructions or diversions (Baker, 1978e). Hills of loess were streamlined by the flood flows (Patton and Baker, 1978a). The basalt bedrock

FIGURE 20.9 GIANT CURRENT RIPPLES
Note the sinuous crests and cuspate troughs covering a large catastrophic flood bar. The town of Malaga, Washington, is in the foreground

was eroded to form a bizarre landscape, known collectively as 'scabland', and consisting of erosional grooves, potholes, rock basins, inner channels, and cataracts. Of hydraulic significance are the giant current ripples, composed predominantly of gravel and commonly over 5 m high and spaced 100 m apart (Figure 20.9). The giant current ripples are directly related to the shear stresses, mean velocities, and stream powers exhibited by the flood flows (Baker, 1973).

Impressive though it is, the Channeled Scabland is but a small part of an immense portion of the Columbia River Basin, much of which was modified by late Pleistocene floods. The greatest floods were released from glacial Lake Missoula, which impounded about 2×10^{12} m^3 of water during its maximum extent. At its peak discharge Lake Missoula yielded approximately 21×10^6 $m^3 s^{-1}$ (Baker, 1973). The lake volume probably could only maintain this peak flow for several hours, but immense discharges could have been maintained for several days (Baker, 1978d). After traversing the Channeled Scabland, the flooding affected the entire downstream portion of the Columbia River system, even extending its influence to the abyssal sea floor off the mouth of the Columbia River (Griggs *et al.*, 1970).

The problem of cataclysmic flooding raises numerous questions that go to the basic assumptions of scientific investigation (Baker, 1978c, 1981). Although not yet fully answered, these questions are central to any scientific discipline. Palaeohydrology may be a new cross-discipline, but it inherits a long tradition of concern with anomalous phenomena.

CONCLUSIONS

In his insightful review of geomorphological processes on planetary surfaces Sharp (1980) observed that one of the lessons from the comparative study of landforms on different planets is to 'think big'. Geomorphologists need to recognize that anomalies of large, rare, and/or ancient phenomena are worthy of study, not for their singularity, but for the large questions that they raise concerning landscape development and the nature of environmental change.

The present lack of adequate hydrologic data on large temporal and spatial scales is being overcome with new systems of global remote sensing and new approaches to accurate geochronology. At the same time, powerful computer models of global atmospheric circulation can predict large-scale hydrologic change. These technological advances will facilitate hydrogeomorphic inquiry at expanded spatial and temporal scales. However, the new methodologies will point to needs for heretofor understudied aspects of hydrogeomorphology. The feedback between method and results will be especially healthy for the growth of cross-disciplinary sciences such as palaeohydrology.

In the broad context of hydroclimatic interactions, the emerging field of palaeohydrology may have found a role for itself. Both long period events and extreme phenomena have remained core problems in hydrology, inadequately explained by the existing research paradigm. Perhaps these puzzles will serve as catalysts for new advances in palaeohydrology and hydrogeomorphology.

It is an irony of our times that many of the ancient geologic 'experiments' in altered fluvial regimen are being redone as a result of large-scale surface mining, forest clearcutting, interbasin water diversions, and other man-made alterations of natural fluvial systems. From this perspective, large-scale fluvial palaeohydrologic studies can serve as analogic models for hydrogeomorphic change. The recognition that methodologic uniformitarianism is equivalent to simplicity or Occam's razor (Albritton, 1967) allows the following inversion of Archibald Geikie's famous maxim: the past is the key to the present (and future).

REFERENCES

Albritton, C. C. 1967. Uniformity, the ambiguous principle. *Geological Society of America Special Paper 89*, 1–2.

Baker, V. R. 1973. Palaeohydrology and sedimentology of Lake Missoula flooding in eastern Washington. *Geological Society of America Special Paper 144*. 79 pp.

Baker, V. R. 1977. Stream channel response to floods with examples from central Texas. *Geological Society of America Bulletin*, **88**,. 1057–1071.

Baker, V. R. 1978a. Adjustment of fluvial systems to climate and source terrain in tropical and subtropical environments. In Miall, A. D. (Ed.), *Fluvial Sedimentology*, Canadian Society of Petroleum Geologists, Calgary, 211–230.

Baker, V. R. 1978b. A preliminary assessment of the fluid erosional processes that shaped the Martian outflow channels. *Proceedings of the Lunar and Planetary Science Conference,* **9**, 3205–3223.

Baker, V. R., 1978c. The Spokane flood controversy and the Martian outflow channels. *Science,* **202**, 1249–1256.

Baker, V. R. 1978d. Palaeohydraulics and hydrodynamics of scabland floods. In Baker, V. R., and Nummedal, D. (Eds.), *The Channeled Scabland*, National Aeronautics and Space Administration, Washington, 59–79.

Baker, V. R. 1978e. Large-scale erosional and depositional features of the Channeled Scabland. In Baker, V. R., and Nummedal, D. (Eds.), *The Channeled Scabland*, National Aeronautics and Space Administration, Washington, 81–115.

Baker, V. R. 1979. Erosional processes in channelized water flows on Mars. *Journal of Geophysical Research,* **84**, 7985–7993.

Baker, V. R. (Ed.) 1981. *Catastrophic Flooding: The Origin of the Channeled Scabland*, Dowden, Hutchinson and Ross, Stroudsburg, Pennsylvania, 360 pp.

Baker, V. R. 1982a. Geology, determinism, and risk assessment. In *Scientific Basis of Water-Resource Management*, National Academy Press, Washington, 109–117.

Baker, V. R. 1982b. *The Channels of Mars*, University of Texas Press, Austin, 198 pp.

Baker, V. R., and Kochel, R. C. 1978. Morphometry of streamlined forms in terrestrial and Martian channels. *Proceedings of the Lunar and Planetary Science Conference,* **9**, 3193–3203.

Baker, V. R., Kochel, R. C., Patton, P. C., and Pickup, G. in press. Palaeohydrologic analysis of Holocene slackwater sediments. In Collinson, J., and Lewin, J. (Eds.), *Modern and Ancient Fluvial Systems*, International Association of Sedimentologists Special Publication.

Baker, V. R., and Milton, D. J. 1974. Erosion by catastrophic floods on Mars and Earth. *Icarus*, **23**, 27–41.

Baker, V. R., and Penteado-Orellana, M. M. 1977. Adjustments to Quaternary climatic change by the Colorado River in central Texas. *Journal of Geology,* **85**, 395–422.

Baker, V. R., and Penteado-Orellana, M. M. 1978. Fluvial sedimentation conditioned by Quaternary climatic change in central Texas. *Journal of Sedimentary Petrology,* **48**, 433–451.

Bowler, J. M. 1978. Quaternary climate and tectonics in the evolution of the Riverine Plain, southeastern Australia. In Davies, J. L., and Williams, M. A. J. (Eds.), *Landform Evolution in Australia*, Australian National University Press, Canberra, 70–112.

Bull, W. B. 1964. Geomorphology of segmented alluvial fans in western Fresno County, California. *U.S. Geological Survey Professional Paper 352-E*, 89–129.

Butler, B. E., Blackburn, G., Bowler, J. M., Lawrence, C. R., Newell, J. W., and Pels, S. 1973. *A Geomorphic Map of Riverine Plain of South-eastern Australia*, Australian National University Press, Canberra.

Carey, S. W. 1962. Scale of geotectonic phenomena. *Geological Society of India,* **3**, 97–105.

Charney, J. G. 1975. Dynamics of deserts and drought in the Sahel. *Quaterly Journal of the Royal Meteorological Society,* **101**, 193–202.

Church, M. 1980. Records of recent geomorphological events. In Cullingford, R. A., Davidson, D. A., and Lewin, J. (Eds.), *Timescales in Geomorphology*, John Wiley and Sons, New York, 13–29.

Church, M., and Mark, D. M. 1980. On size and scale of geomorphology. *Progress in Physical Geography,* **4**, 342–390.

CLIMAP Project Members 1976. The surface of the Ice-Age Earth. *Science,* **191**, 1131–1136.

Currie, L. A. (Ed.) 1982. *Nuclear and Chemical Dating Techniques*, American Chemical Society, Washington, 516 pp.
Damuth, J. E., and Fairbridge, R. W. 1970. Equatorial Atlantic deep-sea arkosic sands and Ice Age aridity in tropical South America. *Geological Society of America Bulletin,* **31**. 189–206.
Dury, G. M. 1965. Theoretic implications of underfit streams. *U.S. Geological Survey Professional Paper 452–C*, 43 pp.
Dury, G. M. 1976. Discharge prediction, present and former, from channel dimensions. *Journal of Hydrology,* **30**, 219–245.
Eagleson, P. S. 1982. Hydrology and climate. In *Sceintific Basis for Water-Resource Management*, National Academy Press, Washington, 31–40.
Elachi, C., Brown, W. E., Cimino, J. B., Dixon, T., Evans, D. L., Ford, J. P., Saunders, R. S., Breed, C., Masursky, H., McCauley, J. F., Schaber, G., Dellwig, L., England, A., MacDonald, H., Martin-Kaye, P., and Sabins, F. 1982. Shuttle imaging radar experiment. *Science,* **218**, 996–1003.
Ethridge, F. G., and Schumm, S. A. 1978. Reconstructing palaeochannel morphologic and flow characteristics: methodology, limitations, and assessment. In Miall, A. D. (Ed.), *Fluvial Sedimentology*, Canadian Society of Petroleum Geologists, Calgary, 703–721.
Friedman, I. 1977. The Amazon Basin, another Sahel? *Science,* **197**, 7.
GARP. 1975. The physical basis of climate and climate modelling. *GARP Publication Series Number,* **16**, 265 pp.
Gibbs, R. J. 1967. The geochemistry of the Amazon River system. Part I. The factors that control the salinity and composition and concentration of the suspended solids. *Geological Society of America Bulletin,* **78**, 1203–1232.
Griggs, G. B., Kulm, L. D., Waters, A. C., and Fowler, G. A. 1970. Deep-sea gravel from Cascadia Channel. *Journal of Geology,* **78**, 611–619.
Hack, J. T. 1957. Studies of longitudinal stream profiles in Virginia and Maryland. *U.S. Geological Survey Professional Paper 294–B*, 45–94.
Haffner, J. 1969. Speciation in Amazonian forest birds. *Science,* **165**, 131–137.
Hansen, J., Johnson, D., Lacis, A., Lebedeff, S., Lee, P., Rind, D., and Russell, G. 1981. Climatic impact of increasing atmospheric carbon dioxide. *Science,* **213**, 957–966.
Heatherwick, G. 1974. Flood forescasting and warnings, Moreton Region. In *January 1974 Flood, Moreton Region*, Institution of Engineers Australia, Queensland Division, Brisbane, 56–79.
Holz, R. K., and Baker, V. R. 1981. An examination of fluvial morphological characteristics of western Amazon streams from Apollo–Soyuz photographs. In Deutsch, M., Wiesnet, D. R., and Rango, A. (Eds.), *Satellite Hydrology*, American Water Resources Association, Minneapolis, Minnesota, 252–259.
Holz, R. K., Baker, V. R., Sutton, S. M., Jr., and Penteado-Orellana, M. M. 1979. South American river morphology and hydrology. In El-Baz, F., and Warner, D. M. (Eds), *Apollo–Soyuz Test Project Summary Science Report, National Aeronautics and Space Administration Special Publication,* **412**, 545–594.
Kilmartin, R. F. 1980. Hydroclimatology — a needed cross-discipline. In *Improved Hydrologic Forecasting, Why and How*, Proceedings of the Engineering Foundation Conference, American Society of Civil Engineers, 160–198.
Klammer, G. 1971. Uber Plio-Pleistozane Terrassen und ihre Sedimehte im unteren Amazonasgebiet. *Zeitschrift für Geomorphologie* **15**, 62–106.
Kochel, R. C., and Baker, V. R. 1982. Paleoflood hydrology. *Science,* **215**, 353–361.
Kochel, R. C., Baker, V. R., and Patton, P. C. 1982. Palaeohydrology of southwestern Texas. *Water Resources Research,* **18**, 1165–1183.

Komar, P. D. 1979. Comparisons of the hydraulics of water flows in Martian outflow channels with flows of similar scale on Earth. *Icarus*, **37**, 156–181.

Lamb, H. H., and Morth, H. T. 1978. Arctic ice, atmospheric circulation and world climate. *Geographical Journal,* **144**, 1–22.

Lettau, H., Lettau, K., and Molion, L. C. B. 1979. Amazonia's hydrologic cycle and the role of atmospheric recycling in assessing deforestation effects. *Monthly Weather Review,* **107**, 227–237.

Long, D. G. F. 1978. Proterozoic stream deposits: Some problems of recognition and interpretation of ancient sandy fluvial systems. In Miall, A. D. (Ed.), *Fluvial Sedimentology*, Canadian Society of Petroleum Geologists, Calgary, 313–341.

Labesoone, J. M., Rolim, J. L., and DeCastro, C. 1977. Late Cretaceous and Cenozoic history of Northeastern Brazil, *Geologie en Mijnbouw,* **56**, 129–139.

Mandelbrot, B. B., and Wallis, J. R. 1968. Noah, Joseph and operational hydrology. *Water Resources Research,* **4**, 909–918.

Mars Channel Working Group, in press. Channels and valleys on Mars. *Geological Society of America Bulletin.*

McCauley, J. F., Schaber, G. G., Breed, C. S., Grolier, M. J., Haynes, C. V., Issawi, B., Elachi, C., and Blom, R. 1982. Subsurface valleys and geoarchaeology of the eastern Sahara revealed by Shuttle Radar. *Science,* **218**, 1004–1020.

Meade, R. H., Nordin, C. F., Jr., Curtis, W. F., Rodrigues, F. M. C., Do Vale, C. M., and Edmond, J. M. 1979. Sediment loads in the Amazon River. *Nature,* **278**, 161–163.

Micklin, P. P. 1978. Environmental factors in Soviet interbasin water transfer policy. *Environmental Management,* **2**, 567–580.

Micklin, P. P. 1981. A preliminary systems analysis of impacts of proposed Soviet river diversions on Arctic sea ice. *EOS,* 62, 489–493.

National Aeronautics and Space Administration. 1982. *Global Change: Impacts on Habitability*, Jet Propulsion Laboratory D–95, Pasadena, California, 15 pp.

National Research Council. 1982. *Scientific Basis of Water-Resource Management*, National Academy Press, Washington, 127 pp.

Newell, R. E. 1971. The Amazon forest and atmospheric general circulation. In Matthews, W. H., Kellogg, W. W., and Robinson, G. D. (Eds.), *Man's Impact on the Climate*, MIT Press, Cambridge, Massachussetts, 457–459.

Patton, P. C., and Baker, V. R. 1978. Origin of the Cheney–Palouse Scabland Tract. In Baker, V. R., and Nummedal, D. (Eds.), *The Channeled Scabland*, National Aeronautics and Space Administration, Washington, D.C., 117–130.

Patton, P. C., and Dibble, D. S. 1982. Archaeologic and geomorphic evidence for the palaeohydrologic record of the Pecos River in west Texas. *American Journal of Science,* **282**, 97–121.

Peterson, G. M., Webb, T., III, Kutzbach, J. E., Van Der Hammen, T., Wijmstra, T. A., and Street, F. A. 1979. The continental record of environmental conditions at 18,000 yr. B.P.: an initial evaluation. *Quaternary Research,* **12**, 47–82.

Potter, P. E. 1974. Sedimentology: Past, present and future. *Die Naturwissenschaften,* **61**,. 461–467.

Potter, P. E. 1978. Significance and origin of big rivers. *Journal of Geology,* **86**, 13–33.

Prance, G. T. 1973. Phytogeographic support for the theory of Pleistocene forest refuges in the Amazon Basin, based on evidence from the distribution patterns in Caryocaraceae, Chrysobalanaceae, Dichapetalceae, and Lecythidaceae. *Acta Amazonica,* **3**, 5–28.

Sagan, C., Toon, O. B., and Pollack, J. B. 1979. Anthropogenic albedo changes and the Earth's climate. *Science,* **206**, 1363–1368.

Savat, J. 1975. Some morphological and hydraulic characteristics of river patterns in the Zaire basin. *Catena,* **2**, 161–180.

Schumm, S. A. 1967. Palaeohydrology: application of modern hydrologic data to problems of the ancient past. In *International Hydrology Symposium,* Proceedings Volume 1, Fort Collins, Colorado, 185–193.

Schumm, S. A. 1968. River adjustment to altered hydrologic regimen, Murrumbidgee River and palaeochannels, Australia. *U.S. Geological Survey Professional Paper 598*, 65 pp.

Schumm, S. A. 1969. River metamorphosis. *American Society of Civil Engineers Proceedings, Journal of the Hydraulics Division,* **95**, 255–273.

Schumm, S. A. 1972. Fluvial palaeochannels. In Rigby, J. K., and Hamblin, W. K. (Eds.), *Recognition of Ancient Sedimentary Environments*, Society of Economic Palaeontologists and Mineralogists, Tulsa, Oklahoma, 98–107.

Schumm, S. A. 1977. *The Fluvial System,* John Wiley and Sons, New York, 338 pp.

Schumm, S. A. 1979. Geomorphic thresholds: the concept and its applications. *Transactions of the Institute of British Geographers, N.S.,* **4**, 485–515.

Sharp, R. P. 1980. Geomorphological processes on terrestrial planetary surfaces. *Annual Reviews of Earth and Planetary Sciences,* **8**, 213–261.

Shukla, J., and Mintz, Y. 1982. Influence of land-surface evapotranspiration on the Earth's climate. *Science,* **215**, 1498–1500.

Starkel, L. (Ed.) 1982. Evolution of the Vistula River Valley during the last 15,000 years. *Polish Academy of Sciences, Geographical Studies Special Issue No. 1*, 169 pp.

Sternberg, H. O. 1975. *The Amazon River of Brazil.* Franz Steiner Verlag GMBH, Wiesbaden, 75 pp.

Stockton, C. W. 1975. Long-term streamflow records reconstructed from tree rings. *Papers of the Laboratory of Tree Ring Research,* **5**, 111 pp.

Tricart, J. 1974. Existence, au Quaternaire, de periods seches en Amazonie et dans les regions voisines. *Revue de Geomorphologie Dynamique*, **23**, 145–158.

Tricart, J. 1975. Influence des oscillations climatiques recentes sur le modele en Amazonie orientale (region de Santarem), d'apres les images de radar lateral. *Zeitschrift für Geomorphologie,* **19**, 140–163.

Tricart, J. 1977. Types de lits fluviaux en Amazonie bresilienne. *Annales de Geographie,* **86**, 1–54.

Twidale, C. R. 1976. On the survival of palaeoforms. *American Journal of Science,* **276**, 77–95.

Van der Hammen, T. 1974. The Pleistocene changes of vegetation and climate in tropical South America. *Journal of Biogeography,* **1**, 3–26.

Wallace, D., and Sagan, C. 1979. Evaporation of ice in planetary atmospheres: ice-covered rivers on Mars. *Icarus,* **39**, 385–400.

Washington, W. M., and Williamson, D. L. 1977. A description of the NCAR global circulation models. In Chang, J. (Ed.), *General Circulation Models of the Atmosphere*, Academic Press, New York, 111–172.

Willeke, G. E. 1980. Myths and uses of hydrometeorology in forecasting. In *Improved Hydrologic Forecasting, Why and How*, Proceedings of the Engineering Foundation Conference, American Society of Civil Engineers, 117–124.

Zonneveld, J. I. S. 1968. Quaternary climatic changes in the Caribbean and northern South America. *Eiszeitalter und Gegenwart,* **19**, 203–208.

Background to Palaeohydrology
Edited by K. J. Gregory

21
Postscript for palaeohydrology

K. J. Gregory

A perspective for palaeohydrology is clearly coordinated by Chapter 20 with an elaboration of the significance of interpretation at the macro scale, an indication of the value of evidence from tropical as well as from temperate latitudes, and a demonstration of the way in which investigation of palaeohydrology can afford results of relevance for understanding future changes in areas such as Amazonia which are experiencing dramatic environmental change. The introductory chapter (pp. 10) identified a continuing need for an interdisciplinary approach to palaeohydrology, and the development of links between components of palaeohydrology and between scales of attention together with new techniques, may all be bringing the background into the foreground.

The background chapters show how a number of links are developing including that from soil properties including micromorphology (Chapter 7) to palaeohydrology, the link between soils and colluvial sequences and the archaeological record (Chapter 8), the connection between understanding channel change and the palaeobotanical record (chapter 15), and between process studies and the sedimentary record (Chapter 6). The need to intensify the link between palaeohydrology and palaeoecology is at the root of the organisation of IGCP Project 158 into two subprojects A and B (Chapter 10) and it is also possible to visualize further benefits from links with other investigations of environmental change such as the evidence from mountain glaciers cited by Starkel (Chapter 9). It is also profitable to explore links at different spatial and temporal scales and Thornes shows (Chapter 3) how temporal scales of attention need to be reconciled and interrelated for palaeodischarge analysis, whereas in the portrayal of patterns of world sediment yield (Chapter 4) Walling and Webb clearly indicate how detailed and small scale investigations can be the basis for insight into, and the architect for, world patterns against which future palaeohydrologies may be viewed. World patterns are also emerging more convincingly for the atmosphere

(Chapter 2) as well and can similarly benefit from information collected at a range of scales.

It is against the background of such scale transforms that the implications of the local derivation of sediments and of the fact that not all palaeohydrological changes are revealed by the characteristics of sediments (Chapter 16) can be viewed. In addition to reconciling data from a number of different scales it is also desirable to extend and to develop the range of techniques available. Thus an analogue approach illustrated by comparing southern Britain with the Arctic (Chapter 18) is one with further scope, and further developments of methodological and theoretical techniques are also possible. Palaeohydraulic analytical techniques now in their early stages (Chapter 5) are capable of further extension, and more sophisticated equations can be developed for flow estimation (Chapter 14). Mineral magnetic measurement is a technique with very considerable potential for application to sediments from a variety of environments (Chapter 7) and there is further scope for techniques which discriminate between composition of stream networks (Chapter 11), of channel patterns (Chapters 13,17), and of deposits (Chapter 18). Emphasis upon the refinement of techniques such as the use of principal components for pollen analysis data can lead towards the elaboration of thresholds (Chapter 19). Other new concepts have emerged and these range from the relation established by Graf between tractive force and biomass (Chapter 12) to the attempts to utilize concepts of stream power (Chapter 13), and both of these theoretical approaches are indicators of further developments that are possible. As indicated in Chapter 10 and detailed in Chapter 2 the refinement of General Circulation Models may eventually provide a background for palaeohydrology and the input from remote sensing (Chapter 20) will continue to stimulate progress.

It may be that many components of the palaeohydrological cycle merit further elucidation and that emphasis has hitherto centred more upon the river channel and upon the vegetation than upon other aspects of the basin which act as controls upon the palaeohydrological cycle. However in palaeohydrology, as in geomorphology (Lewin, 1980), one is limited to the available windows through which information may be obtained at several scales by using an increasing range of methodological and theoretical techniques. Emphasis in palaeohydrological research is increasingly showing that investigators in different disciplines are using the same windows and perhaps this may lead us towards what Francis Bacon (1561–1626) described — 'The inseparable propriety of time, which is ever more and more to disclose truth.'

REFERENCE

Lewin, J. 1980. Available and appropriate timescales in geomorphology. In Cullingford, R. A., Davidson, D. A., and Lewin, J. (Eds.), *Timescales in Geomorphology*, J. Wiley, Chichester, 3–10.

Index

Subjects indexed below are restricted to major references. Material in illustrations and tables is indicated by page references in italics. Locations and authors are not included